INTRODUCTION TO
INFRARED AND RAMAN SPECTROSCOPY

INTRODUCTION TO INFRARED AND RAMAN SPECTROSCOPY

SECOND EDITION

Norman B. Colthup

AMERICAN CYANAMID COMPANY
STAMFORD, CONNECTICUT

Lawrence H. Daly

STATE UNIVERSITY OF
NEW YORK AT ALBANY
ALBANY, NEW YORK

Stephen E. Wiberley

DEPARTMENT OF CHEMISTRY
WALKER LABORATORY
RENSSELAER POLYTECHNIC INSTITUTE
TROY, NEW YORK

1975

ACADEMIC PRESS · New York   San Francisco   London

A Subsidiary of Harcourt Brace Jovanovich, Publishers

ACADEMIC PRESS, INC.
111 Fifth Avenue, New York, New York 10003

United Kingdom Edition published by
ACADEMIC PRESS, INC. (LONDON) LTD.
24/28 Oval Road, London NW1

Library of Congress Cataloging in Publication Data

Colthup, Norman B
      Introduction to infrared and Raman spectroscopy.

      Bibliography: p.
      Includes index.
      1.   Chemistry, Organic.   2.   Spectrum, Infra-red.
3.   Raman spectroscopy.   I.   Daly, Lawrence H., joint
author.   II.   Wiberley, Stephen E., joint author.
III.   Title.
QD272.S6C64 1975   547'.3463   75-1270
ISBN 0-12-182552-3

CONTENTS

CONTENTS

Chapter 10

ETHERS, ALCOHOLS, AND PHENOLS

Chapter 11

AMINES, C=N, AND N=O COMPOUNDS

Chapter 12

COMPOUNDS, CONTAINING BORON, SILICON, PHOSPHORUS, SULFUR, OR HALOGEN 335

Chapter 13

MAJOR SPECTRA–STRUCTURE CORRELATIONS BY SPECTRAL REGIONS 365

Chapter 14

THE THEORETICAL ANALYSIS OF
MOLECULAR VIBRATIONS 459

PREFACE

In this second edition the general plan of the first edition has been retained, but large sections have been rewritten or expanded and many new illustrations have been added. As in the first edition, both infrared and Raman spectroscopy are discussed, covering some theoretical aspects and experimental considerations, but with the greatest emphasis on detailed group frequency correlations.

The chapter on vibrational and rotational spectra has been rewritten and expanded. The theoretical aspects are presented on an introductory level. While there are no detailed quantum mechanical calculations presented, infrared and Raman spectra are discussed from the quantum viewpoint. However, classical analogies, which are usually easier to understand, are often referred to for clarification. While much of the theory in this chapter is introductory and rather general, some details of the vibrational–rotational spectra of diatomic molecules are presented.

The chapter on experimental consideration now includes sections on laser Raman sources, Fourier transform spectroscopy and integrated intensities, with supplemental material on internal reflection spectroscopy and other subjects. As in the first edition, instrumental aspects are covered rather broadly, but some aspects of sample handling techniques are fairly detailed.

The chapter on molecular symmetry and elementary group theory has been expanded somewhat. There is a fuller development of the meaning of the character tables and their uses in spectroscopic problems, including a more complete discussion of the effects of symmetry on infrared and Raman activity.

A large portion of this book is devoted to detailed group frequency correlations. The vibrational origin of group frequencies has been expanded and is emphasized throughout the text. The authors feel that knowledge of vibrational interaction and coupling effects can be very useful in the ordinary applications of spectral interpretation using group frequencies. Pertinent new material has been added to the group frequency discussions, particularly when it applies to new band assignments. Some Raman group frequency correlations have been added, particularly when the bands in question are weak in the infrared spectra. As in the first edition, group frequencies are presented in text form with selected references, in spectra–structure correlation chart form, and in the form of many spectral examples which serve to illustrate group frequencies. These spectra have been expanded with the addition of some high resolution spectra in the carbon–hydrogen stretching region and some new spectra for problem analysis. In addition, a number of new Raman spectra have been included to illustrate Raman group frequencies.

Since the full understanding of group frequency vibrations is based on vibrational analysis, it is helpful to have some understanding of how the vibrational analyses of molecules are achieved so that the spectroscopist can better evaluate the literature on this subject. A method for calculating fundamental frequencies is discussed and a detailed example (chloroform) is worked out. The approach is not new but, as in the first edition, the emphasis lies in developing various steps required in a detailed manner not readily available elsewhere. While some use of group theory and matrix algebra is made, no detailed previous knowledge of these fields is necessary as sufficient material on these fields is presented. This section has been expanded to include the calculation of the form of the normal coordinates, the potential energy distributions, and other related topics. The purpose is not to present a complete coverage of all the ramifications of the normal coordinate method, but rather to outline a guide for the beginner through the various details of the procedures that are covered.

NORMAN B. COLTHUP
LAWRENCE H. DALY
STEPHEN E. WIBERLEY

CHAPTER 1

VIBRATIONAL
AND ROTATIONAL SPECTRA

1.1 Introduction

The energy of a molecule consists partly of translational energy, partly of rotational energy, partly of vibrational energy, and partly of electronic energy. For a first approximation these energy contributions can be considered separately. Electronic energy transitions normally give rise to absorption or emission in the ultraviolet and visible regions of the electromagnetic spectrum. Pure rotation gives rise to absorption in the microwave region or the far infrared. Molecular vibrations give rise to absorption bands throughout most of the infrared region of the spectrum. In this book we shall mainly be concerned with the interaction of electromagnetic radiation with molecular vibrations and rotations.

Electromagnetic radiation is characterized by its wavelength λ, its frequency v, and its wavenumber $\bar{v}$. The wavenumber expressed in cm^{-1} is the number of waves in a 1 cm long wavetrain. The wavenumber is related to the other units by

$$\bar{v} = \frac{v}{c} \qquad \bar{v} = \frac{1}{\lambda} \qquad (1.1)$$

where c is the velocity of light, 2.997925×10^{10} cm/sec, v is the frequency in cycles per second (sec^{-1}) or Hertz (Hz), and λ is the wavelength in cm. In terms of these units, the ultraviolet, visible, infrared, and microwave regions of the electromagnetic spectrum assume the values shown in Table 1.1.

In the infrared region of the electromagnetic spectrum the practical unit for wavelength is 10^{-4} cm or 10^{-6} m. This unit has long been called the micron, μ, but recently has been renamed the micrometer, μm. Wavelength is a property of radiation but not a property of molecules. The properties that radiation and molecules have in common are energy and frequency. The frequency in Hz in the infrared part of the spectrum is an inconveniently large

TABLE 1.1

VALUES FOR λ, $\bar{\nu}$, AND ν

Region	λ (cm)	$\bar{\nu}$ (cm^{-1})	ν (Hz)
Ultraviolet			
(far)	1×10^{-6} to 2×10^{-5}	1×10^{6} to 50,000	3×10^{16} to 1.5×10^{15}
(near)	2×10^{-5} to 3.8×10^{-5}	50,000 to 26,300	1.5×10^{15} to 7.9×10^{14}
Visible	3.8×10^{-5} to 7.8×10^{-5}	26,300 to 12,800	7.9×10^{14} to 3.8×10^{14}
Infrared			
(near)	7.8×10^{-5} to 2.5×10^{-4}	12,800 to 4000	3.8×10^{14} to 1.2×10^{14}
(middle)	2.5×10^{-4} to 5×10^{-3}	4000 to 200	1.2×10^{14} to 6×10^{12}
(far)	5×10^{-3} to 1×10^{-1}	200 to 10	6×10^{12} to 3×10^{11}
Microwave	1×10^{-1} to 1×10^{2}	10 to 0.01	1×10^{12} to 3×10^{8}

number and the wavenumber (cm^{-1}) which is proportional to frequency [see Eq. (1.1)] is more commonly used. Infrared spectra have been presented so that the horizontal coordinate is either linear with wavelength (common in NaCl prism instruments) or linear with wavenumber (common in grating instruments). The wavenumber is sometimes referred to as the "frequency in cm^{-1}." The term "frequency in cm^{-1}" is understood to mean the frequency (in Hz) divided by the velocity of light (in cm/sec) or the reciprocal of the radiation wavelength (in cm).

The vibrational and rotational frequencies of molecules can be studied by Raman spectroscopy as well as by infrared spectroscopy. While they are related to each other, the two types of spectra are not exact duplicates and each has its individual strong points. In Raman spectroscopy, only the wavenumber is used.

1.2   Photon Energy

According to the quantum theory the energy of a photon E_p is given by

$$E_p = h\nu \qquad \text{or} \qquad E_p = hc\bar{\nu} \qquad (1.2)$$

where h is Planck's constant, 6.6256×10^{-27} erg-sec, or 6.6256×10^{-34} joule-sec. This photon energy may be absorbed or emitted by a molecule in which case the rotational, vibrational, or electronic energy of the molecule will be changed by an amount ΔE_m. According to the principle of conservation of energy

$$\Delta E_m = E_p = h\nu = hc\bar{\nu} \tag{1.3}$$

If the molecule gains energy, ΔE_m is positive and a photon is absorbed. If the molecule loses energy, ΔE_m is negative and a photon is emitted. A commonly used form of Eq. (1.3) is

$$\bar{\nu} = \frac{\Delta E_m}{hc} \tag{1.4}$$

which reads: The wavenumber of the absorbed or emitted photon is equal to the change in the molecular energy term expressed in cm^{-1} (E/hc).

1.3 Degrees of Freedom of Molecular Motion

In the study of molecular vibrations we can start with a classical model of the molecule where the nuclei are represented by mathematical points with mass. The internuclear forces holding the molecule together are assumed to be similar to those exerted by massless springs which tend to restore bond lengths or bond angles to certain equilibrium values. Each mass requires three coordinates to define its position, such as x, y, and z in a cartesian coordinate system for example. As a result it has three independent degrees of freedom of motion, in the x, y, or z direction. If there are N atomic nuclei in the molecule, there will be a total of $3N$ degrees of freedom of motion for all the nuclear masses in the molecule.

The center of gravity of the molecule requires three coordinates to define its position and therefore has three independent degrees of freedom of motion which are translations of the center of gravity of the molecule. When a nonlinear molecule is in its equilibrium configuration, it requires three rotational coordinates to specify the molecular orientation about the center of gravity. For example, these can be three angular coordinates specifying rotation about three mutually perpendicular axes, each going through the center of gravity. A nonlinear molecule, therefore, has three independent rotational degrees of freedom. A linear molecule only has two independent rotational degrees of freedom about two mutually perpendicular axes, perpendicular to the molecular axis. Rotation of a linear molecule about the molecular axis is not considered a degree of freedom of motion since no displacements of nuclei are involved. After subtracting the translational and rotational degrees of freedom from the total $3N$ degrees of freedom, we are left with $3N - 6$ *internal* degrees of freedom for a nonlinear molecule and $3N - 5$ *internal* degrees of freedom for a linear molecule. Translations of the center of gravity and rotations about the center of gravity can all take place independently without any change occurring in the shape of the molecule. The internal degrees of freedom change the shape of the molecule without moving the center of gravity and without rotating the molecule.

1.4   Normal Modes of Vibration

It can be shown (Chapter 14) that the $3N - 6$ internal degrees of freedom of motion of a nonlinear molecule correspond to $3N - 6$ independent normal modes of vibration. *In each normal mode of vibration all the atoms in the molecule vibrate with the same frequency and all atoms pass through their equilibrium positions simultaneously.* The relative vibrational amplitudes of the individual atoms may be different in magnitude and direction but the center of gravity does not move and the molecule does not rotate. If the forces holding the molecule together are linear functions of the displacement of the nuclei from their equilibrium configurations, then the molecular vibrations will be harmonic. In this case each cartesian coordinate of each atom plotted as a function of time will be a sine or cosine wave when the molecule performs one normal mode of vibration (see Fig. 1.1).

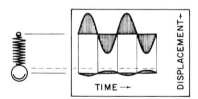

FIG. 1.1.   Normal mode of vibration of a ball and spring model of a diatomic molecule such as HCl. The displacement versus time plot for each mass is a sine wave and the center of gravity (dashed line) is motionless.

1.5   Mechanical Molecular Models

Normal modes of vibration can be demonstrated experimentally using vibrating mechanical models.[1] One such arrangement is illustrated in Fig. 1.2. A source of oscillation is provided by an eccentric shaft on the side of the main shaft of a motor whose rotational speed is adjustable. A ball and spring

[1] C. F. Kettering, L. W. Shutts, and D. H. Andrews, *Phys. Rev.* **36**, 531 (1930).

model of a molecule such as CO_2 is suspended by long threads, one attached to each ball. This leaves the model free to move in the horizontal plane. A short length of thread is used to connect the eccentric to the suspension thread of an end ball about 6 in. above the ball. This horizontal coupling thread is held at an angle to the model axis so that as the motor rotates the eccentric can exert weak periodic forces on the model, both along and across the model axis. As the motor speed is varied, the frequency of the disturbing force is changed. When the disturbing force frequency matches one of the natural frequencies of vibration of the model, resonance occurs. The model responds by performing one of its normal modes of vibration (see Fig. 1.2)

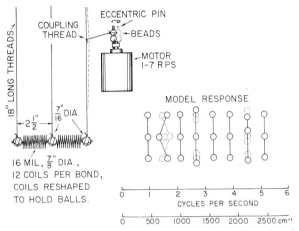

FIG. 1.2. Demonstration of molecular vibrations using ball and spring molecular models, in this case of a molecule such as CO_2.

where it can be seen that all the masses perform simple harmonic motions with the same frequency and go through their equilibrium positions simultaneously. In some modes some masses may vibrate with zero amplitude, that is, they may remain motionless. At a different disturbing force frequency another normal mode of vibration may be activated. When the disturbing force frequency does not coincide with the frequency of a normal mode of vibration, the model remains nearly motionless.

Mechanical molecular models have their limitations because the masses are not mathematical points and the helical springs are not massless and are only an approximate representation of the molecular force field. For demonstration purposes, however, the relative frequencies obtained from a properly constructed model agree sufficiently well with the relative frequencies of the actual molecule. The ratio of the stretching to bending force constants of the spring should be adjusted so that the ratio is about 8 or 10 to 1 which is approximately the ratio found in real molecules. This ratio can be increased if too small by permanently stretching out the spring a little. The model specifications in Fig. 1.2 are not critical. If larger masses are used, stiffer springs can be used. However, for clarity of visual observation, the frequencies should be in the 1–7 cycles/sec range.

If, instead of being excited with the motor, the model is simply struck with a hammer, the resulting motions appear to be erratic and nonsinusoidal. However, closer inspection will show that the motions consist of a simultaneous performance of all the normal modes of vibrations excited previously plus translations and rotations. In the case of the model in Fig. 1.2, free "translations" and "rotations" become pendulum oscillations of the model as a whole because of the suspension system, but these oscillations are made considerably lower in frequency than the lowest vibrational frequency and therefore cause no difficulties.

The linear CO_2 molecule has $3N - 5$ or 4 fundamental modes of vibration. The model discussed here has only three because it is constrained by the suspension system to move in a plane. The fourth fundamental is an out-of-plane bending vibration identical in character and frequency to the in-plane bending vibration performed by the model. Such pairs of vibrations which have the same frequency are called *doubly degenerate* vibrations. A bending vibration in any direction perpendicular to the molecular axis can be described as a linear combination of the two fundamental components.

1.6   Coordinates Used to Describe Molecular Vibrations

The stretching vibrations of the model in Fig. 1.2 representing a molecule like CO_2 are illustrated in Fig. 1.3 along with several types of coordinates used for molecular vibrations. *Cartesian displacement coordinates* X measure displacement from the equilibrium position of each nuclear mass using cartesian coordinates. Each atom has its own cartesian coordinate system with the origin defined by the equilibrium position of the atom. *Internal coordinates* measure the change in shape of a molecule as compared to its equilibrium shape without regard for the molecules' position or orientation in space. For example, there can be a change in a bond length or a bond angle from its equilibrium value. In Fig. 1.3 the bond length internal coordinates are represented by S_1 and S_2, where S_1 is one bond length at any time minus the equilibrium bond length, and S_2 is similarly defined.

Normal modes of vibration may be described using either cartesian displacement coordinates or internal coordinates. For every normal mode of vibration all the coordinates vary periodically with the same frequency and go through equilibrium simultaneously. The form of the normal mode of

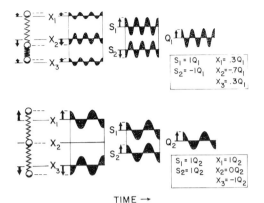

TIME →

FIG. 1.3. The stretching vibrations of a linear triatomic model and types of coordinates used to characterize the vibrations. The cartesian coordinates X are mass displacements. The internal coordinates S are bond length changes. In this case $S_1 = X_1 - X_2$ and $S_2 = X_2 - X_3$. During the performance of one normal mode of vibration, these coordinates have values in specified proportions to the value of one oscillating normal coordinate Q.

vibration is defined by specifying the relative amplitudes (which may be positive or negative) of the various coordinates in the set being used. For example, in Fig. 1.3, S_1/S_2 equals minus one and one for the two vibrations, respectively.

Each normal mode of vibration can also be characterized by a single *normal coordinate*, Q, which varies periodically. As one normal coordinate vibrates, every cartesian displacement coordinate and every internal coordinate vibrates, each with an amplitude in specified proportion (positive or negative) to the amplitude of the normal coordinate, the proportions being such that the resulting motion is a normal mode of vibration. Since normal modes of vibration can be excited independently of each other, normal coordinates are also independent in the sense that each normal coordinate makes a separate and independent contribution to the total vibrational potential and kinetic energy. This simplifies the quantum mechanical formulation where contributions from each normal coordinate can be treated separately.

1.7 Classical Vibrational Frequency Formula for a Diatomic Molecule

Let the diatomic molecule be represented by two masses m_1 and m_2 connected by a massless spring (see Fig. 1.4). For simplicity the masses may be allowed to move only along the molecular axis, like beads on a wire, for example. Let the displacement of each mass from equilibrium along the axis be X_1 and X_2. In this case $(X_2 - X_1)$ is the amount the bond length differs

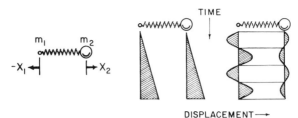

FIG. 1.4. Diatomic molecular model. On the right are the two solutions to the equations of axial motion where mass displacement as a function of time is plotted for translation and vibration.

from the equilibrium length. It will be assumed that the bond "spring" obeys Hooke's law which means that if the bond is not at its equilibrium length, each mass will experience a force equal to a constant, F, times the bond length distortion from equilibrium $(X_2 - X_1)$. We can equate the Hooke's law force, exerted along the spring axis on each mass, to mass m times acceleration (d^2X/dt^2) along this same axis (Newton's law) to obtain two equations, one for each mass.

$$F(X_2 - X_1) = m_1 \frac{d^2X_1}{dt^2} \quad \text{and} \quad -F(X_2 - X_1) = m_2 \frac{d^2X_2}{dt^2} \quad (1.5)$$

When $(X_2 - X_1)$ is positive (stretched spring) the force on mass m_1 is directed in the plus x direction and the force on m_2 is directed in the minus X direction which accounts for the negative sign. The solutions to these equations of motion are

$$X_1 = A_1 \cos(2\pi vt + \alpha) \quad \text{and} \quad X_2 = A_2 \cos(2\pi vt + \alpha) \quad (1.6)$$

which describe simple harmonic motions for both masses, each oscillating as a cosine function of time t. The frequency v is in cycles per second, $2\pi v$ has the units of angular velocity in radians per second, and $2\pi vt$ and α are in radians. We note that while each mass oscillates along the spring axis with a different maximum amplitude A_1 and A_2, each mass oscillates with the *same* frequency v and phase constant α, which means that both masses go through their equilibrium positions simultaneously. Differentiating these equations twice with respect to time t we obtain

$$\frac{d^2X_1}{dt^2} = -A_1 4\pi^2 v^2 \cos(2\pi vt + \alpha)$$

and   (1.7)

$$\frac{d^2X_2}{dt^2} = -A_2 4\pi^2 v^2 \cos(2\pi vt + \alpha)$$

Substituting the values for X_1 and X_2 [Eq. (1.6)] and their second derivatives [Eq. (1.7)] into the original equations of motion [Eq. (1.5)] we obtain, after cancelling the common factor $\cos(2\pi vt + \alpha)$,

$$F(A_2 - A_1) = -m_1 A_1 4\pi^2 v^2 \quad \text{and} \quad F(A_2 - A_1) = m_2 A_2 4\pi^2 v^2 \quad (1.8)$$

These equations can be rearranged to give the amplitude ratios A_1/A_2

$$\frac{A_1}{A_2} = \frac{F}{F - m_1 4\pi^2 v^2} \quad \text{and} \quad \frac{A_1}{A_2} = \frac{F - m_2 4\pi^2 v^2}{F} \quad (1.9)$$

Eliminating A_1/A_2 and solving for v we obtain

$$16\pi^4 m_1 m_2 v^4 - 4\pi^2 F(m_1 + m_2)v^2 + F^2 = F^2$$

so

$$v^4 = v^2 \left(\frac{F(m_1 + m_2)}{4\pi^2 m_1 m_2} \right) \quad (1.10)$$

In one solution we obtain $v = 0$ and substituting this into either amplitude equation [Eq. (1.9)] we obtain $A_1 = A_2$. For any time t [from Eq. (1.6)] we obtain $X_1 = X_2$. This nongenuine zero frequency vibration is actually a translation in the X or spring axis direction. For the other solution we obtain

$$v = \frac{1}{2\pi} \sqrt{F\left(\frac{1}{m_1} + \frac{1}{m_2}\right)} \quad (1.11)$$

where we note the vibrational frequency is independent of amplitude. When this value of v is substituted into either amplitude equation [Eq. (1.9)] we obtain

$$\frac{A_1}{A_2} = \frac{-m_2}{m_1} \quad (1.12)$$

or at any time t [from Eq. (1.6)] $X_1/X_2 = -m_2/m_1$ which indicates that the center of gravity does not move. These two solutions for v correspond to the two degrees of freedom the two masses have in the spring axis direction, (1) translation without vibration and (2) vibration without translation, (see Fig. 1.4). Sometimes [Eq. (1.11)] is written using the reduced mass u as

$$v = \frac{1}{2\pi} \sqrt{\frac{F}{u}} \quad (1.13)$$

where

$$u = \frac{m_1 m_2}{m_1 + m_2} \quad \text{or} \quad \frac{1}{u} = \frac{1}{m_1} + \frac{1}{m_2}$$

Given the frequency and the reduced mass, the force constant can be calculated by

$$F = 4\pi^2 \nu^2 u \qquad (1.14)$$

In Table 1.2 we present some calculated force constants for some diatomic molecules. These are arranged in order of decreasing force constants including triple bonds (N≡N), double bonds (O=O), single bonds (H—Cl) and ionic bonds (Na$^+$Cl$^-$). Reduced masses are given in unified atomic mass units (u)

TABLE 1.2

OBSERVED VIBRATIONAL FREQUENCIES AND CALCULATED FORCE CONSTANTS FOR
SELECTED DIATOMIC MOLECULES

Molecule	$\bar{\nu}$ (cm^{-1})	ν (10^{10} Hz)	u (u)	u (10^{-26} kg)	F (10^2 N m^{-1})
N$_2$	2331	6987	7.002	1.1625	22.4
CO	2143	6425	6.856	1.1384	18.6
NO	1876	5624	7.466	1.2397	15.5
O$_2$	1556	4666	7.997	1.3279	11.4
HF	3958	11,866	0.9430	0.15658	8.7
H$_2$	4159	12,469	0.5039	0.08367	5.1
HCl	2886	8652	0.9800	0.16272	4.8
F$_2$	892	2674	9.499	1.5773	4.5
I$_2$	213	640	63.45	10.536	1.7
NaCl	378	1133	13.950	2.3162	1.2

and kilograms (kg) and force constants are given in newtons per meter (N m^{-1}). In the cgs system, force constants are given in dynes per centimeter. One newton per meter equals 10^3 dyne cm^{-1}. Molecular force constants are often expressed in millidynes per angstrom. One newton per meter equals 10^{-2} mdyne Å^{-1} and 1 dyne cm^{-1} = 10^{-5} mdyne Å^{-1}.

1.8   Infrared Absorption and the Change in Dipole Moment

In the mechanical analogy illustrated in Fig. 1.2 the model analogous to the molecule has certain natural vibrational frequencies. The eccentric on the motor provides an oscillating disturbing force analogous to infrared radiation oscillations which exert forces on the molecule. When the disturbing frequency matches the natural vibrational frequency of the model, the model absorbs energy from the motor and thereby increases its own vibrational energy by vibrating with increased amplitude. At nonresonant frequencies the model is ineffective in absorbing energy from the motor. This is analogous to infrared absorption where the molecule absorbs radiation energy by increasing its own vibrational energy. In a spectrometer the molecule is irradiated with a whole range of infrared frequencies but is only capable of absorbing radiation energy at certain specific frequencies which match the natural vibrational frequencies of the molecule (see Fig. 1.5), and these occur in the infrared region of the electromagnetic spectrum.

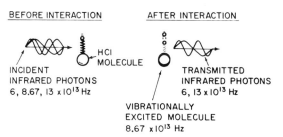

FIG. 1.5. Schematic illustration of infrared absorption. The HCl molecule whose vibrational frequency is 8.67×10^{13} Hz increases its vibrational energy by absorbing the energy of an infrared photon which has this same frequency.

While the absorption *frequency* depends on the molecular vibrational frequency, the absorption *intensity* depends on how effectively the infrared photon energy can be transferred to the molecule, and this depends on the *change in dipole moment* that occurs as a result of molecular vibration. The dipole moment is defined, in the case of a simple dipole (see Fig. 1.6), as the magnitude of either charge in the dipole multiplied by the charge spacing. In a complex molecule this simple picture can be retained if the positive particle represents the total positive charge of the protons concentrated at the center of gravity of the protons, and the negative particle represents the total negative charge of the electrons concentrated at the center of gravity of the electrons. Since the wavelength of infrared radiation is far greater than the size

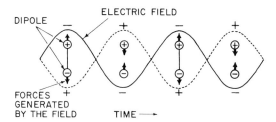

FIG. 1.6. Forces generated on a dipole by an oscillating electric field. These forces tend to alternately increase and decrease the dipole spacing.

of most molecules, the electric field of the photon in the vicinity of a molecule can be considered uniform over the whole molecule. The electric field of the photon exerts forces on the molecular charges and by definition the forces on opposite charges will be exerted in opposite directions (see Fig. 1.6). Therefore the oscillating electric field of the photon will exert forces tending to induce the dipole moment of the molecule to oscillate at the frequency of the photon. At certain frequencies the dipole moment and the nuclei oscillate simultaneously. These are molecular vibrational frequencies where the vibration results in a change in dipole moment. At such a frequency a forced dipole moment oscillation will tend to activate a nuclear vibration. The more the dipole moment changes during a vibration, the more easily the photon electric field can activate that vibration. If a molecular vibration should cause *no* change in dipole moment then a forced dipole moment oscillation cannot activate that vibration. This is summarized in the selection rule that *in order to absorb infrared radiation, a molecular vibration must cause a change in the dipole moment of the molecule*. It can be shown that the intensity of an infrared absorption band is proportional to the square of the change in dipole moment caused by the molecular vibration giving rise to the absorption band.

In the HCl molecule around the electronegative chlorine atom there will be a slight excess of negative charge and around the hydrogen atom there will be a slight excess of positive charge. These two regions can be looked on as constituents of a simple dipole (see Fig. 1.7). During the vibration of the HCl

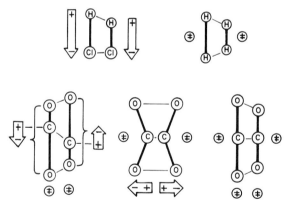

FIG. 1.7.   Dipole moment changes in certain molecular vibrations.

molecule the dipole spacing changes and the excess charge distribution may change also, both of which effects change the dipole moment. Therefore the vibration is infrared active because the electric field of the photon can interact with the changing dipole moment. In the hydrogen molecule, H_2, on the other hand, a center of symmetry is present which implies a zero dipole moment. During the vibration the center of symmetry is retained so the dipole moment does not change since it remains zero. This vibration is infrared inactive, since there is no vibrationally caused change in dipole moment with which the electric field of the photon can interact. In the CO_2 molecule a center of symmetry is present which is retained during the symmetrical, inphase stretching vibration, so this vibration is infrared inactive. If the center of negative charge is assumed to be halfway between the electronegative oxygen atoms and the center of positive charge is assumed to be at the carbon atom, then it can be seen (Fig. 1.7) that the antisymmetrical, out of phase stretching vibration and the bending vibration both cause a change in dipole moment and so are infrared active. It should be clear from the above arguments that if a molecule in its equilibrium configuration has a center of symmetry, then *vibrations during which the center of symmetry is retained will be infrared inactive.*

1.9 *Anharmonicity and Overtones*

Up to now we have been discussing only harmonic vibrations. Mechanical anharmonicity results if the restoring force is not linearly proportional to the nuclear displacement coordinate. Electrical anharmonicity results if the change in dipole moment is not linearly proportional to the nuclear displacement coordinate. If a vibration is mechanically harmonic, the classical picture of a plot of nuclear displacement versus time is a sine or cosine wave (see Fig. 1.8). If mechanical anharmonicity is present this plot will be periodic but not a simple sine or cosine wave. One result of mechanical anharmonicity is that the vibrational frequency will no longer be completely independent of amplitude as it is in the harmonic case.

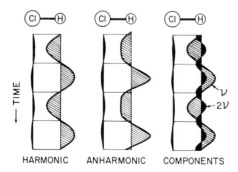

FIG. 1.8. Plots of mass displacement versus time for harmonic and anharmonic vibrations. On the right are the main components of the anharmonic curve in the middle.

If a plot is made of dipole moment versus time for a classical vibration, a periodic but nonsinusoidal wave will result if *either* mechanical *or* electrical anharmonicity is present. However, any such periodic function can be resolved into simple sine or cosine components (see Fig. 1.8) where the frequencies are integral multiples of the fundamental vibrational frequencies (Fourier analysis). This means that if the molecular vibration is anharmonic, the dipole moment will oscillate with the fundamental frequency *and* integral multiples thereof. These are called the fundamental first overtone, second overtone, etc., and these dipole moment oscillations can interact with electromagnetic radiation which has the fundamental frequency and integral multiples thereof. *The intensity of an overtone absorption is dependent on the amount of anharmonicity in the vibration.* Overtones can be detected in the infrared spectrum but they are usually quite weak which implies that although molecular vibrations are measurably anharmonic, the anharmonicity is not great and can be ignored in a reasonably good first approximation.

1.10   Vibrational Potential Function

A vibrational potential function of a diatomic molecule is illustrated in Fig. 1.9 which is a plot of potential energy V versus internuclear distance r. The changes in internuclear potential energy are caused by changes in the electronic and nuclear repulsion energy as a function of internuclear distance. Since vibrating nuclei move much more slowly than electrons, it is assumed that the electronic energy adjusts essentially instantaneously to a change in nuclear position (Born–Oppenheimer approximation). At large internuclear distances the two atomic components attract each other, whereas at small internuclear distances, internuclear repulsion becomes dominant and this results in an equilibrium spacing r_e for the nuclei at some intermediate internuclear

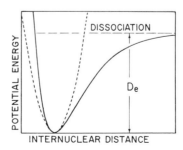

FIG. 1.9.   The potential energy of a diatomic molecule plotted as a function of internuclear distance. The dotted line is the harmonic oscillator potential function.

distance where the energy is at a minimum. At large internuclear distances the energy changes level off at a level of energy which is sufficient to dissociate the molecule.

We usually do not have an exact expression for the forces which hold the nuclei in their equilibrium positions. In point of fact we frequently use infrared and Raman spectroscopy to determine the molecular force field. A commonly assumed first approximation for a diatomic molecule is the harmonic oscillator potential function where the potential energy V as a function of $(r - r_e)$ is

$$V = \tfrac{1}{2}F(r - r_e)^2 \tag{1.15}$$

where r is the internuclear distance, r_e is the equilibrium internuclear distance, and F is a constant. A plot of V vs. $(r - r_e)$ yields a parabola which for sufficiently small values of $(r - r_e)$ is a good approximation for the actual potential energy curve around the minimum (see Fig. 1.9). From classical physics the force is equal to the negative of the derivative of the potential energy with respect to the coordinates. From Eq. (1.15) we get

$$\text{Force} = -\frac{dV}{d(r - r_e)} = -F(r - r_e) \tag{1.16}$$

This can be recognized as Hooke's law where the restoring force is proportional to the bond length distortion or internal coordinate $(r - r_e)$. The force constant F is equal to the second derivative of the potential energy [Eq. (1.15)] with respect to the internal coordinate $(r - r_e)$:

$$F = \frac{d^2V}{d(r - r_e)^2} \tag{1.17}$$

An often used empirical equation called the Morse function[2] (illustrated in Fig. 1.9) is a useful better approximation than the simple quadratic function for the diatomic molecule potential energy, especially at larger than infinitesimal displacements.

$$V = D_e(1 - e^{-\beta(r - r_e)})^2 \tag{1.18}$$

In this equation, V is the potential energy as a function of $(r - r_e)$, β is a constant, and D_e is the energy required to dissociate the molecule measured from the curve minimum (see Fig. 1.9). For very small values for $(r - r_e)$

[2] P. M. Morse, *Phys. Rev.* **34**, 57 (1929).

near equilibrium, the exponent in Eq. (1.18) is approximated by expansion $(e^x \cong 1 + x)$ to give

$$V = D_e \beta^2 (r - r_e)^2 \qquad (1.19)$$

Therefore, near equilibrium, the Morse function resembles the harmonic oscillator function [Eq. (1.15)] where from Eq. (1.17) and Eq. (1.14) $F = 2D_e \beta^2 = 4\pi^2 v_e^2 u$ from which

$$\beta = v_e \sqrt{\frac{2u\pi^2}{D_e}} \qquad (1.20)$$

where v_e is the vibrational frequency for an infinitesimal amplitude and u is the reduced mass. Other slightly more accurate functions have also been proposed.[3]

The potential energy for a diatomic molecule as a function of the internal coordinate S (which is $r - r_e$) can be expanded in a Taylor series about the equilibrium configuration so that

$$V = V_e + \left(\frac{\partial V}{\partial S}\right)_e S + \frac{1}{2}\left(\frac{\partial^2 V}{\partial S^2}\right)_e S^2 \qquad (1.21)$$

plus terms of order S^3 and higher powers. The first term V_e is the energy at equilibrium which can be defined as zero for the vibrational problem. In the second term $(\partial V/\partial S)_e$ is the linear rate of change of potential energy at the equilibrium position and is equal to zero. For sufficiently small amplitudes the cubic and higher order terms can be neglected and we are left with the quadratic terms only which is the harmonic approximation. If we equate $(\partial^2 V/\partial S^2)_e$ with a force constant F, Eq. (1.21) becomes identical with Eq. (1.15).

In the most general case, the potential energy for a polyatomic molecule as a function of the internal coordinates S can be expanded in a similar manner to give

$$V = \frac{1}{2}\sum_i \sum_j F_{ij} S_i S_j \qquad (1.22)$$

where

$$F_{ij} = \left(\frac{\partial^2 V}{\partial S_i \partial S_j}\right)_e$$

[3] E. R. Lippincott and R. Schroeder, *J. Chem. Phys.* **23**, 1131 (1955).

where cubic and higher order terms are neglected in the harmonic approximation. The most important force constants are those where $i = j$ since those where $i \neq j$ are usually small when internal coordinates are used and are frequently set equal to zero in approximate treatments.

1.11   Introduction to the Quantum Effect

The wave-particle duality of matter is expressed in the De Broglie relationship

$$p = \frac{h}{\lambda} \tag{1.23}$$

where p is the momentum mv (normally a characteristic of particles), λ is the wavelength (normally a characteristic of waves), and h is Planck's constant. The wavelengths of macroscopic objects are much too small to give rise to any observable effects. For example, a rifle bullet weighing 2 gm and moving with a velocity of 3.3×10^4 cm/sec has a theoretical De Broglie wavelength of 10^{-31} cm or 10^{-23} Å. However, the hydrogen nucleus with a mass of 1.67×10^{-24} gm, moving as it can in some molecular rotations and vibrations with a velocity on the order of 5×10^5 cm/sec, has a wavelength on the order of 0.8 Å. Since such a wavelength is about the same order of magnitude as various molecular dimensions the wave properties of matter will lead to definite observable effects on the molecular level.

Since a wavelength is involved this means there must be some sort of wavefunction ψ that specifies the amplitude of the wave as a function of the coordinates. Schrödinger presented a generalized wave equation involving ψ which was generally applicable to spectroscopic problems. (See books on quantum mechanics.) The physical interpretation of the amplitude of the wavefunction (ψ) used in the Schrödinger equation was provided by Born. He specified that the square of the amplitude of the wavefunction ($|\psi|^2$ if ψ is complex) gives the *probability* of finding the particle at a given location as a function of the coordinates.

To illustrate the quantum effect let us consider a simple example of *oscillation* where a particle is free to move in one direction only, back and forth between the walls of a box. For example, we could picture an idealized bead moving on a frictionless horizontal wire and rebounding elastically when it strikes the walls (see Fig. 1.10). Within the walls of the box the potential energy is constant and, because of the lack of friction or external energy, the

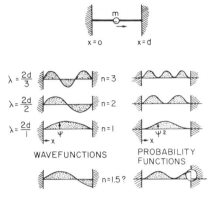

FIG. 1.10. Wavefunctions and probability functions for a small mass oscillating between two walls on a hypothetical "guide wire."

kinetic energy ($\frac{1}{2}mv^2$) must also be constant. This means that the square of the momentum (p^2) of the particle will also be constant because

$$\text{Kinetic energy} = \frac{mv^2}{2} = \frac{m}{m}\frac{mv^2}{2} = \frac{(mv)^2}{2m} = \frac{p^2}{2m} \qquad (1.24)$$

Since p^2 is constant we can substitute for it its De Broglie equivalent h^2/λ^2 to get for the energy E (in excess of the constant potential energy)

$$E = \frac{h^2}{2m\lambda^2} \qquad (1.25)$$

where we note that energy is a function of a wavelength.

One type of wavefunction with a specific wavelength is a sine wave. The wavefunction (ψ) for a stationary sinusoidal wave and its square, are given below.

$$\psi = A\sin(2\pi x/\lambda) \qquad \psi^2 = A^2\sin^2(2\pi x/\lambda) \qquad (1.26)$$

where ψ is the amplitude as a function of the coordinate x, A is the maximum amplitude, and λ is the wavelength. For the oscillating particle in question all appropriate probability functions (ψ^2) within the box must become zero when the particle reaches a wall since the probability function for finding the particle beyond the wall is zero, and there cannot be two different probabilities for finding the particle at the same location (at the wall in this case). This boundary condition restricts the choice of wavefunctions (ψ) within the box which must also become zero at both walls. A suitable choice could be a stationary sinusoidal wavefunction [Eq. (1.26)] where the distance between the walls, d, is an integral multiple, n, of half-wavelengths ($\lambda/2$) of the wave-functions as illustrated in Fig. 1.10. In this case

$$d = n(\lambda/2) \quad \text{and} \quad \lambda = 2d/n \quad \text{when } n = 1, 2, 3, \ldots \quad (1.27)$$

in which case the wavefunction is given by

$$\psi = A \sin(n\pi x/d) \quad (1.28)$$

where, as required, $\psi = 0$ at the walls where $x = 0$ or $x = d$. When the wave-length λ [Eq. (1.27)] is substituted into the energy expression [Eq. (1.25)] we get

$$E = n^2\left(\frac{h^2}{8md^2}\right) \quad (1.29)$$

Here the only variable in the energy equation for a given system is the quantum number n which can have only integral values. This example serves to show that *energy is quantized as a natural consequence of the wave characteristics of particles*. The probability functions (ψ^2) for these quantized energy states are illustrated in Fig. 1.10.

As another example let us consider a simple type of *rotation* where a particle is constrained to move in a circle of radius r. For example, we could consider an idealized bead on a horizontal frictionless circular wire (see Fig. 1.11).

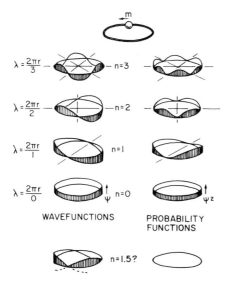

FIG. 1.11. Wavefunctions and probability functions for a small mass rotating in a circle on a hypothetical "guide wire."

We will seek solutions where a probability function (ψ^2) gives a single number for the probability of finding the particle at a particular location on the circumference. This restriction on ψ^2 means that suitable wavefunctions ψ could consist of sine or cosine waves on the circumference which repeat themselves after the full circumference has been traversed, otherwise the wave is destroyed by interference. This means that the circumference ($2\pi r$) must be an integral multiple of the wavelength λ (see Fig. 1.11) so that

$$2\pi r = n\lambda \qquad n = 0, 1, 2, 3, \ldots \qquad (1.30)$$

As in the particle in the box example, the potential energy is constant so the energy in excess of this constant potential energy is the same as that given previously by Eq. (1.25).

$$E = \frac{h^2}{2m\lambda^2} \qquad (1.31)$$

Substitution into this the value for λ from Eq. (1.30) we get

$$E = n^2 \left(\frac{h^2}{8\pi^2 r^2 m} \right) \qquad n = 0, 1, 2, 3, \ldots \qquad (1.32)$$

where again for a given system the only variable in the energy equation is the quantum number n which can have only integral values. The probability functions (ψ^2) for these quantized energy states are illustrated in Fig. 1.11. Note that when $n = 0$, $\lambda = \infty$ [Eq. (1.30)] and the momentum p is zero [Eq. (1.23)] which means the particle is not moving. The uniform probability function implies that the particle may be found with equal probability anywhere on the ring which is acceptable. The angular coordinate of the particle is infinitely uncertain. For the particle in the box, if $n = 0$ then $\lambda = \infty$ [Eq. (1.27)] which is unacceptable because a nonzero amplitude wave with an infinitely long wavelength cannot be drawn whose amplitude will become zero at the walls as required to avoid having two probability values here. Since λ cannot be infinite *the momentum can never be zero* for the oscillating particle which can never stop moving.

The Schrödinger equation is not limited to cases where the potential energy is constant and gives correct solutions for molecular vibration and rotation problems. In these cases the wavefunctions are not simple sinusoidal waves but, as will be seen, bear some resemblance to the artificially simple solutions presented in this section.

1.12   The Quantum Mechanical Harmonic Oscillator

Up to now we have been discussing the molecular vibrational problem in purely classical terms and we should now discuss how quantum mechanics affects the problem. Consider the classical case of a harmonic oscillator consisting of a single mass on a spring free to move horizontally in the X direction only (see Fig. 1.12). If the mass is manually pulled out and held at a distance

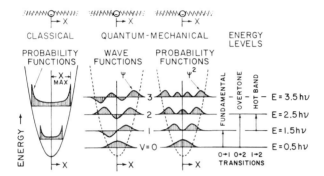

FIG. 1.12.   The harmonic oscillator. The energy levels, wavefunctions, and probability functions for the quantum mechanical oscillator are illustrated along with the energy function and probability functions for the classical oscillator.

X_{max}, measured from the equilibrium position, the potential energy is $FX_{max}^2/2$ and the kinetic energy is zero. If the mass is released at this point, the total energy is

$$E = \tfrac{1}{2}FX_{max}^2 \qquad (1.33)$$

This energy must be conserved subsequently so when the mass reaches equilibrium where the potential energy is zero, the kinetic energy is necessarily $FX_{max}^2/2$. As a result the mass overshoots the equilibrium position and continues until it reaches the position $-X_{max}$ where again the energy will be entirely potential, and the cycle is reversed. In other words, oscillation occurs between the limits plus and minus X_{max}. In the classical oscillator the total vibrational energy $FX_{max}^2/2$ is continuously variable since it depends on X_{max} (the vibrational amplitude), and this may theoretically have any value.

For the diatomic molecule the same results apply except that the X displacement coordinate is replaced by the internal coordinate $(r - r_e)$. For polyatomic molecules, each nonlinear molecule acts as though it consisted of

$3N - 6$ separate harmonic oscillators in each of which the X displacement coordinate is replaced by the appropriate normal coordinate.

The results from quantum mechanical studies show that molecular energy including vibrational energy is not continuously variable but is quantized. This means that vibrational energy may have only certain discrete values called energy levels. The vibrational energy cannot have a value intermediate between these energy levels for any definite period of time but under certain conditions it may jump from one level to another. A quantum mechanical treatment shows that the harmonic oscillator discussed above has energy levels E given by

$$E_{vib} = (v + \tfrac{1}{2})hv \qquad (1.34)$$

where h is Planck's constant, v is the classical vibrational frequency of the oscillator, and v is a quantum number which can have only integer values $(0, 1, 2, 3, \ldots)$. Note that the lowest possible vibrational energy which occurs when $v = 0$ is not zero but $hv/2$ so that the oscillator must always retain a little vibrational energy. If the vibrational energy could become zero then we would simultaneously know both the position (equilibrium position) and the momentum (zero) of the oscillator and this would violate Heisenberg's uncertainty principle, one of the fundamental principles of quantum mechanics.

The energy levels are shown in Fig. 1.12 as horizontal lines. These energy levels have energies equal to those of the classical oscillator when it vibrates with an amplitude indicated by where the dotted parabola crosses the energy level. This classical amplitude can be calculated by equating the classical energy equation [Eq. (1.23)] with quantum mechanical energy equations [Eq. (1.34)]

$$E_{vib} = \tfrac{1}{2}F(r - r_e)^2_{max} = (v + \tfrac{1}{2})hv \qquad (1.35)$$

For the diatomic molecule F is equal to $4\pi^2 v^2 u$ [Eq. (1.14)] so

$$(r - r_e)_{max} = \frac{1}{2\pi}\sqrt{\frac{(2v + 1)h}{vu}} = 5.807\sqrt{\frac{2v + 1}{\bar{v}u}} \qquad (1.36)$$

In the right-hand expression $\bar{v}$ is in cm^{-1}, u is in unified atomic mass units, and $(r - r_e)_{max}$ is in angstroms. For HCl, $\bar{v} = 2886$ cm^{-1}, and $(r - r_e)_{max}$ for the $v = 0$ level is ± 0.11 Å to be compared with the equilibrium bond length 1.275 Å.

The methods of quantum mechanics indicate that we have indefinite knowledge of the position of a mass during the vibration. However, the *probability* of finding the mass at a given position X can be calculated for each vibrational energy level and some of these probability functions are illustrated in Fig. 1.12. It can be seen that these wavelike functions have nodes where at certain values of X the probability of finding the mass is zero. The quantum number v for the level is the same as the number of nodes not counting those at infinity. The quantized levels of energy arise because particles have wavelike properties and time-independent standing waves have an integral number of nodes. For higher quantum numbers the probability function starts to resemble the classical oscillator probability function where the probability of finding the mass is greatest at a position where the velocity is slowest at the extremes of the vibration (see Fig. 1.12).

Also note that the quantum mechanical probability functions indicate a small but finite probability that the mass will be found beyond the classical oscillator amplitude limit (beyond the dotted parabola). In these regions the potential energy is larger than the total energy which implies a negative value for what normally corresponds to classical kinetic energy.

1.13   The Boltzmann Distribution Function

In an assembly of a large number of diatomic molecules of a particular type in thermal equilibrium, the relative populations of the various energy levels can be calculated by the Boltzmann distribution function which for non-degenerate energy states is

$$\frac{n_i}{n_0} = \frac{e^{-E_i/kT}}{e^{-E_0 kT}} \qquad \text{or} \qquad \frac{n_i}{n_0} = e^{-(E_i - E_0)/kT} \tag{1.37}$$

Here n_i and n_0 are the numbers of molecules in the ith and zero energy levels, $(E_i - E_0)$ is the energy difference between the levels, k is Boltzmann constant 1.38054×10^{-16} erg $^\circ K^{-1}$ or 1.38054×10^{-23} J $^\circ K^{-1}$, and T is the absolute temperature. For the vibrational case the lowest possible level is that where $v = 0$ so $(E_i - E_0) = v_i hc\bar{v}$ [see Eq. (1.34)]. At $300^\circ K$, $kT/hc = 209$ cm^{-1} so in Eq. (1.37) the exponent is $-v_i \bar{v}/209$ for $T = 300^\circ$ K. Since for most diatomic molecules $\bar{v}$ is far greater than 209 cm^{-1} it follows that most molecules at $300^\circ K$ will be in the ground vibrational level, $v = 0$, and only a few in the $v = 1$ level and fewer still in the $v = 2$ level. For HCl with $\bar{v} = 2886$ cm^{-1} the population of the $v = 1$ level is about 10^{-6} times as great as the population of the $v = 0$ level at $300^\circ K$.

1.14   Vibrational Transitions and Infrared Absorption

The energy E_p of a photon is equal to hv_p where v_p is the frequency of the photon. Under certain circumstances an unsymmetrical diatomic molecule may be induced to increase its vibrational energy by an amount ΔE_m by absorbing the energy of the photon E_p in which case

$$\Delta E_m = E_p = hv_p \tag{1.38}$$

The only variable in the vibrational energy equation $E_m = (v + \frac{1}{2})hv_m$ is the quantum number v. If the quantum number changes by $+1$ then the diatomic molecule gains energy by an amount ΔE_m equal to the above energy equation subtracted from the same equation where $v + 1$ is substituted for v.

$$\Delta E_m = [(v + 1) + \frac{1}{2}]hv_m - (v + \frac{1}{2})hv_m$$

so $\tag{1.39}$

$$\Delta E_m = hv_m$$

By combining Eqs. (1.38) and (1.39) it is clear that

$$v_p = v_m \tag{1.40}$$

The photon which has the right energy to increase the vibrational quantum number by one has a frequency equal to the classical vibrational frequency of the molecule. If the quantum number should change by -1 then $\Delta E_m = -hv_m$ and the molecule loses energy. In this case a photon with the same frequency will be *emitted* by the molecule to maintain the energy balance. In the quantum mechanical harmonic oscillator *the vibrational quantum number may change only by* ± 1. All other transitions are forbidden if the vibration is harmonic. This quantum mechanical selection rule corresponds to the classical picture where the vibrational energy can be changed only when the electronic field of the photon and the molecular dipole moment vibrate at the same frequency.

If the quantum number changes by $+2$ then

$$\Delta E_m = [(v + 2) + \frac{1}{2}]hv_m - (v + \frac{1}{2})hv_m$$

so $\tag{1.41}$

$$\Delta E_m = 2hv_m$$

When Eq. (1.41) is combined with Eq. (1.38) it is clear that

$$v_p = 2v_m \tag{1.42}$$

so that the photon which has the right energy to increase the quantum number by two has a frequency *twice* that of the molecule. If the vibration is harmonic there is no dipole moment component vibrating at this frequency so there is no way to induce this transition. If the vibration is anharmonic, then as we have seen (Section 1.9) there *will* be a small dipole moment component vibrating at 2, 3, 4, etc., times the molecular frequency. This corresponds to the selection rule where in an *anharmonic* quantum mechanical oscillator the quantum number may change by ±1, ±2, ±3, ...

Since from the Boltzmann distribution function most molecules at normal temperatures exist in the ground vibrational level $v = 0$, the allowed transition $v = 1 \leftarrow v = 0$ will dominate the infrared absorption spectrum. This is called the *fundamental transition* and is responsible for most of the infrared absorption of interest to a chemical spectroscopist. Other allowed transitions such as $(v = 2 \leftarrow v = 1)$, $(v = 3 \leftarrow v = 2)$ and so forth will give rise to absorption at the same frequency as the fundamental transition in a harmonic oscillator. However, the relative intensities will be relatively low, since the $v = 1$ and $v = 2$ levels have a relatively low population. These transitions which do not originate from the $v = 0$ level are sometimes called " hot band " transitions. This is because the population of the nonzero levels will increase if the temperature is increased and this will increase the relative intensity of the " hot band."

1.15 The Anharmonic Oscillator

The harmonic oscillator approximation is adequate for most of the features of the spectrum, but some of the finer details such as the appearance of overtones indicate that real molecules have somewhat anharmonic potential functions. Examples of these discussed already are the Morse potential function or the Taylor series expansion with cubic and higher order terms retained (see Section 1.10). A quantum mechanical treatment of the anharmonic oscillator yields the following equation for the energy levels.

$$E_{\text{vib}} = hc\bar{v}_e(v + \tfrac{1}{2}) - hcx_e\,\bar{v}_e(v + \tfrac{1}{2})^2 + \cdots \tag{1.43}$$

Higher order terms are usually not needed. The energy level terms expressed in cm^{-1} (E/hc) are given by

$$\frac{E_{\text{vib}}}{hc} = \bar{v}_e\left(v + \frac{1}{2}\right) - x_e\bar{v}_e\left(v + \frac{1}{2}\right)^2 \cdots \tag{1.44}$$

Unlike the harmonic oscillator energy levels, these are no longer equally spaced. One result is that hot bands ($v = 2 \leftarrow v = 1$, etc.) will no longer have exactly the same frequency as the fundamental ($v = 1 \leftarrow v = 0$) band. This result corresponds to the classical case where frequency is no longer completely independent of amplitude when mechanical anharmonicity is present.

As was stated before, overtone transitions such as $v = 0$ to $v = 2, 3, 4, \ldots$, are allowed in an anharmonic oscillator. However, the nonuniform spacing of the energy levels in an anharmonic oscillator means that an overtone will not be found at exactly $2, 3, 4, \ldots$, times the frequency of the fundamental. The wavenumber $\bar{v}_v$ of the fundamental and its overtones are given by subtracting from Eq. (1.44), the same equation with $v = 0$, and setting

$$\Delta E_{\mathrm{vib}}/hc = \bar{v}_v$$

[see Eq. (1.4)].

$$\bar{v}_v = \bar{v}_e v - x_e \bar{v}_e (v + v^2) \qquad v = 1, 2, 3, \ldots \tag{1.45}$$

In chloroform the vibration which chiefly involves C—H stretching has a much higher frequency (3019 cm^{-1}) than the other chloroform vibrations. As a result of this separation a series of overtones for this vibration can be recognized in the spectrum and are listed in Table 1.3 (measured by one of the authors). In this table the overtone calculations using the harmonic oscillator approximations are simply integral multiples of the fundamental at 3019 cm^{-1}, and these clearly do not match the experimental values very well. The overtone calculations using the anharmonic oscillator Eq. (1.45) with $\bar{v}_e = 3145$ cm^{-1}

TABLE 1.3

THE C—H STRETCHING VIBRATION FOR CHCl$_3$ LIQUID

Transition $v = 0$ to $v =$	Observed wavenumber (cm^{-1})	Anharmonic oscillator calculation[a] (cm^{-1})	Harmonic oscillator calculation[b] (cm^{-1})	Absorbance for a 1 cm cell
1	3019	3019	3019	170
2	5912	5912	6038	15
3	8677	8679	9057	0.55
4	11318	11320	12076	0.02
5	13850	13835	15095	0.001
6	16270	16224	18114	0.0001

[a] $\bar{v} = 3145 v - 63(v^2 + v)$ $v = 1, 2, 3, \ldots$.
[b] $\bar{v} = 3019 v$ $v = 1, 2, 3, \ldots$.

and $x_e \bar{v}_e = 63$ cm^{-1} are clearly a much better fit. The absorbance values in Table 1.3 indicate the drop in absorption intensity for the higher overtones. The highest frequency bands are measured by passing radiation through about a meter of liquid chloroform.

The energy levels for the C—H stretching vibration for chloroform are illustrated in Fig. 1.13 for both the harmonic and anharmonic cases. The Morse function has been assumed for the anharmonic potential function (treating the C—H bond as a diatomic molecule). The experimental dissociation energy, D_0, for the chloroform C—H bond[4] is 95.7 ± 1 kcal/mole which expressed in cm^{-1} is 33460 cm^{-1}. This represents the energy difference between the $v = 0$ level (the lowest possible vibrational energy) and the energy value at

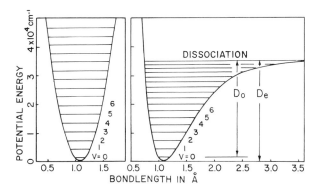

FIG. 1.13. Harmonic and anharmonic potential functions and energy levels for the C—H stretching vibration in CHCl$_3$.

very large bond distances (See Fig. 1.13). In order to get D_e, the spectroscopic dissociation energy measured from the bottom of the curve, the energy of the $v = 0$ level [see Eq. (1.44)] must be added to the experimental dissociation energy D_0. In this case the energy of the $v = 0$ level is 1557 cm^{-1}, which makes the experimental value of D_e about 35,000 cm^{-1}. If the Morse potential function is assumed, an equation can be derived[2]

$$D_e = \frac{\bar{v}_e^2}{4 x_e \bar{v}_e} \tag{1.46}$$

from which approximate values for D_e can be calculated. For the chloroform C—H bond, D_e calculated by this Morse function equation is 39,250 cm^{-1}, which is in rough agreement with experiment.

[4] T. L. Cottrell, "The Strength of Chemical Bonds," p. 186. Butterworth, London, 1958.

1.16   Combination and Difference Bands

In polyatomic molecules, which can be looked on as $3N - 6$ separate oscillators, anharmonicity causes not only the appearance of overtones but weak combination bands and difference bands as well. A combination band appears near the frequency of the sum of two (or sometimes three) fundamental bands. For example, $CHCl_3$, which has a CH stretch band at 3019 cm^{-1} and a CH bend band at 1216 cm^{-1}, also has a weak band at 4217 cm^{-1}, a little lower than the expected summation position at 4235 cm^{-1} because of anharmonicity. This results from a transition from the ground level to a *new* combination level not involved in fundamental transitions. Difference bands like hot bands do not arise from the ground vibrational level and, therefore, are observed less often and will disappear at low temperatures. The transition is from an *excited* level of one vibration, usually of low frequency, to an excited level of another higher frequency vibration. Since a difference band involves no new levels, the wavenumber should be exactly equal to the wavenumber difference of the two fundamentals involved regardless of anharmonicity. In the infrared spectrum of CO_2 the 667 cm^{-1} band, which is the bending vibration fundamental, has weak bands on either side at 721 and 618 cm^{-1}. These are difference bands involving the 667 cm^{-1} band and two bands at 1388 and 1285 cm^{-1} (to be discussed in the next section) active only in the Raman effect, so that $1388 - 667 = 721$ cm^{-1} and $1285 - 667 = 618$ cm^{-1}.

1.17   Fermi Resonance

In a harmonic oscillator-type polyatomic molecule, each normal mode of vibration has at best one spectroscopically active transition from the ground state ($v = 1 \leftarrow v = 0$) associated with it. If anharmonicity is present, new bands involving overtone and combination transitions can appear in the spectrum but usually these are weak. Sometimes, however, the frequency of an overtone or combination band may happen to have nearly the same value as the frequency of another fundamental. In such a case it may happen that two relatively strong bands may be observed where only one strong band for the fundamental was expected. These are observed at somewhat higher and lower frequencies than the expected unperturbed positions of the fundamental and overtone. This effect is called Fermi resonance. Both bands involve the fundamental and both bands involve the overtone. The apparent intensification of the overtone comes from the fact that the fundamental is involved in both bands. In this case the $v = 1$ level of one vibration has an energy close to that of the $v = 2$ level of another vibration. These levels can interact quantum mechanically because of anharmonic terms in the potential energy. These terms may cause a significant perturbation when the energy difference between the two unperturbed levels involved is small. The vibrations involved should be of a type which can be coupled by the anharmonic terms, and the overtone and the fundamental should be of the same symmetry type.

An example of this effect can be seen in the spectra of many aldehydes where the aldehyde C—H in-plane bending vibration has an absorption near 1400 cm^{-1}. The overtone of this vibration is expected near 2800 cm^{-1} and this is close to the expected frequency in cm^{-1} for the aldehyde C—H stretching vibration. In Fig. 1.14 in the benzaldehyde spectrum, strong interaction

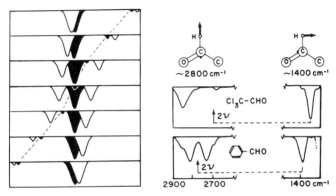

FIG. 1.14. Fermi resonance between an intense fundamental and a normally weak overtone. On the left the theoretical unperturbed bands are shaded and the observed bands are unshaded. On the right $2v$ marks the expected position of the unperturbed overtone of the C—H bending vibration in two aldehydes. Fermi resonance is strongly involved in the splitting of the C—H stretch band in benzaldehyde.

is revealed by the presence of two approximately equal intensity bands near 2800 cm^{-1}, symmetrically displaced about the expected position for the overtone. In the spectrum of chloral the aldehyde C—H bending and stretching vibrations are shifted somewhat and there is no longer such a close coincidence between the bending overtone and stretching fundamental. The two bands near 2800 cm^{-1} are not symmetrically placed about the expected position for the overtone and the intensities are markedly unequal, indicating that much less interaction has occurred.

If it is assumed that the unperturbed overtone has negligible absorption intensity relative to the unperturbed fundamental, the wavenumbers of the observed bands can be corrected for the shift caused by Fermi resonance by

$$\bar{v} = \frac{\bar{v}_1 + \bar{v}_2}{2} \pm \frac{\bar{v}_1 - \bar{v}_2}{2}\left(\frac{\rho - 1}{\rho + 1}\right) \qquad (1.47)$$

Here $\bar{\nu}$ is a corrected wavenumber, $\bar{\nu}_1$ and $\bar{\nu}_2$ are the observed wavenumbers, and ρ is the ratio of the observed absorption intensities of the two bands.[5]

The classic example of Fermi resonance is seen in the Raman spectrum of CO_2. This has two strong bands at 1388 and 1285 cm^{-1} where one strong band was expected for the in-phase stretching vibration fundamental. Interaction occurs between the $v = 1$ level of this vibration and the $v = 2$ (overtone) level of the 667 cm^{-1} bending vibration, the overtone of which is expected near 1334 cm^{-1}.

1.18   Rotation of Linear Molecules

Figure 1.15 shows the hydrogen chloride molecule where m_H is the mass of the hydrogen atom and m_{Cl} is the mass of the chlorine atom. Let G represent the center of gravity of the molecule, d the distance of the chlorine atom

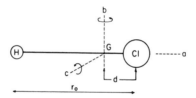

FIG. 1.15.   Hydrogen chloride as a rigid rotator.

from G, and r_0 the equilibrium internuclear distance. According to the lever principle,

$$m_H(r_0 - d) = m_{Cl}d \qquad (1.48)$$

$$d = \frac{m_H r_0}{m_H + m_{Cl}} \qquad (1.49)$$

If it is assumed that the molecule rotates as a rigid rotator about the axis b which passes through the center of gravity, G, and is perpendicular to the internuclear axis, then the moment of inertia I_b about the axis is given by

$$I_b = m_H(r_0 - d)^2 + m_{Cl}d^2 \qquad (1.50)$$

[5] J. Overend, *in* "Infrared Spectroscopy and Molecular Structure" (M. Davies, ed.), Chapter 10, p. 352. Amer. Elsevier, New York, 1963.

Substituting for d from Eq. (1.49)

$$I_b = \frac{m_H m_{Cl}}{m_H + m_{Cl}} r_0^2 \qquad (1.51)$$

and recalling the definition of the reduced mass, u,

$$I_b = u r_0^2 \qquad (1.52)$$

where

$$u_{HCl} = \frac{m_H m_{Cl}}{m_H + m_{Cl}}$$

In other words, the molecule rotating about the axis b has the same moment of inertia I_b as a particle of effective mass u moving in a circle of radius r_0.

The classical rotational energy E_{rot} of a body rotating about an axis is

$$E_{rot} = \tfrac{1}{2} I (2\pi v_r)^2 \qquad \text{or} \qquad E_{rot} = \frac{P^2}{2I} \qquad (1.53)$$

where v_r is the rotational frequency (cycles per second), $2\pi v_r$ is the angular velocity (radians per second), I is the moment of inertia, and P is the angular momentum. The rotational energy and angular momentum may have any value.

The quantum mechanical rotational energy for molecules is not continuously variable. For the rigid diatomic molecule the rotational energy levels E_{rot} are given by

$$E_{rot} = \frac{h^2 J(J+1)}{8\pi^2 I_b} \qquad (1.54)$$

where J is the rotational quantum number which can assume only integer values 0, 1, 2, 3, ... This equation can be written with the energy level terms given in frequency in cm^{-1} (E/hc) as

$$\frac{E_{rot}}{hc} = BJ(J+1) \qquad (1.55)$$

where B, which is called the rotational constant, is defined by

$$B = \frac{h}{8\pi^2 I_b c} \qquad (1.56)$$

As in the vibrational problem the rotating masses cannot be precisely located. According to quantum mechanics one can only calculate the probabilities of finding the masses at certain positions. In a rotating rigid molecule, each mass is found somewhere on a spherical surface of revolution about the center of gravity. An atom of the rotating diatomic molecule has probability functions as a function of bond orientation as illustrated in Fig. 1.16. The regions of higher probability on the spherical surface are represented by darker shading. As seen in Fig. 1.16 there are an integral number of nodal

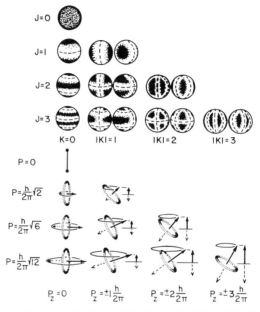

FIG. 1.16. The rigid rotator. Regions of higher probability of finding a mass on the spherical surface of revolution are represented by shading for various values of J and K in the top set of drawings. In the bottom set of drawings a corresponding classical rigid rotator is shown both rotating and precessing in a vertically oriented field. The angular momentum vectors are indicated by arrows (dotted for rotation in the opposite direction).

lines (resembling latitude and longitude lines) where the probability of finding the mass is zero. The total number of nodal lines on the surface is the same as the rotational quantum number J. Again these standing wavefunctions indicate that quantization occurs because of the wave nature of particles. These functions are the same as those for the electron in the hydrogen atom for one particular distance from the nucleus.

The total angular momentum P of the diatomic molecule is given in quantum mechanics by

$$P = \frac{h}{2\pi} \sqrt{J(J+1)} \tag{1.57}$$

However, molecules with identical magnitudes for the total angular momentum vector may have different values for its vector orientation in space. From quantum mechanics the only thing that can be told about the orientation of the angular momentum vector is the angle θ it makes with one specific axis, say the axis of an external electric or magnetic field. This is given by

$$\cos \theta = \frac{K}{\sqrt{J(J+1)}} \qquad (1.58)$$

where $K = 0, \pm 1, \pm 2, \ldots, \pm J$ so that for each value for J there are $2J + 1$ possible values for K. This means that the component of angular momentum in the unique z axis direction, P_z, is quantized so that from the preceding equations

$$P \cos \theta = P_z = \frac{h}{2\pi} K \qquad (1.59)$$

The magnitude of K is the same as the number of "longitude-like" nodal lines (which intersect at the vertical z axis) on the spherical surface in Fig. 1.16. The functions with nonzero K values come in pairs. This corresponds to the fact that the angular momentum component in the z direction may be positive or negative depending on whether the molecule is rotating clockwise or counterclockwise. The time-independent functions illustrated actually result from linear combinations of states where K is an integer with a positive and a negative value. In the bottom of Fig. 1.16 the corresponding classical picture of a rotating diatomic molecule is shown where the angular momentum vector is precessing as a spinning top in a gravitational field. The vertical component of the total angular momentum is time independent and well defined as $0, \pm 1, \pm 2,$ and ± 3 times $h/2\pi$ but the horizontal components are not defined because they are changing with time. The variability of the quantum number K accounts for the fact that there are $2J + 1$ different energy states in each energy level characterized by the quantum number J.

1.19   Rotational Transitions and Infrared Absorption

In Fig. 1.17 is illustrated the classical case of a plane polarized electromagnetic wave whose electric field varies in the x direction. Let this wave approach a rotating molecule with a permanent dipole moment μ. For simplicity allow the x direction to lie in the plane defined by the rotating dipole

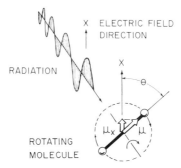

FIG. 1.17.   The radiation's electric field interacting with the rotating dipole moment of a rotating diatomic molecule.

moment. If the angle the dipole moment makes with the x direction is θ, then the component of the dipole moment in the x direction is

$$\mu_x = \mu \cos \theta \qquad (1.60)$$

For the linear molecule θ varies periodically because of rotation and can be written as

$$\theta = 2\pi v_r t \qquad (1.61)$$

where v_r is the rotational frequency, $2\pi v_r$ is the angular velocity (radians per second), and t is time. Combining Eqs. (1.60) and (1.61)

$$\mu_x = \mu \cos(2\pi v_r t) \qquad (1.62)$$

Therefore, the component of the dipole moment μ_x in the direction of the electric field of the radiation oscillates with a frequency v_r. In the more general case the x direction makes an angle ϕ with the plane defined by the rotating dipole moment, in which case the maximum amplitude of μ_x in Eq. (1.62) will not be μ but the constant $\mu \cos \phi$, otherwise the result is the same.

The electric field of the radiation exerts forces on the molecular charges in the x direction and therefore attempts to make μ_x oscillate at the frequency of the photon v_p. When v_p and v_r are nearly the same, the forces exerted by the radiation electric field will remain synchronized with the rotating molecular dipole moment for a sufficiently long time to cause an increase in the rotational energy at the expense of radiation energy. From the above it should be clear that *in order to absorb infrared (or microwave) radiation the rotating molecule must possess a permanent dipole moment.* Homonuclear diatomic molecules such as N_2 or O_2 have no pure rotational infrared spectrum because their permanent dipole moment is zero.

According to the quantum mechanical picture, an increase in molecular rotational energy ΔE_{rot} should equal the energy of the photon E_p whose absorption caused the rotational energy increase.

$$\Delta E_{rot} = E_p = h\nu_p = hc\bar{\nu}_p \tag{1.63}$$

If the rotational quantum number for the linear molecule increases by one, the molecule gains rotational energy ΔE_{rot} given by

$$\Delta E_{rot} = Bhc[(J + 1)(J + 2)] - BhcJ(J + 1) \tag{1.64}$$

This equation is obtained by subtracting Eq. (1.55) from the same equation with $J + 1$ substituted for J. From this one obtains

$$\Delta E_{rot} = 2Bhc(J + 1) \tag{1.65}$$

Combining Eq. (1.63) with Eq. (1.65), one obtains

$$\bar{\nu}_p = 2B(J + 1) \tag{1.66}$$

This equation gives the frequency in cm^{-1} of the photon whose energy has the right value to increase the rotational quantum number from J to $J + 1$. This is the only allowed transition for absorption. Therefore, B can be calculated from the observed frequencies in cm^{-1}. For example, the spacing between two adjacent absorption lines is equal to $2B$ from Eq. (1.66). From B we can calculate the moment of inertia [Eq. (1.56)] and from that the internuclear distance [Eq. (1.51)]. Figure 1.18 shows the energy levels, transitions, and origin of the rotational absorption spectrum of a diatomic molecule.

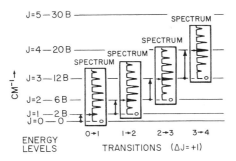

FIG. 1.18. The rotational energy levels and the rotational spectrum of a diatomic molecule. The relation of the spectral lines to the energy levels is illustrated by positioning the spectrum so the zero wavenumber point is on the energy level from which the transition originates.

The quantum mechanical selection rule for absorption, $\Delta J = +1$, corresponds classically to the idea that the electric field radiation oscillating with frequency v_p will be most effective in increasing the molecule rotational frequency v_r'' to some slightly higher frequency v_r' when v_p is intermediate between v_r'' and v_r'. In this case the oscillating electric field of the radiation can remain nearly synchronized in the proper phase with the rotating molecular dipole moment as long as possible while the rotational frequency is changing from v_r'' to v_r'.[6] This rotational frequency increase is a natural consequence of an increase in rotational energy caused by the absorption of the photon energy. This was not the case for vibrational energy changes where the frequency is unaltered.

The classical rotational frequency v_r'' for the linear molecule in the energy level J can be obtained by equating Eq. (1.53), the classical rotational energy, and Eq. (1.54), the quantum mechanical energy,

$$\frac{1}{2}I_b(2\pi v_r'')^2 = \frac{h^2}{8\pi^2 I_b}J(J+1) \tag{1.67}$$

$$(v_r'')_J = 2Bc\sqrt{J(J+1)} \tag{1.68}$$

The classical rotational frequency v_r' for the $J+1$ quantum number is obtained from the above equation by substituting $J+1$ for J.

$$(v_r')_{J+1} = 2Bc\sqrt{(J+1)(J+2)} \tag{1.69}$$

From Eq. (1.66) the frequency of the photon v_p having the right energy to cause transition from J to $J+1$ has a frequency given by

$$v_p = 2Bc(J+1) \tag{1.70}$$

After dividing Eqs. (1.68), (1.69), and (1.70) by $2Bc\sqrt{J+1}$ it can be seen that

$$(v_r'')_J : v_p : (v_r')_{J+1} = \sqrt{J} : \sqrt{J+1} : \sqrt{J+2} \tag{1.71}$$

so that the frequency of the absorbed photon is indeed intermediate between the initial and final classical rotational frequencies of the molecule when the quantum number increases by one.

The reader can verify that the photon which has the proper energy to increase the quantum number by two has a frequency far higher than the initial or final classical rotational frequency involved in such a transition.

[6] C. T. Moyniham, J. Chem. Educ. **46**, 431 (1969).

In fact, it has a frequency a little higher than twice the initial classical rotational frequency but there is no dipole moment component oscillating at this frequency. There is no anharmonicity in rotation. This corresponds to a quantum mechanically forbidden transition.

The pure rotational spectrum of HCl has been measured[7] and the results are shown in Table 1.4. The decreases in the values shown in the last column

TABLE 1.4

FAR INFRARED ABSORPTION BANDS OF HCl AND RELATED CALCULATIONS

Wavenumber, $\bar{\nu}$ (cm^{-1})	Quantum numbers associated with energy levels		$\dfrac{\bar{\nu}}{(J+1)} = 2B$
	Lower	Upper	
	$J \to$	$J+1$	
83.32	3	4	20.830
104.13	4	5	20.826
124.73	5	6	20.788
145.37	6	7	20.767
165.89	7	8	20.736
186.23	8	9	20.692
206.60	9	10	20.660
226.86	10	11	20.624

are caused by centrifugal stretching. In the upper rotational levels the molecule rotates faster and the bond is elongated slightly, resulting in a decrease in the spacing between the levels and a slight increase in the moment of inertia. If the small deviation caused by centrifugal stretching is ignored, then $2B$ equals approximately 20.8 cm^{-1} and $B = 10.4$ cm^{-1}. Now from Eq. (1.56)

$$I_b = \frac{h}{8\pi^2 Bc} = \frac{6.6256 \times 10^{-27}}{8 \times (3.1416)^2 \times 10.4 \times 2.9979 \times 10^{10}} = 2.69 \times 10^{-40} \text{ gm cm}^2 \tag{1.72}$$

The data for HCl was obtained on the gas containing both chlorine isotopes, so the mass based on normal isotope abundance is used here to calculate the reduced mass of HCl.

$$u = \frac{35.5 \times 1}{35.5 + 1} \frac{1}{6.02 \times 10^{23}} = 1.616 \times 10^{-24} \text{ gm/molecule} \tag{1.73}$$

[7] R. L. Hansler and R. A. Oetjen, *J. Chem. Phys.* **21**, 1340 (1953).

From Eq. (1.52) the internuclear distance can be calculated as

$$r_0^2 = \frac{2.69 \times 10^{-40}}{1.616 \times 10^{-24}} = 1.665 \times 10^{-16} \text{ cm}^2$$

and

$$r_0 = 1.29 \times 10^{-8} \text{ cm} = 1.29 \text{ Å}$$

(1.74)

1.20   The Nonrigid Rotator

In a nonrigid rotating diatomic molecule, centrifugal distortion will elongate the bond as the rotational frequency increases. To accurately fit the energy levels, a correction term must be added to Eq. (1.55)

$$E_{rot} = BhcJ(J + 1) - Dhc[J(J + 1)]^2$$

(1.75)

where D is the centrifugal distortion constant. The frequency in cm^{-1}, $\bar{v}_r$ of the photon which has the correct energy to increase the quantum number from J to $J + 1$ is given by subtracting Eq. (1.75) from the same equation with $J + 1$ substituted for J.

$$\frac{\Delta E_{rot}}{hc} = \bar{v}_r = 2B(J + 1) - 4D(J + 1)^3$$

(1.76)

Dividing through by $J + 1$

$$\frac{\bar{v}_r}{J + 1} = 2B - 4D(J + 1)^2$$

(1.77)

This equation is in the same form as that for a straight line of the form $y = a + bx$. If the values of $\bar{v}_r/(J + 1)$ are plotted against $(J + 1)^2$, the intercept equals $2B$ and the slope equals $-4D$. This plot for HCl is shown in Fig. 1.19, yielding $B = 10.44$ cm and $D = 5.2 \times 10^{-4}$ cm.

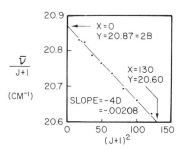

FIG. 1.19.   Plot for the determination of the rotational constants for HCl from the observed rotational frequencies.

1.21 Rotational Line Intensities

In an unsymmetrical diatomic molecule a rotational transition $(J + 1 \leftarrow J)$ will give rise to an individual line in the pure rotational spectrum whose intensity is proportional to the J rotational energy level population n_j as given by the Boltzmann distribution function.

$$\frac{n_j}{n_0} = \frac{g_j e^{-E_j/kT}}{g_0 e^{-E_0/kT}} = \frac{g_j}{g_0} e^{-(E_j - E_0)/kT} \tag{1.78}$$

Here the factor g_j is the number of energy states with the same energy E_j. In the rotational case the lowest level $(J = 0)$ is a single state $(g_0 = 1)$ but the higher levels have a $(2J + 1)$ multiplicity of states in each level (see Fig. 1.14). Therefore, the Boltzmann distribution function for the number n_j of molecules in the J level relative to n_0 (those in the $J = 0$ level) is

$$\frac{n_j}{n_0} = (2J + 1)e^{-J(J+1)Bhc/kT} \tag{1.79}$$

This function is plotted for two temperatures in Fig. 1.20 and can be seen to come to a maximum at a J value given by

$$J_{max} = \sqrt{\frac{kT}{2Bhc} - \frac{1}{2}} \tag{1.80}$$

This equation is obtained by differentiating Eq. (1.79), setting $dn_j/dJ = 0$ and solving for J.

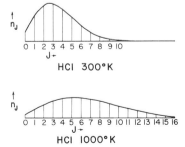

FIG. 1.20.   Relative distribution of HCl molecules in different rotational energy states.

TABLE 1.5

TYPES OF ROTATORS

Type	Moments of inertia	E_{rot}	Structure	Examples
(1) Linear	$I_a = 0, \; I_b = I_c$	$J(J+1)Bhc$	Linear	HCl, HBr, all diatomic molecules, CO_2, C_2H_2, etc.
(2) Spherical top	$I_a = I_b = I_c$	$J(J+1)Bhc$	Tetrahedron Octahedron	CH_4, CCl_4, SiH_4, etc. SF_6, UF_6, etc.
(3) Symmetric top (oblate)	$I_a = I_b \neq I_c$	$J(J+1)Bhc + K^2(C - B)hc$	Structure with one threefold or higher-fold axis	BF_3, C_6H_6, $CHCl_3$
(prolate)	$I_a \neq I_b = I_c$	$J(J+1)Bhc + K^2(A - B)hc$	Structure with one threefold or higher-fold axis	C_2H_6, CH_3Cl, etc.
(4) Asymmetric top	$I_a \neq I_b \neq I_c$	No simple equation	Structure lacking a threefold or higher-fold axis	H_2O, CH_2O, CH_3OH, etc.

1.22 Types of Rotators

In a molecule, the axes about which the moments of inertia are at a minimum and at a maximum are designated the a and c axes, respectively, with b as the axis of intermediate moment of inertia. The moments of inertia about these three mutually perpendicular axes are designated in order of increasing magnitude as I_a, I_b, and I_c. With moments of inertia as a basis, molecules can be classified as rotators into the four distinct types shown in Table 1.5.

Linear molecules have I_a, the moment of inertia about the molecular axis, equal to zero and I_b equal to I_c, which are the moments of inertia about two axes perpendicular to the molecular axis and to each other. Polyatomic linear molecules and diatomic molecules have identical rotational energy equations. However, pure infrared rotational spectra can only be observed for those molecules which possess a permanent dipole moment. In carbon dioxide and acetylene, for example, the permanent dipole moment is zero because of symmetry.

Spherical top molecules which have three equal moments of inertia have a rotational energy equation identical to that for linear molecules. However, since spherical top molecules have no permanent dipole moment (it is equal to zero), no pure infrared rotational spectrum is observed.

Molecules with a threefold or higher axis of symmetry (see Section 3.1) are symmetrical tops and have two equal moments of inertia. The threefold or higher symmetry axis is called the unique axis of the molecule. If the two smaller moments of inertia are equal ($I_a = I_b \neq I_c$), then the molecule is an oblate or coinlike rotator. If the two larger moments of inertia are equal ($I_a \neq I_b = I_c$), then the molecule is a prolate or rodlike rotator. An asymmetric top molecule has three different moments of inertia ($I_a \neq I_b \neq I_c$). The rotational spectrum and the rotational energy expression are complex but some individual molecules such as water have been successfully treated.

1.23 Rotation of Symmetric Top Molecules

The total rotational energy can be expressed as a sum of the rotational energy components about the three principal axes

$$E_{\text{rot}} = \frac{P^2}{2I} = \frac{P_a^2}{2I_a} + \frac{P_b^2}{2I_b} + \frac{P_c^2}{2I_c} \tag{1.81}$$

where P_a, P_b, and P_c are the angular momentum components about the a, b, and c axes. In a prolate symmetrical top $I_b = I_c$ so

$$E_{\text{rot}} = \frac{P_a^2}{2I_a} + \frac{P_b^2 + P_c^2}{2I_b} \tag{1.82}$$

The total angular momentum P is the vector resultant of the angular momentum components, so from the Pythagorean theorem

$$P^2 = P_a^2 + P_b^2 + P_c^2 \quad \text{and} \quad P_b^2 + P_c^2 = P^2 - P_a^2 \tag{1.83}$$

Substituting this value of $P_b^2 + P_c^2$ in Eq. (1.82) and rearranging

$$E_{\text{rot}} = \frac{P^2}{2I_b} + P_a^2\left(\frac{1}{2I_a} - \frac{1}{2I_b}\right) \tag{1.84}$$

which is the classical expression for a symmetric top where a is the unique axis.

For a rigid rotator the results from quantum mechanics were given for the total angular momentum P [Eq. (1.57)] and for the angular momentum about the unique axis P_a [Eq. (1.59)] as

$$P = \frac{h}{2\pi}\sqrt{J(J+1)} \quad \text{and} \quad P_a = \frac{h}{2\pi}K$$

Substituting these quantum mechanical results into Eq. (1.84) for a prolate symmetrical top

$$E_{\text{rot}} = J(J+1)Bhc + K^2(A-B)hc \tag{1.85}$$

where

$$A = \frac{h}{8\pi^2 I_a c} \quad \text{and} \quad B = \frac{h}{8\pi^2 I_b c}$$

For an oblate symmetrical top an identical expression results except that the unique axis is c, and thus A in Eq. (1.85) is replaced by $C = h/(8\pi^2 I_c c)$.

Since the rotation of a symmetric top molecule about its unique or symmetry axis results in no change in dipole moment, infrared radiation cannot change K which characterizes the angular momentum component about the unique axis. The selection rules are $\Delta K = 0$ and $\Delta J = \pm 1$ for a symmetric top with a permanent dipole moment. When the energy difference between two successive energy levels $(J + 1 \leftarrow J)$ is evaluated by subtracting Eq. (1.85) from the same equation where $J + 1$ has been substituted for J, we obtain

$$\Delta E_{rot} = 2Bhc(J + 1) \quad \text{and} \quad \bar{v}_p = 2B(J + 1) \tag{1.86}$$

This rigid rotator result is the same as the comparable equation for the linear molecule [Eq. (1.65)]. Therefore, this analysis of the pure rotational spectrum of a symmetric top molecule can yield only one moment of inertia which is about an axis perpendicular to the unique axis.

Chloroform is a symmetric top molecule and the transition $J = 7 \leftarrow 6$ has been observed in the microwave spectrum[8] at $46,227.2 \times 10^6$ Hz for $CH^{35}Cl_3$ and at $45.502.4 \times 10^6$ Hz for $CD^{35}Cl_3$. From Eq. (1.86) the moment of inertia I_b is calculated as 254.1×10^{-40} gm cm^2 for $CH^{35}Cl_3$ and 258.2×10^{-40} gm cm^2 for $CD^{35}Cl_3$.

In a spherical top molecule $I_a = I_b = I_c$ so $A = B = C$ and Eq. (1.85) reduces to the comparable equation for the linear molecule [Eq. (1.55)].

In a linear molecule I_a becomes vanishingly small so A becomes tremendously large. In Eq. (1.85) if K has any value other than zero, then the factor K^2Ahc will make E_{rot} far larger than the amount of energy available to the molecule at normal temperatures, so K is only equal to zero and Eq. (1.85) reduces to the form of Eq. (1.55).

1.24 Vibrational–Rotational Spectrum, Classical Picture

Imagine a linear molecule vibrating along its axis and also rotating about its center of gravity (see Fig. 1.21a). The dipole moment, μ, in the bond axis direction is oscillating because of the vibration, so

$$\mu = a \sin 2\pi v_v t \tag{1.87}$$

where a is the amplitude, v_v is the vibrational frequency, and t is time. Imagine that this molecule is approached by a plane-polarized infrared ray where the electric field is oscillating in the x direction which (for simplicity) lies in the plane defined by the rotating dipole moment.

At any time t the molecular axis makes an angle θ with the x axis so

$$\theta = 2\pi v_r t \tag{1.88}$$

where v_r is the rotational frequency (cycles per second) and $2\pi v_r$ is the angular velocity (radians per second). The dipole moment component, μ_x, in the

[8] R. R. Unterberger, R. Trambarulo, and W. V. Smith, *J. Chem. Phys.* **18**, 565 (1950).

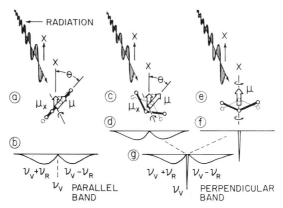

FIG. 1.21. A classical picture of the origin of parallel and perpendicular vibrational–rotational bands in the spectra of linear molecules. The band wings result when the rotation changes the orientation of the vibrationally caused dipole moment oscillation. The central peak results when rotation causes no orientation change in the oscillating dipole moment.

photon electric field direction x, is $\mu_x = \mu \cos \theta$, so making appropriate substitutions we get

$$\mu_x = a \sin(2\pi v_v t)\cos(2\pi v_r t) \tag{1.89}$$

and making use of a trigonometric identity we can write

$$\mu_x = \frac{a}{2} \left[\sin 2\pi(v_v + v_r)t + \sin 2\pi(v_v - v_r)t \right] \tag{1.90}$$

If the direction of the electric field x does not lie in the plane defined by the rotating dipole moment but makes an angle ϕ with it, the right side of Eq. (1.90) is multiplied by the constant ($\cos \phi$) but otherwise it is unchanged. Therefore, the dipole moment component in the photon electric field direction oscillates with the frequencies $(v_v + v_r)$ and $(v_v - v_r)$ and can interact with electromagnetic radiation which has these same frequencies. If we consider a large number of identical molecules in the gaseous state, all will vibrate at the same frequency v_v, which depends on the masses and force constants. However, the rotational frequency v_r is variable and the collection of molecules will have a variety of rotational frequencies. The distribution function of numbers of molecules versus rotational frequencies will reach a maximum at some intermediate rotational frequency (see Fig. 1.20). When these distribution functions are substituted for v_r in $(v_v + v_r)$ and $(v_v - v_r)$ then an absorption band of the broad doublet type is predicted (see Fig. 1.21b). This classical prediction for band shape is fairly well realized in a low resolution infrared spectrum of a gaseous unsymmetrical diatomic molecule where the dipole moment change is along the molecular axis.

A linear polyatomic molecule such as CO_2 has two kinds of infrared active vibrations. When the dipole moment changes along the axis, as in the antisymmetrical stretching vibration, then a broad doublet is predicted using the previous arguments. When the dipole moment changes perpendicular to the molecular axis as in the bending vibration, a different situation arises. A linear molecule can rotate with equal probability about two mutually perpendicular axes. When the rotational axis is perpendicular to the plane in which the molecule is bending (see Fig. 1.21c), then the molecular rotation changes the orientation of the oscillating dipole moment as before with similar spectral results (Fig. 1.21d). However, when the rotational axis direction is parallel to the direction of the dipole moment change (see Fig. 1.21e), then the molecular rotation does not affect the orientation of the changing dipole moment which simply oscillates at the vibrational frequency v_v and gives rise to a single spectral peak (Fig. 1.21f). Since the molecule can rotate with equal probability about both axes, the total band is a combination of a central peak and two broad wings, the peak having the same integrated intensity as the wings (see Fig. 1.21g). In the out-of-phase antisymmetrical stretching vibration (Fig. 1.21a) it is not possible to rotate the molecule without also changing the orientation of the oscillating dipole moment, so a central peak is forbidden. These classical band contour predictions are fairly well realized in a low resolution spectrum of CO_2 gas where the two strong bands both have similar broad doublet structures but only one band has a central peak. It is seen that *the vibrational–rotational band shape can be used to tell whether the dipole moment change is along or across the axis of a linear molecule*, which in this case is sufficient to assign the bands unambiguously to the bending and stretching vibrations.

If the resolution is increased sufficiently, then in the spectrum the two broad wings will be resolved into a series of discrete lines because the angular momentum is quantized and not continuously variable as in the classical picture.

1.25 Vibrational–Rotational Spectrum, Quantum Mechanical Treatment

A vibrational–rotational band in the infrared spectrum of a gas has a complex structure because a change in the vibrational energy, E_v, of a molecule may be accompanied by a change in the rotational energy, E_r. We are assuming initially that these energies are additive. By combining Eqs. (1.34) and (1.55) for the linear molecule, the total energy is given by

$$E_{v+r} = (v + \tfrac{1}{2})hc\bar{v}_0 + BhcJ(J + 1) \qquad (1.91)$$

Let J'' represent the rotational quantum number in the ground vibrational state $v = 0$ and J' represent the rotational quantum number in the first excited vibrational state $v = 1$. Subtracting Eq. (1.91) where $v = 0$, $J = J''$ from the same equation with the substitution $v = 1$, $J = J'$ yields

$$\frac{E_{v+r}}{hc} (\text{cm}^{-1}) = \bar{v}_0 + B[J'(J' + 1) - J''(J'' + 1)] \tag{1.92}$$

A polyatomic linear molecule can have two types of vibrational–rotational bands. A *parallel* band results when the change in dipole moment is parallel to the molecular axis. A *perpendicular* band results when the dipole moment change is perpendicular to the molecular axis.

For the parallel band the selection rule is $\Delta J = \pm 1$. For the perpendicular band the selection rule is $\Delta J = 0, \pm 1$. When $\Delta J = 0$ and $J' = J''$, Eq. (1.92) becomes

$$\frac{E_{v+r}}{hc} = Q(\text{cm}^{-1}) = \bar{v}_0 \tag{1.93}$$

This equation which is independent of the value of J gives the frequency in cm^{-1} of the so-called Q branch of the perpendicular band.

When $\Delta J = 1$ and $J' = J'' + 1$, Eq. (1.92) becomes

$$\frac{\Delta E_{v+r}}{hc} = R(\text{cm}^{-1}) = \bar{v}_0 + 2(BJ'' + 1) \qquad J'' = 0, 1, 2, \ldots \tag{1.94}$$

This equation gives the frequencies in cm^{-1} of the rotational bands which make up the R branch whose values depend on J'' (the rotational quantum number of the lower vibrational state).

When $\Delta J = -1$ and $J' = J'' - 1$, Eq. (1.82) becomes

$$\frac{\Delta E_{v+r}}{hc} = P(\text{cm}^{-1}) = \bar{v}_0 - 2B(J'') \qquad J'' = 1, 2, 3, \ldots \tag{1.95}$$

This is the equation for the frequencies of the rotational bands which make up the P branch whose values depend on J''. In this equation J'' cannot be zero since J' is one integer lower than J'' (see Fig. 1.22).

Parallel and perpendicular bands in linear molecules have identical P and R branches with a spacing of $2B$ between the rotational bands according to these equations. The perpendicular band has a central Q branch which distinguishes it from parallel bands. An unusual exception to this is the NO molecule. In this odd electron molecule there is a resultant electronic angular

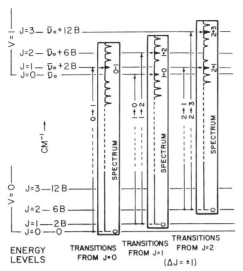

FIG. 1.22. The vibrational–rotational energy levels and the vibrational–rotational spectrum of a diatomic molecule. To show the relations, the spectra are positioned so that the zero wavenumber point is on the energy level from which the transition originates.

momentum about the molecular axis which gives rise to a Q branch in a parallel band.

Under conditions of low resolution the P and R branches of the parallel band appears as a broad doublet where the fine structure is not resolved. The doublet spacing, $\Delta\bar{v}$, of the intensity maxima of the P and R branches is given by

$$\Delta\bar{v} = \frac{1}{\pi c}\sqrt{\frac{kT}{I}} \tag{1.96}$$

where c is the velocity of light, k is the Boltzmann constant, T is the absolute temperature, and I is the moment of inertia. This equation is derived by substituting the value for J_{max} [Eq. (1.80)] into the Eqs. (1.94 and 1.95) for the P and R branches and then subtracting Eq. (1.95) from Eq. (1.94).

In the vibrational–rotational band of hydrogen chloride in Fig. 1.23 the individual rotational bands which make up the overall band are not uniformly spaced as they should be according to the equation presented but become more closely spaced at higher wavenumbers. In the stretching vibration the average internuclear distance and hence the moment of inertia increases as the vibrational quantum number increases. Since the rotational constant is proportional to the reciprocal of the moment of inertia, the rotational constant has a smaller value in the excited state than in the ground state. The rotational constant B is, therefore, not constant as has been assumed.

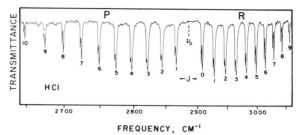

FIG. 1.23. The hydrogen chloride vibrational–rotational band. The P and R branches are labeled. The Q branch (ν_0) is missing since ΔJ cannot equal zero for a parallel band of a linear molecule. Each rotational line is labeled with its rotational quantum number in the ground vibrational state. In this high resolution spectrum each rotational line is a doublet since chlorine has two naturally occurring isotopes. The peaks for $H^{37}Cl$ occur at slightly lower frequencies in cm^{-1} than the peaks for $H^{35}Cl$ [see Eq. (1.11)].

Let B_0 represent the rotational constant in the ground vibrational state and B_1 in the first-excited vibrational state. Equation (1.92) must be modified to yield

$$\frac{\Delta E_{v+r}}{hc} \,(\mathrm{cm}^{-1}) = \bar{\nu}_0 + B_1[J'(J'+1)] - B_0[J''(J''+1)] \qquad (1.97)$$

For the Q branch where $\Delta J = 0$ and $J' = J''$ we obtain, writing J for J'',

$$Q(\mathrm{cm}^{-1}) = \bar{\nu}_0 + (B_1 - B_0)J^2 + (B_1 - B_0)J \qquad J = 0, 1, 2, 3, \ldots$$

For the R branch where $\Delta J = +1$ and $J' = J'' + 1$ we obtain

$$R(\mathrm{cm}^{-1}) = \bar{\nu}_0 + 2B_1 + (3B_1 - B_0)J + (B_1 - B_0)J^2 \qquad J = 0, 1, 2, 3, \ldots$$

For the P branch where $\Delta J = -1$ and $J' = J'' - 1$ we obtain

$$P(\mathrm{cm}^{-1}) = \bar{\nu}_0 - (B_1 + B_0)J + (B_1 - B_0)J^2 \qquad J = 1, 2, 3, \ldots \qquad (1.98)$$

According to these equations there is no requirement for uniform spacing in the P and R branches. The Q branch (when present) no longer consists of a single line.

In the determination of the rotational constants a pair of rotational lines is selected whose separation is dependent on the rotational constant of the ground state (B_0) and a second pair of rotational lines whose separation is dependent on the rotational constant of the excited state (B_1). Two particularly valuable combinations are

$$R_{(J-1)} - P_{(J+1)} = 2B_0(2J+1) \qquad (1.99)$$

$$R_J - P_J = 2B_1(2J+1) \qquad (1.100)$$

Here R_J or P_J stand for the frequency in cm^{-1} of a rotational line in the R or P branch with a quantum number J. These equations come from Eq. (1.98) R and P branches, respectively. The rotational lines of the vibrational–rotational bands of both $H^{35}Cl$ and $H^{37}Cl$ have been accurately measured by several investigators.[9,10] For simplicity only $H^{35}Cl$ bands will be considered and Table 1.6 is constructed listing values for Eqs. (1.99) and (1.100).

TABLE 1.6

TREATMENT OF DATA ON VIBRATIONAL–ROTATIONAL BAND OF $H^{35}Cl$

J	R_J	P_J	$R_J - P_J$	$R_{(J-1)} - P_{(J+1)}$	$2J+1$	$B_1{}^a$	$B_0{}^b$
0	2906.24						
1	2925.90	2865.10	60.80	62.62^c	3	10.13	10.37
2	2944.90	2843.62	101.28	104.34	5	10.13	10.43
3	2963.29	2821.56	141.73	145.96	7	10.12	10.43
4	2981.00	2798.94	182.06	187.53	9	10.11	10.42
5	2998.04	2775.76	222.28	228.96	11	10.10	10.41
6	3014.41	2752.04	262.37	270.26	13	10.09	10.39
7	3030.09	2727.78	302.31	311.40	15	10.08	10.38
8	3045.06	2703.01	342.05	352.36	17	10.06	10.36
9	3059.32	2677.73	381.59	393.10	19	10.04	10.34
					Average	10.10	10.39

^{a} B_1 = column four $\div$ 2 $\times$ column six.
^{b} B_0 = column five $\div$ 2 $\times$ column six.
^{c} If $J = 1$, then $R_{(J-1)} = R_0$ and $P_{(J+1)} = P_2$; thus, $2906.24 - 2843.62 = 62.62$.

It can be seen in Table 1.6 that the rotational constants B_0 and B_1 vary somewhat as the quantum number changes, and again this variation can be explained by centrifugal stretching (see Section 1.20). The average values for the rotational constants in the upper and lower states are 10.10 and 10.39 cm^{-1}, respectively. A better average value could be obtained by plotting $R_J - P_J$ and $R_{(J-1)} - P_{(J+1)}$ versus $2J+1$ on a large scale graph. As illustrated in Fig. 1.24 the slopes of these two plots give $2B_1$ and $2B_0$, respectively. The values of B_1 and B_0 obtained by this method for HCl are 10.06 and 10.36, respectively. The moment of inertia in the upper state is then 2.78×10^{-40} gm cm^2 and in the lower state 2.70×10^{-4} gm cm^2. The latter value agrees well with that calculated from the pure rotational spectrum [see Eq. (1.72)].

[9] E. K. Plyler and D. Tidwell, *Z. Electrochem.* **64**, 717 (1960).
[10] "Table of Wavenumbers for the Calibration of Infrared Spectrometers," *Pure Appl. Chem.* **1**, No. 4 (1961).

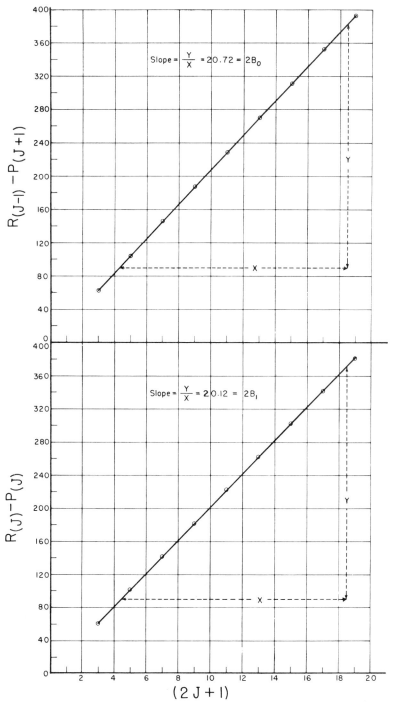

FIG. 1.24. Plots of $R_{(J)} - P_{(J)}$ and $R_{(J-1)} - P_{(J+1)}$ versus $2j + 1$ for determination of rotational constants B_1 and B_0, respectively.

1.26 Vibrational–Rotational Spectrum Nonrigid Rotator

If the nonrigid rotator energy equation [Eq. (1.75)] is used in place of the rigid rotator energy equation in Eq. (1.91) then Eq. (1.97) is modified to read

$$\frac{\Delta E_{v+r}}{hc} = \bar{v}_0 + B_1[J'(J' + 1)] - D_1[J'(J' + 1)]^2$$
$$- B_0[J''(J'' + 1)] + D_0[J''(J'' + 1)^2] \quad (1.101)$$

Proceeding as before we can calculate, writing J for J'',

$$R_{(J-1)} - P_{(J+1)} = (2B_0 - 3D_0)(2J + 1) - D_0(2J + 1)^3 \quad (1.102)$$

$$R_j - P_j = (2B_1 - 3D_1)(2J + 1) - D_1(2J + 1)^3 \quad (1.103)$$

Dividing by $(2J + 1)$ yields

$$\frac{R_{(J-1)} - P_{(J+1)}}{2J + 1} = (2B_0 - 3D_0) - D_0(2J + 1)^2 \quad (1.104)$$

$$\frac{R_j - P_j}{2J + 1} = (2B_1 - 3D_1) - D_1(2J + 1)^2 \quad (1.105)$$

These are equations for straight lines of the form $y = a + bx$ so that if the left-hand functions in these equations are plotted as a function of $(2J + 1)^2$ the line slope will give $-D$ and the intercept will give $2B - 3D$ for the two vibrational states 0 and 1 (see Fig. 1.25). From this study D_1, D_0, B_1, and B_0 can be evaluated. Then B_e, the rotational constant at the bottom of the potential energy curve, can be evaluated by assuming a linear extrapolation of the form

$$B_v = B_e - \alpha\left(v + \frac{1}{2}\right) \quad \text{or} \quad B_e = B_0 + \frac{B_0 - B_1}{2} \quad (1.106)$$

Using the data in Table 1.6 for the $H^{35}Cl$ molecule, D_0 and D_1 are found to have about the same value, 5.2×10^{-4} cm^{-1}. The rotational constants B_e, B_0, and B_1 are 10.592, 10.440, and 10.136 cm^{-1}, respectively. The internuclear distances r_e, r_0, and r_1 are 1.275, 1.284, and 1.303 Å, respectively.

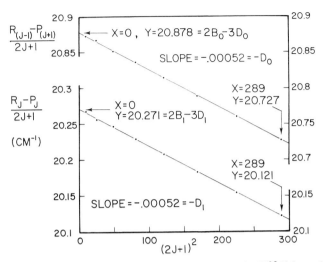

FIG. 1.25.   Plots to determine the four rotational constants for $H^{35}Cl$ from the observed vibrational–rotational spectrum.

Using similar type of data[10] for $D^{35}Cl$ a similar treatment shows D_0 and D_1 are about 1.36×10^{-4}, B_e, B_0, and B_1 are 5.448, 5.391, and 5.279 cm^{-1}, respectively, and r_e, r_0, and r_1 are 1.275, 1.281, and 1.295 Å, respectively. Note that only r_e is same for the two isotopic molecules and that r_0 is slightly different for the two (see Fig. 1.26).

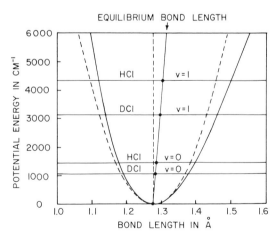

FIG. 1.26.   The equilibrium bond lengths for the $v = 0$ and $v = 1$ states for HCl and DCl. The solid curve is the anharmonic potential function and the dotted curve is the harmonic potential function for both molecules.

1.27 Spherical Top Molecules

Spherical top molecules such as methane and carbon tetrachloride have three equal moments of inertia. All infrared active bands have the same selection rule $\Delta J = 0, \pm 1$. Since the rotational energy equation is the same as that for linear molecules (see Table 1.5) the vibrational–rotational bands resemble the perpendicular bands of a linear molecule with simple P, Q, and R branches (see Fig. 1.27).

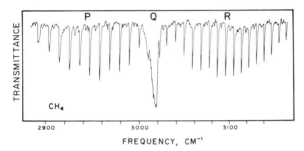

FIG. 1.27.   Vibrational–rotational band of methane.

1.28 Symmetrical Top Molecules

Molecules in this class have two equal moments of inertia which differ from the third moment. The normal modes of a symmetrical top molecule can be divided into parallel and perpendicular vibrations. In the parallel vibrations the dipole moment change occurs parallel to the molecule's major symmetry axis which must be a threefold or higher symmetry axis (see Section 3.1). In the perpendicular vibrations the dipole moment change occurs perpendicular to the major symmetry axis.

The rotational energy equations for symmetrical top molecules (see Table 1.5) involve two different moments of inertia and two rotational quantum numbers, namely, J, characterizing the total angular momentum of the molecule and K, the angular momentum about the major symmetry axis. For *parallel* bands the selection rule is $\Delta J = 0, \pm 1$ with $\Delta K = 0$, except when $K = 0$, in which case $\Delta J = \pm 1$, $\Delta K = 0$. When a difference in energy levels is taken, only the quantum number J is involved and as a result only one moment of inertia is involved, the one not about the symmetry axis. The parallel bands of a symmetric top molecule resemble the perpendicular bands of the linear molecule having P, Q, and R branches corresponding to $\Delta J = -1$

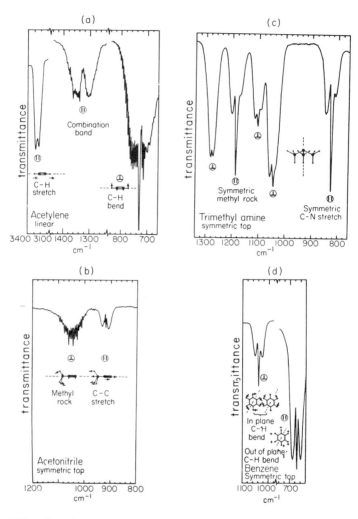

FIG. 1.28. Gas phase contours of absorption bands of acetylene, acetonitrile, tri-methylamine, and benzene measured in a 10-cm cell with a rock salt prism. (a) Acetylene[11] at 81 mm pressure. The parallel and perpendicular bands are marked. The perpendicular band has a central peak which is absent in the parallel bands. There is partial resolution of some rotational fine structure. (b) Acetonitrile[11] at 67 mm Hg pressure. The moment of inertia about the symmetry axis is relatively small. The perpendicular band has a round contour with some fine structure resolved. The parallel band is a triplet. (c) Trimethylamine[12] at 27 mm Hg pressure. The parallel bands are a little wider than the perpendicular bands and show a more distinct central peak. The triplet bands are asymmetrical because of centrifugal distortion. (d) Benzene[13] at 26 mm Hg pressure. The parallel and perpendicular bands are not very different. The references cited refer to the band assignments; the spectra shown were obtained by the authors of this text.

(Poorer), $\Delta J = 0$ (Equal), and $\Delta J = +1$ (Richer). The unresolved band has a broad doublet and a sharp central peak (see Fig. 1.28).[11-13]

For the *perpendicular* bands the selection rule is $\Delta J = 0, \pm 1$ with $\Delta K = \pm 1$. Since two quantum numbers and two moments of inertia are involved, the band structure is more complex than the parallel bands. The unresolved band shape will vary depending on the relative magnitudes of the moments of inertia. When the moment of inertia about the symmetry axis is relatively small, as in acetonitrile, the perpendicular band contour may be broad and round, differing distinctly in this case from the parallel band. When the moments of inertia are more nearly equal, the perpendicular and parallel bands are more similar in shape and may be difficult to distinguish (see Fig. 1.28).

1.29 Asymmetrical Top Molecules

Molecules in this class have three unequal moments of inertia $I_a \neq I_b \neq I_c$ and therefore have the most complex band contours. If, during a vibration, the dipole moment changes parallel to the a, b, or c axis a band contour results which is designated as an A, B, or C type band, respectively.

The contours of the pure A, B, and C type bands are dependent on the relative values of the moments of inertia.[14] The A and C bands both have a sharp central peak and symmetrical broad wing bands. The B type band is unique in that it has no central peak. In molecules where the moment of inertia about the c axis is relatively large the central peak of the C type band is comparatively stronger than the central peak of the A type band (see Fig. 1.29). The wing bands of the A, B, and C bands are made up of two components, $\Delta J = 0$ which causes a doublet of narrow spacing and $\Delta J = \pm 1$ which causes a doublet of somewhat wider spacing. This doubling of each wing band contour is not always apparent but it is clearly illustrated by the B type band in ethylene oxide in Fig. 1.29.

If the dipole moment does not change exactly parallel to the a, b, or c axis a mixed contour will result. Band asymmetry can also result from centrifugal distortion.

[11] G. Herzberg, "Infrared and Raman Spectra of Polyatomic Molecules." Van Nostrand-Reinhold, Princeton, New Jersey, 1945.

[12] J. R. Barceló Matutano and J. Bellanato, *Spectrochim. Acta* **8**, 27 (1956).

[13] C. R. Bailey, J. B. Hale, C. K. Ingold, and J. W. Thompson, *J. Chem. Soc., London*, p. 931 (1936).

[14] R. M. Badger and L. R. Zumwalt, *J. Chem. Phys.* **6**, 711 (1938).

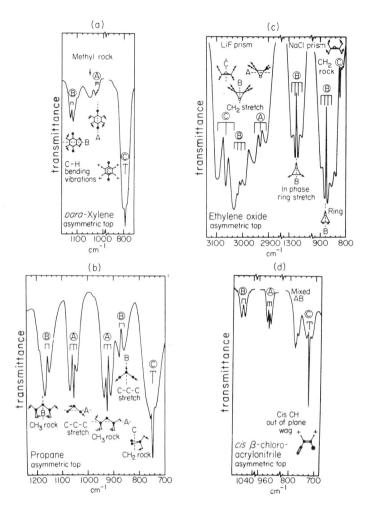

FIG. 1.29. Gas phase contours of absorption bands of *para*-xylene, propane, ethylene oxide, and *cis*-β-chloroacrylonitrile measured in a 10-cm cell (unless noted otherwise) with a rock salt prism. (a) *para*-Xylene[15] at 7 mm Hg pressure in a 100-cm cell. The *B* type bands have a doublet structure and the *A* and *C* bands have a triplet structure. The *C* type band has a relatively strong central peak. (b) Propane[16] at 679 mm Hg pressure. The *A*, *B*, and *C* type bands have the same structure as discussed for *para*-xylene. (c) Ethylene oxide[17] at 72 mmHg pressure except for the 800–900 cm^{-1} band which is at 41 mm Hg pressure. The *A* and *C* bands have a triplet structure. The *B* type bands show four components. (d) *cis*-β-Chloroacrylonitrile at its vapor pressure at 25° C. In this planar molecule the out-of-plane CH wag band has a *C* type contour with the prominent central peak. This band is easily distinguished from nearby in-plane vibrations.

[15] K. S. Pitzer and D. W. Scott, *J. Amer. Chem. Soc.* **65**, 803 (1943).
[16] H. L. McMurry and V. Thornton, *J. Chem. Phys.* **19**, 1014 (1951).
[17] R. C. Lord and B. Nolin, *J. Chem. Phys.* **24**, 656 (1956).

1.30 The Raman Effect

When electromagnetic radiation of energy content hv irradiates a molecule, the energy may be transmitted, absorbed, or scattered. In the Tyndall effect the radiation is scattered by particles (smoke or fog, for example). In Rayleigh scattering the molecules scatter the light. No change in wavelength of individual photons occurs in either Tyndall or Rayleigh scattering. In 1928, C. V. Raman described another type of scattering known as the Raman effect. This effect had been theoretically predicted by Smekal before the successful experimental demonstration of the effect by Raman and is therefore sometimes referred to as the Smekal–Raman effect in the German literature.

In a Raman spectrometer the sample is irradiated with an intense source of monochromatic radiation usually in the visible part of the spectrum. Generally this radiation frequency is much higher than the vibrational frequencies but is lower than the electronic frequencies. The radiation scattered by the sample is analyzed in the spectrometer. Rayleigh scattering can be looked on as an elastic collision between the incident photon and the molecule. Since the rotational and vibrational energy of the molecule is unchanged in an elastic collision, the energy and therefore the frequency of the scattered photon is the same as that of the incident photon. This is by far the strongest component of the scattered radiation. The Raman effect can be looked on as an inelastic collision between the incident photon and the molecule where as a result of the collision the vibrational or rotational energy of the molecule is changed by an amount ΔE_m. In order that energy may be conserved, the energy of the scattered photon, hv_s, must be different from the energy of the incident photon hv_i, by an amount equal to ΔE_m:

$$hv_i - hv_s = \Delta E_m \tag{1.107}$$

If the molecule gains energy then ΔE_m is positive and v_s is smaller than v_i, giving rise to Stokes lines in the Raman spectrum. This terminology arose from Stokes rule of fluorescence which stated that fluorescent radiation always occurs at lower frequencies than that of the exciting radiation. If the molecule loses energy, then ΔE_m is negative and v_s is larger than v_i, giving rise to anti-Stokes lines in the Raman spectrum.

In Fig. 1.30 the lines designated $v = 0$ and $v = 1$ represent vibrational energy levels of a molecule such as HCl. The energy difference between these levels is given by Eq. (1.39) as $\Delta E_m = hv_m$. A transition directly between these two levels causes the absorption of an infrared photon whose frequency is the same as the molecular frequency v_m. In Rayleigh and Raman scattering,

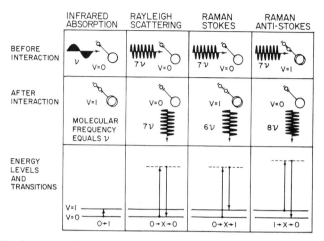

FIG. 1.30. Schematic illustration of Raman and Rayleigh scattering and infrared absorption. In infrared absorption the incident photon has the same frequency as the molecular vibration. In Rayleigh and Raman scattering, the incident photon has much higher frequency (7ν in this figure). The scattered photon is like the incident photon in Rayleigh scattering but in Raman scattering the scattered photon has a lower or higher frequency ($7\nu \pm \nu$). The photon frequency difference is the same as the molecular vibrational frequency.

the frequency of the incident photon is usually much greater than ν_m. When the incident photon interacts with a molecule in the ground vibrational state $v = 0$, the molecule absorbs the photon energy and is raised momentarily to some high level of energy (dashed line) which is not a stable energy level. Therefore, the molecule immediately loses energy and, most probably, returns to the ground vibrational level, emitting a scattered photon (to make up for its energy loss) whose energy and frequency is the same as that of the incident photon. This is Rayleigh scattering. However, a small proportion of the molecules in the unstable high level of energy may fall, not to the ground vibrational level but to the $v = 1$ energy level. The scattered photon in this case has *less* energy than the exciting photon, the difference being

$$h\nu_i - h\nu_s = \Delta E_m = h\nu_m \qquad (1.108)$$

so

$$\nu_s = \nu_i - \nu_m \qquad \text{when} \qquad \Delta v = +1$$

This scattered photon gives rise to a Stokes line in the Raman spectrum. According to quantum mechanics the allowed change in the vibrational quantum number for a Raman transition is $\Delta v = \pm 1$ for a harmonic vibration. The final possibility is that the molecule initially is in the excited state

$v = 1$, absorbs the incident photon energy, and is raised to an unstable high level of energy. When the molecule falls to the ground vibrational level $v = 0$ the energy loss is made up for by the emission of a photon whose energy is *greater* than that of the incident photon by hv_m. This scattered photon gives rise to an anti-Stokes line in the Raman spectrum. According to the Boltzmann distribution function the ratio of the number of molecules in the $v = 1$ state to the number in the $v = 0$ state for a given vibration is

$$\frac{n_1}{n_0} = e^{-(hv_m/kT)} \tag{1.109}$$

At ordinary temperatures most of the molecules exist in the ground state and therefore Stokes lines have greater intensities than anti-Stokes lines which originate from an excited level with lower population. This difference increases as the vibrational frequency increases.

1.31 Polarizability

While infrared and Raman spectrum both involve vibrational and rotational energy levels, they are not duplicates of each other but rather complement each other (see Fig. 1.31). This is because the *intensity* of the spectral band depends on how effectively the photon energy is transferred to the molecule and the mechanism for photon energy transfer differs in the two techniques. This will be shown below.

If a molecule is placed in the electric field of electromagnetic radiation then the electrons and protons will experience oppositely directed forces exerted by the electric field. As a result the electrons are displaced relative to the protons and the polarized molecule has an induced dipole moment caused by the

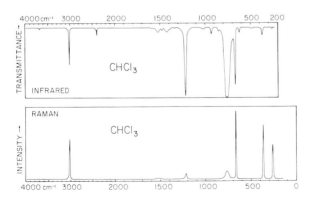

FIG. 1.31.   Infrared and Raman spectra of chloroform.

external field. The induced dipole moment, μ, divided by the strength of the electric field E causing the induced dipole moment, is the polarizability α

$$\mu = \alpha E \tag{1.110}$$

The polarizability can be looked on as the deformability of the electron cloud of the molecule by the electric field. *In order for a molecular vibration to be Raman active, the vibration must be accompanied by a change in the polarizability of the molecule.* This will be shown below.

The electric field of the electromagnetic radiation in the vicinity of the molecule varies with time so that

$$E = E_0 \sin 2\pi v t \tag{1.111}$$

where E_0 is a constant, the maximum value of the field, v, is the frequency of the radiation, and t is time. This oscillating electric field will induce in the molecule an oscillating dipole moment μ whose frequency will be the same as that of the external electric field. Combining Eqs. (1.110) and (1.111)

$$\mu = \alpha E_0 \sin 2\pi v t \tag{1.112}$$

From classical theory this oscillating dipole moment will emit radiation in all directions which has the same frequency as that of the oscillating dipole moment (and, therefore, that of the exciting radiation), and whose intensity is proportional to the square of the maximum value for μ, which is $\alpha^2 E_0^2$ in Eq. (1.112).

In molecules the polarizability α does not have a constant value since certain vibrations and rotations can cause α to vary. For example, during the vibration of a diatomic molecule, the molecular shape is alternately compressed and extended. Since the election cloud is not identical at the extremes of the vibration, a change in polarizability results. For small displacements the polarizability can be expanded in a Taylor series as

$$\alpha = \alpha_0 + \frac{\partial \alpha}{\partial Q} Q + \cdots \tag{1.113}$$

where α_0 is the equilibrium polarizability, Q is a normal coordinate (which is $r - r_e$ in the diatomic case), and $\partial \alpha / \partial Q$ is the rate of change of polarizability with respect to Q measured at the equilibrium configuration. Higher order terms are neglected in the harmonic approximation. The normal coordinate Q varies periodically

$$Q = Q_0 \sin 2\pi v_v t \tag{1.114}$$

where v_v is the frequency of the normal coordinate vibration and Q_0 is a constant, the maximum value for Q. Combining Eqs. (1.114) and (1.115)

$$\alpha = \alpha_0 + \frac{\partial \alpha}{\partial Q} Q_0 \sin 2\pi v_v t \qquad (1.115)$$

The substitution of this value for α into Eq. (1.112) yields

$$\mu = \alpha_0 E_0 \sin 2\pi v t + \frac{\partial \alpha}{\partial Q} Q_0 E_0 (\sin 2\pi v_v t)(\sin 2\pi v t) \qquad (1.116)$$

Making use of a trigonometric identity we can write

$$\mu = \alpha_0 E_0 \sin 2\pi v t + \frac{\partial \alpha}{\partial Q} \frac{Q_0 E_0}{2} [\cos 2\pi (v - v_v)t - \cos 2\pi (v + v_v)t] \qquad (1.117)$$

It can be seen from this equation that the induced dipole moment μ varies with three component frequencies v_v, $v - v_v$, and $v + v_v$ and can therefore give rise to Rayleigh scattering, Stokes, and anti-Stokes Raman scattering, respectively. This classical prediction for these frequencies corresponds to the quantum mechanical result for Raman transitions when $\Delta v = \pm 1$. If the vibrations causes *no* change in polarizability so that $\partial \alpha / \partial Q = 0$, then Eq. (1.117) shows that the Raman component frequencies of the induced dipole moment have zero amplitudes and therefore no radiation with Raman frequencies can be generated.

This result is illustrated graphically in Fig. 1.32b where the induced dipole moment generated by the oscillating electric field of the radiation is plotted as a function of time, for a molecule performing a vibration which changes the polarizability. The molecular vibration (which is much slower than the external field vibration) causes the externally induced dipole moment oscillation to be amplitude modulated as shown. This occurs because the induced dipole moment for a given value of the external field is different when the molecule is stretched or contracted as illustrated in Fig. 1.32a. The steady amplitude oscillation components of the induced dipole moment are illustrated in Fig. 1.32c which shows a large amplitude component with the external field frequency and smaller amplitude components with Raman frequencies. If there were no vibrationally caused change in polarizability, then there would be no amplitude modulation of the induced dipole moment oscillation, and the Raman frequency components would have zero amplitudes.

From classical theory it is known that the intensity of scattered radiation is proportional to the fourth power of the frequency and to the square of the

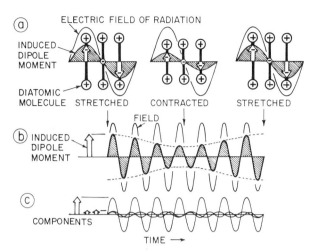

FIG. 1.32. Classical theory for the origin of the Raman effect. In (a) we see a diatomic molecule where "+" represents protons and "−" represents the center of gravity of the electrons. The electrons are displaced by the external field of the photon and an induced dipole moment is generated which changes when the bond length changes during the molecular vibration. The induced dipole moment is plotted in (b) as an amplitude modulated wave with steady amplitude components shown in (c) from which scattered radiation is generated.

maximum value of the excitation electric field (E_0^2). The intensity of Rayleigh scattered radiation is proportional to α_0^2 and the intensity of Raman scattered radiation is proportional to $(\partial\alpha/\partial Q)^2$. The main deficiency of the classical treatment of the Raman effect is that it predicts an incorrect value for the intensity ratio of anti-Stokes to Stokes bands. The correct ratio is

$$\frac{\text{anti-Stokes intensity}}{\text{Stokes intensity}} = \frac{(v + v_v)^4}{(v - v_v)^4}\, e^{-(hv_v/kT)} \tag{1.118}$$

In the classical treatment the exponential term is not present. The exponential term comes from purely quantum mechanical arguments [see Eq. (1.109)] and is in accord with experiment.

1.32   The Tensor Character of the Polarizability

The dipole moment is a vector whose direction is that of the line between the center of gravity of the protons and the center of gravity of the electrons in a molecule. The polarizability is a tensor whose form will be shown below.

In a molecule which has tetrahedral symmetry (CCl_4) or octahedral symmetry (SF_6) an external electric field will generate an induced dipole moment whose direction is the same as that of the field regardless of the orientation of the molecule (see Fig. 1.33). Such a molecule is said to be isotropic. In such a case when the electric field vector is resolved into cartesian coordinate components, Eq. (1.110) becomes

$$\mu_x = \alpha E_x \qquad \mu_y = \alpha E_y \qquad \mu_z = \alpha E_z \qquad (1.119)$$

where μ_x is a component of the induced dipole moment vector. The polarizability is the same in all directions.

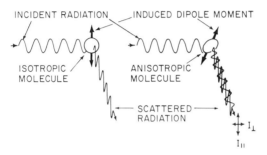

FIG. 1.33. Scattered radiation generated by the induced dipole moment in an isotropic and an anisotropic molecule.

For other molecules the polarizability α may be different in the x, y, and z directions, and as a result the induced dipole moment may not be parallel to the field. Such a molecule is said to be anisotropic. In this case the electric field component E_x will induce a dipole moment which has a component in the x direction but also may have a component in the y and z directions (see Fig. 1.33). For this general case the following equations apply:

$$
\begin{aligned}
\mu_x &= \alpha_{xx} E_x + \alpha_{xy} E_y + \alpha_{xz} E_z \\
\mu_y &= \alpha_{yx} E_x + \alpha_{yy} E_y + \alpha_{yz} E_z \\
\mu_z &= \alpha_{zx} E_x + \alpha_{zy} E_y + \alpha_{zz} E_z
\end{aligned}
\qquad (1.120)
$$

The overall polarizability is then the whole system of these α constants or coefficients. Such a system of coefficients which establishes a linear relationship between vectors (μ and E) is termed a tensor and therefore the polarizability is a tensor.

The polarizability tensor is symmetric, that is, $\alpha_{xy} = \alpha_{yx}$, $\alpha_{yz} = \alpha_{zy}$, and $\alpha_{xz} = \alpha_{yx}$. Symmetric tensors have the property that a particular set of coordinates x', y', and z' can be chosen so that only $\alpha_{x'x'}$, $\alpha_{y'y'}$, and $\alpha_{z'z'}$ are different from zero. These special axes are three mutually perpendicular directions in the molecule for which the induced dipole moments are parallel to the electric field. Equation (1.120) then reduces to

$$\mu_{x'} = \alpha_{x'x'} E_{x'} \qquad \mu_{y'} = \alpha_{y'y'} E_{y'} \qquad \mu_{z'} = \alpha_{z'z'} E_{z'} \qquad (1.121)$$

The three axes in these three directions are called the principle axes of polarizability.

The locus of points formed by plotting $1/\alpha^{1/2}$ in any direction from the origin yields a surface called the polarizability ellipsoid whose axes are x', y', and z'. For a molecule which is completely anisotropic $\alpha_{x'x'} \neq \alpha_{y'y'} \neq \alpha_{z'z'}$, and the ellipsoid has three axes of unequal length. If the polarizability is the same in two directions, the ellipsoid becomes a rotational ellipsoid with two equal axes. If the molecule is isotropic, the ellipsoid becomes a sphere.

While the polarizability ellipsoid may have higher symmetry than the molecule, all the symmetry elements possessed by the molecule will also be possessed by the ellipsoid. If the polarizability ellipsoid is changed in size, shape, or orientation as a result of molecular vibration or rotation, a Raman spectrum will result.

As an example refer to Fig. 1.34, the vibrations of acetylene. In vibrations 1 and 2 the molecular shape is different at the vibrational extremes illustrated so the polarizability ellipsoid will change in size or shape. In vibration 3 the shape is the same at the vibrational extremes but the polarizability ellipsoid will be changed in orientation because the effect of the rotation of the C≡C bond is not compensated for by the counterrotation effect of the two CH bonds which are not symmetrically equivalent to the C≡C bond. Vibrations 1, 2, and 3 are Raman active. Vibrations 4 and 5 are Raman inactive. The polarizability ellipsoid in 4 and 5 cannot change in size or shape because the vibrational extremes have the same shape. The ellipsoid cannot rotate because in vibration 5 the effect of the rotation of one CH bond is nullified by the counter-rotation of the other equivalent CH bond.

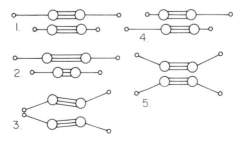

FIG. 1.34.   The vibrations of acetylene, H—C≡C—H.

1.33 Depolarization Ratio

Let the direction of propagation of the incident radiation be the z axis and the direction of observation be perpendicular to the z axis in the xy plane (see Fig. 1.33). The depolarization ratio ρ is defined as the ratio of the intensity of scattered light polarized perpendicular to the xy plane $I_\perp$ to that polarized parallel to the xy plane $I_\parallel$.

$$\rho = \frac{I_\perp}{I_\parallel} \tag{1.122}$$

In order to relate the intensity ratio to theory it is necessary to resolve the polarizability into two parts, an isotropic part α_i and the anisotropic part α_a so that

$$\alpha_i = \tfrac{1}{3}(\alpha_{xx} + \alpha_{yy} + \alpha_{zz}) \tag{1.123}$$

$$(\alpha_a)^2 = \tfrac{1}{2}[(\alpha_{xx} - \alpha_{yy})^2 + (\alpha_{yy} - \alpha_{zz})^2 + (\alpha_{zz} - \alpha_{xx})^2 + 6(\alpha_{xy}^2 + \alpha_{yz}^2 + \alpha_{zx}^2)] \tag{1.124}$$

The values of α_i and $(\alpha_a)^2$ have the property of being invariant to a change in molecular orientation relative to the space-fixed coordinate system. In liquids, the molecular orientation is not fixed. By averaging over all orientations of the polarizability ellipsoid it can be shown that for natural unpolarized incident light the depolarization ratio ρ_n is

$$\rho_n = \frac{I_\perp}{I_\parallel} = \frac{6(\alpha_a)^2}{45(\alpha_i)^2 + 7(\alpha_a)^2} \tag{1.125}$$

and that when the incident radiation is plane polarized (as in a laser source) the depolarization ratio ρ_p is

$$\rho_p = \frac{I_\perp}{I_\parallel} = \frac{3(\alpha_a)^2}{45(\alpha_i)^2 + 4(\alpha_a)^2} \tag{1.126}$$

These equations are for Rayleigh scattering. For Raman scattering the polarizability components α_{xx}, α_{yy}, and so forth, in Eqs. (1.123) and (1.124) should be replaced by the *change* in polarizability with respect to the coordinate $\partial\alpha_{xx}/\partial Q$, and so forth, and the new values α_i' and α_a' are used in Eqs. (1.125) and (1.126).

The value of this formulation lies in the fact that the isotropic part of the polarizability only changes ($\alpha_i' \neq 0$) when the molecule is performing a totally symmetric vibration. In this kind of vibration the vibrationally distorted molecule has all the symmetry elements possessed by the molecule in the equilibrium configuration. Since for this kind of vibration $\alpha_i' \neq 0$, then ρ_n [Eq. (1.125)] will have a value between 0 and $\frac{6}{7}$ and ρ_p [Eq. (1.126)] will have a value between 0 and $\frac{3}{4}$. For the special case of an isotropic molecule α_a' is zero so ρ_n and ρ_p are both zero for totally symmetric vibrations. If the vibrationally distorted molecule is antisymmetric to any symmetry element possessed by the molecule in the equilibrium configuration, then $\alpha_i' = 0$ and ρ_n is $\frac{6}{7}$ and ρ_p is $\frac{3}{4}$ in the Raman spectrum. Therefore, *a measurement of the*

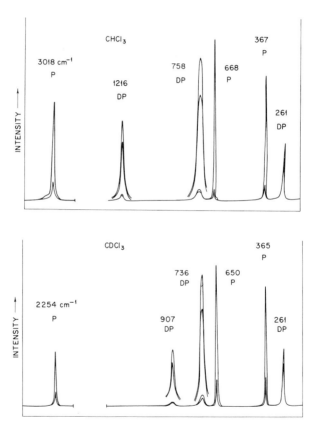

FIG. 1.35. Polarized Raman spectra of $CHCl_3$ and $CDCl_3$. The less intense spectra are run with perpendicular polarization and the more intense spectra are run with parallel polarization. The depolarization ratio is $\frac{3}{4}$ for the depolarized bands (marked DP) and is between 0 and $\frac{3}{4}$ for the polarized bands (marked P), the latter belonging to the totally symmetric species. The weaker bands (1216, 758, 907, and 736 cm^{-1}) have been rerun with higher gain. The incident radiation is from a laser source and is plane polarized.

depolarization ratio provides a means of distinguishing totally symmetrical vibrations from the rest. See Fig. 1.35 for a polarized Raman spectrum of chloroform.

1.34 Pure Rotational Raman Spectra

The pure rotation of a molecule will give rise to a pure rotational Raman spectrum if the polarizability of the molecule varies in different directions at right angles to the axis of rotation. Therefore even symmetrical molecules such as H_2 and CO_2 have pure rotational Raman spectra. The selection rule for pure rotational Raman energy changes for linear molecules is $\Delta J = 0, \pm 2$. This selection rule differs from that for infrared transitions ($\Delta J = \pm 1$). One way to rationalize this is that in an unsymmetrical diatomic. molecule the permanent dipole moment rotates with the molecule but, because of its symmetry, the polarizability ellipsoid after only *one-half* a molecular revolution is indistinguishable from the original ellipsoid. Applying this selection rule to the rotational energy equation (1.55) for linear molecules, we obtain for the wavenumbers of the Stokes lines ($J + 2 \leftarrow J$)

$$\bar{v} = \frac{\Delta E_{rot}}{hc} = B(4J + 6) \qquad J = 0, 1, 2, \ldots \qquad (1.127)$$

and for anti-Stokes lines ($J \rightarrow J - 2$)

$$\bar{v} = \frac{\Delta E_{rot}}{hc} = B(4J - 2) \qquad J = 2, 3, 4, \ldots \qquad (1.128)$$

where J is the quantum number of the initial rotational state. Both sets of lines have a spacing equal to $4B$ with the space between the first line and the exciting line equal to $6B$. See Fig. 1.36 [18] for the rotational Raman spectrum of N_2O. The frequencies in cm^{-1} of this unsymmetrical linear molecule can be reproduced by using Eq. (1.127) with $B = 0.4190$ cm^{-1}. For example, the most intense line in Fig. 1.36, where $J = 15$, has a calculated value of 27.7 cm^{-1}. The relative line intensities compare favorably with the n_j/n_0 values calculated from Eq. (1.79) with $T = 300°K$. The intensity maximum, calculated using Eq. (1.80), is at $J = 15.3$, just above the $J = 15$ line. In *symmetrical*

[18] H. J. Sloane, *Appl. Spectrosc.* **25**, 430 (1971).

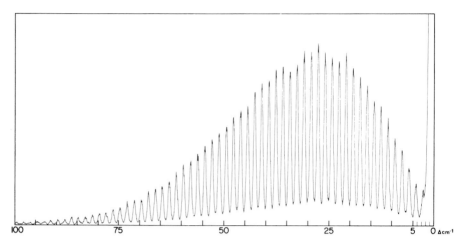

Fig. 1.36. Rotational Raman spectrum of nitrous oxide gas (NNO) run on the Cary Model 82 Raman spectrophotometer with argon ion 5145 Å excitation. From Sloane,[18] used with permission.

linear molecules, alternate line intensity variations can occur because of the influence of nuclear spin.[11]

All symmetric top molecules should give rise to pure rotational Raman spectra. However, the polarizability is unaffected by rotation about the unique symmetry axis. The selection rule is $\Delta J = \pm 1$, ± 2 with $\Delta K = 0$ and can be applied to Eq. (1.85). The spacing between the lines is equal to $2B$. The moment of inertia about the unique symmetry axis cannot be obtained.

Spherical top molecules which are isotropic cannot have a pure rotational Raman spectrum. Asymmetric top molecules have no simple rotational energy equation.

CHAPTER 2

EXPERIMENTAL CONSIDERATIONS

In this chapter the experimental techniques and equipment needed to measure infrared absorption spectra will be discussed first and then the methods and instrumentation required to determine Raman spectra.

The techniques and equipment in the infrared region are dependent on the characteristics of each of the following: the source, the monochromator, and the receptor. Each of these items shall now be considered in the order mentioned.

2.1 Source of Infrared Radiation

Molecules in a gas, a liquid, or a solid are in constant thermal agitation except at the temperature of absolute zero ($-273°C$). As the temperature of a body is increased, the thermal motion and amplitude increase. All the molecules are not excited to the same degree of thermal agitation but follow a Boltzmann distribution curve. Although most of the molecules have approximately the same degree of thermal agitation, smaller numbers of molecules have lower or higher degrees of thermal agitation. Since molecules consist of atoms containing electrical charges, these oscillating centers of electrical charge give off electromagnetic radiation. The distribution of this energy corresponds to the distribution of the number of molecules in the various states of thermal excitation, and hence a hot body emits electromagnetic radiation covering a wide range of wavelengths.

Infrared radiation is normally considered as heat radiation because in approaching an infrared source, a person is certainly aware of its heat aspect. Heat or thermal energy can be transferred by infrared radiation or by conduction and convection. These latter two methods require a physical medium so that the heat can be carried directly from point to point. By contrast infrared transfers heat indirectly as radiation and does not require a direct physical medium. When infrared radiation is absorbed by an object, the radiation is converted to heat.

An object which is white-hot emits considerable energy in the infrared region of the electromagnetic spectrum but a great deal of the radiation is also in the visible region. The fact that the object appears white proves that a continuous spectrum of all visible wavelengths is present. As the object cools, it becomes red in color, indicating that the visible radiation is now confined to the red end or the long wavelength end of the visible spectrum. On cooling further, the radiation decreases and is confined to the infrared region where it is no longer visible to the human eye. The total amount of radiation decreases with temperature and the energy peak shifts toward longer wavelengths.

In this discussion it is well to introduce the concept of a blackbody, which may be defined as a body which absorbs all radiation striking it. A blackbody has the maximum possible absorption and emission and is a theoretical entity since no perfect blackbody exists. The first attempt to predict the spectral distribution of the energy E of a blackbody radiator in a quantitative fashion was made by Wien using the laws of classical physics. Although the equation derived by Wien holds for frequencies in the ultraviolet and visible regions, it fails for infrared frequencies. Lord Rayleigh and Jeans derived a distribution law also based on classical theory which holds for long wavelengths but fails when applied to the spectral region in which Wien's equation is valid. Using the quantum hypothesis, Planck derived a distribution law for blackbody radiation which holds over all wavelengths. Planck's distribution law giving the radiant energy between the wavelengths λ to $\lambda + d\lambda$ may be expressed in the form

$$E_\lambda \, d\lambda = \frac{c_1 \lambda^{-5}}{e^{(c_2/\lambda T)} - 1} \, d\lambda \qquad (2.1)$$

where e = the Naperian base, 2.718, and the constants c_1 and c_2 are given by

$$c_1 = 2\pi c^2 h = 3.740 \times 10^{-12} \text{ W cm}^2 \qquad (2.2)$$

and

$$c_2 = ch/k = 1.438 \text{ cm } ^\circ\text{K} \qquad (2.3)$$

where c is the speed of light, 2.998×10^{10} cm/sec, h is Planck's constant, 6.625×10^{-34} W sec^2, and k is the Boltzmann constant, 1.380×10^{-23} W sec/$^\circ$K. $E_\lambda \, d\lambda$ is the radiant energy emitted per unit area per unit increment of wavelength. With the units shown and $d\lambda$ in μm, the units of $E_\lambda \, d\lambda$ are W/cm^2/μm.

For $c_2 > \lambda T$, the exponential factor in the denominator of Eq. (2.1) is very large compared to unity and consequently this equation becomes

$$E_\lambda \, d\lambda = c_1 \lambda^{-5} e^{-(c_2/\lambda T)} \, d\lambda \qquad (2.4)$$

which is Wien's radiation law.

For $\lambda T \gg c_2$, the exponential factor which can be expressed as an infinite series of the type

$$e^x = 1 + x + \frac{x^2}{2!} + \frac{x^3}{3!} + \cdots$$

becomes approximately

$$e^{c_2/\lambda T} = 1 + c_2/\lambda T \qquad (2.5)$$

and

$$E_\lambda \, d\lambda = (c_1/c_2)\lambda^{-5} \lambda T \, d\lambda \qquad (2.6)$$

$$= 2\pi c k T \lambda^{-4} \, d\lambda \qquad (2.7)$$

which is the Rayleigh-Jeans radiation law.

The wavelength λ_m at which the energy is at a maximum at any temperature is found by differentiating E_λ with respect to λ and setting the derivative $dE_\lambda/d\lambda$ equal to zero. The solution[1] of the resulting equation yields Wien's displacement law

$$\lambda_m = \frac{b}{T} \qquad (2.8)$$

where b is Wien's displacement constant, 2897 μm °K, and T is absolute temperature, °K.

The actual energy distribution of a blackbody radiator at three different temperatures is shown in Fig. 2.1. The energy at any wavelength is given by the area of the vertical strip under the curve at that wavelength. If the height of the strip is E_λ and the width is $d\lambda$, then the area is $E_\lambda \, d\lambda$ as indicated on the 1000°C curve in Fig. 2.1.

Figure 2.1 also shows how the peak of the energy curve is temperature dependent. For example, at a temperature of 1000°C the wavelength at which the energy is at a maximum, $E_{\lambda m}$, is given by Eq. (2.8) and would be

[1] F. W. Sears, "Optics," p. 314. Addison-Wesley, Reading, Massachusetts, 1949.

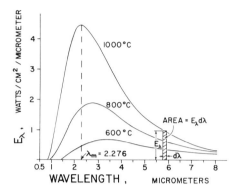

FIG. 2.1.   Energy distribution of a blackbody radiator at 600°, 800°, and 1000°C.

$(2.897 \times 10^3)/(1000 + 273) = 2.276$ μm (see Fig. 2.1). At a lower temperature of 600°C the energy peak shifts to a longer wavelength, namely, 3.318 μm. It is also important to note that the energy output falls off very rapidly at long wavelengths. In the usual infrared spectrometer a slit located between the source and the receptor is programmed to open as the wavelength increases, resulting in a constant energy level as the spectrum is traversed. Nevertheless the low energy output of the source at long wavelengths presents problems in the design of instruments for the far infrared.

Integrating of $E_\lambda \, d\lambda$ [Eq. (2.1)] between the limits 0 and ∞ yields[2]

$$\int_0^\infty E_\lambda \, d\lambda = \sigma T^4 \tag{2.9}$$

which is the Stefan-Boltzmann law and σ is the Stefan-Boltzmann constant. The Stefan-Boltzmann law states that the total radiation from a blackbody is proportional to the fourth power of its absolute temperature. When a source has less radiation than a theoretical blackbody, the ratio of its emission to that of a blackbody at the same temperature is called the emissivity, ρ. For a theoretical blackbody, the emissivity ρ equals 1.00. The radiant energy emitted by a body then becomes

$$E = \rho \sigma T^4 \tag{2.10}$$

where in practical units, E is radiant energy, W cm^{-2}, ρ is emissivity factor, σ is 5.672×10^{-12} W cm^{-2} degree^{-4} (Kelvin), and T is the temperature in degrees Kelvin.

[2] I. Kaplan, "Nuclear Physics," p. 81. Addison-Wesley, Reading, Massachusetts, 1955.

With this equation either an unknown emissivity or an unknown temperature can be calculated when the other terms are determinable. Emissivity factors range from 0.02 for highly polished silver to 0.95 for lampblack.

The ideal infrared source would be one that would give a continuous and high radiant energy output over the entire infrared region. It has been shown, however, that the total amount of energy radiated and the spectral distribution of this energy are dependent on the temperature of the source. A blackbody slit cavity at high temperature would be ideal from the continuous high radiant energy standpoint, but is impractical for spectrometers. However, several infrared sources have been developed which follow the characteristics of the blackbody radiator rather closely in the infrared region. Two examples are the Nernst Glower and the Globar.

The Nernst Glower is a source which is composed mainly of oxides of the rare earths such as zirconium, yttrium, and thorium. These oxides are molded in the shape of a rod about 20 mm long and 1 mm wide. The ends of the rod are attached to ceramic tubes for mounting and have platinum leads for the electrical connection. The source is nonconducting at room temperature and must be preheated to bring it to a conducting state. Like most semiconductors, it has a negative coefficient of resistivity and must be operated with a ballast resistance in a constant voltage circuit. Nernst Glowers are operated generally at 1900°C using 1.2 A and 75 V alternating current.

The Globar is a silicon carbide rod, usually about 2 in. long and $\frac{3}{16}$ in. in diameter. The rod is silvered at the ends to ensure better electrical contact. It is easily mounted and has about the same radiator properties as the Nernst Glower and has the advantage of being more rugged. It is operated at about 50 V and 4–5 A at 1200°C. The resistance of the Globar increases with the length of time it is operated and provision must be made for increasing the voltage by means of a variable transformer. The main disadvantage is that it must have a water-cooled jacket to protect the electrical contacts. The life of the Globar decreases as the temperature is increased because the bonding material boils out, causing high resistance hot spots and eventual failure.

Comparison of the spectral energy distribution of the Globar and the Nernst Glower[3] indicates that the Glower has an advantage in the short wavelength region but at 15 μm the two sources are comparable.

Another infrared source is the invention of Kurt H. Opperman of the Perkin-Elmer Corporation and dates to 1956. Its principle is the utilization of the infrared-radiative properties of incandescent aluminum oxide. A rhodium high-temperature heating coil is inserted in a sintered aluminum oxide tube which is then filled with a slurry of zirconium silicate and aluminum oxide powders and water to prevent deterioration of the rhodium coil

[3] R. A. Friedel and A. G. Sharkey, Jr., *Rev. Sci. Instrum.* **18,** 928 (1947).

in the atmosphere. Under vacuum, all air entrapments are removed. The slurry becomes a tightly packed dry powder mass. Electrical contacts are soldered into the rhodium wire. The solder also caps the aluminum oxide tube at both ends. In use, the source is partially enclosed in a metal shield and ceramic (porcelain) caps to prevent temperature transience due to air currents.

The source is powered by 2.8 V, 30 W, and develops an external temperature of 1200°C. It provides radiant infrared energy from 2.5 to 25 μm.

For all the infrared sources discussed the radiant energy is low in the far infrared, and to obtain sufficient energy the slit width has to be opened considerably with a corresponding decrease in resolution. One might compensate for this low emissivity in the far infrared by raising the temperature of the radiator, say from 2000° to 3000°K. It can be seen from the Stefan-Boltzmann law [Eq. (2.9)] that the total emissivity would increase about fivefold:

$$\frac{E_2}{E_1} = \frac{\sigma T_2^4}{\sigma T_1^4} = \frac{81 \times 10^{12}}{16 \times 10^{12}} = \sim 5 : 1 \qquad (2.11)$$

If the radiation is calculated using Planck's law [Eq. (2.1)] there is only a slight increase in the long wavelength radiation compared with a much greater increase in the short wavelength radiation. Since stray high frequency radiation can cause a considerable error in the absorption measurements, raising the temperature has just as bad an effect as increasing the slit width and it reduces the life of the source.

2.2 Infrared Monochromators and Filters

Prior to discussing infrared monochromators, some mention should be made of the transmittance of optical materials in the infrared region. Table 2.1 lists approximate transmission limits in wavenumbers for some optical materials.

The substances in Table 2.1 are not soluble in water except for sodium chloride, potassium bromide, cesium bromide, and cesium iodide which are very soluble in water. Irtran-2 is a trademark of the Eastman Kodak Company and is an optical material which can stand elevated temperature and can be polished to good optical tolerances. KRS-5 is a synthetic optical crystal consisting of about 42% thallium bromide and 58% thallium iodide. Polyethylene is used below about 600 cm^{-1} as it has absorption bands at higher frequencies.

TABLE 2.1

APPROXIMATE TRANSMISSION LIMITS FOR OPTICAL MATERIALS (in cm^{-1})

Glass	3000
Quartz	2500
Mica	2000
Lithium fluoride	1500
Calcium fluoride	1200
Barium fluoride	850
Irtran-2	750
Sodium chloride	600
Potassium bromide	350
Silver chloride	350
KRS-5	250
Cesium bromide	250
Cesium iodide	200
Polyethylene (high density)	30

(a) PRISM MONOCHROMATORS

Although infrared spectrometers being made today are of the grating type, older infrared spectrometers contain prisms of alkali halides.

For spectroscopic work a prism must be transparent to the particular wavelength region of interest and the dispersion of the prism must be as large as possible. These two requirements are not entirely unrelated as the following discussion will show.

The angular dispersion of a prism can be represented mathematically as $d\theta/d\lambda$ which is the rate of change of the angle of emergence from the prism with respect to wavelength. The angular dispersion may be considered as the product of two terms.

$$\frac{d\theta}{d\lambda} = \frac{d\theta}{dn}\frac{dn}{d\lambda} \qquad (2.12)$$

The first term, $d\theta/dn$, the rate of change of angular dispersion with index of refraction n, is a function of the angle between the two refracting faces. For the case of minimum deviation, where the incident and emergent rays make equal angles with the normals to the prism faces at which the rays enter and emerge, the evaluation of $d\theta/dn$ yields[4]

$$\frac{d\theta}{dn} = \frac{2\sin(\alpha/2)}{\cos i} \qquad (2.13)$$

[4] F. A. Jenkins and H. E. White, "Fundamentals of Optics," p. 465. McGraw-Hill, New York, 1957.

where, as illustrated in Fig. 2.2, α equals the apex angle of the prism and i equals the incident or emergent angle.

From this expression it can be seen that a prism must have a large apex angle to make the first term of the angular dispersion large. Prisms usually have a 60° angle as a compromise between smaller angles which give less dispersion and larger angles which cause greater reflection loss and require more material.

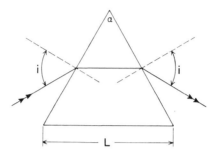

FIG. 2.2.   Passage of a ray at minimum deviation through a prism.

In the expression for angular dispersion, the second term, $dn/d\lambda$, is a function of the prism material and is commonly called the dispersion. This term can be evaluated for a substance by plotting n, the index of refraction against wavelength, λ. Such a plot is shown in Fig. 2.3 for sodium chloride from approximately 2 to 18 μm. Values of $dn/d\lambda$ can be taken at suitable wavelength intervals and plotted to yield the dispersion curve shown. For reference, transmission curves are also plotted at an approximate thickness of 1 mm.

The index of refraction of a given material decreases as the wavelength increases. Second the dispersion, $dn/d\lambda$, approaches a maximum as the substance reaches its wavelength boundary of effective transmission. In other words, a prism made of a given alkali halide will show maximum

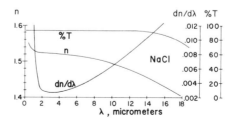

FIG. 2.3.   Variation of refractive index, dispersion, and percent transmission with wavelength of sodium chloride.

dispersion in the infrared wavelength region just below that wavelength where the prism becomes opaque. In order to obtain maximum dispersion over the entire infrared region it is therefore necessary to have several prisms of different materials. Since most data are obtained in the infrared region from 2.5 to 15 μm, a compromise is made by using a sodium chloride prism because it has adequate dispersion in this region.

TABLE 2.2

REGIONS OF OPTIMUM DISPERSION FOR VARIOUS PRISMS

Prism	Wavelength (μm)	Wavenumbers[a] (cm^{-1})
Lithium fluoride	2.0–5.3	5000–1885
Calcium fluoride	5.3–8.5	1885–1175
Sodium chloride	8.5–15.4	1175–650
Potassium bromide	15.4–25	650–400
KRS-5	20–35	500–285
Cesium bromide	25–40	400–250

[a] Values rounded to the nearest 5 cm^{-1}.

The resolving power of a prism can be defined mathematically as $\lambda/d\lambda$ and is a convenient measure of the ability of the prism to separate two spectral bands with a wavelength difference of $d\lambda$. The ultimate resolving power is equal to the product of the length of the prism base, L, and $dn/d\lambda$, the dispersion.[5] Thus, the resolution can be increased by increasing the size of the prism, by using N prisms in series in which case L becomes LN and, as previously discussed, by choosing a prism with high dispersion in the desired wavelength region.

Most infrared prism monochromators have a Littrow arrangement similar to the one shown in Fig. 2.4. In a Littrow arrangement, radiation from the entrance slit is rendered parallel by an off-axis paraboloid. It then goes through the prism to the Littrow mirror which reflects the radiation back through the prism. The paraboloid then focuses the spectrum (which is a dispersed image of the entrance slit) onto the exit slit. The exit slit allows only a small part of the spectrum to go through, thus isolating essentially monochromatic radiation which is then focused onto the detector. The wavelength region which goes through the exit slit is varied by changing the Littrow mirror angle. As was previously discussed, the radiation energy from the source decreases as the wavelength increases. Therefore the slits are usually programmed to open up as the wavelength increases to keep the

[5] F. W. Sears, "Optics," p. 278. Addison-Wesley, Reading, Massachusetts, 1949.

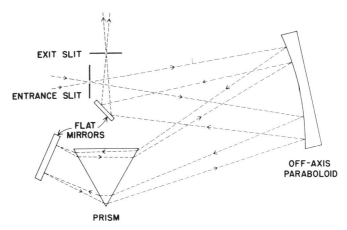

EXIT SLIT

ENTRANCE SLIT

FLAT MIRRORS

OFF-AXIS
PARABOLOID

PRISM

FIG. 2.4.   Diagram of a Littrow prism arrangement.

energy going through relatively constant. Unfortunately opening the slits allows a wider wavelength region to reach the detector, thus decreasing the resolution, but fortunately this effect is more than compensated for by the increase in dispersion $dn/d\lambda$ with increasing wavelength. Sometimes a double monochrometer is used, chiefly to control scattered or stray radiation (radiation of the wrong wavelength reaching the detector). In this arrangement the exit slit of the first monochrometer is the entrance slit of the second monochrometer.

(b)   GRATING MONOCHROMATORS

In a grating monochromator the arrangement is usually similar to that in Fig. 2.4 but with the prism removed and the Littrow mirror replaced by a grating. Usually, more than one grating (or grating order) is required to cover the entire spectral region.

The function of a grating, like that of a prism, is to provide monochromatic radiation from radiation composed of many wavelengths. A diffraction grating consists of a number of equally spaced slits, which diffract light by interference. The theoretical resolving power of a grating may be expressed as mN, where m is the order of the spectrum and N is the number of grooves or rulings on the grating. It is apparent that the highest resolution is obtained by having gratings with a large number of grooves or using large gratings to increase N and by working with higher order spectra. However, the number of grooves is restricted by several factors including the longest wavelength which is to be measured. The size of the grating is limited by cost and space available while for higher order spectra the grating efficiency falls off rapidly.

For monochromatic light of wavelength, λ, the grating law is

$$m\lambda = d(\sin i \pm \sin r) \qquad m = 1, 2, 3, \ldots \qquad (2.14)$$

i.e., the order of the spectra, and as illustrated in Fig. 2.5, i = angle of incidence, r = angle of reflectance or diffraction, d = distance between grooves—that is, the grating constant. The plus sign applies when i and r are on the same side of the grating normal, and the minus sign when they are on opposite sides. The two basic types of diffraction gratings are the transmission and the reflectance types. For the transmission grating, the grooves, which are ruled by an elaborate ruling engine on a plane glass or quartz surface with a diamond needle, are opaque and scatter any incident radiation. The clear undisturbed portions between the grooves transmit the incident radiation and act as slits. Reflecting diffraction gratings are usually made by ruling from 50 to 50,000 equally spaced grooves per millimeter on borosilicate glass coated with aluminum by vacuum evaporation. The diamond does not mark the glass at all but burnishes grooves into the aluminum. This method has the advantage that the diamond does not wear as fast and that the groove shape can be controlled better.

From the master grating replica gratings are made, having their grooves formed in a thin layer of clear plastic that adheres to the surface of a glass

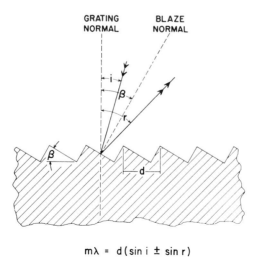

$$m\lambda = d(\sin i \pm \sin r)$$

FIG. 2.5.   Cross-sectional diagram of a plane diffraction grating.

backing plate. The optical surface of the reflectance grating is coated with aluminum. These gratings have an advantage over the master grating in that they can be re-aluminized in the event the aluminum is attacked or stained.

As previously discussed, the number of grooves per millimeter or its reciprocal, called the grating constant, is important in determining the resolving power of the grating. There is a finite limit to the number of grooves per millimeter for transmission gratings because adjacent grooves are so close that the maximum angle that the radiation can leave the grating is limited. In addition, since transmission gratings are formed on glass or quartz plates, they are only valuable in the near infrared region. For infrared instruments plane reflectance type gratings are most valuable.

Another factor to be considered is that the longest wavelength in the spectral region under consideration should be less than the width of the groove face on which the incident radiation falls. If the wavelength of the incident light is materially longer than the width of groove face, the grating acts as a mirror and most of the radiation goes into the zero order rather than being diffracted into the desired orders. By this means visible light can be separated from infrared radiation with the diffraction grating acting as a mirror for the infrared region.

A potential disadvantage of a grating for infrared instrumentation is that the grating, unlike the prism, distributes energy into a number of orders. However this disadvantage can be largely overcome by controlling the groove shape to a specific blaze angle (see Fig. 2.5) thereby concentrating the spectral energy in the desired region of the spectrum for a given order. For reflectance gratings the blaze angle is calculated on the basis that the angle of incidence, i, equals the angle of reflectance, r. Undesired overlapping orders can be eliminated with a fore prism or by appropriate filters, to be discussed next.

(c) FILTERS

Most infrared instruments have filters which pass infrared radiation within a particular wavelength range. Appropriate filters can reduce stray radiation in a prism spectrometer and remove the undesired higher orders obtained with a grating spectrometer. Four types of filters involved in infrared measurements are the transmission, reflectance, interference, and scattering filters.

Transmission filters consist of a thin single filter coating evaporated on a substrate. They have high transmission above and below the wavelength where high absorption produces a sharp cutoff or cut-on. (A cutoff is the wavelength where the filter has almost complete absorption of incident radiation; a cut-on, where absorption ceases to be high.) Many crystalline materials show a preferential reflection at certain wavelengths corresponding in frequency with the predominant vibration rate of ions in the particular crystal lattice. For example, quartz is valuable at 9 μm, calcium fluoride at 23 μm, and cesium bromide at 125 μm. The spectral purity of such a selective reflection increases with the number of reflections. Reflectance filters are of main value in the far infrared region.

An interference filter consists of a series of thin, multilayer films of various castings on a substrate. These layers are either deposited on the substrate or evaporated on in a vacuum. By use of various coatings and substrates and a variation in film thickness, the cutoff or cut-on wavelength can be accurately controlled over a fairly wide range. The wavelengths which are to be left out of the spectrum are reflected; the rest are transmitted through the filter.

A great deal of stray radiation in prism spectrometers falls between 1.5 and 3.5 μm. If a mirror or KBr plate is roughened, this stray radiation can be reduced and these filters are called scattering filters.

Infrared filters can be further conveniently classified in the following manner. The short wavelength pass filter absorbs or reflects wavelengths longer than the cutoff wavelengths. There is a sharp drop in transmission at the cutoff which is less than 5% of the maximum transmission. The long wavelength pass filter will transmit all wavelengths longer than a particular cut-on, absorbing or reflecting those below this wavelength. The third classification is the band-pass filter which has high transmittance between two wavelengths. Transmission is low at both ends of this range, which may be wide or narrow depending on the filter characteristics. In choosing a suitable filter, transmission curves, cutoff sharpness, optical efficiency, cost, availability, stability (how the filter will hold up under prolonged radiation), and surface hardness are important considerations.

2.3 Infrared Detectors

Infrared detectors can be conveniently divided into two general groups, selective and nonselective. Selective detectors include those whose response is markedly dependent on the wavelength of incident radiation. Photographic plates, photocells, photoconductive cells, and infrared phosphors are in this group. The second group (nonselective) includes detectors whose response is directly proportional to incident energy and relatively independent of wavelength. These detectors are better suited for spectroscopy work, and among the more common types are thermocouples, bolometers (metal strip, thermistor, and superconducting), and the pneumatic cell.

The most important type of selective detector is the photoconductive cell, which has a rapid response and high sensitivity. These cells, usually made of materials such as thallous sulfide, lead sulfide, lead telluride, lead selenide, or selenium, show an increase in conductivity when illuminated by infrared light.

A thermocouple is an excellent detector for measuring infrared radiation. By blackening the junction of dissimilar metals, incident radiation is absorbed, causing a temperature rise at the junction and a resultant increase in the electromotive potential developed across the junction leads. Thermocouples are often made by evaporating metals, such as bismuth, antimony, or semiconductor alloys, on a thin film of cellulose nitrate or other supporting base.[6] The thermocouple is placed in an evacuated chamber with a window of NaCl and KBr, or CsI and crystal quartz for long wavelengths, i.e., 50 μm. Common commercial units have sensitivities of about 6–8 $\mu V/\mu W$.

A bolometer is a detecting device which depends on a change of resistance of the material with temperature. The receiving element is made one arm of a Wheatstone bridge circuit, thus providing an easy method of measuring the resistance of the element. Bolometers have an advantage in infrared detection in that they generally have a faster response time than thermocouples.[7] Jones[8] has given a good discussion of the theoretical aspects of their behavior.

In recent years, thermistor bolometers have come into prominence largely because of their increased sensitivities. Thermistors are made of semiconducting material in which the resistance decreases exponentially with increasing temperature. The thermal coefficient of resistivity is about 4 %/°C, or about ten times that for metal bolometers. Thermistors are prepared in the form of flakes, 1–5 mm long, 0.1–1 mm wide, and 0.01–0.02 mm thick, composed of sintered oxides of nickel, cobalt, and manganese. Metal contacts are placed on each end, the assembly is cemented to a backing of glass or quartz, and the whole assembly placed in an evacuated metal case, with an infrared-transparent window.[9] Thermistors have also been prepared of other materials, such as the oxides of uranium.

Commercial thermistor bolometers consist of a matched pair of thermistor flakes, having similar thermal and electrical properties. The second flake serves as a compensating element to eliminate the effect of background radiation. Time constants for thermistors are about the same as for metal bolometers (3 μsec), but the sensitivity is much greater (up to 400 V/W).

Superconducting bolometers use metals or compounds such as columbium nitride at cryogenic temperatures. They offer the advantages of extremely high sensitivity and very fast response.

[6] V. Z. Williams, *Rev. Sci. Instrum.* **19**, 135 (1948).

[7] E. B. Baker and C. D. Robb, *Rev. Sci. Instrum.* **14**, 356 (1943).

[8] R. C. Jones, *J. Opt. Soc. Amer.* **36**, 448 (1946).

[9] W. H. Brattain and J. A. Becker, *J. Opt. Soc. Amer.* **36**, 354 (1946).

The pyroelectric detector consists of a thin pyroelectric crystal such as triglycine sulfate. If such a material is electrically polarized in an electric field, it retains a residual electric polarization after the field is removed. The residual polarization is sensitive to changes in the temperature. Electrodes on the crystal faces collect the charges so the device acts as a capacitor across which a voltage appears, the amount of which is sensitive to the temperature of the device. The pyroelectric detector operates at room temperature. Being a thermal device, it possesses essentially flat wavelength response ranging from the near infrared through the far infrared. It can handle signal frequencies of up to several thousand Hertz and hence is well suited for the Fourier spectrometer since the interferometer can be scanned at a moderately rapid velocity. The principle of the Fourier spectrometer is discussed in more detail in Section 2.9.

A comparison of commercially available far infrared detectors has been made by R. C. Milward.[10]

One of the more ingenious detectors was the pneumatic cell, developed primarily by Golay.[11] The Golay cell consisted of a gas-filled 3 mm cell, which was connected by means of a tube to a second small cell, one side of which was closed by a curved diaphragm which acts as a mirror. A change in the radiant energy incident on the first cell caused a change in gas pressure, which, in turn, moved the mirror on the second cell. A steady beam of light was reflected off the mirror and passed through a grid network to a photocell, which was, in turn, connected to an amplifier and recorder. A slight movement of the mirror would result in a large change in intensity of the light passing through the grid thereby yielding great sensitivity.

It is beyond the scope of this text to consider the amplifiers used with the various detectors discussed. There are several excellent references[6,12,13] which cover this subject adequately.

2.4 *Commercial Instruments*

Major manufacturers of infrared spectrometers are Beckman Instruments, Inc. and the Perkin-Elmer Corporation in the United States and Unicam Instruments Limited in England. Because of the constant changes being

---

[10] R. C. Milward, "The Spex Speaker," Vol. XVII, No. 3, p. 5. Spex Industries, Inc., Metuchen, New Jersey, 1972.

[11] M. J. E. Golay, *Rev. Sci. Instrum.* **18**, 357 (1947).

[12] R. A. Smith, F. E. Jones, and R. P. Chasmar, "The Detection and Measurement of Infra-red Radiation," Chapter XII. Oxford Univ. Press, London and New York, 1957.

[13] G. K. T. Conn and D. G. Avery, "Infrared Methods," Chapter IV. Academic Press, New York, 1960.

made in commercial infrared instruments, it is not advisable to describe the models available.

The first commercial infrared spectrometers were single beam instruments; however, the interference of water vapor and carbon dioxide bands severely limited their application. In 1947 a double beam instrument was described in the literature[14] and made available commercially by Baird Associates, Inc. Since that time single beam instruments have gradually lost their importance except for special applications.

Most commercial instruments are of the double beam type where the radiation from the source is divided into two parts, a sample beam and a reference beam. The sample beam goes through the sample, and the reference beam goes through air. If solutions are being run, the solvent can be put into the reference beam to compensate for solvent bands. By means of a rotating semicircular mirror, first one beam and then the other beam are alternately focused onto the entrance slit of the monochromator. If, at a certain wavelength, the two beams are equal in intensity, no alternating component of radiation reaches the detector. But if the beams are unequal because of sample absorption, then an alternating component of radiation reaches the detector and its energy is selectively amplified. This signal is used to equalize the two beams, using some kind of optical attenuation in the reference beam. The recorder pen is coupled to the motions of the attenuator position as a function of wavelength, thus recording the spectrum.

The newest type of instrument is the infrared Fourier transform spectrometer. This type will be discussed later in Section 2.9.

2.5 Instrument Calibration

It is occasionally necessary to check the calibration of an infrared spectrometer by measuring the spectra of gases whose absorption bands are accurately known.

The spectra of gases measured on a prism spectrometer often have a different appearance as compared to those obtained with a grating spectrometer of much higher resolution. Two problems can result from this difficulty. Identification of the absorption bands in the calibration traces may be an arduous task, and when a group of bands coalesce, if the bands are not symmetrically placed, the frequency of the absorption maximum may be different from that of the strongest band. Care should be taken, therefore, in making such comparisons. Articles showing the spectra of many gases with

[14] W. S. Baird, H. M. O'Bryan, G. Ogden, and D. Lee, *J. Opt. Soc. Amer.* **37**, 754 (1947).

their band positions tabulated have been published,[15,16] and one text[17] deals exclusively with calibration data. For a routine check in the 2–15 μm region a film of polystyrene is convenient since calibration data on polystyrene are available in most manufacturers' manuals. Polystyrene bands used for calibration include 3027.1, 2924, 2850.7, 1944.0, 1601.4, 1154.3, 1028.0, 906.7, and 698.9 cm^{-1}.

2.6 Sample Handling Techniques

Since most sample holders have some type of alkali halide window a brief discussion of polishing rock salt crystals as an example is in order. Block rock salt is prepared for polishing by first cleaving the pieces roughly to shape with a single-edged razor blade and a small hammer. The salt crystal is placed on a metal block or firm surface covered with a few layers of paper. The razor blade is held on the crystal edge parallel to the cleavage planes and the sides and then tapped with the hammer. When a cleavage is started, it is followed until complete. Pieces may be split in halves or thirds safely, but if much smaller sections are cleaved off, these smaller pieces may break. If small pieces are desired, the big pieces should be cleaved in half and the halves in half, etc. The crystals may be finally cut into desired sizes by appropriate cuts perpendicular to the edges.

The cleaved faces can be sanded flat with No. 220 silicon carbide paper such as WETORDRY TRI-M-ITE paper* and fine-sanded with No. 600 paper. Used crystals can be sanded if they are badly scratched or etched. If they have had very light use, they should be repolished without sanding.

One technique for polishing is as follows: two laps are prepared and mounted side by side. One lap is made from a 6-in. diameter metal disc with an indentation in the rim to take a rubber O-ring. Two thicknesses of fine nylon cloth are stretched over the disc and held in place with the O-ring. The other lap is like the first but covered with ordinary diaper cloth.

The nylon lap is moistened completely with water from a medicine dropper and a little Aloxite buffing powder A No. 1 fine† or Cerium Oxide Polishing

---

[15] A. R. Downie, M. C. Magoon, T. Purcell, and B. Crawford, Jr., *J. Opt. Soc. Amer.* **43**, 941 (1953).

[16] E. K. Plyler, L. R. Blaine, and M. Nowak, *J. Res. Nat. Bur. Stand.* **58**, 195 (1957).

[17] International Union of Pure and Applied Chemistry Commission on Molecular Structure and Spectroscopy, "Tables of Wavenumbers for the Calibration of Infra-red Spectrometers." Butterworth, London, 1961.

* Available from the 3M Company, St. Paul, Minnesota.

† Available from the Carborundum Company.

Powder* is sprinkled on and spread evenly with the fingers. The lap should be moist—not wet or unevenly damp. The rock salt crystal is held by the ends and polished about 30 strokes on the wet lap and immediately buffed about 10 strokes on the dry lap. The operation is repeated until a clear crystal is obtained. If the lap is too dry, a frosted effect results. If the lap is too wet, excessively rounded edges and a ripple "orange peel" effect results.

Using fingers for holding the crystal usually results in marks on the plate ends but, if this area is not in the region of the infrared beam, it can be ignored. If a completely clear crystal is desired, a holder or rubber fingers must be used.

Rock salt can also be well polished with alcohol and rouge as a fine abrasive.

Cesium bromide is used when transmission down to 250 cm^{-1} is desired. Cesium bromide does not cleave like NaCl and is usually bought preshaped. It is softer and more easily bent or scratched than NaCl. When sanding damaged CsBr plates the abrasive paper is best wet with an alcohol. The polishing process is similar to that described for NaCl but with water replaced with an alcohol. It is hard to obtain as good a polish with CsBr as one gets with NaCl.

(a) GASES

Absorption spectra of gases can be measured in a wide variety of gas cells ranging from a few centimeters to several meters in path lengths. For compactness the longer path length cells have mirrors which reflect the beam through the cell several times. In cells of one or more meters such a procedure results in a considerable energy loss in the infrared beam. Both glass and metal cells are used although for either very small or very large volumes metal cells are more convenient. To hold a vacuum of 10^{-5} mm or better the glass cells must have carefully polished ends on which the halide windows are cemented with materials such as epoxy resins. Metal cells of many types have been described in the literature. The windows in metal cells with O-ring gaskets can be removed more readily for polishing than from glass cells with cemented windows. In addition, metal cells are more suitable for studies at high temperatures. Freezeout traps or manometers can be connected to such cells with glass to metal seals. If a suitable valve is placed between the cell and the freezeout trap, the gas pressure can be readily controlled with suitable freezing mixtures on the trap, particularly if several different gases are in the cell. The manometer serves to monitor the pressure of the gas. For measurements at room temperature neoprene O-rings are suitable, but for temperatures above 100°C silicone rubber O-rings are more durable.

* Available from Davison Chemical Co., Pompton Plains, New Jersey.

Narrow contoured gas cells with a path length of 10 cm and a volume of only 26 cm^3 are particularly valuable if only small samples of gas are available. Most gases yield suitable spectra at approximately 50 mm pressure. For qualitative analysis an excellent catalog of the infrared spectra of 66 gases has been published by Pierson et al.[18]

(b) LIQUIDS

In most instances the spectra of liquids are measured in either a demountable type cell similar to the one shown in Fig. 2.6 or in fixed thickness or sealed cells similar to that shown in Fig. 2.7. Both types of cell are commercially available from a number of sources.

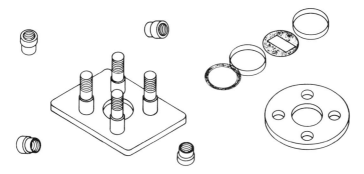

FIG. 2.6.   Demountable type infrared cell for liquids.

If the cell windows are sufficiently flat and parallel the cell thickness of a fixed cell may be measured by observing the interference fringes or percent transmittance undulations which result when the spectrum of the empty cell is taken. Interference fringes result because the transmitted beam finds itself accompanied by the infrared beam which has been twice reflected inside the cell and is retarded by twice the cell thickness relative to the unreflected beam. The thickness t in cm may be calculated by

$$2t = \frac{N}{\bar{v}_1 - \bar{v}_2} \tag{2.15}$$

where N is the number of fringes found between $\bar{v}_1$ and $\bar{v}_2$ (the frequencies in cm^{-1}).

[18] R. H. Pierson, A. N. Fletcher, and E. St. Clari Gantz, *Anal. Chem.* **28**, 1218 (1956).

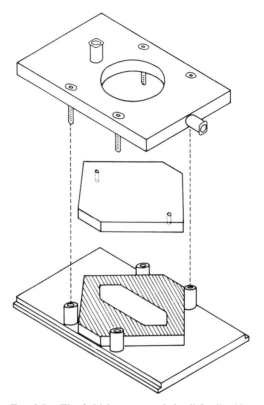

FIG. 2.7.   Fixed thickness or sealed cell for liquids.

Liquids which are not too viscous or corrosive to metal or rock salt are normally measured in sealed cells to prevent undue evaporation. These cells may be filled and emptied with a hypodermic syringe. Either corks or plugs made of an inert plastic, such as Teflon, are used to cap the cells. Viscous or corrosive liquids (including water) are placed in demountable cells. An aluminum spacer foil is placed on the bottom crystal in the metal holder, and a drop of sample is placed in the center of the spacer. The second rock salt crystal is placed on top and fastened gently in place. Usually rubber gaskets are placed between the metal holder and the crystal to prevent cracking the crystal. In many instances with very viscous liquids or heavily absorbing liquids the spacer is omitted. Aqueous solutions may be measured in a limited wavelength region with calcium fluoride plates or in most of the rock salt region with barium fluoride or Irtran-2 and in the complete rock salt region using silver chloride plates. Silver chloride, however, is reactive and deformable and darkens on exposure to light which causes a gradual reduction in its transmission in the infrared region also. For qualitative aqueous solution analysis, it is often simpler to use a demountable cell with rock salt plates and resand and polish the plates after the spectrum has been obtained.

Cavity cells first developed by R. N. Jones of the National Research Council of Canada are made by ultrasonic machining rock salt crystals with a blade of the desired shape which is driven into the block at a frequency of about 20,000 cycles/sec. These cells are economical in cost and suitable for measuring the spectra of most solutions but are less useful for pure liquids since the smallest thickness available at present is 0.05 mm.

For qualitative and quantitative analysis when it is desirable to determine a solute in a solution, the absorption bands of the solvent can be canceled out by filling a second cell of the same thickness as the sample cell with the solvent and placing the second cell in the reference beam. To perform the same function, variable space cells which may be quickly set to give any path length within such ranges as 0.015 mm or 0.08 to 1.0 mm are also commercially available.

(c) SOLIDS

Techniques for obtaining the infrared spectra of solids are more diversified than for either gases or liquids. If a suitable solvent is available, the solid may be dissolved and measured in the manner described for liquids. However, suitable solvents are limited in number and none are totally transparent. Those solvents which are less transparent require relatively high concentrations of the solid before adequate spectra can be obtained. Probably carbon tetrachloride, chloroform, and carbon disulfide are the most valuable solvents from the standpoint of transparency.

An alternate technique most useful for soluble polymers is to prepare a film of the solid by evaporating a solution of the material directly on a rock salt plate or on a material from which the dried film can be peeled. Water solutions can be readily evaporated on silver chloride or Irtran-2 plates.

Another method involves melting the sample on the salt plate and allowing it to resolidify. This method is not recommended for crystalline materials because of possible orientation effects or for samples which might be decomposed by heat. It is most applicable for amorphous materials of a tarry or waxy nature.

Orientation effects result from the fact that radiation is a transverse vibration and only interacts with the components of dipole moment changing in a direction normal to the beam direction. For example, if a benzene ring compound is oriented with the ring coplanar with the salt plate, out-of-plane vibrational bands will be weakened and in-plane bands will be strengthened relative to an unoriented preparation. Oriented films are prepared deliberately to observe just such effects for vibrational analysis studies but for qualitative and quantitative analysis orientation is not desirable.

Crystalline powders strongly scatter radiation and are prepared in a manner to reduce the scattering. Scattering effects are seen as a gradual lowering of the "no absorption" background line at the short wavelength end of the spectrum and as a nonsymmetrical band shape distortion. To reduce the scattering, first, the powder is finely ground, preferably to a particle size smaller than the wavelength being used. Second, the particles are surrounded by a material whose index of refraction is relatively similar to that of the sample. The mull method and the pressed disc technique, to be discussed next, both follow this procedure.

An extremely valuable technique for solid samples is the mineral oil (Nujol) mull method. A small amount of solid sample is mulled in a mortar with a small amount of Nujol to yield a paste which is then transferred to a rock salt plate or the sample may be mulled directly between the salt plates. The small scratches that result from this latter method (which saves time and sample) do no damage and are completely removed when the plate is resurfaced.

The mineral oil technique has the advantage of nonreactivity with samples, reproducibility of results, minor obscuration because of the medium, and rapidity of preparation. The obscured aliphatic carbon-hydrogen region does not always yield definitive information and its loss can usually be tolerated. When information is needed about the carbon-hydrogen region, a second mull can be prepared using either totally fluorinated or chlorinated hydrocarbons such as perfluorokerosene, Halocarbon Oil* or hexachlorobutadiene as the mulling agent. These liquids have strong bands elsewhere but are free of carbon-hydrogen absorption bands. With this complementary preparation, the whole spectral region is available unobscured.

Where there is insufficient sample to fill the sample beam at the desired thickness, the mulled sample should not be thinned further. Instead the regions without sample should be masked off with tape even if this leaves only about 3 mm^2 clear. Modern double beam instruments will yield a suitable spectrum if the slits are widened a little and a compensating wire screen or attenuator is placed in the reference beam.

Another valuable method involves the potassium bromide disc technique. In this method a very small amount of finely ground solid sample is intimately mixed with powdered potassium bromide and then pressed in an evacuated die under high pressure. The resulting discs are transparent and yield excellent

* Halocarbon Products Corp., 82 Burlews Court, Hackensack, New Jersey.

spectra. The only infrared absorption by the potassium bromide matrix is from small amounts of adsorbed water in the powder. However, this can be confused with OH impurity in the sample.

There are a wide variety of dies commercially available for the preparation of potassium bromide discs. In general, a die which can be evacuated prior to pressing the disc is preferable but not mandatory for routine work. Pressures of approximately 8 tons for a pressing time of 5 min or 10 tons for 1 min are suitable for dies which produce 13-mm discs. For larger diameter dies, higher pressures are required. A Carver Laboratory Press of 10-ton capacity is suitable or an economical press can be made from an ordinary hydraulic truck jack and a suitable pressure gauge.

Potassium bromide of infrared quality ground to at least 400 mesh is commercially available from the Harshaw Chemical Company, Cleveland, Ohio. Only two weak water bands should be evident in the spectrum of a disc containing pure potassium bromide, the H_2O stretching vibration at 3400 cm^{-1} and the H_2O bending vibration at approximately 1640 cm^{-1}. Their intensities vary with sample preparation.

The optimum sample concentration usually varies from 0.1% to as much as 3% in some aromatic compounds. The proper concentration is best determined by trial and 0.5% is a good starting point. Other important variables in obtaining good spectra are the particle size and homogeneity of the sample in the potassium bromide matrix. Both these variables are related to the grinding time. With the same concentration of sample in the potassium bromide the absorption bands increase in intensity as the grinding time is increased and usually reach a maximum constant value in from 3 to 5 min. Proper particle size and distribution can be achieved by mechanically grinding the sample in a plastic capsule in a commercially available* dental accessory known as the "Wig-L-Bug."

A technique for handling polymers that are difficult to mull, dissolve, or melt consists of placing the polymer and a suitable amount of KBr in a metal Wig-L-Bug capsule with a metal ball inside. The closed capsule is sealed with masking tape and immersed in liquid air. When bubbling stops, the capsule is placed in the Wig-L-Bug and shaken. After the capsule has reached room temperature a normal KBr pellet is made of the contents. The normally tough polymer becomes brittle at low temperatures.

Another technique involves dissolving the sample in a volatile solvent (even water is suitable), adding the solution to the potassium bromide, and then evaporating the solvent before grinding. Samples taken from the outlet of a gas chromatograph can be conveniently trapped in a porous plug of

* Available from the Crescent Dental Manufacturing Co., Chicago, Illinois.

potassium bromide and the spectra then obtained by pressing a disc in the usual manner.[19]

An alternate procedure involves the so-called freeze drying technique. In this method solutions of weighed amounts of sample and potassium bromide are cooled rapidly in an acetone–Dry Ice mixture. Such rapid cooling yields submicroscopic crystals of the organic compound. The resulting mixture is then dried over phosphorus pentoxide at reduced pressures (approximately 10 mm) and slightly elevated temperatures (50° C).

It has been found on occasion that the spectra obtained on powders prepared by the freeze drying method differ from those obtained by grinding. In addition, there have been differences found between the spectrum of a solid when measured in a Nujol mull as compared to its spectrum in potassium bromide. Several authors have reported changes in the spectra of solids in alkali discs caused either by metathesis[20,21] or the formation of mixed crystals.[22,23] Padgett[24] has discussed the different spectra obtained for cyanuric acid in potassium bromide, potassium chloride, and for a sublimed sample, and concluded that the effect is probably caused by ion interchange. Baker[25] has summarized solid state anomalies in infrared spectroscopy with emphasis on organic compounds. He lists the factors which influence the changes in pellet spectra as follows: (a) crystal energy of the organic phase, (b) total amount of energy used in grinding sample and matrix, (c) lattice energy of matrix, (d) particle size of matrix, (e) stress relaxation involving either the temperature of pellet or powder mixture or the time lapse between grinding the sample mixture and obtaining the spectrum, and (f) occurrence of polymorphic transitions.

Although the potassium bromide method has the disadvantages mentioned for crystalline samples, it is still well suited for obtaining spectra of microsamples and of amorphous polymers or resins. It should not displace the mineral oil mull technique but rather act as a complementary method.

(d)   INTERNAL REFLECTION SPECTROSCOPY

An important method called internal reflection spectroscopy (IRS) or attenuated total reflection (ATR) for obtaining the infrared spectra of solids

[19] H. W. Leggon, *Anal. Chem.* **33**, 1295 (1961).

[20] W. A. Pliskin and R. P. Eischens, *J. Phys. Chem.* **59**, 1156 (1955).

[21] J. A. A. Ketelaar, C. Haas, and J. van der Elsken, *J. Chem. Phys.* **24**, 624 (1956).

[22] V. C. Farmer, *Chem. Ind. (London)* p. 586 (1955).

[23] L. H. Jones and M. M. Chamberlain, *J. Chem. Phys.* **25**, 365 (1956).

[24] W. M. Padgett, II, J. M. Talbert, and W. F. Hamner, *J. Chem. Phys.* **26**, 959 (1957).

[25] A. W. Baker, *J. Phys. Chem.* **61**, 450 (1957).

or films has been described by Fahrenfort.[26] Harrick has also discussed the theory in considerable detail.[27-29]

If a beam of monochromatic radiation traverses a prism so that it is internally reflected from the back face of the prism as shown in Fig. 2.8(a), three effects may be observed depending on the angle of incidence θ_1 of the beam within the dense prism material, and the ratio of the indexes of refraction of the rare medium n_2 to the dense material n_1, as seen in Fig. 2.8(b).

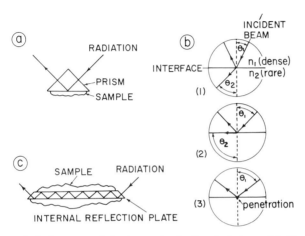

FIG. 2.8. Optical diagrams of the internal-reflection technique. (a) Single reflection from prism. (b) Reflections depending on angle of incident ray. (c) Multiple-reflection method.

(1) If $\sin \theta_1 < n_2/n_1$ some reflection will occur but the larger part of the beam will be refracted, the angle of refraction θ_2 being larger than θ_1.

(2) If $\sin \theta_1 = n_2/n_1$, then the angle of refraction, $\theta_2 = 90°$, and the radiation will travel along the interface. In this case θ_1 is called the critical angle θ_c.

(3) If $\sin \theta_1 > n_2/n_1$, then the angle of refraction θ_2 is imaginary and all the radiation is reflected back into the dense medium.

This last condition is that of total internal reflection. Both experiments and theory have established that in this case an evanescent (i.e., vanishing)

[26] J. Fahrenfort, *Spectrochim. Acta* **17**, 698 (1961).

[27] N. J. Harrick, *Ann. N.Y. Acad. Sci.* **101**, 928 (1963).

[28] N. J. Harrick, *J. Opt. Soc. Amer.* **55**, 851 (1965).

[29] N. J. Harrick, "Internal Reflection Spectroscopy." Wiley (Interscience), New York, 1967.

wave having the same frequency as the totally reflected radiation and whose amplitude diminishes logarithmically persists beyond the interface into the rare medium. Both Fahrenfort and Harrick independently recognized that such an evanescent wave was capable of interacting with an absorbing medium just beyond the interface leading to a reduction or " attenuation " of the totally reflected radiation. Thus, if an absorbing substance is placed on the reflecting surface of the prism, the energy that escapes temporarily from the prism is selectively absorbed. The spectrum of the internally reflected radiation is similar to the conventional infrared absorption spectrum of the substance. The infrared band intensities are the equivalent of a penetration of a few microns into the substance and are independent of the sample thickness (beyond a certain minimum). Because of this, sample preparation is simplified for many materials (such as rubbery polymers, for example) and it is invaluable for obtaining the spectrum of surface films. A further advantage is the absence of interference fringes which can cause errors in the normal transmission method.[30]

To obtain suitable spectra the angle of incidence of the infrared beam on the internally reflecting face of the prism should be adjustable. In addition, the prism should have a higher index of refraction than the substance to be examined. In this respect KRS-5 with an index of refraction of 2.380 is most suitable. Initial ATR devices used a single reflection and a variable angle of incidence to produce the desired spectra. The amount of absorption can be markedly increased by multiplying the number of reflections as much as 25 times as shown in Figure 2.8(c). The amount of absorption is dependent on the area and efficiency of contact between the sample and reflector.

Internal reflection spectra are not exact duplicates of transmission spectra. First of all, the effective penetration of the evanescent wave into the sample is a direct function of the wavelength of the radiation. This means that as the wavelength increases, bands get deeper relative to those in a transmission spectrum. Secondly, the effective penetration of the evanescent wave is an inverse function of the quantity $[\sin^2 \theta_1 - (n_2/n_1)^2]^{1/2}$. This means that band intensities can be increased by decreasing θ_1. However, a complication occurs because the index of refraction of the sample n_2 fluctuates sharply in the region of an absorption band. It is usually higher and lower than normal, respectively, on the long and short wavelength side of the band center. This causes an index of refraction distortion of band shapes. This distortion effect of fluctuations of n_2 on the quantity in the brackets above can be minimized by keeping θ_1 large and well away from the critical angle. This means that single reflection undistorted spectra may be weak so multiple reflections are usually used.

[30] J. V. White and W. M. Ward, *Anal. Chem.* **37**, 268 (1965).

(e) HIGH AND LOW TEMPERATURE METHODS

It is often desirable to obtain infrared spectra of samples at elevated or low temperatures in addition to room temperature. If a sample has a melting point near room temperature, the spectrum is often measured in the liquid phase for comparison with simpler related compounds which are normally liquids. Other examples include the study of the temperature dependence of bands as well as the study of infrared spectra obtained at different temperatures of a chemical reaction which can be carried out directly in the cell. In order to obtain the vapor spectrum of a material with a relatively low vapor pressure either a multiple reflection cell of several meters or a heated gas cell of conventional length can be used. However, in the case of a compound with very low vapor pressure a heated multiple reflection cell is needed.

One type of heated cell for liquids or solids is a small metal box with windows to transmit the infrared beam and a back flange which fits in the cell holder plate on the instrument. Inside the box a conventional cell is mounted into another cell holder plate identical to the one on the instrument. The box is wound with resistance wire and covered with asbestos. A Variac controls the current while the temperature is measured either with a thermometer or a thermocouple which for best results should be mounted into the body of the salt plates of the cell.

Commercial pyrolyzers are available for the controlled thermal degradation of materials which are difficult to prepare for transmission spectroscopy because of toughness, surface texture, or composition, including certain polymers and rubbers.[31,32] Carbon-filled rubbers are best identified by this technique.

Infrared spectra are often measured at low temperature to verify the existence of rotational isomers. The more complex spectra at room temperature can usually be simplified at reduced temperatures yielding the spectrum of the more stable rotational configuration. Absorption bands which are broad and indistinct and overlap in liquids often become sharper as the liquid becomes a solid. Difference bands can sometimes be distinguished from overtone and combination bands by the disappearance of the difference bands at reduced temperatures.

Most designs for low temperature cells[33] include a liquid reservoir for coolants, such as Dry Ice and acetone or liquid nitrogen near the sample mounting. A typical design[34] is shown in Fig. 2.9. A circular insert can also be placed in the copper block so that potassium bromide discs can be studied at low temperatures.

[31] D. L. Harms, *Anal. Chem.* **25**, 1140 (1953).
[32] B. Cleverley and R. Herrmann, *J. Appl. Chem.* **10**, 192 (1960).
[33] A. Walsh and J. B. Willis, *J. Chem. Phys.* **18**, 552 (1950).
[34] E. L. Wagner and D. F. Hornig, *J. Chem. Phys.* **18**, 296 (1950).

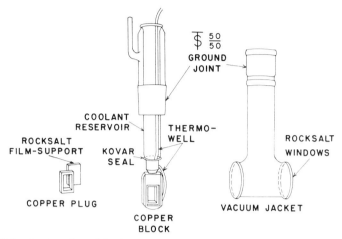

FIG. 2.9.   Design of a cell for low temperature measurements. [From E. L. Wagner and D. F. Hornig. Reprinted by courtesy of the *J. Chem. Phys.* **18**, 296 (1950).]

2.7   *Quantitative Analysis*

(a)   THE ABSORPTION LAW

If P_0 is the incident radiant power impinging on a sample and P is the radiant power after passing through the sample, then the ratio P/P_0 is called the transmittance of the sample and given the symbol T (see Fig. 2.10). Multiplying by 100 gives the percent transmittance ($\% T$) of the sample. If the sample is in a cell of thickness b in cm at a concentration of c in gm/liter then the fundamental equation governing the absorption of radiation is

$$\frac{P}{P_0} = 10^{-abc} \qquad (2.16)$$

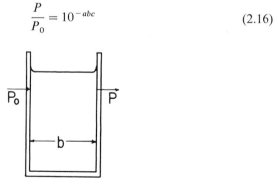

FIG. 2.10.   Graphical illustration of transmission terms.

All the terms in this equation have been defined except a which is a constant characteristic of the sample at a given frequency or wavelength and is called the absorptivity.

Equation (2.16) is cumbersome for calculations and can be transformed by taking the logarithm to the base 10 of both sides of the equation and replacing P/P_0 by P_0/P to eliminate the negative sign. The result is

$$\log_{10} \frac{P_0}{P} = abc \tag{2.17}$$

The term $\log_{10} P_0/P$ is given the symbol A and called the absorbance.* Thus,

$$A = abc \tag{2.18}$$

Historically, the study of the variables in the previous Eq. (2.16) yielded Beer's law denoting conformation with respect to changes in concentration, c, and Bouger's or Lambert's law denoting conformation with respect to changes in thickness, b. Equation (2.16) is obtained by combining[35] the laws just mentioned yielding the Beer-Lambert or Bouger-Beer law. However, for the sake of simplicity the law is now called Beer's law[36] and written as shown in Eq. (2.18). It is seen that the absorbance, A, is a function of three factors, a constant specific for the substance at a particular frequency, the concentration, and the thickness. The product of the concentration and the thickness is a measure of the relative number of molecules in the infrared beam.

A condition implied in this equation is that the radiant energy must be monochromatic, that is, of a single discrete wavelength. This condition is difficult to obtain in practice but is sufficiently approached so that quantitative analyses based on this law may be carried out.

In absorption measurements in the ultraviolet and visible wavelength regions the slit width of the instrument is relatively narrow compared to the widths of the bands in the absorption spectrum. It is worthwhile, therefore, to measure molar absorptivities or ε values which are a product of the absorptivity, a, and the molecular weight, M, of the material, i.e., $\varepsilon = aM$. ε values are measured at band maxima and furnish valuable adjunct information as to the nature of the material. There is little problem in obtaining reproducible ε values from one instrument to another. In the infrared region, however, ε values or absorptivities are seldom measured and reported for a

---

[35] F. H. Lohman, *J. Chem. Educ.* **32**, 155 (1955).

[36] H. K. Hughes *et al.*, *Anal. Chem.* **24**, 1349 (1952).

* The ratio of I_0 and I representing the intensity of the incident and transmitted radiant energy rather than P_0 and P are frequently employed. For current spectroscopic nomenclature, see L. May, *Appl. Spectrosc.* **27**, 419 (1973).

given compound because of lack of reproducibility from one instrument to another. This problem arises because the slit width of the instrument is relatively wide compared to the band widths observed. Hence, the slit width of the instrument at a given frequency is an important variable that must be considered in quantitative analysis or high resolution work. Robinson[37] in a systematic outline of some errors in infrared analysis has pointed out that deviations from Beer's law are to be expected whenever the slits are wider than the width of the absorption band. When the ratio of slit width to band width was 0.4 there was no appreciable deviation from Beer's law. However, at a ratio of 0.8 the deviations became noticeable, being the largest for measurements made at a wavelength taken on the band where the steepest slope occurred.

(b) INTEGRATED BAND INTENSITIES

Peak height measurements $(A = \log_{10} P_0/P)$ used in most quantitative analyses are sensitive to changes in instrumental resolution and can vary considerably from instrument to instrument. The integrated intensity is a measure of the total intensity of the band and shows much less sensitivity to instrumental resolution. It is not widely used for routine quantitative analyses but has much greater theoretical significance in that it is proportional to the square of the change in dipole moment with respect to the normal coordinate $(\partial\mu/\partial Q)^2$.

The integrated intensity[38-42] has been expressed in a variety of units. It can be expressed as the integrated absorptivity, $\mathscr{A}$, defined as

$$\mathscr{A} = \int_{\text{band}} a \, d\bar{v} \tag{2.19}$$

where a is the absorptivity measured with a spectrometer with unlimited resolution, integrated over the whole band. In practice this can be measured for solutions by plotting or recording the band as a function of linear coordinates of absorbance $(A = \log_{10} P_0/P)$ versus wavenumber $(\bar{v})$ and measuring

[37] D. Z. Robinson, *Anal. Chem.* **23**, 273 (1951).

[38] K. S. Seshadri and R. N. Jones, *Spectrochim. Acta* **19**, 1013 (1963).

[39] T. L. Brown, *Chem. Rev.* **58**, 581 (1958).

[40] G. M. Barrow, "Introduction to Molecular Spectroscopy," Chapter 4. McGraw-Hill, New York, 1962.

[41] J. Overend, *in* "Infrared Spectroscopy and Molecular Structure" (M. Davies, ed.), Chapter 10. Amer. Elsevier, New York, 1963.

[42] D. Steele, "Theory of Vibrational Spectroscopy," Chapter 8. Saunders, Philadelphia, Pennsylvania, 1971.

the band area $[\int \log_{10}(P_0/P)\, d\bar{v}]$ with a planimeter or by weighing paper cutouts of the band, from which

$$\mathscr{A} = \frac{1}{bc} \int_{\text{band}} \log_{10} \frac{P_0}{P}\, d\bar{v} \tag{2.20}$$

If the cell length, b, is measured in cm, and the concentration, c, is measured in moles/cm^3, and the spectrum is plotted using wavenumber (cm^{-1}) versus absorbance using $A = \log_{10} P_0/P$, then the integrated absorbance $\mathscr{A}$ is in cm/mole. In these units, $\mathscr{A}$ is related to the dipole moment change with respect to the normal coordinate $(\partial\mu/\partial Q)$ by

$$\mathscr{A} = \frac{\pi N_A}{3c^2(\log_e 10)} \left(\frac{\partial\mu}{\partial Q}\right)^2 \tag{2.21}$$

$$\mathscr{A} = 304.75 \left(\frac{\partial\mu}{\partial Q}\right)^2 \text{ cm mole}^{-1} \tag{2.22}$$

where N_A is Avogadro's number, c is not the concentration but the velocity of light (cm/sec) and $(\partial\mu/\partial Q)$ is in (esu cm)/(cm gm$^{1/2}$). The concentration is often measured in moles/liter rather than in moles/cm^3 in which case the constant 304.75 should be multiplied by 10^{-3} and $\mathscr{A}$ is in IUPAC practical units,[38] 1 mole^{-1} cm^{-2}. The dipole moment is usually expressed in Debye units D where 1 D equals 10^{-18} esu cm. Dipole moment derivatives with respect to interatomic distances have usually been expressed in Debyes per angstrom, D/Å.

Ramsay[43] has shown that measurements made with spectral slit widths one-half the true width of an absorption band yield true and apparent molar absorptivities differing by approximately 20% whereas the difference in the integrated intensity measurement is approximately 2–3%. In the integrated intensity measurements the decrease in peak intensity produced by the finite slit width is roughly compensated by the increase in band width. Ramsay further proved that it was not feasible to characterize band shapes by any one equation. He characterized them by an apparent peak intensity $\ln(T_0/T)_{v_{\max}}$* and an apparent half-intensity width $(\Delta v^a_{1/2})$ and related these quantities to the true peak intensity $\ln(I_0/I)_{v_{\max}}$, the true half-intensity width $(\Delta v^t_{1/2})$, and the slit width s. For convenience, the ratios

$$\frac{\ln (I_0/I)_{v_{\max}}}{\ln (T_0/T)_{v_{\max}}} \quad \text{and} \quad \frac{\Delta v^a_{1/2}}{\Delta v^t_{1/2}}$$

[43] D. A. Ramsay, *J. Amer. Chem. Soc.* **74**, 72 (1952).
* Ramsay used ln rather than $\log_{10}$ as has been adopted in this text.

were calculated as functions of apparent peak intensity and of the ratio of slit width s to the apparent half-band width $\Delta v_{1/2}^a$. The ratio was found to be at an optimum when $s/\Delta v_{1/2}^a = 0.65$, which is approximately the condition for the slit to be equal to $\Delta v_{1/2}^t$. The quantity

$$\frac{\ln(I_0/I)_{v_{\max}}}{\ln(T_0/T)_{v_{\max}}}$$

depends mainly on the ratio of slit width to the half-band width and to a much lesser extent on the peak intensity of the band, the variation being greater for larger values of $s/\Delta v_{1/2}^a$.

Using Ramsay's method, Jones et al.[44] calculated integrated band intensities for a large number of steroids and successfully determined the number of carbonyl groups and their positions in these complex compounds.

(c) SINGLE COMPONENT ANALYSIS

In the determination of either a liquid or solid component in a given solvent the spectrum of the pure component is measured and a frequency (cm^{-1}) or wavelength is selected at which the solvent has little or no absorption. A series of known concentrations of the component in the solvent is prepared and the absorbance or percent transmittance values are measured at the chosen frequency.

Several methods for measuring P_0/P are available. One of these methods[45] is that of "base line absorbance." In this method the spectrum of the mixture is scanned through the analytical band having maximum absorption at frequency v as shown in Fig. 2.11. A base line is drawn between v_1 and v_2 which are the points selected after the most favorable analytical peaks have been determined for all the compounds in question by a detailed consideration of their spectra. In general, the base line is drawn to approximate the "no absorbance" condition. The base line absorbance is calculated from the equation

$$A_b = \log_{10} \frac{P_b}{P} \tag{2.23}$$

where A_b = the base line absorbance; P_b = the distance from zero line to the base line; P = the distance on the spectrum from zero line to the selected absorption peak. P_b and P are measured at the same frequency, v.

[44] R. N. Jones, D. A. Ramsay, D. S. Keir, and K. Dobriner, *J. Amer. Chem. Soc.* **74**, 80 (1952).

[45] J. J. Heigl, M. F. Bell, and J. U. White, *Anal. Chem.* **19**, 293 (1947).

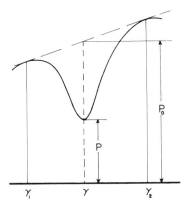

FIG. 2.11.   Base line method for determining absorbance or transmittance.

Another technique for making quantitative measurements involves the so-called "cell in, cell out" method. With the spectrophotometer set at the desired wavelength, P_0 and P are measured closely and consecutively in time by recorder deflection. In one modification P_0 is measured by placing in the infrared beam an empty cell containing a single window of thickness equal to that of the combined windows in the cell containing the sample to be analyzed. P is measured with the sample in the cell and $\log_{10} P_0/P$ is plotted versus the concentrations to obtain an analytical working curve. In the other modification of the "cell in, cell out" method a P_0^a is measured for air with no cell in the spectrophotometer, and then a P^a, for the analytical cell containing the least absorbing component at the frequency involved. Subsequently, a P_0^b for air and P^b for the mixture are measured. The ratio P_0^b/P^b divided by the ratio P_0^a/P^a is then substituted for P_0/P in Eq. (2.17).

If matched cells are available, both cells can be filled with pure solvent and placed in the reference and sample beams, respectively, and the recorder carefully adjusted to an absorbance reading of 0.0 or a 100% T value. The cell in the sample beam can then be removed and used to measure the solutions of known concentration measured in the sample beam versus the solvent in the reference beam. If percent transmittance values are measured on the recorder paper or scale, these values are plotted on the log axis of semilog paper versus the concentration on the linear coordinate scale. If absorbance values are measured rather than percent transmittance, the absorbance readings are plotted on linear coordinate paper versus the concentration. In both instances straight line plots should result if Beer's law holds. As has been previously mentioned, deviations from Beer's law can result if the slit width of the instrument at the given frequency exceeds the band width. Or deviations can occur if the molecular environment is not the same over the whole concentration range studied because of association, dissociation, complex formation, or a similar process.

Another quantitative technique involves an internal standard. Barnes and associates[46] developed a quantitative infrared method using an internal standard in Nujol mulls and Wiberley et al.[47] applied the method to potassium bromide pellets. The internal standard method eliminates the need to determine the sample thickness. This assumption is readily justified by consideration of Beer's law.[47] The absorbance of the known material to be assayed at v_k will be given by

$$A_k = a_k b c_k \qquad (2.24)$$

and the absorbance of the internal standard at frequency v_s by

$$A_s = a_s b c_s \qquad (2.25)$$

Now, dividing the first equation by the second,

$$\frac{A_k = a_k b c_k}{A_s = a_s b c_s} \qquad (2.26)$$

the b's cancel and because a_k and a_s are both constants at the wavenumbers at which the measurements are made, and c_s, the concentration of the internal standard, is constant, these constants can be accumulated in an overall constant, K, and

$$\frac{A_k}{A_s} = K c_k \qquad (2.27)$$

Hence, a plot of A_k/A_s vs. c_k will give a straight line. In this method it is not necessary to determine a_k and a_s, or even to know c_s exactly, to obtain an empirical working curve.

For universal application in quantitative infrared work an internal standard should have a simple spectrum with a few sharp bands, be stable to heat and not absorb excessive moisture, be easily ground to a small particle size, be readily available, and nontoxic. Substances recommended for internal standards in Nujol mulls have been D-alanine, calcium carbonate, lead thiocyanate and naphthalene and in potassium bromide pellets, potassium

[46] R. B. Barnes, R. C. Gore, E. F. Williams, S. G. Linsley, and E. M. Peterson, Ind. Eng. Chem., Anal. Ed. 19, 620 (1947).

[47] S. E. Wiberley, J. W. Sprague, and J. E. Campbell, Anal. Chem. 29, 210 (1957).

thiocyanate, and sodium azide.[48] Sodium azide has a very strong peak at 2140 cm^{-1} and a weak peak at 1309 cm^{-1}. In the determination of sodium fluoroacetate in dried residues from soil dispersion, sodium azide along with potassium bromide yielded Nujol mulls of medium absorbance values. The addition of the potassium bromide not only decreases the amount of Nujol required, thereby increasing the transmittance of the mull, but also facilitates the grinding of the solids before mulling.

In the quantitative analysis of gases, the absorption law is

$$A = abp \qquad (2.28)$$

where p is the pressure of the gas (or partial pressure in a mixture of gases) and replaces c, the concentration, in Eq. (2.18). Thus, a plot of absorbance versus pressure should yield a straight line. However, severe deviation from linearity can result from pressure broadening.[49] Pressure broadening is characterized by unusual changes in band intensity with increasing pressure, resulting in different absorbance values than predicted by Beer's law. Pressure broadening may be caused by self-broadening as well as by foreign gas broadening. Self-broadening occurs for single component analysis, i.e., when the vapor pressure of the gas equals the total pressure in the gas cell. Foreign gas broadening occurs when some foreign gas is mixed with the infrared active gas being determined. Self-broadening is not of too great concern in quantitative analysis because empirical curves can be constructed to eliminate the error. However, foreign gas broadening presents serious problems in the analysis of complex gas mixtures.

(d) MULTICOMPONENT ANALYSIS

The analysis of complex mixtures is based on the fact that absorbances are additive, i.e., if the absorbance of component A at 2860 cm^{-1} is 0.400 and that of component B at the same frequency is 0.250, then the total absorbance of the mixture at 2860 cm^{-1} is 0.650. This assumption is usually justified unless the components react with one another, or dissociate, associate, or form complexes, etc. The effect of spectral slit width previously discussed must also be considered.

For example, consider the analysis of a mixture containing three components all of which absorb in the infrared region. None of these compounds

[48] R. T. M. Frazer, *Anal. Chem.* **31**, 1602 (1959).
[49] M. G. Mellon, "Analytical Absorption Spectroscopy," p. 498. Wiley, New York, 1950.

will absorb the same amount at all frequencies. Now the absorbance A of component 1 at frequency v_1 is given by

$$A = a_1 bc_1$$

and the absorbance of the three components at v_1 is then

$$A_1 = a_1 bc_1 + a_2 bc_2 + a_3 bc_3$$

and at v_2

$$A_2 = a_1' bc_1 + a_2' bc_2 + a_3' bc_3 \tag{2.29}$$

and at v_3

$$A_3 = a_1'' bc_1 + a_2'' bc_2 + a_3'' bc_3$$

As a rule b is not measured but the same cell is used for the complete analysis and the b term included in the determination of a. It is of course necessary to experimentally determine the combined absorptivity constant and cell length constant by measurement of pure samples of the individual components separately at v_1, v_2, and v_3. Substitution of these values then yields

$$A_1 = k_1 c_1 + k_2 c_2 + k_3 c_3$$
$$A_2 = k_1' c_1 + k_2' c_2 + k_3' c_3 \tag{2.30}$$
$$A_3 = k_1'' c_1 + k_2'' c_2 + k_2'' c_3$$

where k_1, k_2, k_3, k_1', etc., are now known. Measurement of A_1, A_2, and A_3 on the mixture then yields three equations containing three unknown concentrations, c_1, c_2, and c_3, which can be readily calculated.

If more components are involved, it is simpler to transpose the above equation by electrical computers or reciprocal matrix methods[50] to the form

$$
\begin{Vmatrix} A_1 \\ A_2 \\ A_3 \end{Vmatrix} =
\begin{Vmatrix} k_1 & k_2 & k_3 \\ k_1' & k_2' & k_3' \\ k_1'' & k_2'' & k_3'' \end{Vmatrix} \times
\begin{Vmatrix} c_1 \\ c_2 \\ c_3 \end{Vmatrix} \rightarrow
\begin{Vmatrix} c_1 \\ c_2 \\ c_3 \end{Vmatrix} =
\begin{Vmatrix} \text{inverse} \\ k \\ \text{matrix} \end{Vmatrix} \times
\begin{Vmatrix} A_1 \\ A_2 \\ A_3 \end{Vmatrix} \tag{2.31}
$$

where the terms are as previously defined.

[50] R. P. Bauman, "Absorption Spectroscopy," p. 404. Wiley, New York, 1962.

2.8 Polarized Infrared Radiation

The intensity of an infrared absorption band is proportional to the square of the transition moment (or changing dipole moment) of the molecular vibration causing the band. The band intensity also depends on the relative directions of the transition moment and the electric field vector of the incident radiation. The component of the transition moment in the direction of the electric field vector is equal to the transition moment multiplied by the cosine of the included angle. The square of this quantity is proportional to the absorbance.

Polarized infrared radiation can be used to reveal information about the direction of transition moments of normal modes of vibration in solid oriented compounds. When the molecular orientation in a solid is known, polarization studies can be of help in making band assignments. The measured direction of the transition moment of the vibration causing a band must agree with the direction predicted from the known structure if the assignment is correct. Conversely, if the band assignment is known and the molecular orientation in the solid is not known, some information may be deduced about the molecular orientation.

In liquids and gases, the normal modes of the *molecule* interact with radiation. One complication of the solid phase over the liquid phase is that in the crystalline solid state, the normal modes of the *unit cell* are the ones which interact with the radiation. A normal mode has been defined as a vibration where all the atoms are motionless or perform simple harmonic motion with the same frequency, all atoms going through their equilibrium positions simultaneously. Therefore the normal modes of a unit cell consist of the normal modes of the individual molecules taken in as many different phases as there are molecules per unit cell. In a unit cell containing two molecules, each with a carbonyl group, the two carbonyl stretching modes would consist of in-phase and out-of-phase stretching of the two carbonyl groups.

(a) POLARIZERS

Polarized infrared radiation is obtained with transmission polarizers using silver chloride[51] or selenium.[52] A thin plate of silver chloride or thin film of selenium is placed in the beam of unpolarized infrared radiation and is tilted at the polarizing angle relative to the beam.

[51] R. Newman and R. S. Halford, *Rev. Sci. Instrum.* **19**, 270 (1948).
[52] A. Elliott, E. J. Ambrose, and R. B. Temple, *J. Opt. Soc. Amer.* **38**, 212 (1948).

The polarizing angle of incidence, i, is related to the index of refraction, n, of the material by the equation

$$n = \tan i \qquad (2.32)$$

The tilted plate will reflect some radiation to the side which consists of radiation completely polarized in the plane perpendicular to the plane containing the angle of incidence. Some of this polarization component is also transmitted, however, so that the transmitted beam is only partially polarized. Additional plates remove more and more of the unwanted component by reflection, thus improving the degree of polarization of the transmitted beam. Common commercial polarizers consist of about six silver chloride plates (see Fig. 2.12).

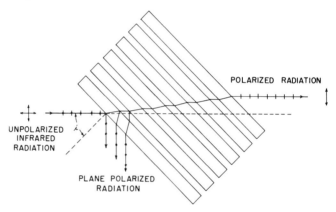

FIG. 2.12.   Plane polarization of infrared radiation by several plates. If I is the incident parallel beam of radiation striking at the angle of the incidence, i, the reflected beam (R) is polarized. The incident plane is the plane of the paper, therefore beam R is vibrating normal to the paper. This reflection-polarization process occurs at each face of the stack of plates. After a sufficient number of reflections the transmitted beam (T) has been depleted of vibrations normal to the plane of incidence and consists almost completely of vibrations in this plane. Thus, there are two beams of polarized light produced.

The newest commercial infrared polarizer is a wire grid construction that is compact in size and transmitts polarized infrared radiation of very high quality. The wire grid polarizer is formed of uniformly spaced parallel fine strips of vapor deposited metal on a transparent substrate.

(b) SAMPLE PREPARATION

The molecules in the sample must be oriented in some way. In crystalline compounds, a single crystal of suitable thinness and area or a group of crystals similarly oriented may be prepared. These may be obtained by allowing molten material to crystallize between salt plates separated with a spacer under conditions where a temperature gradient exists on the plate.[53] They may be obtained under certain conditions from the vapor phase[54] or from solution.[55] Single crystals may be cut and polished to the required size, if possible, the cuts being made parallel to one of the principal axes to avoid double refraction effects.

Thin polymer films may be oriented by stretching, sometimes at elevated temperatures. Such treatment results in uniaxial orientation where some crystallographic axis, usually the long chain axis, will tend to line up parallel to the direction of stretch but there is no preferred orientation of crystallites about this axis. Another technique for polymer orientation consists of rolling the film between rollers which can be heated. The film may also be rolled after sandwiching between silver chloride plates,[56] the silver chloride being removed, if desired, by sodium thiosulfate solution. Rolling results in orientation along the rolling direction (like stretching) but may also result in the orientation of particular crystallographic planes parallel to the film plane (double orientation). X-ray examination can determine the enhancement of orientation.

(c) SAMPLE MEASUREMENT

The oriented sample and the polarizer are placed in the spectrometer beam and spectra are taken with the electric vector of the radiation parallel and perpendicular to the orientation direction of the sample. If the sample is not uniform in thickness or shape, it may be left stationary while the polarizer is rotated so that the same part of the sample is used. Errors are caused by this method because of the inherent polarization in the spectrometer, chiefly caused by the prism.[57] To avoid the above difficulty when rotating the polarizer, measurements may be made with the polarizer oriented 45° with respect to the slit, and then rotated 90° to a position 45° on the other side. If the sample is uniform enough, the polarizer is left stationary in the direction of maximum transmission while the sample is rotated 90°. This is the most satisfactory way to make polarization measurements. Beam condensers may be used for small samples but the additional convergence of the beam has some effect on the results.[58,59]

[53] F. Halverson and R. J. Francel, *J. Chem. Phys.* **17**, 694 (1949).

[54] S. Zwerdling and R. S. Halford, *J. Chem. Phys.* **23**, 2221 (1955).

[55] H. Hallman and M. Pope, "Technical Report on Preparation of Thin Anthracene Single Crystals," ONR Contract No. 285 (25). 1958.

[56] A. Elliott, E. J. Ambrose, and R. B. Temple, *J. Chem. Phys.* **16**, 877 (1948).

[57] E. Charney, *J. Opt. Soc. Amer.* **45**, 980 (1955).

[58] D. L. Wood, *Ann. N.Y. Acad. Sci.* **69**, Art. 1, 194 (1957).

[59] R. D. B. Fraser, *J. Chem. Phys.* **21**, 1511 (1953).

Dichroism is observed when an absorption band intensity changes when the relative orientation of the polarizer and the sample is altered. The experimentally determined dichroic ratio of a particular mode of vibration is equal to $A_{\parallel}/A_{\perp}$, where A is equal to the absorbance of the band with the electric vector parallel and perpendicular to the orientation direction. The experimentally determined dichroic ratio may differ from the true ratio for various reasons[57] such as imperfect polarization by the polarizer, polarization by the spectrometer, imperfectly oriented samples, the presence in partially oriented polymers of some randomly oriented crystallites, birefringence, or the effects of beam convergence in condensing systems.

(d) APPLICATIONS

An application of the use of polarized infrared spectroscopy is found in the study of *ortho*-substituted nitrobenzenes.[60] If the *ortho*-substituted group is large enough, it may sterically rotate the nitro groups (which are normally coplanar with the ring) out of the ring plane. The oriented crystals were prepared from the molten salt by crystallization between salt plates where a temperature gradient was set up. It was established that in the oriented crystals the planes of the rings are seen edge on and are perpendicular to the direction of crystal growth. The transition moment of the symmetrical NO_2 stretching frequency near 1350 cm^{-1} is parallel to the C—N bond and is in the plane of the ring regardless of the rotation of the NO_2 group. This planar vibration band shows perpendicular dichroism, the band being more intense when the electric vector is perpendicular to the direction of crystal growth. The transition moment of the asymmetric NO_2 stretching frequency near 1530 cm^{-1} is parallel to the O—O direction. If the NO_2 group is coplanar with the ring, the group will have perpendicular dichroism. If the NO_2 group is rotated 180° out of the ring plane, this band will have parallel dichroism. Thus it was established that *o*-nitrophenol and *o*-nitroaniline had the NO_2 group coplanar with the ring, while *o*-nitrochlorobenzene and *o*-nitrobromobenzene had a nonplanar NO_2 group due to the presence of the adjacent halogen atoms.[60]

Polarization measurements were made on crystalline acrylamide to verify assignments and to verify a proposed crystal structure[61] which was not as yet known in detail from X-ray work. The model was derived from hydrogen bonding considerations and was in accord with the symmetry elements and unit cell dimensions known from X-ray measurements.[62]

    [60] R. J. Francel, *J. Amer. Chem. Soc.* **74**, 1265 (1952).

    [61] N. B. Colthup, Paper 98, presented at the Pittsburgh Conference on Analytical Chemistry and Applied Spectroscopy (1959).

    [62] B. Post and L. A. Siegle, private communication.

An acetone solution of acrylamide was put on a rock salt plate. After evaporation of the solvent, acrylamide crystallized in patterns on the plate. Under the polarizing microscope extinction was noted, parallel and perpendicular to the direction of crystal growth. An area of uniform extinction and therefore uniform orientation was selected and masked off. It was decided from the above and from the examination of the spectrum that the *BC* plane (the preferred cleavage plane) was parallel to the rock salt plate face and the direction of crystal growth was parallel to the *C* axis.

Acrylamide has four molecules per unit cell so that for every normal mode of vibration of the acrylamide molecule there were four modes for the unit cell (see Fig. 2.13). Two modes are infrared inactive because acrylamide crystallizes into two pairs of dimers, each with a center of symmetry.[61] The

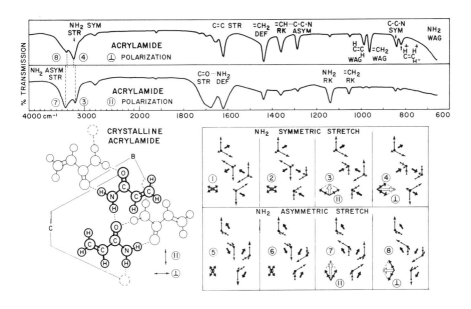

FIG. 2.13. The polarized infrared spectrum of acrylamide. The illustration shows the four repeating molecules of crystalline acrylamide, arranged in two pairs of planar dimers, tilted out of the *BC* plane (heavy outline molecules above the plane, others below). The *B* and *C* units of the unit cell are shown as well as the direction of the electric vector of the radiation with which the polarized spectra were obtained. The unit cell has four modes of vibration for every molecular mode. These are illustrated schematically for the asymmetric and symmetric NH_2 stretching frequencies. The arrows on the NH_2 groups represent atomic motion. The heavy black arrows represent molecular transition moments. The white arrows represent unit cell transition moments which are the vector resultants of the molecular transition moments. Only modes 3, 4, 7, and 8 are infrared active and these are labeled on the spectra. The relative magnitudes of the two infrared active unit cell transition moments in each case depend on the orientation of the molecular transition moments.

two infrared active modes will have transition moments parallel and perpendicular to the direction of crystal growth as seen from a direction normal to the BC plane. The parallel and perpendicularly polarized spectra thus isolate the two active modes which have nearly the same but not identical frequencies. The relative band intensities of the two modes yield information about the direction of the vibrational transition moment of the molecule, projected onto the BC plane (normal to the incident beam). The ratio of the parallel to perpendicular transition moments of the unit cell for a given molecular vibration in this case is equal to the ratio of cos α to sin α where α is the acute angle which the BC plane projection of the molecular transition moment makes with the C axis or parallel direction (the direction of crystal growth). The band absorbance ratio $A_{\parallel}/A_{\perp}$ equals the ratio of the squares of the transition moments or $\cos^2 \alpha/\sin^2 \alpha$ or $A_{\parallel}/A_{\perp} = \cot^2 \alpha$.

The angle α obtained indicates the orientation of the BC plane projection of the molecular transition moment. The transition moment does not necessarily lie exactly along the involved molecular bond; for example, the $C{=}O$ stretching, due to interaction with the $C{-}N$ stretching and NH_2 deformation vibrations, has a transition moment direction which may differ appreciably from the $C{=}O$ direction. The NH_2 stretching bands probably give the best idea of the molecular orientation. As the $C{=}C$ stretching band is partially obscured by amide bands, the CH_2 deformation band is the best indicator of the $C{=}C$ orientation. The out-of-plane CH wagging vibrational bands are strong in the perpendicular spectrum, indicating that the planar dimers are rotated out of the BC plane, about the C axis. In this fashion it was decided that the proposed assignments and the proposed structure were mutually compatible, though not necessarily uniquely defined. The Raman spectrum and deuteration studies provide additional information on assignments.[63] A few of the assignments made below 1000 cm^{-1} differ from the data in the literature, because correlations of vinyl frequencies were involved in the present author's assignments.

In polymers which are completely uniaxially oriented by stretching, the dichroic ratio is given by $A_{\parallel}/A_{\perp} = 2 \cot^2 \alpha$, where $A_{\parallel}$ and $A_{\perp}$ are the absorbances of the spectral band in radiation polarized parallel and perpendicular to the direction of stretch, and α is the angle the transition moment makes with the stretching direction.[59] This treatment takes into account that in a long chain polymer, the chains usually lie parallel to the direction of stretch but the crystallites about this axis take all possible orientations. If the parallel polarized absorbance is greater than the perpendicular (parallel dichroism), the angle between the transition moment and the direction of stretch is less than 54°44'. If this angle is greater than 54°44', the perpendicularly polarized

[63] N. Jonathan, *J. Mol. Spectrosc.* **6**, 205 (1961).

absorbance will be the stronger (perpendicular dichroism). This relationship will remain true in the case of incomplete orientation. If quantitative use is to be made of the dichroic ratio, account must be taken of the fact that polymer preparations are usually incompletely oriented.[56,59]

Polarized infrared radiation can distinguish the folded and extended forms of proteins and polypeptides.[64,65] In the extended form, the backbone chains are lined up side by side and are connected with cross-chain hydrogen bonds, the $N-H\cdots O=C$ unit lying approximately perpendicular to the chain direction. The NH stretching band exhibits perpendicular dichroism. In the folded form, the NH of one amide group hydrogen bonds to the $C=O$ of another amide group of the same chain, the $N-H\cdots O=C$ unit lying approximately parallel to the chain progression axis. The NH stretching band exhibits parallel dichroism.

2.9 Infrared Fourier Transform Spectroscopy

This technique[66-68] is based on the blending of a Michelson interferometer with a sensitive infrared detector and a digital minicomputer.

The principle of the Michelson interferometer is illustrated in Fig. 2.14. Radiation from the source is passed through a beam splitter which transmits half the beam to a movable mirror and reflects the other half to a fixed mirror. These two beams are reflected from their respective mirrors and recombined either constructively or destructively at the beam splitter depending

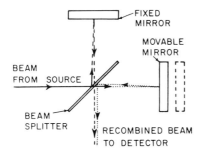

FIG. 2.14.   Diagram of the Michelson interferometer.

[64] E. J. Ambrose, A. Elliott, and R. B. Temple, *Nature (London)* **163**, 859 (1949).

[65] A. Elliott, *J. Appl. Chem.* **6**, 341 (1956).

[66] W. J. Hurley, *J. Chem. Educ.* **43**, 236 (1966).

[67] M. J. D. Low, *J. Chem. Educ.* **47**, A163, A255, and A415 (1970).

[68] R. O. Kagel and S. T. King, *Ind. Res.* **15**, 54 (1973).

on the position of the movable mirror. As the path difference between the two beams is altered, the detector signal pattern obtained is the interferogram of the radiation entering the interferometer. For monochromatic radiation the amplitude of the detector signal is a cosine function of mirror position. For polychromatic radiation the detector signal is a summation of all the constructive or destructive interferences of each wavelength interacting with every other wavelength and results in an interferogram. A cosine Fourier transform yields the mathematical relationship between the intensity of the interferogram as a function of mirror travel, $I(x)$ and the intensity of the frequency $I(v)$ of the infrared radiation,

$$I(x) = \int_{-\infty}^{\infty} I(v) \cos(2\pi x v) \, dv \qquad (2.33)$$

and the inverse transform, calculated by a computer,

$$I(v) = \int_{-\infty}^{\infty} I(x) \cos(2\pi v x) \, dx \qquad (2.34)$$

relates the interferogram obtained to the infrared spectrum.

It should be emphasized there is no monochromator in the system and radiation of many frequencies passes through the sample. All frequencies not selectively absorbed reach the detector. Unlike prism and grating spectrometers in which only one small frequency range impinges on the detector at one time, there are many frequencies yielding a very large signal-to-noise ratio. This characteristic is known as the multiplex or Fellgett advantage.[69] The interferometer signal is directly proportional to the total observation time (T) while the noise is proportional to $T^{1/2}$. Hence, the signal-to-noise ratio is proportional to $T^{1/2}$. In a prism or grating instrument the effective observation time is the time each small fractional part ($1/M$) of the whole spectrum is being measured. This is equal to the total observation time T times $1/M$. Therefore, in a prism or grating instrument the signal-to-noise ratio is proportional to $(T/M)^{1/2}$, with the net result that the interferometer has an advantage of $M^{1/2}$ over the grating instrument operated at an equivalent signal-to-noise ratio.

Another advantage is the light throughput of the interferometer which is about 40 times greater than for a dispersive instrument. Therefore, Fourier transform spectroscopy has an important application for analysis of micro-samples[68] and for far infrared spectroscopy. There are some disadvantages such as the equipment cost, time delay between experiment and reception of results, and the fact that interferometers are single beam instruments.

[69] P. Fellgett, *J. Phys. Radium* **19**, 187 and 237 (1958).

2.10 Raman Sources

In the earliest work on Raman spectra the primary radiation source was the mercury arc. Mercury has its most intense line at 4358 Å with other lines at approximately 2537, 3650, 4057, 5461, 5770, and 5790 Å. To obtain relatively monochromatic radiation at 4358 Å these other wavelengths were eliminated with appropriate glass filters or solutions.

The most successful mercury source was a lamp developed by Welsh et al.[70] It consisted of a four-turn helix of Pyrex tubing with a cooling coil passing through the center of the helix. Annular electrodeless lamps powered by microwaves at 2450 Mcycles were also used as a Raman source.[71] With the advent of the laser in 1960, the implications of this source for Raman spectroscopy was obvious and the sources previously mentioned became obsolete.

Laser stands for light amplification by stimulated emission of radiation. Stimulated emission is initiated by a photon whose energy equals the exact difference in energy between two energy states or levels of an electron. Such a photon can cause an electron in the upper excited level to fall to the lower level and radiate its energy as a new photon so there are now two photons instead of one (see Fig. 2.15). Most excited states of atoms have lifetimes of about 10^{-8} sec. However, there are some excited states called metastable states in which electrons cannot easily decay by giving up their energy. Electrons in such energy states have lifetimes of 10^{-3} sec or more. Thus,

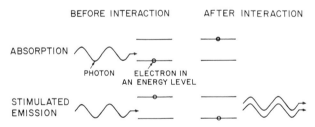

BEFORE INTERACTION   AFTER INTERACTION

ABSORPTION

PHOTON   ELECTRON IN
AN ENERGY LEVEL

STIMULATED
EMISSION

FIG. 2.15. The stimulated emission process compared with the absorption process. In both cases the electron energy level transition is triggered by the incident photon whose energy exactly equals the energy difference between the two levels involved. Normally, most electrons are in the lower unexcited level and absorption dominates. In a laser an abnormally high number of electrons are in the upper excited level and stimulated emission can occur.

[70] H. L. Welsh, M. F. Crawford, T. R. Thomas, and G. R. Love, Can. J. Phys. 30, 577 (1952).

[71] N. S. Ham and A. Walsh, Spectrochim. Acta 12, 88 (1958).

energy can be stored if a sufficient number of electrons can be brought to the metastable state. This energy can be released as an intensity increase in a given wavelength of light in a laser amplifier under suitable conditions. If a light beam having a wavelength corresponding to the difference between the upper and lower excited levels of an atom is passed through the medium, the energy of the light beam will be amplified if there are more atoms in the upper state but absorbed if there are more atoms in the lower state (see Fig. 2.15). The situation of having more atoms in the upper state is called a population inversion because the natural tendency is for the electrons to be at a lower rather than a higher energy state.

The first solid state laser developed by Maiman[72] in 1960 was a ruby laser (1 part Cr^{3+} : 2000 parts Al_2O_3) with an output at 6943 Å. The original ruby laser was followed by other solid state lasers such as the neodymium laser and semiconductor lasers such as gallium–arsenide. Shortly thereafter it was shown that a population inversion could be achieved by means of a gas discharge. Hence, gas lasers such as the helium–neon laser (6328 Å), the argon ion laser (4579, 4658, 4727, 4765, 4880, 4965, 5017, and 5145 Å), the CO_2 laser (10.6 μm), and the krypton ion laser (4762, 4825, 5208, 5308, 5682, 6471, and 6764 Å) were also developed. Other type lasers include liquid lasers (of which the dye laser has attracted the most attention), chemical lasers, and metal–vapor lasers.[73] The ruby laser as a Raman source was first used by Porto and Wood.[74] However, it was only useful for photographic recording. In 1965, Weber and Porto[75] discussed the helium–neon laser as a light source for high resolution Raman spectroscopy.

The helium–neon laser at 6328 Å has been a commonly used source with its main limitation being detector sensitivity. Also popular are the argon ion laser at 4880 and 5145 Å and the krypton ion laser at 6764 Å with some use being made of the helium–cadmium laser at 4416 Å.

2.11 Raman Spectrographs

There have been a number of custom built Raman spectrographs described in the literature. Initially, Raman spectra were recorded on photographic plates with exposure times ranging from minutes to days. Present commercial instruments use photoelectric recording and can be connected to a computer for curve averaging. A spectrograph of high light gathering power with low

[72] T. H. Maiman, *Nature (London)* **187**, 493 (1960).
[73] W. T. Silvast, *Sci. Amer.* **228**, 89 (1973).
[74] S. P. S. Porto and D. L. Wood, *J. Opt. Soc. Amer.* **52**, 251 (1962).
[75] A. Weber and S. P. S. Porto, *J. Opt. Soc. Amer.* **55**, 1033 (1965).

stray light is a requirement. Usually a double or triple monochromator system is employed with the exit slit from the first monochromator serving as an entrance slit to the second. The gratings are moved by a single cosecant bar drive which yields a spectrum linear in wavenumber.

2.12 Raman Measurements

(a) SAMPLE TECHNIQUES

Gases, liquids, and solids can be studied by Raman spectroscopy with liquids being the easiest to handle.

Liquids can be placed in sample holders of a few millimeters outside diameter. For microsamples, Bailey, Kint, and Scherer[76] have used aluminized capillary tubes with a cell volume as small as 0.04 μl.

Solids can be measured with a special holder which allows the solid samples to be suspended in the center of the laser beam. Powders can be held in capillary tubes.

Solids can also be dissolved in suitable solvents and handled in the same manner as liquids. Water, which transmits so poorly in the infrared region, is an ideal solvent for inorganic compounds. The selection of an organic solvent is dependent on its solvent power, its Raman spectrum, and possible chemical interaction with the solute.

Kiefer and Bernstein[77] have described a rotating sample system (0–3000 rpm) for highly absorbing liquids or solutions and a rotating device for colored crystal powders.[78] A modified optical arrangement was subsequently used by Kiefer[79] for difference Raman spectroscopy, resulting in cancellation of unwanted Raman lines of the solvent in the spectrum of the solution and accurate determination of small environmental frequency shifts.

Koningstein and Gächter[80] have described a surface-scanning technique for highly colored crystals. In this method the surface is scanned by placing a rotating refractor plate between the sample and the lens used to focus the laser beam on the crystal surface.

Raman spectra of gases are measured in gas cells of the multiple-reflection type with pressures adjustable from millimeters to atmospheres.

[76] G. F. Bailey, S. Kint, and J. R. Scherer, *Anal. Chem.* **39**, 1040 (1967).

[77] W. Kiefer and H. J. Bernstein, *Appl. Spectrosc.* **25**, 500 (1971).

[78] W. Kiefer and H. J. Bernstein, *Appl. Spectrosc.* **25**, 609 (1971).

[79] W. Kiefer, *Appl. Spectrosc.* **27**, 253 (1973).

[80] J. A. Koningstein and B. F. Gächter, *J. Opt. Soc. Amer.* **63**, 892 (1973).

(b)   DETERMINATION OF DEPOLARIZATION RATIO

It has been pointed out that the determination of the depolarization ratio, ρ_n or ρ_p is an important aid in the assignment of Raman lines to certain vibrations (see Section 1.34). For example, totally symmetrical vibrations yield polarized Raman lines (i.e., with values of ρ_p less than $\frac{3}{4}$) and unsymmetrical vibrations yield unpolarized Raman lines. For example, as shown in the Raman spectrum of carbon tetrachloride in Fig. 2.16 the Raman band at 459 cm^{-1} is polarized and therefore corresponds to the totally symmetrical vibration,

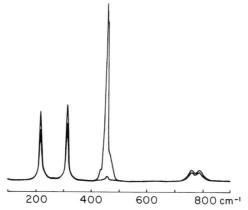

FIG. 2.16.   The polarized Raman spectrum of CCl$_4$. The upper and lower curves correspond to $I_\parallel$ and $I_\perp$, respectively. For the 459 cm^{-1} band the depolarization value is close to zero, expected for totally symmetrical vibrations in isotropic molecules. For the other bands the depolarization ratio is close to 0.75; expected for antisymmetrical vibrations using polarized incident radiation from a lazer source. Courtesy of J. R. Downey, Jr. and G. J. Janz, Rensselaer Polytechnic Institute, Troy, N.Y.

ν_1. In contrast, the Raman bands at 218, 314, 762, and 791 cm^{-1} are depolarized and assigned to antisymmetrical vibrations.

With a laser source the measurement of depolarization ratios is much simpler than was true for the mercury source. There are quite a few techniques available for measuring depolarization ratios.[81,82] One method is as follows: The laser beam (which is plane polarized) is passed through a half-wave plate then through the sample cell, an analyzer (rotated to obtain parallel and perpendicular measurements), a calcite wedge called a scrambler,[83] and finally focused on the entrance slit of the monochromator.

[81] W. F. Murphy, M. V. Evans, and P. Bender, *J. Chem. Phys.* **47**, 1836 (1967).

[82] W. Proffitt and S. P. S. Porto, *J. Opt. Soc. Amer.* **63**, 77 (1973).

[83] A. Weber, S. P. S. Porto, L. E. Cheesman, and J. J. Barrett, *J. Opt. Soc. Amer.* **57**, 19 (1967).

The scrambler depolarizes the radiation before it enters the monochromator, and thus avoids the necessity of correcting for instrumental polarization effects. In an alternate procedure the analyzer is left in a fixed orientation and the polarization plane of the incident radiation is rotated by rotating the half-wave plate.

(c) INTENSITY MEASUREMENTS

The determination of the absolute intensities of Raman lines is even more difficult than the determination of the absolute intensity of infrared absorption bands. Bernstein and Allen[84] as well as Rea[85] have considered the variables which affect observed Raman intensities. Variables dependent on the nature of the sample are the refractive index of the sample and the absorption of both the exciting and scattered radiation by the sample. Variables independent of the sample include the spectral variation in the transmittance of the instrument and in the frequency response of the detector, the variation caused by the convergence of the exciting and Raman radiation, and such instrumental conditions as slit width, scanning speed, and time constant.

For these reasons the intensity of a Raman line is usually measured in terms of the arbitrarily chosen 459 cm^{-1} line of carbon tetrachloride (see Fig. 2.16). This ratio has been called the scattering coefficient.

A reasonably satisfactory standard intensity scale for the interchange of intensity data has been defined by Danti.[86]

(d) HIGH AND LOW TEMPERATURE METHODS

There have been a large number of methods described for high temperature Raman measurements. For molten salt studies techniques have been described by Janz and James[87] using a Toronto type mercury lamp and by Clarke et al.[88] and Melveger et al.[89] using laser excitation. A high temperature multiple-reflection cell as well as a capillary cell have been described by Krasser and Nürnberg.[90]

[84] H. J. Bernstein and G. Allen, *J. Opt. Soc. Amer.* **45**, 237 (1955).

[85] D. G. Rea, *J. Opt. Soc. Amer.* **49**, 90 (1959).

[86] A. Danti, "Information for Contributors to the Catalogs of Raman Spectral Data," Chemical Thermodynamic Properties Center. Agricultural and Mechanical College of Texas, College Station, Texas.

[87] G. J. Janz and D. W. James, *J. Chem. Phys.* **35**, 739 (1961).

[88] J. H. R. Clarke, C. Solomons, and K. Balasubrahmanyam, *Rev. Sci. Instrum.* **38**, 655 (1967).

[89] A. J. Melveger, R. K. Khanna, B. R. Guscott, and E. R. Lippincott, *Inorg. Chem.* **7**, 1630 (1968).

[90] W. Krasser and H. W. Nürnberg, *Chem. Instrum.* **1**, 47 (1968).

Low temperature measurements do not present as many experimental problems as high temperature measurements. Rice *et al.*[91] have described a relatively simple low temperature procedure. A copper rod was drilled to fit the Raman tube closely and kept immersed in a Dewar flask containing liquid nitrogen or a Dry Ice–acetone mixture. The exciting light reached the Raman tube through two large windows cut into the copper jacket. The formation of interfering layers of condensed atmospheric moisture on the Raman tube was prevented by a jet of either nitrogen gas or cold alcohol. Bryant[92] has described a cryostat for the measurement of low temperature Raman spectra of crystals.

[91] B. Rice, J. M. Gonzalez Barredo, and T. F. Young, *J. Amer. Chem. Soc.* **73**, 2306 (1951).

[92] J. I. Bryant, *Spectrochim. Acta, Part A* **24**, 9 (1968).

CHAPTER 3

MOLECULAR SYMMETRY

3.1 Symmetry Properties

Symmetry plays an important role in the structure of molecules: Some are highly symmetrical, some less so, and many have no symmetry at all. It would thus seem reasonable to set up some system whereby molecules could be classified by their symmetry characteristics. Crystallographers were the first to study symmetry properties in connection with crystal studies. When spectroscopists realized that the symmetry of a molecule played an important role in which vibrations were permitted and which excluded, they also became concerned with symmetry properties and the establishment of a convenient reference system.

In this chapter symmetry will be discussed from the standpoint of the spectroscopist rather than the crystallographer. For a more general and detailed treatment of this subject the reader should consult additional references.[1-11]

[1] P. J. Wheatley, "The Determination of Molecular Structure." Oxford Univ. Press, London and New York, 1959.

[2] J. C. D. Brand and J. C. Speakman, "Molecular Structure." Arnold, London, 1960.

[3] H. Eyring, J. Walter, and G. E. Kimball, "Quantum Chemistry." Wiley, New York, 1960.

[4] F. A. Cotton, "Chemical Applications of Group Theory." Wiley (Interscience), New York, 1963.

[5] R. P. Bauman, "Absorption Spectroscopy." Wiley, New York, 1963.

[6] K. Nakamoto, "Infrared Spectra of Inorganic and Coordination Compounds." Wiley, New York, 1963.

[7] G. M. Barrow, "Introduction to Molecular Spectroscopy." McGraw-Hill, New York, 1962.

[8] J. E. Rosenthal and G. M. Murphy, Rev. Mod. Phys. 8, 317 (1936).

[9] A. G. Meister, F. F. Cleveland, and M. J. Murray, Amer. J. Phys. 11, 239 (1943).

[10] G. Herzberg, "Infrared and Raman Spectra of Polyatomic Molecules." Van Nostrand-Reinhold, Princeton, New Jersey, 1945.

[11] E. B. Wilson, Jr., J. C. Decius, and P. C. Cross, "Molecular Vibrations." McGraw-Hill, New York, 1955.

From the standpoint of infrared and Raman spectroscopy molecules can be conveniently classified using the following five symmetry elements:

(1) A center of symmetry designated by i.

(2) A p-fold rotation axis of symmetry designated by C_p, where C stands for cyclic and rotation through $2\pi/p$ or $360°/p$ produces an orientation indistinguishable from the original molecule.

(3) Planes of symmetry usually designated by σ with subscripts v, h, or d depending on whether the plane is a vertical, horizontal, or diagonal plane of symmetry.

(4) A p-fold rotation-reflection axis of symmetry designated by S_p where rotation through $2\pi/p$ or $360°/p$ followed by a reflection at a plane perpendicular to the axis of rotation produces an orientation indistinguishable from the original molecule.

(5) The identity, I, a trivial symmetry element possessed by all molecules and introduced for the purposes of mathematical group theory.

Each of these symmetry elements will now be examined in more detail.

(a) CENTER OF SYMMETRY

A molecule has a center of symmetry, i, if by reflection at the center (inversion) the molecule is transformed into a configuration indistinguishable from the original one. For every atom with x, y, and z coordinates from the center there must be an identical atom with $-x$, $-y$, and $-z$ coordinates. Some examples of molecules possessing a center of symmetry are shown in Fig. 3.1(a, b, c). In the case of carbon dioxide there is an atom located at the center. However, dibromodichloroethane and benzene do not have an atom located at the center. In the case of dibromodichloroethane only the *trans* isomer has a center of symmetry. If one hydrogen atom has the coordinates x, y, and z from the center, then the other hydrogen atom has the cordinates $-x$, $-y$, and $-z$ and similar conditions apply to the bromine and chlorine atoms. However, for the *cis* isomer such conditions would not be fulfilled. It should also be quite obvious that a molecule may only possess one center of symmetry.

(b) ROTATION AXIS OF SYMMETRY

A molecule has a rotation axis of symmetry, C_p, if rotation by $360°/p$ yields a molecular configuration indistinguishable from the original one. Again referring to Fig. 3.1(d) rotation about the axis shown through $180°$

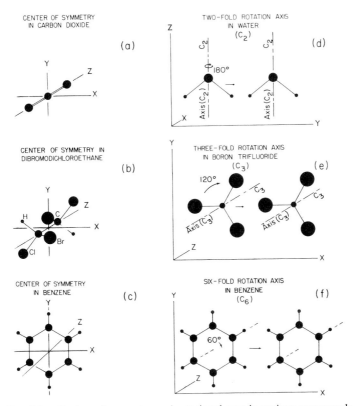

FIG. 3.1.   Centers of symmetry and rotational axes in various compounds.

yields a water molecule indistinguishable from the molecule upon which the rotation has been performed. In this case $p = 2$, and the axis is designated a C_2 axis and often spoken of as a twofold or diad axis. In the case of boron trichloride [Fig. 3.1(e)] rotation through $120°$ or $360°/3$ yields a similar result. Likewise chloroform shown in Fig. 3.2(b) has a C_3 axis of rotation. For $n = 3$, 4, ..., the axes are designated as threefold, fourfold, ..., or triad, tetrad, ... axes. In the case of benzene shown in Fig. 3.1(f) there is a sixfold axis of rotation passing through the center of symmetry. In addition, coincident with this axis there are a C_2 axis and a C_3 axis since rotation through $180°$ or $120°$ will also yield a benzene molecule indistinguishable from the original one on which the rotation has been performed. Perpendicular to this sixfold axis of rotation are six C_2 axes, three of which pass through the carbon atoms and three of which pass through the carbon—carbon bonds. Rotation about each of these axes through $180°$ yields an indistinguishable benzene molecule. By convention, the axis of highest symmetry is designated as the z axis.

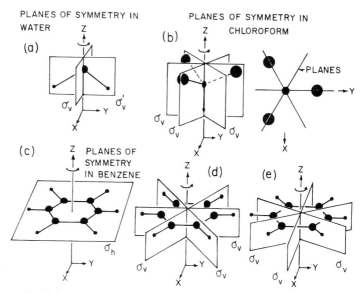

FIG. 3.2.   Planes of symmetry in water, chloroform, and benzene.

(c)   PLANES OF SYMMETRY

A molecule has a plane of symmetry, σ, if by reflection at the plane the molecule is transformed into a configuration indistinguishable from the original one. In other words, a plane of symmetry bisects the molecule into two equivalent parts, one part being the mirror image of the other. For instance, consider the water molecule shown in Fig. 3.2(a). A plane is shown which bisects the H–O–H angle of 105°. If every portion of the water molecule on the left-hand side of this plane were translated across the plane to the right-hand side to a new position equidistant from the plane and vice versa (i.e., the right-hand portion translated to the left) the resulting figure could not be distinguished from the original one. The water molecule, therefore, has a plane of symmetry bisecting the 105° angle. It is evident that this is a special plane in the xz dimension. For any other planes parallel to this particular plane a corresponding operation of reflection similar to that just described would not yield the same water molecule. The plane bisecting the 105° angle is spoken of as a symmetry element. It should be clear that for every symmetry element there is a corresponding symmetry operation, that is, a coordinate transformation (in this case a reflection) which produces a configuration of atoms indistinguishable from the original molecule.

It might be well to point out that if the hydrogen atom on the left-hand side had been labeled 1 and the one on the right-hand side labeled 2, then after performing the symmetry operation the final molecule would have been distinguishable from the starting molecule. However, such a procedure imposes a restriction that does not apply in the usual symmetry considerations. It is not necessary that each atom or nucleus end up after the symmetry operation in its exact original position but only that the resulting total molecule be indistinguishable from the original molecule.

Again considering Fig. 3.2(a) it is quite apparent that the plane in which the water molecule lies is also a plane of symmetry in the yz dimension. Both planes of symmetry are designated as σ_v because they are both vertical to the major symmetry axis, i.e., the C_2 axis shown in Fig. 3.2(a) in the z direction. To distinguish the two planes, one is designated as σ_v'. The choice is arbitrary but in planar molecules of this type it is conventional to have the x axis perpendicular to the molecular plane so the atoms are in the yz or σ_v' plane.[12] Likewise the chloroform molecule shown in Fig. 3.2(b) has three planes of symmetry vertical to the threefold rotation axis. However, in the case of benzene the plane in which the carbon and hydrogen atoms lie [Fig. 2.3(c)] is horizontal to the major symmetry axis, i.e., the C_6 axis [refer to Fig. 3.1(f)]. The remaining six planes of symmetry [Fig. 3.2(d) and (e)] are vertical to the C_6 axis. Hence, benzene has one horizontal plane (σ_h) and six vertical planes ($6\sigma_v$) of symmetry.

(d) p-FOLD ROTATION-REFLECTION AXIS OF SYMMETRY

A molecule has a p-fold rotation-reflection axis of symmetry, S_p, if rotation through $360°/p$ followed by reflection at a plane perpendicular to the axis of rotation yields a configuration indistinguishable from the starting molecule. This is the most difficult symmetry operation to vizualize but with the aid of Fig. 3.3 perhaps it can be made clear. In the case of dibromodichloroethane, rotation through $180°$ does not yield the original configuration. However, by reflection at a plane perpendicular to the axis of rotation [see Fig. 3.3(a)] the original structure is obtained. Such an axis is designated as an S_2 axis. Likewise boron trifluoride has an S_3 and allene an S_4 rotation-reflection axis [see Fig. 3.3(b) and (c)]. It may be interesting to compare boron trifluoride with chloroform. Boron trifluoride has a C_3 axis coincident with the S_3 axis. However, chloroform being nonplanar has only a C_3 axis. Hence, boron trifluoride is a molecule of higher symmetry. In the case of benzene [see Fig. 3.1(f)] there is an S_6 axis coincident with the C_6 axis previously discussed.

---

[12] R. S. Mulliken, *J. Chem. Phys.* **23**, 1997 (1955).

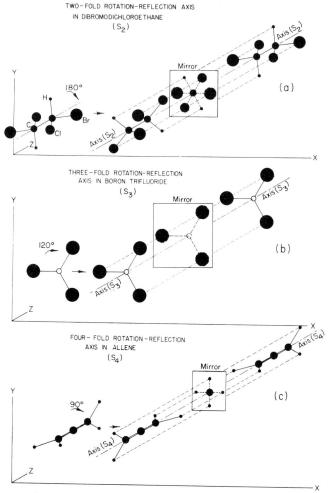

FIG. 3.3.   Rotation-reflection axes in various compounds.

(e) THE IDENTITY, I

The fifth and last symmetry element is the identity, I, sometimes designated in the literature by the symbol E. All molecules possess the identity even if they possess none of the other four symmetry elements. The symmetry operation corresponding to this symmetry element involves leaving the molecule unaltered. Hence, the resulting molecule cannot be distinguished from the original. This symmetry element is introduced for mathematical reasons which will become apparent in the following sections.

3.2 Point Groups

In the previous section it was mentioned that all molecules possess the symmetry element, the identity. Also it was shown that a large number of molecules have additional symmetry elements. For example, the water molecule has two vertical planes of symmetry and a twofold rotation axis of symmetry. With a little thought it will be realized that any molecule which has two vertical planes must of necessity have a twofold rotation axis. However, the reverse is not true, for a molecule may possess a twofold rotation axis and not the two mirror planes of symmetry. Likewise in the case of either chloroform or boron trifluoride the three vertical planes of symmetry require that there be a threefold rotation axis of symmetry in each of these molecules. Hence, it can be seen that the presence of certain symmetry elements means that others are of necessity required. In the case of benzene the symmetry can be adequately described by the following elements, a sixfold axis, a plane of symmetry in the rotation axis, and a plane of symmetry perpendicular to this axis. Yet as was shown in the previous section benzene has additional symmetry elements, namely, a center of symmetry, and (coincident with the sixfold axis) a sixfold rotation-reflection axis and a twofold and threefold axis as well as a two-, three-, and sixfold rotation-reflection axis, six twofold axes, and seven planes of symmetry. Thus, the existence of certain symmetry elements requires per se the presence of others. It might be assumed that any possible combination of symmetry elements is permitted. However, such an assumption can be readily proved false. For example, a molecule cannot possess a threefold and fourfold axis in the same direction. Likewise it can be proved by mathematical group theory that only certain combinations of symmetry elements are possible. Such a restricted combination of symmetry elements that leaves at least one point unchanged is called a *point group*. A point group is also a group in a mathematical sense as will be demonstrated in detail in the next section. For the present the main concern will be with the classification of molecules into distinct point groups and the shorthand notations describing these groups. Point groups can be classified by either the Hermann-Mauguin or the Schoenflies notation.[1] In this text the Schoenflies notation has been employed in the previous section and will be naturally continued here.

In principle molecules may belong to any of the possible point groups permitted by mathematical group theory. In reality, however, the actual chemical examples of specific point groups are somewhat restricted. In dealing with crystals only one-, two-, three-, four-, and sixfold axes are permissible. With such a restriction, there are only thirty-two possible combinations of symmetry elements yielding the thirty-two crystal point groups.

Table 3.1 lists these thirty-two groups plus four additional point groups (marked with a superscript "^{a}") for which there are actual molecular examples.

In the following discussion all the individual point groups will not be considered. Rather the comments will be restricted to those particular point groups where it is felt some questions regarding the symmetry elements might arise.

Point group C_p. A molecule possessing only a C_p rotation axis of symmetry falls in this group. Those molecules with no symmetry at all except for the identity, I, fall in group C_1. A molecule with only a twofold axis of symmetry would be a member of the C_2 point group, etc.

Point group S_p. A molecule having only a p-fold rotation reflection axis of symmetry belongs to the S_p point group. This point group is designated also by C_i for those molecules with a center of symmetry, such as the *trans* form of dibromodichloroethane [see Figs. 3.1(b) and 3.3(a)], and having as a necessary consequence an S_2 axis.

Point group C_{pv}. Since in the Schoenflies notation, rotation axes are designated as being vertical, the point group C_{pv} includes those molecules with a vertical rotation axis of order p with p vertical planes of symmetry lying in the rotation axis. The water molecule belongs to the C_{2v} point group while chloroform belongs to the C_{3v} point group (see Fig. 3.2). Boron trifluoride, however, does not belong to the C_{3v} group because it possesses an S_3 element and therefore has higher symmetry [see Fig. 3.1(e)]. The point group C_{1v} is usually designated by C_s and can just as logically be designated by C_{1h} since the Schoenflies notation of designating the rotation axis as being vertical is an arbitrary one. Linear heteronuclear diatomic molecules belong to the point group $C_{\infty v}$ since they possess an infinite-fold axis and an infinite number of planes through the axis.

Point group C_{ph}. Molecules belonging to this point group have a rotation axis of order p and a horizontal plane of symmetry perpendicular to the axis. As mentioned C_{1h} is equivalent to C_{1v} or C_s. The point group C_{2h} has an S_2 axis coincident with the C_2 axis.

Point group D_p. The D stands for dihedral. Molecules belonging to this point group have a p-fold axis, C_p, and perpendicular to this axis p-twofold axes at equal angles to each other. No D_1 group is listed in Table 3.1 since such a group is identical with C_2.

Point group D_{pd}. Similar to the point group D_p, molecules in this group have a p-fold axis, C_p, and perpendicular to this axis p-twofold axes plus

TABLE 3.1
Important Point Groups and Their Symmetry Elements

Symbol (General)	Symbol (Specific)	Center i	C_2	C_3	C_4	C_5	C_6	S_2	S_4	S_6	S_8	σ_v	σ_h	σ_a
C_p	C_1													
	C_2		1											
	C_3			1										
	C_4				1									
	C_6						1							
S_p	S_2	1						1						
	S_4		1						1					
	S_6	1		1						1				
C_{pv}	C_{1v}		(equivalent to C_s or C_{1h})									1		
	C_{2v}		1									2		
	C_{3v}			1								3		
	C_{4v}				1							4		
	C_{6v}						1					6		
	$C_{\infty v}{}^a$		(C_∞ ; any C_p)									∞		
C_{ph}	C_{2h}	1	1					1					1	
	C_{3h}			1									1	
	C_{4h}	1			1								1	
	C_{6h}	1					1						1	
D_p	D_2		3											
	D_3		3	1										
	D_4		4		1									
	D_6		6				1							
D_{pd}	D_{2d}		3						1					2
	D_{3d}	1	3	1						1				3
	$D_{4d}{}^a$		4		1						1			4
D_{ph}	D_{2h}	1	3									(3 undesignated planes)		
	D_{3h}		3	1								3	1	
	D_{4h}	1	4		1				1			4	1	
	$D_{5h}{}^a$		5			1						5	1	
	D_{6h}	1	6				1			1		6	1	
	$D_{\infty h}{}^a$	1	∞ ; (C_∞)									∞	1	
T	T		3	4										
	T_d		3	4					3			(6 undesignated planes)		
	T_h	1	3	4						4		(3 undesignated planes)		
O	O		6	4	3									
	O_h	1	6	4	3				3	4		(9 undesignated planes)		

^{a} See text preceding table.

p-planes of symmetry (σ_d) passing through the p-fold axis and bisecting the angles between the two consecutive twofold axes. The point group D_{2d} has three mutually perpendicular C_2 axes and one S_4 axis coincident with one of the C_2 axes and two diagonal planes of symmetry passing through the S_4 axis. Allene shown in Fig. 3.3(c) belongs to the point group D_{2d}.

In the point group D_{3d} the S_6 axis is coincident with the C_3 axis.

The point group D_{4d} has one C_4 axis (coincident with one C_2 and one S_8 axis), four C_2 axes, and four diagonal planes of symmetry. In Table 3.1 the coincident C_2 axis is not listed.

Point group D_{ph}. Molecules with a p-fold axis and p-vertical planes of symmetry lying in the rotation axis plus a horizontal plane of symmetry perpendicular to the C_p axis fall in this point group. The D_{1h} group is identical with C_{2v} and therefore not listed as such. Boron trichloride [Fig. 3.1(e)] is an example of the D_{3h} point group.

In the point group D_{4h} the S_4 axis is coincident with the C_4 axis. On this same axis there is a coincident C_2 axis which is not included in Table 3.1. The four C_2 axes which are perpendicular to the C_4 axis are of course listed.

In the point group D_{6h} the S_6 axis which is coincident with the C_6 axis is listed but not the C_2 and C_3 or S_2 and S_3 axes which are also coincident with the C_6 axis. Benzene shown in Fig. 3.1(c) and (f) and in Fig. 3.2(c) is the only known example of this group.

Point group T. The T stands for tetrahedral. Molecules belong to this group if they have four threefold axes and three mutually perpendicular twofold axes. The T_h group has the same symmetry elements plus a center of symmetry.

Point group T_d. The classic example of this point group is methane shown in Fig. 3.4. Methane has four threefold axes, three mutually perpendicular twofold axes (coincident with S_4 axes), and two mutually perpendicular planes of symmetry through each twofold axis, or a total of six planes of symmetry.

Point group O. The O stands for octahedral. Molecules in this group have three mutually perpendicular fourfold axes and four threefold axes.

Point group O_h. This group has the same symmetry elements as point group O plus a center of symmetry and nine planes of symmetry. As a consequence of this center of symmetry there are also three S_4 axes coincident with the three C_4 axes. Some typical chemical examples of selected point groups are tabulated in Table 3.2.

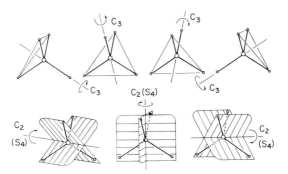

FIG. 3.4. Symmetry elements of methane (point group, T_d).

TABLE 3.2

CHEMICAL EXAMPLES OF SELECTED POINT GROUPS

General	Specific	Chemical examples
C_p	C_1	CHFClBr, N_2H_5Cl
	C_2	H_2O_2, CH_2Cl—CH_2Cl (gauche)
C_{pv}	C_{1v}	ONCl, OSCl$_2$, HFCO, CH_3OH
	C_{2v}	H_2O, F_2CO, CH_2Cl_2, F_2CCl_2
	C_{3v}	NH_3, CHCl$_3$, F_3CCN, CCl_3—CF_3
	C_{4v}	SbCl$_5$, Fe$_5$(CO)$_5$, B_5H_9
	$C_{\infty v}$	HCN, OCS, OCN$^-$, HC$\equiv$CCN
C_{ph}	C_{2h}	C_3O_2, BrH$_2$C—CH_2Br (trans)
	C_{3h}	H_3BO_3, B(OCH$_3$)$_3$
D_{pd}	D_{2d}	B_2Cl_4, H_2C=C=CH_2
	D_{3d}	C_2H_6, C_6H_{12} (cyclohexane)
	D_{4d}	S_8, S_2F_{10}
D_{ph}	D_{3h}	BF$_3$, C_3H_6(cyclopropane)
	D_{4h}	C_4H_8(cyclobutane), (PNCl$_2$)$_4$
	D_{6h}	C_6H_6, C_6Cl_6
	$D_{\infty h}$	CO_2, CS_2, NCC$\equiv$CCN
T	T_d	P_4, CH$_4$, CCl$_4$, SO$_4{}^{2-}$
O	O_h	SF$_6$, UF$_6$, WF$_6$

3.3 Group Theory

The important point groups to which most molecules can be assigned have been discussed; so it is now pertinent to consider how mathematical group theory is involved in the treatment of point groups.

A set of so-called elements I, A, B, C, ..., is said to be a group if the following four conditions apply:

(1) There is an identity element, I, for which $AI = IA = A$ where A represents any element in the group.
(2) The elements obey a law of combination which states that the product of any two elements of the group is also an element of the group, i.e., $AB = C$.
(3) The elements obey a law of association, e.g., $A(BC) = (AB)C$.
(4) Every element in the group has an inverse in the group, i.e., for every element A there is an inverse A^{-1} such that $A \times A^{-1} = I = A^{-1} \times A$ and $B \times B^{-1} = I = B^{-1} \times B$, etc. In other words the product of the element and its inverse yields the identity element.

As a concrete example, consider the four elements 1, -1, i, and $-i$, where $i = -1^{1/2}$, and apply the above four conditions assuming that $I = 1$, $A = -1$, $B = i$, and $C = -i$. It can be seen that these conditions can be more readily checked if a multiplication table is arranged as follows. Each element is arranged in a horizontal column and in a vertical column. By cross multiplication the table can be completed; thus for the first vertical column, $1 \times 1 = 1$, $-1 \times 1 = -1$, $i \times 1 = i$, and $-i \times 1 = -i$.

TABLE 3.3

MULTIPLICATION TABLE FOR A GROUP OF FOUR ELEMENTS

	1	-1	i	$-i$
1	1	-1	i	$-i$
-1	-1	1	$-i$	i
i	i	$-i$	-1	1
$-i$	$-i$	i	1	-1

The remaining columns can be completed in a similar manner. Now the table can be examined to see if the four requirements of a group are fulfilled.

(1) Since the identity is 1 and A is -1 then $-1 \times 1 = 1 \times -1 = -1$. For the element B, $i \times 1 = 1 \times i = i$ and for the element C, $-i \times 1 = 1 \times -i = -i$ and the first condition holds for all elements.
(2) Simple inspection of Table 3.3 shows that the product of any two elements of the group is also an element. In mathematical terms the elements of the group constitute a set; so it would be stated that the set is closed with respect to the binary operation.
(3) To obey the law of association $A(BC) = AB(C)$ or in this example $-1 \times (i \times -i) = -1 \times i(-i)$ and hence this condition is satisfied.

(4) Examination of the elements with respect to the inverse requirement reveals that this condition is fulfilled since the inverse of 1 is 1, the inverse of -1 is -1, the inverse of $+i$ is $-i$, and the inverse of $-i$ is $+i$. Also substitution into $A \times A^{-1} = I = A^{-1} \times A$ yields $(-1)(-1) = 1 = (-1)(-1)$ and into $B \times B^{-1} = I = B^{-1} \times B$ yields $(i)(-i) = 1 = (-i)(i)$, etc. Thus, it has been shown explicitly that the four elements I, A, B, and C do constitute a group in a mathematical sense.

3.4 Group Theory Applied to Point Groups

In dealing with a given molecule or, to be more general, a given point group, the elements of the group are the *symmetry operations* or so-called covering operations. As discussed earlier a symmetry or covering operation is an operation performed on the molecule which changes it into a configuration indistinguishable from the original. A symmetry operation essentially is equivalent to the renumbering of symmetrically related nuclei. It should be mentioned at this point that chloroform will be the specific example which will be treated in this text. As previously indicated in Table 3.2, chloroform is a member of the C_{3v} point group.

The symmetry operations for the chloroform molecule are shown in Fig. 3.5 and are as follows:

(1) Rotation through $360°/3$ or $2\pi/3$ radians and designated by C_3. There are two such elements, C_3^+ and C_3^-. A clockwise rotation through

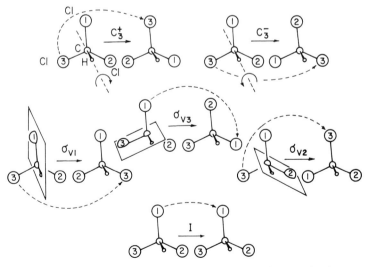

FIG. 3.5. The symmetry operations for the chloroform molecule.

120° is designated C_3^+. (Chlorine No. 3 moves to the position originally held by chlorine No. 1.) If this operation is performed twice ($C_3^+ \times C_3^+$ or C_3^2) a clockwise rotation of 240° results which is equivalent to a counterclockwise rotation of $-120°$ which is designated C_3^-. (Chlorine No. 3 moves to the position originally held by chlorine No. 2.)

(2) Reflection in a plane through the rotation axis and one of chlorine atoms and designated by σ_v. Since there are three such planes, there are three elements, σ_{v_1}, σ_{v_2}, and σ_{v_3}.

(3) Rotation of zero degrees about the C_3 axis and designated by I. The I stands for the identity operation and there is only one such element.

It can be seen that there are a total of six elements for any molecule belonging to the C_{3v} point group. By preparing a multiplication table (Table 3.4) similar to Table 3.3, it can be shown that these elements also fulfill the requirements for a group in a mathematical sense.

TABLE 3.4

MULTIPLICATION TABLE FOR C_{3v} SYMMETRY OPERATIONS[a]

	I	C_3^+	C_3^-	σ_{v_1}	σ_{v_2}	σ_{v_3}
I	I	C_3^+	C_3^-	σ_{v_1}	σ_{v_2}	σ_{v_3}
C_3^+	C_3^+	C_3^-	I	σ_{v_3}	σ_{v_1}	σ_{v_2}
C_3^-	C_3^-	I	C_3^+	σ_{v_2}	σ_{v_3}	σ_{v_1}
σ_{v_1}	σ_{v_1}	σ_{v_2}	σ_{v_3}	I	C_3^+	C_3^-
σ_{v_2}	σ_{v_2}	σ_{v_3}	σ_{v_1}	C_3^-	I	C_3^+
σ_{v_3}	σ_{v_3}	σ_{v_1}	σ_{v_2}	C_3^+	C_3^-	I

[a] In using such a table it is important to note that the operation performed first is at the top and the operation performed second is at the left.

If Table 3.4 is examined for the four requirements for a group, it is seen that the first requirement is fulfilled because there is an identity element.

The second requirement regarding the law of combination is obeyed because the product of any two elements of the group is also an element of the group. For example, consider the multiplication of σ_{v_3} (side row) by σ_{v_1} (top row) to yield C_3^+. The product $\sigma_{v_3}\sigma_{v_1}$ means that the symmetry operation σ_{v_1} should be carried out first and then followed by σ_{v_3}. (The convention is similar to algebraic operations; the operation applied first is written on the right.) The result of carrying out these two operations should be the same as performing the single operation C_3^+. Figure 3.6(a) demonstrates that such a result is obtained. Starting with the configuration shown, reflection n the σ_{v_1} plane interchanges atoms 2 and 3. Reflection in the σ_{v_3} plane then

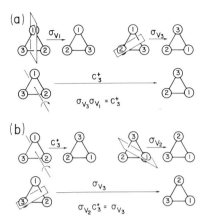

FIG. 3.6. Examples of multiplication of symmetry elements. It is important to note that the planes are numbered with the molecule in the equilibrium position, i.e., with atom 1 at the top, followed by atom 2 on the right and atom 3 on the left. Thus, in the equilibrium position the plane bisecting atom 1 is σ_{v_1} and that bisecting atom 2 is σ_{v_2}, etc. These planes remain fixed and do *not* change with the symmetry operation. Note that in example (a), after the σ_{v_1} operation, the σ_{v_3} plane now bisects atom 2. Atoms 3 and 2 have been interchanged by σ_{v_1} but the σ_{v_3} plane is unaltered.

interchanges atoms 1 and 3. As shown, the single operation of rotation through 120° yields the same result. Hence, $\sigma_{v_3}\sigma_{v_1} = C_3^+$. Again, considering another example, Fig. 3.6(b) demonstrates that rotation through 120° followed by a reflection in the σ_{v_2} plane equals the single operation of reflection in the σ_{v_3} plane. It can be readily demonstrated in a similar fashion that the product of any two of the six elements yields an element of the group.

The third requirement is that the law of association must be obeyed, for example, $A(BC) = AB(C)$. For the C_{3v} group consider the case of

$$\sigma_{v_1}(\sigma_{v_2}\sigma_{v_3}) = (\sigma_{v_1}\sigma_{v_2})\sigma_{v_3} \tag{3.1}$$

Application of Table 3.3 yields

$$(\sigma_{v_2}\sigma_{v_3}) = C_3^+ \qquad \text{and} \qquad (\sigma_{v_1}\sigma_{v_2}) = C_3^+ \tag{3.2}$$

and substituting into Eq. (3.1)

$$\sigma_{v_1}C_3^+ = C_3^+\sigma_{v_3} \tag{3.3}$$

but from Table 3.3

$$\sigma_{v_1}C_3^+ = \sigma_{v_2} \qquad \text{and} \qquad C_3^+\sigma_{v_3} = \sigma_{v_2} \tag{3.4}$$

so

$$\sigma_{v_2} = \sigma_{v_2} \tag{3.5}$$

Note that the order of performing the operation must be consistent. The σ_{v_3} is taken from the top row and the σ_{v_2} from the side column of Table 3.3, thereby yielding C_3^+. If the reverse procedure had been followed, the σ_{v_2} being taken from the top row and the σ_{v_3} from the side column, the result would have been C_3^-. It can also be shown for any other like combination of elements that the law of association holds.

The fourth and final condition requiring that there be an inverse of each element of the group may be satisfied by noting in Table 3.3 the pairs of operations which yield the identity I, when multiplied together, for example, $\sigma_{v_1} \times \sigma_{v_1} = I$, $C_3^+ \times C_3^- = 1$. To summarize, then, it has been shown that the six elements do constitute a group as mathematically defined.

3.5 Representations of Groups

In this section some elementary use of matrices will be made. Matrix properties and operations are discussed in Section 14.21 but in this chapter it is only necessary to know how to multiply two matrices as in the example below.

$$\left\| \begin{matrix} 1 & 2 \\ 3 & 4 \end{matrix} \right\| \times \left\| \begin{matrix} 5 & 6 \\ 7 & 8 \end{matrix} \right\| = \left\| \begin{matrix} (1 \times 5 + 2 \times 7) & (1 \times 6 + 2 \times 8) \\ (3 \times 5 + 4 \times 7) & (3 \times 6 + 4 \times 8) \end{matrix} \right\| = \left\| \begin{matrix} 19 & 22 \\ 43 & 50 \end{matrix} \right\|$$

The elements of a group such as the symmetry operations just discussed can be represented by a set of numbers or, more generally, by matrices. Any set of such matrices which, when multiplied, satisfies the group multiplication table is said to be a representation of the group. The purpose is to enable us to express geometrical relationships in algebraic form. This can be made clear with an example. In the first column of Fig. 3.7 we illustrate the form of the normal coordinates for H_2O including translations and rotations. In the second, third, and fourth column we illustrate the results of performing the C_2, σ_v or σ_v' symmetry operations on the normal coordinates in column one. It can be seen that a symmetry operation either leaves the form of the normal coordinate unchanged or else multiplies it by minus one, which reverses every displacement vector.

$$Q_i \xrightarrow{\text{symmetry operation}} (\pm 1) Q_i$$

This means that the form of a nondegenerate normal coordinate is either symmetrical $(+1)$ or antisymmetrical (-1) to each symmetry element possessed by the equilibrium configuration. As we shall see, the transformation number (± 1) in Fig. 3.7 is actually a one-by-one matrix. In Fig. 3.7 it

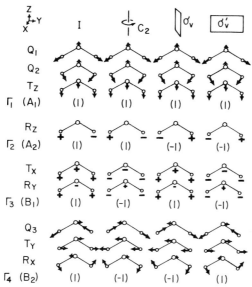

FIG. 3.7.   The form of the normal coordinates (schematic) of H_2O and their transformation by the symmetry operations of the C_{2v} group. The transformation either leaves the normal coordinate unchanged or else reverses its sign.

is seen that Q_1, Q_2, and T_z each have the same transformation number (1) for each symmetry operation. This set of four numbers is a representation (Γ_1) for the symmetry operations of the group where each symmetry operation is represented by (1).

	I	C_2	σ_v	σ_v'
Γ_1	(1)	(1)	(1)	(1)

It is clear that any two of these numbers multiplied together will equal any one of the numbers, so this representation satisfies the group multiplication table for the C_{2v} group. The complete set of representations of transformation numbers illustrated in Fig. 3.7 is given in Table 3.5.

TABLE 3.5

REPRESENTATIONS FOR THE C_{2v} POINT GROUP

	I	C_2	σ_v	σ_v'
Γ_1	1	1	1	1
Γ_2	1	1	-1	-1
Γ_3	1	-1	1	-1
Γ_4	1	-1	-1	1

We must now illustrate that the numbers in each of these representations obey the C_{2v} group multiplication table given in Table 3.6. From the C_{2v}

TABLE 3.6

MULTIPLICATION TABLE FOR C_{2v} SYMMETRY OPERATIONS

	I	C_2	σ_v	σ_v'
I	I	C_2	σ_v	σ_v'
C_2	C_2	I	σ_v'	σ_v
σ_v	σ_v	σ_v'	I	C_2
σ_v'	σ_v'	σ_v	C_2	I

group multiplication table we have, for example,

$$\sigma_v C_2 = \sigma_v' \tag{3.6}$$

The normal coordinates in column two of Fig. 3.7 result from the C_2 operation. When these are transformed in accordance with Eq. (3.6) by the application of the σ_v operation, the result will be equivalent to the normal coordinates in column four of Fig. 3.7 which result from the σ_v' operation in agreement with Eq. (3.6). Likewise, in Table 3.5 for each representation Γ, the number in the C_2 column multiplied by the number in the σ_v column equals the number in the σ_v' column in agreement with Eq. (3.6). Other such relationships can be verified, indicating we have indeed written in Table 3.5 four separate representations for the C_{2v} group as defined.

The use of normal coordinates as a basis for the representations leads to simple results such as Table 3.5 because of the following reasons. If a symmetrical molecule is distorted in some manner from its equilibrium configuration or if its nuclei are moving in some manner, the potential and kinetic energies cannot be changed by the performance of a symmetry operation on the molecule, since any such operation is equivalent to the renumbering of symmetrically related atoms. The potential energy V and the kinetic energy T in terms of the normal coordinates Q are given mathematically by

$$V = \tfrac{1}{2}\sum_i \lambda_i Q_i^2 \quad \text{and} \quad T = \tfrac{1}{2}\sum_i \dot{Q}_i^2 \tag{3.7}$$

where λ_i is a constant involving the vibrational frequency and $\dot{Q}$ is the time derivation of Q (see Chapter 14). The main point is that the total energy involves only *squared* functions of the normal coordinates. Therefore it is clear that if the energy is to remain constant after the performance of a symmetry operation, then $(Q_i)^2$ must remain constant and a nondegenerate normal coordinate must be left unchanged or simply be changed in sign by any symmetry operation.

The representations in Table 3.5 developed using normal coordinates as a basis are not the only possible representations. An important set of representations can be developed using cartesian displacement coordinates as a basis. In Fig. 3.8 the arrows indicate x, y, and z components of arbitrary

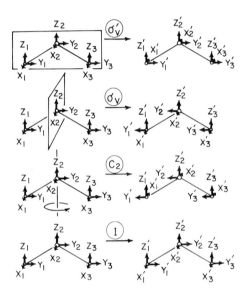

FIG. 3.8. The effect of symmetry operations on cartesian displacement coordinates for H_2O.

displacements from equilibrium of the H_2O atoms. These displacements are transformed by the symmetry operations of the C_{2v} point group where the transformed cartesian displacement coordinates are primed. In Fig. 3.8 for the σ'_v transformation (reflection in the plane of the page), the set of primed displacement coordinates is related to the original coordinates by the following linear equations:

$$
\begin{array}{lll}
X'_1 = -X_1 & X'_2 = -X_2 & X'_3 = -X_3 \\
Y'_1 = Y_1 & Y'_2 = Y_2 & Y'_3 = Y_3 \\
Z'_1 = Z_1 & Z'_2 = Z_2 & Z'_3 = Z_3
\end{array}
\tag{3.8}
$$

The nine equations of this set can be written using a nine-by-nine matrix as follows:

$$
\begin{Vmatrix} X_1' \\ Y_1' \\ Z_1' \\ X_2' \\ Y_2' \\ Z_2' \\ X_3' \\ Y_3' \\ Z_3' \end{Vmatrix} =
\begin{Vmatrix}
-1 & 0 & 0 & 0 & 0 & 0 & 0 & 0 & 0 \\
0 & 1 & 0 & 0 & 0 & 0 & 0 & 0 & 0 \\
0 & 0 & 1 & 0 & 0 & 0 & 0 & 0 & 0 \\
0 & 0 & 0 & -1 & 0 & 0 & 0 & 0 & 0 \\
0 & 0 & 0 & 0 & 1 & 0 & 0 & 0 & 0 \\
0 & 0 & 0 & 0 & 0 & 1 & 0 & 0 & 0 \\
0 & 0 & 0 & 0 & 0 & 0 & -1 & 0 & 0 \\
0 & 0 & 0 & 0 & 0 & 0 & 0 & 1 & 0 \\
0 & 0 & 0 & 0 & 0 & 0 & 0 & 0 & 1
\end{Vmatrix} \times
\begin{Vmatrix} X_1 \\ Y_1 \\ Z_1 \\ X_2 \\ Y_2 \\ Z_2 \\ X_3 \\ Y_3 \\ Z_3 \end{Vmatrix} =
\begin{Vmatrix} -X_1 \\ Y_1 \\ Z_1 \\ -X_2 \\ Y_2 \\ Z_2 \\ -X_3 \\ Y_3 \\ Z_3 \end{Vmatrix} \quad (3.9)
$$

The equivalence of Eqs. (3.8) and (3.9) is verified by multiplying the nine-by-nine matrix of Eq. (3.9) as shown, following the rules for matrix multiplication (see Section 14.21). The nine-by-nine matrix is the transformation matrix for the σ_v' operation. The transformation matrix for the identity operation I is like the above but with only $+1$ on the diagonal.

The transformation matrices for the σ_v and C_2 operations are as follows:

$$
\sigma_v \qquad\qquad\qquad\qquad\qquad C_2
$$

$$
\begin{Vmatrix}
0 & 0 & 0 & 0 & 0 & 0 & 1 & 0 & 0 \\
0 & 0 & 0 & 0 & 0 & 0 & 0 & -1 & 0 \\
0 & 0 & 0 & 0 & 0 & 0 & 0 & 0 & 1 \\
0 & 0 & 0 & 1 & 0 & 0 & 0 & 0 & 0 \\
0 & 0 & 0 & 0 & -1 & 0 & 0 & 0 & 0 \\
0 & 0 & 0 & 0 & 0 & 1 & 0 & 0 & 0 \\
1 & 0 & 0 & 0 & 0 & 0 & 0 & 0 & 0 \\
0 & -1 & 0 & 0 & 0 & 0 & 0 & 0 & 0 \\
0 & 0 & 1 & 0 & 0 & 0 & 0 & 0 & 0
\end{Vmatrix}
\quad
\begin{Vmatrix}
0 & 0 & 0 & 0 & 0 & 0 & -1 & 0 & 0 \\
0 & 0 & 0 & 0 & 0 & 0 & 0 & -1 & 0 \\
0 & 0 & 0 & 0 & 0 & 0 & 0 & 0 & 1 \\
0 & 0 & 0 & -1 & 0 & 0 & 0 & 0 & 0 \\
0 & 0 & 0 & 0 & -1 & 0 & 0 & 0 & 0 \\
0 & 0 & 0 & 0 & 0 & 1 & 0 & 0 & 0 \\
-1 & 0 & 0 & 0 & 0 & 0 & 0 & 0 & 0 \\
0 & -1 & 0 & 0 & 0 & 0 & 0 & 0 & 0 \\
0 & 0 & 1 & 0 & 0 & 0 & 0 & 0 & 0
\end{Vmatrix}
$$

$$
(3.10)
$$

If each of these matrices is to represent a specific symmetry operation then they must satisfy the group multiplication table. From the C_{2v} multiplication Table 3.5, the C_2 operation followed by σ_v is

$$
\sigma_v C_2 = \sigma_v' \tag{3.11}
$$

If the C_2 matrix is postmultiplied by the σ_v matrix in Eq. (3.10), then from matrix multiplication, the σ_v' matrix in Eq. (3.9) will result. Other such relationships can be verified indicating that these large matrices are indeed proper representations for the C_{2v} group.

An important property of these transformation matrices is their *character*. The character of a matrix of a representation is symbolized by χ and is equal to the sum of the elements on the diagonal. The characters of the transformation matrices for H_2O, Eqs. (3.9) and (3.10), using cartesian coordinates as a basis are

	I	C_2	σ_v	σ_v'
χ_{H_2O}	$+9$	-1	$+1$	$+3$

$$(3.12)$$

It should be noticed that in this case only atoms left unmoved by a symmetry operation can contribute to the character for that operation. This result will be used later.

3.6 Irreducible Representations

Actually there are an infinite number of matrices which can form a representation of a group. These can be generated by a similarity transformation as follows. If X is a square matrix with an inverse X^{-1} and I, A, B, and C are large matrices which are representations of a group, then new matrices A', B', and C' can be generated by

$$A' = X^{-1}AX \qquad B' = X^{-1}BX \qquad C' = X^{-1}CX \qquad (3.13)$$

Let us say the group multiplication table requires that

$$AB = C \qquad (3.14)$$

In matrix multiplication, a matrix times its inverse such as XX^{-1} equals the identity matrix (ones on the diagonal, zero elsewhere), and in matrix algebra the identity matrix has the same role that the integer one has in algebraic multiplication which means that XX^{-1} can be inserted or removed at will in a matrix equation. If we insert XX^{-1} between A and B in Eq. (3.14) and premultiply and postmultiply each side of this equation by X^{-1} and X, respectively, we obtain

$$X^{-1}AXX^{-1}BX = X^{-1}CX \qquad \text{or} \qquad A'B' = C' \qquad (3.15)$$

This means that A', B', and C', which have been generated by a similarity transformation, also obey the group multiplication table and therefore also form a valid representation for the group. It is important to note that *the character is not changed by a similarity transformation.*[6, 7]

It may be possible to find a similarity transformation which reduces matrices with off-diagonal terms such as those in Eq. (3.10) into matrices which have smaller submatrices centering on the diagonal and zeros elsewhere such as

$$
\begin{array}{ccc}
A & B & C \\
\left\|
\begin{array}{c|c|cc}
a & 0 & 0 & 0 \\
\hline
0 & b & 0 & 0 \\
\hline
0 & 0 & c_1 & c_2 \\
0 & 0 & c_3 & c_4
\end{array}
\right\|
&
\left\|
\begin{array}{c|c|cc}
d & 0 & 0 & 0 \\
\hline
0 & e & 0 & 0 \\
\hline
0 & 0 & f_1 & f_2 \\
0 & 0 & f_3 & f_4
\end{array}
\right\|
&
\left\|
\begin{array}{c|c|cc}
g & 0 & 0 & 0 \\
\hline
0 & b & 0 & 0 \\
\hline
0 & 0 & i_1 & i_2 \\
0 & 0 & i_3 & i_4
\end{array}
\right\|
\end{array}
\qquad (3.16)
$$

If a transformation can be found which will put *all* the matrices of a given representation into this general form, the representation is said to be reducible. If no transformation can further diagonalize all submatrices such as c, f, and i in Eq. (3.16) then the set of matrices of a given representation is said to be completely reduced and the sets of submatrices are called the *irreducible representations.*[8] The fact that the submatrices are also representations can be clarified by referring to Eq. (3.16). If the group multiplication table requires that $AB = C$ then, following the rules for matrix multiplication, it is clear that $ad = g$, $be = h$, and $cf = i$ so that the set of submatrices a, d, and g, for example, form part of a representation which as described above cannot be further reduced.

The matrices of the representation using cartesian displacement coordinates as a basis can be reduced by some transformation (unspecified here) into matrices of the representation using normal coordinates as a basis. For example, the matrix characterizing the C_2 operation using H_2O normal coordinates as a basis can be written

$$
C_2
$$

$$
\begin{Vmatrix} Q_1' \\ Q_2' \\ T_z' \\ R_z' \\ T_x' \\ R_y' \\ Q_3' \\ T_y' \\ R_x' \end{Vmatrix}
=
\begin{Vmatrix}
1 & 0 & 0 & 0 & 0 & 0 & 0 & 0 & 0 \\
0 & 1 & 0 & 0 & 0 & 0 & 0 & 0 & 0 \\
0 & 0 & 1 & 0 & 0 & 0 & 0 & 0 & 0 \\
0 & 0 & 0 & 1 & 0 & 0 & 0 & 0 & 0 \\
0 & 0 & 0 & 0 & -1 & 0 & 0 & 0 & 0 \\
0 & 0 & 0 & 0 & 0 & -1 & 0 & 0 & 0 \\
0 & 0 & 0 & 0 & 0 & 0 & -1 & 0 & 0 \\
0 & 0 & 0 & 0 & 0 & 0 & 0 & -1 & 0 \\
0 & 0 & 0 & 0 & 0 & 0 & 0 & 0 & -1
\end{Vmatrix}
\times
\begin{Vmatrix} Q_1 \\ Q_2 \\ T_z \\ R_z \\ T_x \\ R_y \\ Q_3 \\ T_y \\ R_x \end{Vmatrix}
=
\begin{Vmatrix} Q_1 \\ Q_2 \\ T_z \\ R_z \\ -T_x \\ -R_y \\ -Q_3 \\ -T_y \\ -R_x \end{Vmatrix}
\qquad (3.17)
$$

When multiplied out as shown, this reads $Q'_1 = (1)Q_1$, $Q_1 = (1)Q_2$, $Q'_3 = (-1)Q_3$, and so forth for the C_2 operation which can be verified by referral to Fig. 3.7.

If this matrix characterizing C_2 is compared with that obtained for the C_2 operation in cartesian displacement coordinates [Eq. (3.10)], it can be seen that the character is -1 as before but now there are no off-diagonal elements. We have only one-by-one matrices on the diagonal which obviously cannot be further reduced. Since the matrices for the other operations will be in this same form, the representation for the water molecule using normal coordinates as a basis is a completely reduced one. Each of the nine one-by-one matrices on the diagonal of the large matrix in Eq. (3.17) represents the C_2 operation in one of the four possible irreducible representations illustrated in Fig. 3.7 and tabulated in Table 3.5.

3.7 The Character Table

TABLE 3.7

CHARACTERS OF THE IRREDUCIBLE REPRESENTATIONS OF THE C_{2v} POINT GROUP

		I	$C_2(2)$	$\sigma_v(xz)$	$\sigma'_v(yz)$	
(Γ_1)	A_1	1	1	1	1	T_z
(Γ_2)	A_2	1	1	-1	-1	R_z
(Γ_3)	B_1	1	-1	1	-1	T_x, R_y
(Γ_4)	B_2	1	-1	-1	1	T_y, R_x

In Table 3.7 the *characters* of the four possible different irreducible representations for the C_{2v} group are listed where in this case the character is the same as the one-by-one transformation matrix. Table 3.7 is called a character table for the C_{2v} group. Along the top are given the symmetry elements. Along the left are the vibration types or species which correspond to the irreducible representations. It so happens that the character under the identity symmetry element I is the same as the degeneracy of the vibration of the indicated type, 1 for singly degenerate (nondegenerate), or 2 for doubly degenerate. For nondegenerate vibrations the characters under the other symmetry elements are $+1$ for symmetrical vibrations and -1 for for antisymmetrical vibrations. The final column of Table 3.7 contains the normal coordinates for translations and rotations. Since cartesian coordinates are specified for these, cartesian planes and axes are specified (in parenthesis) after the symmetry element symbols.

In Fig. 3.7, Q_1 and Q_2 are A_1 vibration types or species and are designated according to the symmetry operations which they represent as follows:

A	symmetric with respect to the principal axis of symmetry
B	antisymmetric with respect to the principal axis of symmetry
E	doubly degenerate vibrations, the irreducible representation is two dimensional, i.e., a 2×2 matrix (see Section 3.6)
F	triply degenerate vibrations, i.e., a three-dimensional representation
g and u (subscripts)	symmetric or antisymmetric with respect to a center of symmetry*
1 and 2 (subscripts)	symmetric or antisymmetric with respect to a rotation axis (C_p) or rotation-reflection axis (S_p) other than the principal axis or in those point groups with only one symmetry axis with respect to a plane of symmetry
prime and double prime (superscript)	symmetric or antisymmetric with respect to a plane of symmetry

These abbreviations are used except for linear molecules belonging to the point groups $C_{\infty v}$ and $D_{\infty h}$. For these two point groups the designations chosen are the same as for the electronic states of homonuclear diatomic molecules. Large Greek letters are used as follows:

Σ^+	symmetric with respect to a plane of symmetry through the molecular axis
Σ^-	antisymmetric with respect to a plane of symmetry through the molecular axis
Π, Δ, Φ	degenerate vibrations with a degree of degeneracy increasing in this order.

In designating symmetry species the order just listed is followed. For example, in the case of the C_{2v} point group the designations are A_1, A_2, B_1 and B_2 rather than A', A'', B', and B''.

In the case of the point group C_{3v} the characters of the irreducible representations are given in Table 3.8.

---

* g and u are taken from the German words *gerade* and *ungerade* meaning even and uneven, respectively.

TABLE 3.8

CHARACTERS OF THE IRREDUCIBLE REPRESENTATION OF THE C_{3v} POINT GROUP

	I	$C_3^+(z)$	$C_3^-(z)$	σ_{v1}	σ_{v2}	σ_{v3}	
(Γ_1) A_1	1	1	1	1	1	1	T_z
(Γ_2) A_2	1	1	1	-1	-1	-1	R_z
(Γ_3) E	2	-1	-1	0	0	0	T_x, T_y
							R_x, R_y

As in the C_{2v} case a given symmetry operation changes a nondegenerate normal coordinate (A_1 and A_2 types) into $+1$ times itself (symmetrical) or -1 times itself (antisymmetrical). For example, in the CH stretching vibrations of chloroform (C_{3v}) the carbon and hydrogen move along the symmetry axis which is coincident with the CH bond (see Fig. 3.9). This normal coordinate (A_1) is symmetrical with respect to all the symmetry elements and is unchanged by all symmetry operations shown in Fig. 3.5.

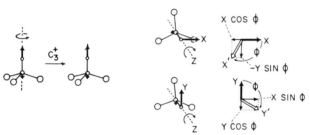

FIG. 3.9.   The effect of the C_3^+ symmetry operation on the C—H stretching and bending vibrations of $CHCl_3$. The stretching vibration remains unchanged but the displacement vectors of the doubly degenerate bending vibrations are each rotated $120°$ and are transformed to new vectors (white arrows) with components as indicated.

The situation for the doubly degenerate E vibrational type is best illustrated by an example. In the doubly degenerate CH, bending vibrations (E) in chloroform the hydrogen motions have two mutually perpendicular components (x and y) at right angles to the CH bond or symmetry axis z. As seen in Fig. 3.9, the C_3^+ operation, for example, rotates each component $120°$, thus transforming each component into a linear combination of both components so that

$$x \xrightarrow{\ C_3^+\ } x' = x \cos \phi - y \sin \phi \qquad (\cos \phi = -\tfrac{1}{2}) \qquad (3.18)$$

$$y \longrightarrow y' = x \sin \phi + y \cos \phi \qquad (\sin \phi = \sqrt{3}/2) \qquad (3.19)$$

It is seen that a normal coordinate which is a member of a degenerate set is transformed by a symmetry operation into a linear combination of the members of the set. In Eq. (3.7) the kinetic and potential energies were expressed in terms of normal coordinates, and it was shown that a symmetry operation must leave the energy unchanged and therefore cannot change the square of a nondegenerate normal coordinate. For two degenerate normal coordinates Q_{1a} and Q_{1b}, the frequency parameter λ_1 is the same for both in which case from Eq. (3.7)

$$T + V = \tfrac{1}{2}(\dot{Q}_{1a}^2 + \dot{Q}_{1b}^2) + \tfrac{1}{2}\lambda_1(Q_{1a}^2 + Q_{1b}^2) + \cdots \qquad (3.20)$$

Now it is only required that $(Q_{1a}^2 + Q_{1b}^2)$ be unchanged by a symmetry operation. Such a requirement can be met, for example, if Q_{1a} is changed by a symmetry operation into $(Q_{1a} \cos \phi - Q_{1b} \sin \phi)$, where ϕ is some angle, and Q_{1b} is changed into $(Q_{1a} \sin \phi + Q_{1b} \cos \phi)$ since, as the reader can verify,

$$(Q_{1a}^2 + Q_{1b}^2) = [(Q_{1a} \cos \phi - Q_{1b} \sin \phi)^2 + (Q_{1a} \sin \phi + Q_{1b} \cos \phi)^2] \quad (3.21)$$

Therefore we can write for all the doubly degenerate coordinates such as Q_{4a} and Q_{4b} (the CH bending coordinate)

$$Q_{4a} \xrightarrow{C_3^+} Q'_{4a} = Q_{4a} \cos \phi - Q_{4b} \sin \phi \qquad (\cos \phi = -\tfrac{1}{2}) \qquad (3.22)$$

$$Q_{4b} \longrightarrow Q'_{4b} - Q_{4a} \sin \phi + Q_{4b} \cos \phi \qquad (\sin \phi = \sqrt{3}/2) \qquad (3.23)$$

See Fig. 3.10 for a similar treatment for the CCl_3 bending coordinates. A two-by-two matrix is needed to characterize the transformation of pairs of E type normal coordinates for the C_3^+ and other operations. Equations (3.22) and (3.23) can be written in matrix form as

$$\begin{Vmatrix} Q'_a \\ Q'_b \end{Vmatrix} = \begin{Vmatrix} -\tfrac{1}{2} & -\sqrt{3/2} \\ +\sqrt{3/2} & -\tfrac{1}{2} \end{Vmatrix} \times \begin{Vmatrix} Q_a \\ Q_b \end{Vmatrix} \qquad (3.24)$$

and as discussed above this transformation matrix cannot be further reduced. The complete set of matrices forming the irreducible representation for E type normal coordinates for the C_{3v} group is given by

$$
\begin{array}{ccc}
I & C_3^+ & C_3^- \\[4pt]
\begin{Vmatrix} 1 & 0 \\ 0 & 1 \end{Vmatrix} &
\begin{Vmatrix} -\tfrac{1}{2} & -\sqrt{3/2} \\ +\sqrt{3/2} & -\tfrac{1}{2} \end{Vmatrix} &
\begin{Vmatrix} -\tfrac{1}{2} & +\sqrt{3/2} \\ -\sqrt{3/2} & -\tfrac{1}{2} \end{Vmatrix}
\end{array}
$$

$$(3.25)$$

$$
\begin{array}{ccc}
\sigma_{v_1} & \sigma_{v_2} & \sigma_{v_3} \\[4pt]
\begin{Vmatrix} 1 & 0 \\ 0 & -1 \end{Vmatrix} &
\begin{Vmatrix} -\tfrac{1}{2} & +\sqrt{3/2} \\ +\sqrt{3/2} & +\tfrac{1}{2} \end{Vmatrix} &
\begin{Vmatrix} -\tfrac{1}{2} & -\sqrt{3/2} \\ -\sqrt{3/2} & +\tfrac{1}{2} \end{Vmatrix}
\end{array}
$$

The characters of these matrices are listed in Table 3.8 in the E row.

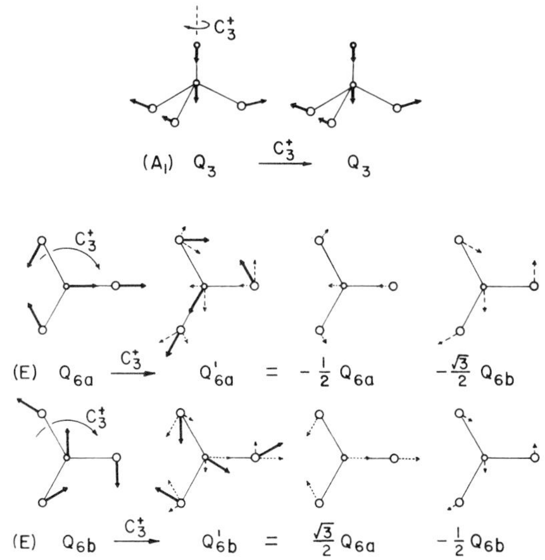

$(A_1) \quad Q_3 \xrightarrow{C_3^+} Q_3$

$(E) \quad Q_{6a} \xrightarrow{C_3^+} Q'_{6a} = -\frac{1}{2}Q_{6a} \quad -\frac{\sqrt{3}}{2}Q_{6b}$

$(E) \quad Q_{6b} \xrightarrow{C_3^+} Q'_{6b} = \frac{\sqrt{3}}{2}Q_{6a} \quad -\frac{1}{2}Q_{6b}$

Fig. 3.10. CCl$_3$ bending normal coordinates (schematic) for chloroform transformed by the C_3^+ operation. The A_1 bending coordinate Q_3 is transformed into $+1$ times itself by the C_3^+ operation. Each of the doubly degenerate E bending coordinates, Q_{6a} and Q_{6b}, is transformed by the C_3^+ operation into a linear combination of both coordinates. Top views of the doubly degenerate coordinates are illustrated where the central arrow represents both carbon and hydrogen displacements.

It can be noticed that the *character* of the matrix for the operation C_3^+ is always the same as that for C_3^- and also that the characters of the matrices for the operations σ_{v_1}, σ_{v_2}, and σ_{v_3} are always the same. Operations of this symmetry related sort are said to belong to the same *class*. The C_{3v} point group therefore has three different classes of symmetry operations I, C_3, and σ_v. In group theory a class is defined as a set of elements say A and B where

$$X^{-1}AX = A \text{ or } B \qquad X^{-1}BX = A \text{ or } B \qquad (3.26)$$

where X is, in turn, all the elements of the group. Since a similarity transformation does not change the character, the matrix representations of all the members of a class have the same characters. The point group character table is usually simplified by only listing the characters of the *classes* of symmetry operations rather than all the individual symmetry operations. The number of possible different irreducible representations is equal to the number of classes in the group which makes the character table a square array as seen in Table 3.9.

TABLE 3.9

CHARACTER TABLE FOR THE C_{3v} POINT GROUP

	(1) I	(2) C_3	(3) σ_v	
A_1	1	1	1	T_z
A_2	1	1	-1	R_z
E	2	-1	0	(T_x, T_y) (R_x, R_y)

3.8 Irreducible Representation Components in a Representation

From group theory there is a systematic method for calculating how many times a given irreducible representation appears in any reducible representation.[6,7] This is given by

$$N_i = \frac{1}{N_g} \sum_{\substack{\text{all} \\ \text{classes}}} N_e \chi(R) \chi_i(R) \tag{3.27}$$

Here $\chi_i(R)$ is the character of the irreducible representation type i for the operation (R) found in the character table; $\chi(R)$ is the character of the reducible representation for the operation (R) which is to be broken down into irreducible representation components; N_e is the number of elements in each class (the number in parenthesis in the first row of Table 3.9); N_g is the number of elements in the group, namely, the sum of the N_e values and N_i is the number of times the irreducible representation type i appears in the reducible representation being analyzed.

As an example let us consider the characters for the representation for the molecular motion of H_2O using cartesian coordinates as a basis in Eq. (3.12).

	I	C_2	σ_v	σ_v'	
χ_{H_2O}	$+9$	-1	$+1$	$+3$	(3.28)

This can be broken down using Eq. (3.27) and the C_{2v} character Table 3.7.

$$N_{A_1} = \tfrac{1}{4}[(1 \cdot 9 \cdot 1) + (1 \cdot -1 \cdot 1) + (1 \cdot 1 \cdot 1) + (1 \cdot 3 \cdot 1)] = 3$$
$$N_{A_2} = \tfrac{1}{4}[(1 \cdot 9 \cdot 1) + (1 \cdot -1 \cdot 1) + (1 \cdot 1 \cdot -1) + (1 \cdot 3 \cdot -1)] = 1$$
$$N_{B_1} = \tfrac{1}{4}[(1 \cdot 9 \cdot 1) + (1 \cdot -1 \cdot -1) + (1 \cdot 1 \cdot 1) + (1 \cdot 3 \cdot -1)] = 2$$
$$N_{B_2} = \tfrac{1}{4}[(1 \cdot 9 \cdot 1) + (1 \cdot -1 \cdot -1) + (1 \cdot 1 \cdot -1) + (1 \cdot 3 \cdot 1)] = 3$$

This result is summarized by saying that the reducible representation equation (3.28) contains irreducible representations in the following distribution:

$$\Gamma = 3A_1 + A_2 + 2B_1 + 3B_2 \tag{3.29}$$

This result can be checked by adding the appropriate products of characters from the C_{2v} character in Table 3.7 and comparing the sum with Eq. (3.28).

	I	C_2	σ_v	σ_v'
$3A_1$	3	3	3	3
A_2	1	1	-1	-1
$2B_1$	2	-2	2	-2
$3B_2$	3	-3	-3	3
χ_{H_2O}	9	-1	1	3

(add)

In principle a similarity transformation can be found which will change the cartesian coordinate transformation matrices [Eq. (3.10)] into the completely reduced normal coordinate transformation matrices which include translations and rotations [Eq. (3.17)]. Since a similarity transformation does not change the character, the character of the matrices using normal coordinates as a basis will be the same as Eq. (3.28), and the analysis of the representation will be the same as Eq. (3.29). Equation (3.17) also illustrates the fact that for each irreducible representation present in the total representation, there is one nondegenerate normal coordinate. Therefore the results from Eq. (3.29) indicates that for H_2O there are three normal coordinates of the A_1 type, one of the A_2 type, two of the B_1 type, and three of the B_2 type included in the total representation for the motions of N particles which is the representation using the $3N$ cartesian coordinates as a basis in Eq. (3.28). These numerical results can be verified by referring to Fig. 3.7. Since translation and rotation normal coordinates are included, the number of vibrational normal coordinates for each vibrational type can be obtained by subtracting the number of translations and rotations given in the character table from the above results. When this is done we are left with two A_1 and one B_2 vibrational types of normal coordinates for H_2O.

3.9 Transformation Properties of a Vector

The translation of the center of gravity of a molecule is a vector, and so is the dipole moment. Let us consider a vector with x, y, and z components and investigate its transformation properties when symmetry operations are performed on it. If such a vector is rotated about the z axis, the z component will remain unchanged. In Fig. 3.11 is illustrated the projection of this vector on the xy plane. The vector projection r starts in position A and is rotated through an angle ϕ to position B. From Fig. 3.11 we can write the following relationships:

$$r = r' \qquad x = r \cos \alpha \qquad y = r \sin \alpha$$
$$x' = r \cos(\alpha + \phi) \qquad y' = r \sin(\alpha + \phi) \qquad z' = z \tag{3.30}$$

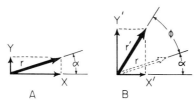

FIG. 3.11.   Rotation of a vector about a symmetry axis through an angle ϕ.

The following equations are trigonometric identities:

$$\cos(\alpha + \phi) = \cos \alpha \cos \phi - \sin \alpha \sin \phi \tag{3.31}$$

$$\sin(\alpha + \phi) = \cos \alpha \sin \phi + \sin \alpha \cos \phi \tag{3.32}$$

If both sides of both of these equations are multiplied by r, then by substituting the values from Eq. (3.30) we obtain

$$x' = x \cos \phi - y \sin \phi$$
$$y' = x \sin \phi + y \cos \phi \tag{3.33}$$
$$z' = z$$

These are the transformation equations for a proper rotation of a vector through an angle ϕ about a symmetry axis such as C_3. The identity operation is a proper rotation through zero degrees.

An improper rotation is a rotation followed by a reflection in a plane perpendicular to the axis of rotation. An example is the S_2 symmetry operation. The inversion through a center of symmetry is the same as an improper rotation through an angle of $180°$. A reflection by a plane of symmetry is an improper rotation through an angle of zero degrees. The transformation equations for an improper rotation are the same as those for a proper rotation except that $z' = -z$. Equation (3.33) can be written in matrix form as

$$\left\| \begin{array}{c} x' \\ y' \\ z' \end{array} \right\| = \left\| \begin{array}{ccc} \cos \phi & -\sin \phi & 0 \\ \sin \phi & \cos \phi & 0 \\ 0 & 0 & \pm 1 \end{array} \right\| \times \left\| \begin{array}{c} x \\ y \\ z \end{array} \right\| \tag{3.34}$$

where $+1$ and -1 are for proper and improper rotations, respectively. The character for the transformation of a vector for each symmetry operation is therefore

$$\chi = \pm 1 + 2 \cos \phi \qquad (3.35)$$

3.10 The Number of Fundamentals of Each Type

A method for deducing the number of normal coordinates of each type was outlined previously using the character table and the characters for the complete representation for molecular motion using cartesian coordinates as a basis. It is not necessary to construct the full transformation matrices to deduce the required characters. The first step is to recognize that in the representations using cartesian coordinates as a basis [Eqs. (3.9) and (3.10)] only atoms left undisplaced by a symmetry operation R can contribute to the character for that operation. Let the number of such atoms be U_R. While such an atom is undisplaced, its cartesian coordinate displacement vector components will be rotated through an angle ϕ by a symmetry operation as described in the previous section so each such atom contributes $(\pm 1 + 2 \cos \phi)$ to the character, and the total character for the operation R is

$$\chi = U_R (\pm 1 + 2 \cos \phi) \qquad (3.36)$$

This character includes translations and rotations. The translation of the center of gravity is a vector with x, y, and z components and therefore its character for the operation (R) is $(\pm 1 + 2 \cos \phi)$ as described in the previous section. The character for a rotation will not be developed here but is given by $(1 \pm 2 \cos \phi)$, where plus and minus are for proper and improper rotations, respectively.[5,8] The character in Eq. (3.36) minus the character for translation and rotation given above is the character $\chi_v(R)$ for the representation for true vibrations for the operation (R) and is given by

$$\chi_v(R) = (U_R - 2)(1 + 2 \cos \phi) \qquad (3.37)$$

for proper rotations and

$$\chi_v(R) = (U_R)(-1 + 2 \cos \phi) \qquad (3.38)$$

for improper rotations. Since $\chi_v(R)$ is not an irreducible representation, it must be reduced with the aid of Eq. (3.27), where $\chi_v(R)$ is substituted for $\chi(R)$ to give

$$N_i(V) = \frac{1}{N_g} \sum N_e \, \chi_v(R) \chi_i(R) \tag{3.39}$$

where $N_i(V)$ is the number of vibrational normal coordinates (and vibrational fundamentals) of vibration type i.

In $CHCl_3$, for example, by referring to Fig. 3.5 we can see that the number of atoms unshifted by the symmetry operations I, C_3, and σ_v are

$$U_I = 5 \qquad U_{C_3} = 2 \qquad U_{\sigma_v} = 3 \tag{3.40}$$

The identity operation is a proper rotation with $\phi = 0$, the C_3 operation is a proper rotation with $\phi = 120°$, and the σ_v operation is an improper rotation with $\phi = 0°$. The values needed are summarized in Table 3.10. These values

TABLE 3.10

CALCULATION OF $\chi_v(R)$

C_{3v}	(1) I	(2) C_3	(3) σ_v
ϕ	$0°$	$120°$	$0°$
$2\cos\phi$	2	-1	2
$\pm 1 + 2\cos\phi$	3	0	1
U_R	5	2	3
$\chi_v(R)$	9	0	3

for $\chi_v(R)$ are used with $\chi_i(R)$ values from the C_{3v} character Table 3.8 in Eq. (3.39) to get

$$\begin{aligned}
N_{A_1}(V) &= \tfrac{1}{6}[(1 \cdot 9 \cdot 1) + (2 \cdot 0 \cdot 1) + (3 \cdot 3 \cdot 1)] = 3 \\
N_{A_2}(V) &= \tfrac{1}{6}[(1 \cdot 9 \cdot 1) + (2 \cdot 0 \cdot 1) + (3 \cdot 3 \cdot -1)] = 0 \\
N_E(V) &= \tfrac{1}{6}[(1 \cdot 9 \cdot 2) + (2 \cdot 0 \cdot -1) + (3 \cdot 3 \cdot 0)] = 3 \\
\chi_v(R) &= 3\chi_{A_1}(R) + 0\chi_{A_2}(R) + 3\chi_E(R)
\end{aligned} \tag{3.41}$$

The $CHCl_3$ molecule therefore has three A_1 vibrations, no A_2 vibrations, and three doubly degenerate E vibrations giving a total of nine fundamental vibrations. It should be remembered that $N_E(V)$ is the number of times the irreducible representation of E type appears in the complete representation, and since the E type representation consists of two-by-two matrices there are two fundamental components in each E type representation.

3.11 Selection Rules

It has been stated that a molecular vibration will be infrared active only if the vibration causes a change in the dipole moment μ, and that the intensity of the infrared band will be proportional to the square of the change in dipole moment with respect to the normal coordinate $(\partial\mu/\partial Q)^2$. In quantum mechanical terms the intensity of an infrared band is proportional to the square of a related quantity called the *transition moment* whose x, y, and z components are[10,11]

$$\int_{-\infty}^{+\infty} \psi_f^* \mu_x \psi_i \, d\tau \qquad \int_{-\infty}^{+\infty} \psi_f^* \mu_y \psi_i \, d\tau \qquad \int_{-\infty}^{+\infty} \psi_f^* \mu_z \psi_i \, d\tau \qquad (3.42)$$

Here ψ_i is the wavefunction of the initial vibrational state and ψ_f^* is the wavefunction of the final vibrational state involved in the transition. (The asterisk superscript means that if ψ_f is complex, the complex conjugate of ψ_f is used). The quantities μ_x, μ_y, and μ_z are the components of the dipole moment and $d\tau$ is the volume element. In order for the transition $\psi_f \leftarrow \psi_i$ to be infrared active, at least one of the above integrals must be nonzero.

If a molecule has some symmetry, then for certain vibrational symmetry species all three integrals may be zero as a necessary consequence of the symmetry. Since the transition moment components in Eq. (3.42) are definite integrals over the whole configuration space of the molecule, they will be unchanged by an symmetry operation, since such operations are equivalent to the renumbering of symmetrically related atoms. Furthermore, the band intensity, which is proportional to the square of the transition moment, is an observable quantity and must have the same value for all indistinguishable orientations of the molecule. This means that if the integrals are to have nonzero values, then the product $\psi_f \mu \psi_i$ must be *totally symmetric*, that is, it must transform according to the totally symmetric irreducible representation. One advantage of this formulation is that the results are independent of whether or not anharmonicity is present.

The vibrational wavefunctions ψ as a function of a normal coordinate Q are illustrated in Fig. 3.12 (see also Fig. 1.12). These wavefunctions have symmetry properties related to those of the normal coordinate. For example, if a symmetry operation leaves the normal coordinate unchanged, the wavefunction will remain unchanged (regardless of anharmonicity). If a nondegenerate antisymmetrical normal coordinate Q is reversed in sign by a symmetry operation ($Q \rightarrow -Q$) then, as seen in Fig. 3.12, the wavefunction is left unchanged if the quantum number is even ($\psi_0 \rightarrow \psi_0$, $\psi_2 \rightarrow \psi_2$) or is reversed in sign *like the normal coordinate* when the quantum number is odd ($\psi_1 \rightarrow -\psi_1$, $\psi_3 \rightarrow -\psi_3$). This holds true even when anharmonicity is present.

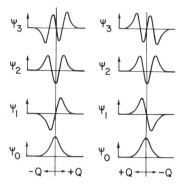

FIG. 3.12. The vibrational wavefunctions ψ as a function of a normal coordinate Q. The right-hand set shows the effect when an antisymmetrical nondegenerate normal coordinate is reversed in sign by a symmetry operation. When $Q \rightarrow -Q$ then $\psi_{even} \rightarrow \psi_{even}$ and $\psi_{odd} \rightarrow -\psi_{odd}$.

In the ground state the total vibrational wavefunction is a product of normal coordinate wavefunctions where all the quantum numbers are $v = 0$ (which is even). This ground state wavefunction is not changed by any symmetry operation and must belong to the totally symmetric species. In a fundamental transition, the molecule goes from the ground vibrational state to an excited state where the quantum number is $v = 1$ (which is odd) for one normal coordinate only and $v = 0$ (which is even) for all the other normal coordinates. In this case the wavefunction of the excited state must have the *same symmetry properties as that one excited normal coordinate.* Although not obvious from the above discussion, this holds true even when degenerate vibrations are present. Because of these symmetry properties, the wavefunctions can be used as bases for irreducible representations for the symmetry group of the molecule.

In order to investigate the symmetry properties of the product $(\psi_f \mu \psi_i)$ we can start by investigating the symmetry properties of the product $(\psi_f \psi_i)$ and then comparing these to the symmetry properties of the dipole moment μ. (These quantities commute.) In order to do this it is necessary to define the character of the *direct product* of representations. This is most easily done with an example where the direct product of the characters for the A_1 species (totally symmetric ground state) and the B_2 species (one type of excited state) for the C_{2v} group is obtained as shown below.

	I	C_2	σ_v	σ_v'	
$\chi_{A_1}(R)$	1	1	1	1	
$\chi_{B_2}(R)$	1	-1	-1	1	(multiply)
$\chi_{A_1 \times B_2}(R)$	1	-1	-1	1	

where it is seen that the character of the direct product $\chi_{A_1 \times B_2}(R)$ is the same as $\chi_{B_2}(R)$. The significance of this in quantum mechanics is that the direct product of the characters of the irreducible representations representing two wavefunctions (ψ_f and ψ_i) will yield the character of a representation (not necessarily irreducible) which represents the wavefunction product ($\psi_f \psi_i$). If ψ_i is the totally symmetric ground state (where all the characters are $+1$) then the wavefunction product ($\psi_f \psi_i$) belongs to the same species as that for the excited state ψ_f, which for a fundamental transition is the same as that for the excited normal coordinate.

In order for the transition moment [Eq. (3.42)] to be nonzero it was pointed out that the product ($\psi_f \psi_i \mu$) must belong to the totally symmetric species. This product will be totally symmetric only if the wavefunction product ($\psi_f \psi_i$) belongs to the *same* species as at least one component of the dipole moment μ. This is because the direct product of the characters of an irreducible representation (representing $\psi_f \psi_i$) times an *identical* set of characters (representing μ) yields or contains the characters of the totally symmetric representation. For example, in a nondegenerate species, the square of any character $(\pm 1)^2$ equals $+1$, the character of the totally symmetric species. These arguments are summarized in a general seleection rule[10] as follows.

A vibrational transition between energy levels is allowed in the infrared spectrum when at least one component of the dipole moment vector has the same species as the product of the quantum mechanical wavefunctions which characterize the molecular state for the two levels involved. For example, in order for a fundamental transition to be infrared active, at least one component of the dipole moment should belong to the same species as ($\psi_f \psi_i$) which is the same as that of the excited normal coordinate.

In order for a transition to be Raman active, an integral of the type $\int \psi_f^* \alpha_{gg'} \psi_i \, d\tau$ must be nonzero where $\alpha_{gg'}$ is one component of the polarizability g and g' are x, y, or z. This leads to a selection rule for Raman activity identical to the one above for infrared activity, but with the polarizability tensor substituted for the dipole moment vector. The species of the dipole moment vector and the polarizability tensor will be examined next.

3.12 Infrared Activity of Fundamentals

The dipole moment is a vector whose transformation under symmetry operations was studied in Section 3.9. The character $\chi_M(R)$ for the transformation of the dipole moment vector for the operation R is given by Eq. (3.35) as

$$\chi_M(R) = \pm 1 + 2 \cos \phi \qquad (3.43)$$

The character of the representation of the dipole moment vector for each symmetry operation in the C_{3v} point group, for example, is given in Table 3.11.

TABLE 3.11

CALCULATION OF $\chi_M(R)$

C_3	(1)I	(2)C_3	(3)σ_v	
ϕ	0	120°	0	
$2\cos\phi$	2	-1	2	
± 1	1	1	-1	
$\pm 1 + 2\cos\phi$	3	0	1	$= \chi_M(R)$

Since $\chi_M(R)$ is not like any of the irreducible representations in the C_{2v} character table (Table 3.8) it must be reduced with the aid of Eq. (3.27) where $\chi_M(R)$ is substituted for $\chi(R)$.

$$N_i(M) = \frac{1}{N_g} \sum N_e \chi_M(R)\chi_i(R) \tag{3.44}$$

In this equation $N_i(M)$ is the number of times the character $\chi_i(R)$ of the vibration type from the character table appears in $\chi_M(R)$. In the C_{3v} case we use the character table (Table 3.8) and $\chi_M(R)$ values from Table 3.10 to get

$$
\begin{aligned}
N_{A_1}(M) &= \tfrac{1}{6}[(1\cdot 3\cdot 1) + (2\cdot 0\cdot 1) + (3\cdot 1\cdot 1)] &= 1\\
N_{A_2}(M) &= \tfrac{1}{6}[(1\cdot 3\cdot 1) + (2\cdot 0\cdot 1) + (3\cdot 1\cdot -1)] &= 0\\
N_E(M) &= \tfrac{1}{6}[(1\cdot 3\cdot 2) + (2\cdot 0\cdot -1) + (3\cdot 1\cdot 0)] &= 1\\
\chi_M(R) &= \chi_{A_1}(R) + \chi_E(R) &
\end{aligned}
\tag{3.45}
$$

Only those vibration types whose characters are contained in $\chi_M(R)$ are infrared active. Therefore, fundamental vibrations of type A_1 and E are allowed and type A_2 are forbidden in the infrared spectrum of any molecule which belongs to the C_{3v} point group.

In the last column of the character table the three components of translation (T_x, T_y, T_z) are listed. A vibrational species will be infrared active if one or more of the three components of translation are listed in the row for that species in the character table. This is because the dipole moment and the translation of the center of gravity are both vectors and are transformed by symmetry operations in the same way. Furthermore, the direction of the translation listed $(T_x$, for example) indicates the direction (x) of the change in dipole moment for that vibrational species.

3.13 Raman Activity of Fundamentals

The polarizability is a tensor which, like a vector, has transformation properties under symmetry operations. The character $\chi_\alpha(R)$ for the transformation of the polarizability tensor for the operation R will not be derived here but is given by[5,11]

$$\chi_\alpha(R) = 2 \pm 2 \cos \phi + 2 \cos 2\phi \qquad (3.46)$$

where the plus sign is used for proper rotations and the minus sign for improper rotations. The character of the representation of the polarizability tensor for each symmetry operation in the C_{3v} point group, for example, is given in Table 3.12.

TABLE 3.12

CALCULATION OF $\chi_\alpha(R)$

C_3	$(1)I$	$(2)C_3$	$(3)\sigma_v$	
ϕ	0	120°	0	
$\pm 2 \cos$	2	−1	−2	
$2 \cos 2\phi$	2	−1	2	
$2 \pm \cos \phi + 2 \cos 2\phi$	6	0	2	$= \chi_\alpha(R)$

Since $\chi_\alpha(R)$ is not like any of the irreducible representations in the C_{3v} character table it must be reduced with the aid of Eq. (3.27) where $\chi_\alpha(R)$ is substituted for $\chi(R)$.

$$N_i(\alpha) = \frac{1}{N_g} \sum N_e \chi_\alpha(R)\chi_i(R) \qquad (3.47)$$

In this equation $N_i(\alpha)$ is the number of times the character $\chi_i(R)$ of the vibration type from the character table (Table 3.8) appears in $\chi_\alpha(R)$. In the C_{3v} case we use the character table (Table 3.8) and $\chi_\alpha(R)$ values from Table 3.11 to get

$$
\begin{aligned}
N_{A_1}(\alpha) &= \tfrac{1}{6}[(1 \cdot 6 \cdot 1) + (2 \cdot 0 \cdot 1) + (3 \cdot 2 \cdot 1)] &= 2 \\
N_{A_2}(\alpha) &= \tfrac{1}{6}[(1 \cdot 6 \cdot 1) + (2 \cdot 0 \cdot 1) + (3 \cdot 2 \cdot -1)] = 0 \\
N_E(\alpha) &= \tfrac{1}{6}[(1 \cdot 6 \cdot 2) + (2 \cdot 0 \cdot -1) + (3 \cdot 2 \cdot 0)] = 2 \\
\chi_\alpha(R) &= 2\chi_{A_1}(R) + 2\chi_E(R)
\end{aligned}
\qquad (3.48)
$$

Only those vibration types whose characters are contained in $\chi_\alpha(R)$ are active in the Raman effect. Therefore, fundamental vibrations of type A_1 and E are allowed, and type A_2 are forbidden in the Raman spectrum of any molecule which belongs to the C_{3v} point group.

In the last column of some character tables, the components of the polarizability (α_{xx}, α_{xy}, ...) which change during a vibration are listed in the row for Raman active vibrational species. In character tables of groups with degenerate species, linear combinations such as $\alpha_{xx} + \alpha_{yy}$ or $\alpha_{xx} - \alpha_{yy}$ may be listed. Degenerate pairs of polarizability components are enclosed in parenthesis.

3.14 Overtone and Combination Bands

Since molecular vibrations in general are slightly anharmonic, both the infrared and Raman spectrum may contain weak overtone and combination bands. A *combination* energy level is one which involves two or more normal coordinates with different frequencies that have vibrational quantum numbers greater than zero. For example, a combination band which appears at the sum of the wavenumbers of two different fundamentals involves a transition from the ground vibrational level (belonging to the totally symmetric species) to an excited combination level where two different normal coordinates each have a quantum number of one and all the others have a quantum number zero. To obtain the spectral activity of the combination band transition it is necessary to determine the symmetry species of the excited wavefunction. In quantum mechanics the total vibrational wavefunction is equal to the product of the wavefunctions for each normal coordinate. For the combination band transition this means that the symmetry species of the excited wave function will be the same as that of the *direct product* of the characters of the vibrational symmetry species of the two (or more) normal coordinates involved. An example for the C_{3v} group is evaluated as follows:

	I	C_3	σ_v
$\chi_{A_1}(R)$	1	1	1
$\chi_E(R)$	2	-1	0
$\chi_{A_1 \times E}(R)$	2	-1	0

(multiply)

In this case the character of the direct product $\chi_{A_1 \times E}(R)$ is the same as $\chi_E(R)$. Whenever an E type fundamental transition is allowed in the infrared and Raman spectrum then the combination band transition involving A_1 and E type normal coordinates will also be allowed.

The selection rules for the remaining binary combinations for the C_{3v} group may be obtained except for $E \times E$. In this case

	I	C_3	σ_v
$\chi_E(R)$	2	-1	0
$\chi_E(R)$	2	-1	0
$\chi_{E \times E}(R)$	4	1	0

(multiply)

Since $\chi_{E \times E}$ does not have the character of A_1, A_2, or E, it contains more than one component and the reduction formula [Eq. (3.27)] is used with the character of the direct product replacing $\chi(R)$, and where $N_i(EE)$ is the number of times $\chi_i(R)$ appears in $\chi_{E \times E}(R)$.

$$N_{A_1}(EE) = \tfrac{1}{6}\{(1) \cdot 4 \cdot 1 + (2) \cdot 1 \cdot 1 + (3) \cdot 0 \cdot 1\} = 1$$
$$N_{A_2}(EE) = \tfrac{1}{6}\{(1) \cdot 4 \cdot 1 + (2) \cdot 1 \cdot 1 + (3) \cdot 0 \cdot 1\} = 1$$
$$N_E(EE) = \tfrac{1}{6}\{(1) \cdot 4 \cdot 2 + (2) \cdot 1 \cdot -1 + (3) \cdot 0 \cdot 0\} = 1$$

Therefore, $E \times E = A_1 + A_2 + E$ which can be verified by noting that the sum of the characters of each class in Table 3.9 equals $\chi_{E \times E}(R)$ found above. If any of the three components A_1, A_2, or E is allowed, then the $E \times E$ combination is allowed. In this case the $E \times E$ combination is allowed in both the infrared and Raman spectrum. The possible binary combinations for the C_{3v} group are summarized in Table 3.13.

An overtone transition originates from the ground vibrational level and goes to an overtone level where, for one normal coordinate only, the quantum number is greater than one and all the other quantum numbers are zero. As was stated earlier the symmetry species of the excited state for a nondegenerate normal coordinate will belong to the totally symmetric species if the vibrational quantum number is even, and will belong to the same species as the excited normal coordinate if its quantum number is odd (see Fig. 3.12). For nondegenerate normal coordinates, an overtone transition where the final quantum number is odd will be allowed in the spectrum if the fundamental for that normal coordinate is allowed. If the final quantum number is even, an overtone transition for a nondegenerate normal coordinate will be allowed in the spectrum if the fundamental of the totally symmetric species is allowed. In the C_{3v} case, for example, the selection rules for both the infrared and Raman spectrum for the overtones of the type A_2 vibration can be summarized as

$$A_2^n = \begin{cases} n & \text{even—allowed} \\ n & \text{odd—forbidden} \end{cases}$$

In a similar manner it is obvious that the A_1 overtones are allowed for both even and odd values for n.

The spectroscopic activity for overtones of *degenerate* vibrations involve a more complex analysis and we will simply present an appropriate procedure. For doubly degenerate vibrations, the following equation is applied.[11]

$$\chi_E^n(R) = \tfrac{1}{2}[\chi_E^{n-1}(R)\chi_E(R) + \chi_E(R^n)] \qquad (3.49)$$

where $\chi_E(R)$ is the character corresponding to the operation R performed n times in succession. In order to evaluate χ_E^2 for the C_{3v} group one must determine $\chi_E(I^2)$, $\chi_E(C_3^2)$, and $\chi_E(\sigma_v^3)$. In this case $\chi_E(I^2)$ is the same as $\chi_E(I)$ and equals 2. Carrying out the operation C ($\pm 120°$) twice is the same as only carrying it out once in the opposite direction, therefore $\chi_E(C_3^2) = \chi_E(C_3) = -1$. The operation σ_v performed twice is the same as performing the identity operation, and therefore $\chi_E(\sigma_v^2) = \chi_E(I) = 2$. The values of $\chi_E(R)^n$ up to $n = 5$ are as follows:

	I	C_3	σ_v
$\chi_E(R)^2$	2	-1	2
$\chi_E(R)^3$	2	2	0
$\chi_E(R)^4$	2	-1	2
$\chi_E(R)^5$	2	-1	0

Using Eq. (3.49), the character of the first overtone $\chi_E^2(R)$ is obtained as follows:

	I	C_3	σ_v	
$\chi_E(R)$	2	-1	0	
$\chi_E(R)$	2	-1	0	
				(multiply)
$\chi_E(R)^2$	4	1	0	
$\chi_E(R)^2$	2	-1	2	
				(add)
	6	0	2	
				(divide by 2)
$\chi_E^2(R)$	3	0	1	

Since $\chi_E^2(R)$ does not have the character of A_1, A_2, or E, it contains more than a single component, and the reduction formula of Eq. (3.27) is used with $\chi_E^2(R)$ replacing $\chi(R)$ and

$$N_{A_1}(E^2) = \tfrac{1}{6}\{(1) \cdot 3 \cdot 1 + (2) \cdot 0 \cdot 1 + (3) \cdot 1 \cdot 1\} = 1$$
$$N_{A_2}(E^2) = \tfrac{1}{6}\{(1) \cdot 3 \cdot 1 + (2) \cdot 0 \cdot 1 + (3) \cdot 1 \cdot -1\} = 0$$
$$N_{E}(E^2) = \tfrac{1}{6}\{(1) \cdot 3 \cdot 2 + (2) \cdot 0 \cdot -1 + (3) \cdot 1 \cdot 0\} = 1$$

Therefore $E^2 = A_1 + E$, which can be verified by noting that the sum of the A_1 and E characters in Table 3.9 yield $\chi_E^2(R)$. Since A_1 and E type fundamentals are allowed in both infrared and Raman spectroscopy, the first overtone of E type normal coordinates are also allowed.

A difference band appears at a wavenumber equal to the difference between the wavenumber of two different fundamentals. In this case the initial state is not the ground state but is one where for one normal coordinate the quantum number is one and the other quantum numbers are zero. The final state is one where some other normal coordinate with a higher frequency has a quantum number of one and the other quantum numbers are all zero. Each of the two levels involved has a wavefunction with the same symmetry as its normal coordinate whose quantum number is equal to one. From the selection rule given earlier, the spectral activity of the difference band transition is evaluated by determining the symmetry species of the direct product of the characters of the vibrational symmetry species of the two normal coordinates involved. Mechanically, this procedure is identical to that given

TABLE 3.13

SELECTION RULES FOR C_{3v} MOLECULES

Vibration type	Infrared[a]	Raman[a]
A_1	a	a
A_2	f	f
E	a	a
$A_1 \times A_1$	a	a
$A_1 \times A_2$	f	f
$A_1 \times E$	a	a
$A_2 \times A_2$	a	a
$A_2 \times E$	a	a
$E \times E$	a	a
A_1^n	a	a
$A_2^n (n$ even$)$	a	a
$A_2^n (n$ odd$)$	f	f
E^n	a	a

[a] a, allowed; f, forbidden.

in the example for the combination band transition, which means that any time a combination band transition is allowed the difference band transition involving the same two normal coordinates will also be allowed.

The selection rules which apply to any molecule belonging to the C_{3v} point group are summarized in Table 3.13.

3.15 Symmetry Coordinates

The internal coordinates of H_2O are illustrated in the top of the first column of Fig. 3.13. These are S_1 and S_2, each of which is one bond length minus the equilibrium bond length, and ϕ which is the bond angle minus the equilibrium bond angle. Symmetry coordinates are appropriate linear combinations of symmetry related internal coordinates. The symmetry related internal coordinates in H_2O are S_1 and S_2 and trial combinations of these coordinates are $(S_1 + S_2)$ and $(S_1 - S_2)$. These are illustrated in the bottom of the first column of Fig. 3.13 along with the unique coordinate ϕ. In the second, third, and fourth column of Fig. 3.13 we show the results of performing the C_2, σ_v, or σ_v' symmetry operations of the C_{2v} group on the coordinates in column one. In the top two rows it can be seen that the I and σ_v' operations leave S_1 and S_2 unchanged while the C_2 and σ_v' operations change S_1 into S_2 and S_2 into S_1. In the bottom row it can be seen that the combination coordinate $(S_1 - S_2)$ is unchanged by the I and σ_v operations

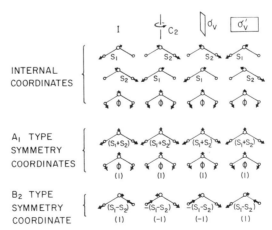

FIG. 3.13. Internal coordinates and symmetry coordinates for H_2O and their transformations by the symmetry operations of the C_{2v} group. These nondegenerate symmetry coordinates are either left unchanged or are changed in sign by the symmetry operation.

but is reversed in sign by the C_2 and σ_v operations so that $(S_1 - S_2)$ is changed to $(\pm 1)(S_1 - S_2)$. The factor (± 1) is like the appropriate character in the B_2 irreducible representation of the C_{2v} group $(1, -1, -1, 1)$. Except for a normalization factor (see Chapter 14), the coordinate $(S_1 - S_2)$ is a B_2 type symmetry coordinate of the C_{2v} group. It can be seen that the coordinates $(S_1 + S_2)$ and ϕ are unchanged by all the symmetry operations so that $(S_1 + S_2)$ is changed to $(1)(S_1 + S_2)$. The factor (1) is like the characters in the A_1 irreducible representation of the C_{2v} group. Except for a normalization factor these are A_1 type symmetry coordinates of the C_{2v} group.

If a molecule is performing one normal mode of vibration, then each internal coordinate is oscillating with an amplitude in specified proportion to the amplitude of the normal coordinate for that vibration (see Fig. 1.3). When a vibration belongs to a particular symmetry species, it is also true that each symmetry coordinate of that particular symmetry species oscillates with an amplitude in specified proportion to the amplitude of the normal coordinate of that same symmetry species. The normal coordinate contains no contributions from symmetry coordinates belonging to symmetry species other than its own. This is the main advantage in using symmetry coordinates.

For example, the in-phase stretch symmetry coordinate and the bending coordinate of water belong to the same symmetry species, which means the "in-phase stretch" vibration may bend the molecule somewhat and vice versa. However, the out-of-phase stretch vibration (for small amplitudes) does not change the bond angle at all since these coordinates belong to different symmetry species. Furthermore, since there is only one B_2 symmetry coordinate for water, this means that the B_2 normal coordinate for water is identical to the B_2 out-of-phase stretch symmetry coordinate for water.

3.16 Isotope Effects and the Product Rule

The relationship between nuclear mass and vibrational frequency has been introduced earlier. If we assume, as is generally possible, that the force constants are unaltered by isotope substitution, the shift in observed frequency can be attributed principally to mass effects. Qualitatively this shift will aid in the assignment of frequencies simply by observing what bands are altered and to what extent they are altered by isotopic substitution. Vibrations which involve large amplitudes of the atom of the substituted species should shift appreciably, and the larger the isotopic mass difference the greater will be the shift. Deuterium substitution gives a particularly large shift.

Consider a diatomic molecule such as HCl. The wavenumber ratio of the normal and isotopic species (primed) is

$$\frac{\bar{v}'_e}{\bar{v}_e} = \frac{1303\sqrt{F/u'}}{1303\sqrt{F/u}} = \sqrt{\frac{u}{u'}} \qquad \text{where} \qquad u = \frac{m_1 m_2}{m_1 + m_2} \qquad (3.50)$$

$$\frac{\bar{v}'_e}{\bar{v}_e} = \sqrt{\frac{M'}{M}} \sqrt{\frac{m_1 m_2}{m'_1 m'_2}} \qquad \text{where} \qquad M \text{ (mol. wt.)} = m_1 + m_2 \qquad (3.51)$$

Harmonic zero order wavenumbers $(\bar{v}_e)$ are used because the diatomic formula was derived assuming a harmonic potential function. For $^1H^{35}Cl$ and the deuterated isotope $^2H^{35}Cl$ the wavenumber ratio calculated from the above equation is

$$\left(\frac{\bar{v}'_e}{\bar{v}_e}\right)_{\text{calc}} = 0.717 \qquad (3.52)$$

The observed wavenumbers $(\bar{v})$ are 2886 and 2091 cm^{-1} and the derived harmonic zero order wavenumbers $(\bar{v}_e)$ [see Eq. (1.44)] are 2990 and 2144 m^{-1} for $^1H^{35}Cl$ and $^2H^{35}Cl$, respectively, so

$$\frac{\bar{v}'}{\bar{v}} = 0.725 \qquad \text{and} \qquad \frac{\bar{v}'_e}{\bar{v}_e} = 0.717 \qquad (3.53)$$

These results can be compared with the calculated value in Eq. (3.52). The observed wavenumber ratio $(\bar{v}'/\bar{v})$ which is uncorrected for anharmonicity departs slightly from the predicted value.

A general quantitative relation involving the isotope effect, similar in form to Eq. (3.53), is the Teller-Redlich product rule. It is based on the assumption that the product of the zero order wavenumber ratios $(\bar{v}'_e/\bar{v}_e)$ for all vibrations of a given symmetry type is independent of the potential constants and depends only on the relative masses of the atoms and on the geometrical structure of the molecule. The general relation is

$$\frac{\bar{v}'_1 \bar{v}'_2 \cdots \bar{v}'_i}{\bar{v}_1 \bar{v}_2 \cdots \bar{v}_i} = \left[\left(\frac{M'}{M}\right)^t \left(\frac{I'_x}{I_x}\right)^{r_x} \left(\frac{I'_y}{I_y}\right)^{r_y} \left(\frac{I'_z}{I_z}\right)^{r_z} \left(\frac{m_1}{m'_1}\right)^{N_1} \left(\frac{m_2}{m'_2}\right)^{N_2} \cdots \left(\frac{m_j}{m'_j}\right)^{N_j}\right] \qquad (3.54)$$

where all factors pertaining to isotopic species are primed and $\bar{v}_1, \bar{v}_2, \cdots, \bar{v}_i$ represent the harmonic wavenumbers of the i genuine vibrations of one symmetry type. If the symmetry type is doubly degenerate the doubly degenerate wavenumber $\bar{v}$ is still entered only once in Eq. (3.54). M is the molecular weight, and I_x, I_y, and I_z are the three moments of inertia of the molecule (equivalent to I_a, I_b, and I_c used earlier in this book). The t and r_x, r_y, and r_z values are the number of nongenuine vibrations (translations and rotations) which belong to the vibrational species considered and may be obtained directly from the character table. The value for t is the number of times T_x, T_y, or T_z appears in the species considered except for degenerate species. In the doubly degenerate E species, for example, T_x and T_y are considered as *one* degenerate pair and only one is counted. The r_x value is either 1 or 0 depending on whether or not R_x belongs to the vibrational species considered, and the same relationship holds for r_y and R_y and also r_z and R_z. However if there are degenerate species such as E then $(I_x'/I_x)^{r_x} = (I_y'/I_y)^{r_y}$ and only one of these moment of inertia ratios is used.

The m_j value is the mass of one representative atom j, of a set of identical symmetry related atoms, whose positions are exchanged by symmetry operations. The N_j value can be looked on as the number of degrees of freedom that the atoms of a symmetrically related set can contribute to the vibrational species considered. For example, for the A_1 species for $CHCl_3$, the C and H atoms are constrained to move along the symmetry axis so each atom has only one degree of freedom of motion for this species. Each chlorine atom is constrained to move only in the plane defined by the H—C—Cl atoms and therefore contributes two degrees of freedom for this species. Once the motion of one chlorine is specified, the motions of the other chlorines are fixed because of the symmetry relationships so they do not contribute additional degrees of freedom. For the E species, in one of the degenerate components the C and H atoms are constrained to move perpendicular to the symmetry axis in the x direction say, and each atom contributes one degree of freedom to one component of the E species. The chlorine set contributes three degrees of freedom to one component of the E species.

Group theory provides a systematic procedure for calculating the N_j values. The N_j value can be looked on as the number of vibrations (including nongenuine ones) which each set of atoms contributes to the vibrational species considered and are given by the reduction formula [Eq. (3.27)]

$$N_j = \frac{1}{N_g} \sum N_e \chi_j(R)\chi_i(R) \tag{3.55}$$

where $\chi_j(R)$ values are given by

$$\chi_j(R) = U_j(R)(\pm 1 + 2 \cos \phi) \tag{3.56}$$

where $+1$ or -1 is used for proper or improper rotations, respectively. The $U_j(R)$ values are the number of atoms j of each set left unchanged by each symmetry operation R. Equation (3.56) may be compared to Eq. (3.36) to which it is related.

Table 3.14 contains the C_{3v} character table and related data needed for the isotope effect for $CHCl_3$.

TABLE 3.14

C_{3v} CHARACTER TABLE AND DATA FOR $CHCl_3$ ISOTOPE EFFECT

C_{3v}	I	$(2)C_3$	$(3)\sigma_v$	
A_1	1	1	1	T_z
A_2	1	1	-1	R_z
E	2	-1	0	$(T_x, T_y); (R_x, R_y)$
ϕ	0	120	0	
$\cos\phi$	1	$-\frac{1}{2}$	1	
$U_C(R)$	1	1	1	
$U_H(R)$	1	1	1	
$U_{Cl}(R)$	3	0	1	
$\chi_C(R)$	3	0	1	
$\chi_H(R)$	3	0	1	
$\chi_{Cl}(R)$	9	0	1	

The N_j values can now be obtained from Eq. (3.57). For the carbon atom, the only one in the set, the N_j value for the A_1 species is

$$N_C = \tfrac{1}{6}[(1 \cdot 3 \cdot 1) + (2 \cdot 0 \cdot 1) + (3 \cdot 1 \cdot 1)] = 1 \qquad (3.57)$$

and in the same way the other N_j values for the A_1 species are

$$N_H = 1 \quad \text{and} \quad N_{Cl} = 2 \qquad (3.58)$$

The corresponding values for the E species (one component) are

$$N_C = 1 \quad N_H = 1 \quad \text{and} \quad N_{Cl} = 3 \qquad (3.59)$$

The three A_1 vibrations for $CHCl_3$ are 3019, 668, and 368 cm^{-1} and for $CDCl_3$ are 2255, 650, and 366 cm^{-1}. The product rule [Eq. (3.54)] can now be applied to the A_1 vibrations using the observed wavenumbers

$$\frac{\bar{v}_1' \bar{v}_2' \bar{v}_3'}{\bar{v}_1 \bar{v}_2 \bar{v}_3} \cong \left[\left(\frac{M'}{M}\right)\left(\frac{m_H}{m_H'}\right)\right]^{1/2} \qquad (3.60)$$

$$\frac{(2255)(650)(366)}{(3019)(668)(368)} \cong \left[\left(\frac{120.4}{119.4}\right) \left(\frac{1.008}{2.015}\right) \right]^{1/2} \tag{3.61}$$

$$0.723 \cong 0.710 \tag{3.62}$$

The agreement is not exact because the observed wavenumbers were used rather than zero order harmonic wavenumbers. In a like manner for the E vibrations Eq. (3.54) becomes

$$\frac{\bar{v}_4' \, \bar{v}_5' \, \bar{v}_6'}{\bar{v}_4 \, \bar{v}_5 \, \bar{v}_6} \cong \left[\left(\frac{M'}{M}\right) \left(\frac{I_x'}{I_x}\right) \left(\frac{m_H}{m_H'}\right) \right]^{1/2} \tag{3.63}$$

$$\frac{(905)(740)(262)}{(1216)(757)(261)} \cong \left[\left(\frac{120.4}{119.4}\right) \left(\frac{258.2}{254.1}\right) \left(\frac{1.008}{2.015}\right) \right]^{1/2} \tag{3.64}$$

$$0.730 \cong 0.715 \tag{3.65}$$

where 258.2 and 254.1 (both times 10^{-40} gm cm^2) are one of the two equal moments of inertia for $CDCl_3$ and $CHCl_3$, respectively (see Section 1.24). The good agreement in both cases helps confirm the given assignment for the six fundamental vibrations. The moments of inertia may be calculated by the method of Hirshfelder.[13]

3.17 Character Tables and Selection Rules

Table 3.15 lists the character tables and selection rules for important point groups. Polyatomic linear molecules which belong to the $C_{\infty v}$ and $D_{\infty h}$ groups are a special case. There are an infinite number of rotations about the axis on which the nuclei lie which yield an equivalent configuration of the molecule. The determination of the selection rules is based on the application of a reduction formula which for nonlinear molecules involves a summation over all the elements of a finite group. It is not possible to make these calculations in the same way for linear molecules because one would have to sum over an infinite number of elements. Therefore, for continuous groups such as $C_{\infty v}$ and $D_{\infty h}$ it is necessary to replace the summation with an integration. See Ferigle and Meister for this treatment.[14]

[13] J. O. Hirshfelder, *J. Chem. Phys.* **8**, 431 (1940).
[14] S. M. Ferigle and A. G. Meister, *Amer. J. Phys.* **20**, 421 (1952).

TABLE 3.15

SYMMETRY ELEMENTS, SYMMETRY TYPES (OR SPECIES), AND CHARACTERS AND SELECTION RULES

(The vibrations belonging to the totally symmetric species give rise to polarized Raman bands. E = doubly degenerate; F = triply degenerate)

Point Group C_1

Symmetry elements
   I

Symmetry types and characters

C_1	I
A	1

Selection rules (forbidden vibrations)
   Infrared, none
   Raman, none

Point Group C_2

Symmetry elements
   I, $C_2(z)$

Symmetry types and characters

C_2	I	$C_2(z)$		
A	1	1	T_z, R_z	α_{xx}, α_{yy}, α_{zz}, α_{xy}
B	1	-1	T_x, T_y; R_x, R_y	α_{yz}, α_{xz}

Selection rules (forbidden vibrations)
   Infrared, none
   Raman, none

Point Group C_{1v}, C_{1h}, or C_s

Symmetry elements
   I, $\sigma(xy)$

Symmetry types and characters

$C_{1v} \equiv C_{1h} \equiv C_s$	I	$\sigma(xy)$		
A'	1	1	T_x, T_y; R_z	α_{xx}, α_{yy}, α_{zz}, α_{xy}
A''	1	-1	T_z; R_x, R_y	α_{yz}, α_{xz}

Selection rules (forbidden vibrations)
   Infrared, none
   Raman, none

TABLE 3.15 (continued)

Point Group C_{2v}

Symmetry elements
  I, $C_2(z)$, $\sigma_v(xz)$, $\sigma_v(yz)$

Symmetry types and characters

C_{2v}	I	$C_2(z)$	$\sigma_v(xz)$	$\sigma_v'(yz)$		
A_1	1	1	1	1	T_z	$\alpha_{xx}, \alpha_{yy}, \alpha_{zz}$
A_2	1	1	-1	-1	R_z	α_{xy}
B_1	1	-1	1	-1	T_x ; R_y	α_{xz}
B_2	1	-1	-1	1	T_y ; R_x	α_{yz}

Selection rules (forbidden vibrations)
  Infrared, A_2, $A_1 \times A_2$, $B_1 \times B_2$, A_2^n (n odd)
  Raman, none

Point Group C_{3v}

Symmetry elements
  I, $2C_3(z)$, $3\sigma_v$

Symmetry types and characters

C_{3v}	I	$2C_3(z)$	$3\sigma_v$		
A_1	1	1	1	T_z	$\alpha_{xx} + \alpha_{yy}, \alpha_{zz}$
A_2	1	1	-1	R_z	
E	2	-1	0	$(T_x, T_y); (R_x, R_y)$	$(\alpha_{xx} - \alpha_{yy}, \alpha_{xy}), (\alpha_{yz}, \alpha_{xz})$

Selection rules (forbidden vibrations)
  Infrared, A_2, $A_1 \times A_2$, A^n (n odd)
  Raman, A_2, $A_1 \times A_2$, A_2^n (n odd)

Point Group C_{4v}

Symmetry elements
  I, $2C_4(z)$, $C_4^2 \equiv C_2''$, $2\sigma_v$, $2\sigma_d$

Symmetry types and characters

C_{4v}	I	$2C_4(z)$	$C_4^2 \equiv C_2''$	$2\sigma_v$	$2\sigma_d$		
A_1	1	1	1	1	1	T_z	$\alpha_{xx} + \alpha_{yy}, \alpha_{zz}$
A_2	1	1	1	-1	-1	R_z	
B_1	1	-1	1	1	-1		$\alpha_{xx} - \alpha_{yy}$
B_2	1	-1	1	-1	1		α_{xy}
E	2	0	-2	0	0	$(T_x, T_y); (R_x, R_y)$,	$(\alpha_{yz}, \alpha_{xz})$

Selection rules (forbidden vibrations)

Infrared	Raman
A_2, B_1, B_2, $A_1 \times A_2$, $A_1 \times B_1$, $A_1 \times B_2$, $A_2 \times B_1$, $A_2 \times B_2$, $B_1 \times B_2$, A_2^n (n odd), B_1^n (n odd), B_2^n (n odd)	A_2, $A_1 \times A_2$, $B_1 \times B_2$, A_2^n (n odd)

TABLE 3.15 (continued)

Point Group $C_{\infty v}$

Symmetry elements

$I, 2C_\infty^\varphi, 2C_\infty^{2\varphi}, 2C_\infty^{3\varphi}, \ldots, \infty\,\sigma_v$

Symmetry types and characters

$C_{\infty v}$	I	$2C_\infty^\varphi$	$2C_\infty^{2\varphi}$	$2C_\infty^{3\varphi}$	$\ldots$	$\infty\,\sigma_v$		
$A_1 \equiv \Sigma^+$	1	1	1	1	$\ldots$	1	T_z	$\alpha_{xx}+\alpha_{yy},\ \alpha_{zz}$
$A_2 \equiv \Sigma^-$	1	1	1	1	$\ldots$	-1	R_z	
$E_1 \equiv \Pi$	2	$2\cos\varphi$	$2\cos 2\varphi$	$2\cos 3\varphi$	$\ldots$	0	$(T_x, T_y);$ (R_x, R_y)	$(\alpha_{yz},\ \alpha_{xz})$
$E_2 \equiv \Delta$	2	$2\cos 2\varphi$	$2\cos 2\cdot 2\varphi$	$2\cos 3\cdot 2\varphi$	$\ldots$	0		$(\alpha_{xx}-\alpha_{yy},\ \alpha_{xy})$
$E_3 \equiv \Phi$	2	$2\cos 3\varphi$	$2\cos 2\cdot 3\varphi$	$2\cos 3\cdot 3\varphi$	$\ldots$	0		
$\ldots$	$\ldots$				$\ldots$	$\ldots$	$\ldots$	

Selection rules (forbidden vibrations) (See Hirshfelder[13] for selection rules for combination bands and overtones)

Infrared, Σ^-, Δ, Φ

Raman, Σ^-, Φ

TABLE 3.15 (*continued*)

Point Group C_{2h}

Symmetry elements
I, $C_2(z)$, $\sigma_h(xy)$, i

Symmetry types and characters

C_{2h}	I	$C_2(z)$	$\sigma_h(xy)$	i		
A_g	1	1	1	1	R_z	$\alpha_{xx}, \alpha_{yy}, \alpha_{zz}, \alpha_{xy}$
A_u	1	1	-1	-1	T_z	
B_g	1	-1	-1	1	$R_x; R_y$	α_{yz}, α_{xz}
B_u	1	-1	1	-1	$T_x; T_y$	

Selection rules (forbidden vibrations)

Infrared	Raman
$A_g, B_g, A_g \times A_g, A_g \times B_g,$ $B_g \times B_g, A_u \times A_u, A_u \times B_u,$ $B_u \times B_u, A_g^n, B_g^n, A_u^n$ (n even), B_u^n (n even)	$A_u, B_u, A_g \times A_u, A_g \times B_u,$ $B_g \times A_u, B_g \times B_u, A_u^n$ (n odd), B_u^n (n odd)

Point Group C_{3h}

Symmetry elements
I, C_3, σ_h, S_3

Symmetry types and characters

C_{3h}	I	C_3	σ_h	S_3		
A'	1	1	1	1	R_z	$\alpha_{xx} + \alpha_{yy}, \alpha_{zz}$
A''	1	1	-1	-1	T_z	
E'	2	-1	2	-1	T_x, T_y	$(\alpha_{xx} - \alpha_{yy}, \alpha_{xy})$
E''	2	-1	-2	1	R_x, R_y	$(\alpha_{yz}, \alpha_{xz})$

Selection rules (forbidden vibrations)

Infrared	Raman
$A', E'', A' \times A', A' \times E'',$ $A'' \times A'', A'' \times E', A'^n, A''^n$ (n even)	$A'', A' \times A'', A''^n$ (n odd)

TABLE 3.15 *(continued)*

Point Group D_{2d} or V_d

Symmetry elements

$I, 2S_4(z), S_4^2 \equiv C_2'', 2C_2, 2\sigma_d$

Symmetry types and characters

$D_{2d} \equiv V_d$	I	$2S_4(z)$	$S_4^2 \equiv C_2''$	$2C_2$	$2\sigma_d$		
A_1	1	1	1	1	1		$\alpha_{xx} + \alpha_{yy}, \alpha_{zz}$
A_2	1	1	1	-1	-1	R_z	
B_1	1	-1	1	1	-1		$\alpha_{xx} - \alpha_{yy}$
B_2	1	-1	1	-1	1	T_z	α_{xy}
E	2	0	-2	0	0	$(T_x, T_z); (R_x, R_y)$	$(\alpha_{yz}, \alpha_{xz})$

Selection rules (forbidden vibrations)

Infrared	Raman
$A_1, A_2, B_1, A_1 \times A_1, A_1 \times A_2,$ $A_1 \times B_1, A_2 \times A_2, A_2 \times B_2,$ $B_1 \times B_1, B_1 \times B_2, B_2 \times B_2,$ $A_1^n, A_2^n, B_1^n, B_2^n$ (*n* even)	$A_2, A_1 \times A_2, B_1 \times B_2, A_2^n$ (*n* odd)

Point Group D_{3d} or S_{6v}

Symmetry elements

$I, 2S_6(z), 2S_6^2 \equiv 2C_3, S_6^3 \equiv S_2 \equiv i, 3C_2, 3\sigma_d$

Symmetry types and characters

$D_{3d} \equiv S_{6v}$	I	$2S_6(z)$	$2S_6^2 \equiv 2C_3$	$S_6^3 \equiv S_2 \equiv i$	$3C_2$	$3\sigma_d$		
A_{1g}	1	1	1	1	1	1		$\alpha_{xx} + \alpha_{yy}, \alpha_{zz}$
A_{1u}	1	-1	1	-1	1	-1		
A_{2g}	1	1	1	1	-1	-1	R_z	
A_{2u}	1	-1	1	-1	-1	1	T_z	
E_g	2	-1	-1	2	0	0	(R_x, R_y)	$(\alpha_{xx} - \alpha_{yy}, \alpha_{xy}),$
E_u	2	1	-1	-2	0	0	(T_x, T_y)	$(\alpha_{yz}, \alpha_{xz})$

Selection rules (forbidden vibrations)

Infrared	Raman
$A_{1g}, A_{1u}, A_{2g}, E_g, A_{1g} \times A_{1g},$ $A_{1g} \times A_{1u}, A_{1g} \times A_{2g}, A_{1g} \times E_g,$ $A_{1u} \times A_{1u}, A_{1u} \times A_{2u}, A_{1u} \times E_u,$ $A_{2g} \times A_{2g}, A_{2g} \times A_{2u}, A_{2g} \times E_g,$ $A_{2u} \times A_{2u}, A_{2u} \times E_u, E_g \times E_g,$ $E_u \times E_u, A_{1g}^n, A_{1u}^n, A_{2g}^n, A_{2u}^n$ (*n* even), $E_g^n, E_u^2, E_u^4, \cdots$	$A_{1u}, A_{2g}, A_{2u}, E_u, A_{1g} \times A_{1u},$ $A_{1g} \times A_{2g}, A_{1g} \times A_{2u}, A_{1g} \times E_u,$ $A_{1u} \times A_{2g}, A_{1u} \times A_{2u}, A_{1u} \times E_g,$ $A_{2g} \times A_{2u}, A_{2g} \times E_u, A_{2u} \times E_g,$ $E_g \times E_u, A_{1u}^n,$ (*n* odd), A_{2g}^n (*n* odd), A_{2u}^n (*n* odd), $E_u^3, E_u^5, \cdots$

TABLE 3.15 (continued)

Point Group D_{4d} or S_{8u}

Symmetry elements
I, $2S_8(z) \equiv 2C_4$, $2S_8^2 \equiv 2C_4$, $2S_8^3$, $S_8^4 \equiv C_2'' \equiv C_2$, $4C_2$, $4\sigma_d$

Symmetry types and characters

$D_{4d} \equiv S_{8v}$	I	$2S_8(z)$	$2S_8^2 \equiv 2C_4$	$2S_8^3$	$S_8^4 \equiv C_2$	$4C_2$	$4\sigma_d$		
A_1	1	1	1	1	1	1	1		$\alpha_{xx} + \alpha_{yy}$, α_{zz}
A_2	1	1	1	1	1	-1	-1	R_z	
B_1	1	-1	1	-1	1	1	-1		
B_2	1	-1	1	-1	1	-1	1	T_z	
E_1	2	$\sqrt{2}$	0	$-\sqrt{2}$	-2	0	0	(T_x, T_y)	
E_2	2	0	-2	0	2	0	0		$(\alpha_{xx} - \alpha_{yy}, \alpha_{xy})$
E_3	2	$-\sqrt{2}$	0	$\sqrt{2}$	-2	0	0	(R_x, R_y)	$(\alpha_{yz}, \alpha_{xz})$

Selection rules (forbidden vibrations)

Infrared

A_1, A_2, B_1, E_2, E_3, $A_1 \times A_2$, $A_1 \times B_1$, $A_1 \times E_2$, $A_1 \times E_3$,
$A_2 \times B_2$, $A_2 \times E_2$, $A_2 \times E_3$, $B_1 \times B_2$, $B_1 \times E_1$, $B_1 \times E_2$,
$B_2 \times E_1$, $B_2 \times E_2$, A_1^n (n even), A_2^n (n even), B_1^n (n even),
B_2^n (n even), E_1^n (n even), E_3^n (n even)

Raman

A_2, B_1, B_2, E_1, $A_1 \times B_1$, $A_1 \times B_2$, $A_1 \times E_1$, $A_2 \times B_1$,
$A_2 \times B_2$, $A_2 \times E_1$, $B_1 \times B_2$, $B_1 \times E_3$, $B_2 \times E_3$, A_1^2 (n odd),
B_1^n (n odd), B_2^n (n odd), E_1^n (n odd)

TABLE 3.15 (*continued*)

Point Group D_{2h} or V_h

Symmetry elements

I, $\sigma(xy)$, $\sigma(xz)$, $\sigma(yz)$, i, $C_2(z)$, $C_2(y)$, $C_2(x)$

Symmetry types and characters

$D_{2h} \equiv V_h$	I	$\sigma(xy)$	$\sigma(xz)$	$\sigma(yz)$	i	$C_2(z)$	$C_2(y)$	$C_2(x)$		
A_g	1	1	1	1	1	1	1	1		$\alpha_{xx}, \alpha_{yy}, \alpha_{zz}$
A_u	1	-1	-1	-1	-1	1	1	1		
B_{1g}	1	1	-1	-1	1	1	-1	-1	R_z	α_{xy}
B_{1u}	1	-1	1	1	-1	1	-1	-1	T_z	
B_{2g}	1	-1	1	-1	1	-1	1	-1	R_y	α_{xz}
B_{2u}	1	1	-1	1	-1	-1	1	-1	T_y	
B_{3g}	1	-1	-1	1	1	-1	-1	1	R_x	α_{yz}
B_{3u}	1	1	1	-1	-1	-1	-1	1	T_x	

Selection rules (forbidden vibrations)

Infrared	Raman
$A_{1g}, A_{1u}, B_{1g}, B_{2g}, B_{3g}, A_{1g} \times A_{1g}$,	$A_{1u}, B_{1u}, B_{2u}, B_{3u}, A_{1g} \times A_{1u}$,
$A_{1g} \times A_{1u}, A_{1g} \times B_{1g}, A_{1g} \times B_{2g}$,	$A_{1g} \times B_{1u}, A_{1g} \times B_{2u}, A_{1g} \times B_{3u}$,
$A_{1g} \times B_{3g}, A_{1u} \times A_{1u}, A_{1u} \times B_{1g}$,	$A_{1u} \times B_{1g}, A_{1u} \times B_{2g}, A_{1u} \times B_{3g}$,
$A_{1u} \times B_{1u}, A_{1u} \times B_{2u}, A_{1u} \times B_{3u}$,	$B_{1g} \times B_{1u}, B_{1g} \times B_{2u}, B_{1g} \times B_{3u}$,
$B_{1g} \times B_{1g}, B_{1g} \times B_{1u}, B_{1g} \times B_{2g}$,	$B_{1u} \times B_{2g}, B_{1u} \times B_{3g}, B_{2g} \times B_{2u}$,
$B_{1g} \times B_{2u}, B_{1g} \times B_{3g}, B_{1u} \times B_{1u}$,	$B_{2g} \times B_{3u}, B_{2u} \times B_{3g}, B_{3g} \times B_{3u}$,
$B_{1u} \times B_{2u}, B_{1u} \times B_{3u}, B_{2g} \times B_{2g}$,	A_{1g}^n (n odd), B_{1u}^n (n odd), B_{2u}^n (n odd),
$B_{2g} \times B_{3g}, B_{2u} \times B_{2u}, B_{2u} \times B_{3u}$,	B_{3u}^n (n odd)
$B_{3g} \times B_{3g}, B_{3u} \times B_{3u}, A_{1g}^n, A_{1u}^n$,	
B_{1g}^n, B_{1u}^n (n even), B_{2g}^n, B_{2u}^n (n even),	
B_{3g}^n, B_{3u}^n (n even)	

Point Group D_{3h}

Symmetry elements

I, $2C_3(z)$, $3C_2$, σ_h, $2S_3$, $3\sigma_v$

Symmetry types and characters[a]

D_{3h}	I	$2C_3(z)$	$3C_2$	σ_h	$2S_3$	$3\sigma_v$		
A_1'	1	1	1	1	1	1		$\alpha_{xx} + \alpha_{yy}, \alpha_{zz}$
A_1''	1	1	1	-1	-1	-1		
A_2'	1	1	-1	1	1	-1	R_z	
A_2''	1	1	-1	-1	-1	1	T_z	
E'	2	-1	0	2	-1	0	(T_x, T_y)	$(\alpha_{xx} - \alpha_{yy}, \alpha_{xy})$
E''	2	-1	0	-2	1	0	(R_x, R_y)	$(\alpha_{yz}, \alpha_{xz})$

[a] For D_{3h}' (free rotation), A_1', A_1'', A_2', A_2'', E', and E'' are replaced by A_1, $\overline{A}_1$, A_2, $\overline{A}_2$, E, and $\overline{E}$, respectively. The selection rules are the same.

TABLE 3.15 (*continued*)

Selection rules (forbidden vibrations)

Infrared	Raman
$A'_1, A'_2, A''_2, E'', A'_1 \times A'_1,$	$A'_2, A''_1, A''_2, A'_1 \times A'_2, A'_1 \times A''_1,$
$A'_1 \times A'_2, A'_1 \times A''_1, A'_1 \times E'',$	$A'_1 \times A''_2, A'_2 \times A''_1, A'_2 \times A''_2,$
$A'_2 \times A'_2, A'_2 \times A''_2, A'_2 \times E'',$	$A''_1 \times A''_2, (A'_2)^n$ (n odd), $(A''_1)^n$
$A''_1 \times A''_1, A''_1 \times A''_2, A''_1 \times E',$	(n odd), $(A''_1)^n$ (n odd)
$A''_2 \times A''_2, A''_2 \times E', (A'_1)^n, (A'_2)^n,$	
$(A''_1)^n, (A''_2)^n$ (n even)	

Point Group D$_{4h}$

Symmetry elements
  $I, 2C_4(z), C''_2, 2C_2, 2C'_2, \sigma_h, 2\sigma_v, 2\sigma_d, 2S_4, S_2, i$

Symmetry types and characters

D_{4h}	I	$2C_4(z)$	C''_2	$2C_2$	$2C'_2$	σ_h	$2\sigma_v$	$2\sigma_d$	$2S_4$	i		
A_{1g}	1	1	1	1	1	1	1	1	1	1		$\alpha_{xx}+\alpha_{yy}, \alpha_{zz}$
A_{1u}	1	1	1	1	1	-1	-1	-1	-1	-1		
A_{2g}	1	1	1	-1	-1	1	-1	-1	1	1	R_z	
A_{2u}	1	1	1	-1	-1	-1	1	1	-1	-1	T_z	
B_{1g}	1	-1	1	1	-1	1	1	-1	-1	1		$\alpha_{xx}-\alpha_{yy}$
B_{1u}	1	-1	1	1	-1	-1	-1	1	1	-1		
B_{2g}	1	-1	1	-1	1	1	-1	1	-1	1		α_{xy}
B_{2u}	1	-1	1	-1	1	-1	1	-1	1	-1		
E_g	2	0	-2	0	0	-2	0	0	0	2	(R_x, R_y)	$(\alpha_{yz}, \alpha_{xz})$
E_u	2	0	-2	0	0	2	0	0	0	-2	(T_x, T_y)	

Selection rules (forbidden vibrations)

Infrared	Raman
$A_{1g}, A_{1u}, A_{2g}, B_{1g}, B_{1u}, B_{2g}, B_{2u},$	$A_{1u}, A_{2g}, A_{2u}, B_{1u}, B_{2u}, E_u,$
$E_g, A_{1g} \times A_{1g}, A_{1g} \times A_{1u}, A_{1g} \times A_{2g},$	$A_{1g} \times A_{1u}, A_{1g} \times A_{2g}, A_{1g} \times B_{1u},$
$A_{1g} \times B_{1g}, A_{1g} \times B_{1u}, A_{1g} \times B_{2g},$	$A_{1g} \times B_{2u}, A_{1g} \times A_{2u}, A_{1u} \times B_{1g},$
$A_{1g} \times B_{2u}, A_{1g} \times E_g, A_{1u} \times A_{1u},$	$A_{1u} \times B_{2g}, A_{2g} \times A_{2u}, A_{2g} + B_{1u},$
$A_{1u} \times A_{2u}, A_{1u} \times B_{1g}, A_{1u} \times B_{1u},$	$A_{2g} \times B_{2u}, A_{2u} \times B_{1g}, A_{2u} \times B_{2g},$
$A_{1u} \times B_{2g}, A_{1u} \times E_u, A_{2g} \times A_{2g},$	$B_{1g} \times B_{1u}, B_{1g} \times B_{2g}, B_{1u} \times B_{2g},$
$A_{2g} \times A_{2u}, A_{2g} \times B_{1u}, A_{2g} \times B_{2g},$	$B_{2g} \times B_{2u}, A_{1g} \times E_u, A_{2g} \times A_{1u},$
$A_{2g} \times B_{2u}, A_{2g} \times E_g, A_{2u} \times A_{2u},$	$A_{2g} \times E_u, B_{1g} \times B_{2u}, B_{1g} \times E_u,$
$A_{2u} \times B_{1g}, A_{2u} \times B_{1u}, A_{2u} \times B_{2g},$	$B_{2g} \times E_u, E_g \times A_{1u}, E_g \times A_{2u},$
$A_{2u} \times B_{2u}, B_{1g} \times B_{1g},$	$E_g \times B_{1u}, E_g \times B_{2u}, E_g \times E_u,$
$B_{1g} \times B_{1u}, B_{1g} \times B_{2g}, B_{1g} \times E_g,$	A^n_{1u} (n odd), A^n_{2g} (n odd), B^n_{1u} (n odd),
$B_{1u} \times B_{1u}, B_{1u} \times B_{2u}, B_{1u} \times E_u,$	B^n_{2u} (n odd), E^n_u (n odd)
$B_{2g} \times B_{2g}, B_{2g} \times B_{2u}, B_{2g} \times E_g,$	
$B_{2u} \times B_{2u}, B_{2u} \times E_u, E_g \times E_g,$	
$E_u \times E_u, A_{2g} \times B_{1g}, A_{1u} \times B_{2u},$	
$A^n_{1g}, A^n_{1u}, A^n_{2g}, A^n_{2u}$ (n even), $B^n_{1g},$	
$B^n_{1u}, B^n_{2g}, B^n_{2u}, E^n_g, E^n_u$ (n even)	

TABLE 3.15 (continued)

Point Group D$_{6h}$

Symmetry elements

I, $2C_6(z)$ $2C_6^2 \equiv 2C_3$, $C_6^3 \equiv C_2''$, $3C_2$, $3C_2'$, σ_h, $3\sigma_v$, $3\sigma_d$, $2S_6$, $2S_3$, $S_6^3 \equiv S_2 \equiv i$

Symmetry types (or species)

D_{6h}	I	$2C_6(z)$	$2C_6^2 \equiv 2C_3$	$C_6^3 \equiv C_2''$	$3C_2$	$3C_2'$	σ_h	$3\sigma_v$	$3\sigma_d$	$2S_6$	$2S_3$	$S_6^3 \equiv S_2 \equiv i$		
A_{1g}	1	1	1	1	1	1	1	1	1	1	1	1		$\alpha_{xx} + \alpha_{yy}$, α_{zz}
A_{1u}	1	1	1	1	1	1	-1	-1	-1	-1	-1	-1		
A_{2g}	1	1	1	1	-1	-1	1	-1	-1	1	1	1	R_z	
A_{2u}	1	1	1	1	-1	-1	-1	1	1	-1	-1	-1	T_z	
B_{1g}	1	-1	1	-1	1	-1	-1	-1	1	1	-1	1		
B_{1u}	1	-1	1	-1	1	-1	1	1	-1	-1	1	-1		
B_{2g}	1	-1	1	-1	-1	1	-1	1	-1	1	-1	1		
B_{2u}	1	-1	1	-1	-1	1	1	-1	1	-1	1	-1		
E_{1g}	2	1	-1	-2	0	0	-2	0	0	-1	1	2	(R_x, R_y)	$(\alpha_{yz}, \alpha_{xz})$
E_{1u}	2	1	-1	-2	0	0	2	0	0	1	-1	-2	(T_x, T_y)	
E_{2g}	2	-1	-1	2	0	0	2	0	0	-1	-1	2		$(\alpha_{xx} - \alpha_{yy}, \alpha_{xy})$
E_{2u}	2	-1	-1	2	0	0	-2	0	0	1	1	-2		

TABLE 3.15 (continued)

Selection rules (forbidden vibrations)

Infrared

A_{1g}, A_{1u}, A_{2g}, B_{1g}, B_{1u}, B_{2g}, B_{2u}, E_{1g}, E_{2g}, E_{2u}, $A_{1g} \times A_{1g}$,
$A_{1g} \times A_{2g}$, $A_{1g} \times B_{1g}$, $A_{1g} \times B_{1u}$, $A_{1g} \times B_{2u}$, $A_{1g} \times B_{2u}$,
$A_{1g} \times E_{1g}$, $A_{1g} \times E_{2g}$, $A_{1g} \times E_{2u}$, $A_{1u} \times B_{1g}$, $A_{1u} \times B_{2g}$,
$A_{1u} \times A_{2u}$, $A_{1u} \times E_{1u}$, $A_{1u} \times E_{2g}$, $A_{1u} \times E_{2u}$,
$A_{2g} \times A_{2g}$, $A_{2u} \times A_{2g}$, $A_{2g} \times E_{1g}$, $A_{1u} \times A_{2g} \times B_{2g}$,
$A_{2g} \times B_{2u}$, $A_{2g} \times E_{1g}$, $A_{2g} \times E_{2g}$, $A_{2g} \times E_{2u}$,
$B_{1g} \times B_{1g}$, $B_{1g} \times B_{1u}$, $B_{1g} \times E_{1g}$, $B_{1g} \times E_{1u}$,
$B_{1g} \times E_{2g}$, $A_{2u} \times A_{2u}$, $B_{2g} \times B_{2u}$, $A_{2u} \times E_{1u}$,
$A_{2u} \times B_{2u}$, $A_{2u} \times A_{2u} \times E_{2g}$, $A_{2u} \times E_{2u}$, $B_{1u} \times E_{1u}$,
$B_{1u} \times B_{2u}$, $B_{1u} \times E_{1g}$, $B_{1u} \times E_{2u}$, $B_{1u} \times B_{2g}$,
$B_{2g} \times B_{2u}$, $B_{2g} \times E_{1g}$, $B_{1g} \times E_{1u}$, $B_{2u} \times B_{2g}$,
$B_{2u} \times E_{1u}$, $B_{2u} \times E_{1g}$, $B_{2u} \times E_{2u}$, $B_{2u} \times B_{2g}$,
$E_{1u} \times E_{1u}$, $E_{1u} \times E_{2u}$, $E_{2g} \times E_{2u}$, $E_{2u} \times E_{2u}$, A_{1g}^n, A_{1u}^n,
A_{2g}^n, A_{2u}^n (n even) B_{1g}^n, B_{1u}^n, B_{2g}^n, B_{2u}^n, E_{1g}^n, E_{1u}^n (n even), E_{2g}^n,
E_{2u}^n (n even)

Raman

A_{1u}, A_{2g}, A_{2u}, B_{1u}, B_{1u}, B_{2g}, B_{2u}, E_{1u}, E_{2u}, $A_{1g} \times A_{1u}$,
$A_{1g} \times A_{2g}$, $A_{1g} \times A_{2u}$, $A_{1g} \times B_{1g}$, $A_{1g} \times B_{1u}$, $A_{1g} \times B_{2g}$,
$A_{1g} \times B_{2u}$, $A_{1g} \times E_{1u}$, $A_{1g} \times E_{2u}$, $A_{1u} \times A_{2g}$, $A_{1u} \times A_{2u}$,
$A_{1u} \times B_{1g}$, $A_{1u} \times B_{1u}$, $A_{1u} \times B_{2g}$, $A_{1u} \times B_{2u}$, $A_{1u} \times E_{1g}$,
$A_{1u} \times E_{2g}$, $A_{2g} \times A_{2u}$, $A_{2g} \times B_{1g}$, $A_{1u} \times B_{2g} \times B_{2g}$,
$A_{2g} \times B_{2u}$, $B_{1g} \times B_{1u}$, $A_{2g} \times E_{1u}$, $B_{1g} \times B_{2g}$,
$B_{1g} \times B_{2u}$, $B_{1g} \times E_{1u}$, $A_{2u} \times E_{2g}$, $B_{1u} \times A_{2u}$,
$A_{2u} \times B_{1g}$, $A_{2u} \times B_{2g}$, $A_{2u} \times E_{1g}$, $A_{2u} \times E_{2g}$, $B_{1u} \times E_{1u}$,
$B_{1u} \times B_{2u}$, $B_{1u} \times E_{1g}$, $B_{1u} \times E_{2g}$, $B_{2g} \times B_{2u}$, $B_{2g} \times E_{1u}$,
$B_{2g} \times E_{2u}$, $B_{2u} \times E_{1g}$, $B_{2u} \times E_{2g}$, $E_{1g} \times E_{1u}$, $E_{1u} \times E_{2u}$,
$E_{1u} \times E_{2u}$, $E_{2g} \times E_{2u}$, A_{1u}^n (n odd), A_{2g}^n (n odd), A_{2u}^n (n odd),
B_{1g}^n (n odd), B_{1u}^n (n odd), B_{2g}^n (n odd), B_{2u}^n (n odd), E_{1u}^n (n odd),
E_{2u}^n (n odd)

TABLE 3.15 (continued)

Point Group $D_{\infty h}$

Symmetry elements

$I,\ 2C_\infty^\varphi,\ 2C_\infty^{2\varphi},\ 2C_\infty^{3\varphi}, \ldots,\ \sigma_h,\ \infty C_2,\ \infty\sigma_v,\ 2S_\infty^\varphi,\ 2S_\infty^{2\varphi}, \ldots,\ S_2\equiv i$

Symmetry types and characters

$D_{\infty h}$	I	$2C_\infty^\varphi$	$2C_\infty^{2\varphi}$	$2C_\infty^{3\varphi}$	$\ldots$	σ_h	∞C_2	$\infty\sigma_v$	$2S_\infty^\varphi$	$2S_\infty^{2\varphi}$	$\ldots$	$S_2\equiv i$		
$A_{1g}\equiv\Sigma_g^+$	1	1	1	1	$\ldots$	1	1	1	1	1	$\ldots$	1		$\alpha_{xx}+\alpha_{yy},\ \alpha_{zz}$
$A_{1u}\equiv\Sigma_u^+$	1	1	1	1	$\ldots$	-1	1	-1	-1	-1	$\ldots$	-1	T_z	
$A_{2g}\equiv\Sigma_g^-$	1	1	1	1	$\ldots$	1	-1	-1	1	1	$\ldots$	1	R_z	
$A_{2u}\equiv\Sigma_u^-$	1	1	1	1	$\ldots$	-1	-1	1	-1	-1	$\ldots$	-1		
$E_{1g}\equiv\Pi_g$	2	$2\cos\varphi$	$2\cos 2\varphi$	$2\cos 3\varphi$	$\ldots$	-2	0	0	$-2\cos\varphi$	$-2\cos 2\varphi$	$\ldots$	2	(R_x, R_y)	$(\alpha_{xz},\ \alpha_{yz})$
$E_{1u}\equiv\Pi_u$	2	$2\cos\varphi$	$2\cos 2\varphi$	$2\cos 3\varphi$	$\ldots$	2	0	0	$2\cos\varphi$	$2\cos 2\varphi$	$\ldots$	-2	(T_x, T_y)	
$E_{2g}\equiv\Delta_g$	2	$2\cos 2\varphi$	$2\cos 4\varphi$	$2\cos 6\varphi$	$\ldots$	2	0	0	$2\cos 2\varphi$	$2\cos 4\varphi$	$\ldots$	2		$(\alpha_{xx}-\alpha_{yy},\ \alpha_{xy})$
$E_{2u}\equiv\Delta_u$	2	$2\cos 2\varphi$	$2\cos 4\varphi$	$2\cos 6\varphi$	$\ldots$	-2	0	0	$-2\cos 2\varphi$	$-2\cos 4\varphi$	$\ldots$	-2		
$E_{3g}\equiv\Phi_g$	2	$2\cos 3\varphi$	$2\cos 6\varphi$	$2\cos 9\varphi$	$\ldots$	-2	0	0	$-2\cos 3\varphi$	$-2\cos 4\varphi$	$\ldots$	2		
$E_{3u}\equiv\Phi_u$	2	$2\cos 3\varphi$	$2\cos 6\varphi$	$2\cos 9\varphi$	$\ldots$	2	0	0	$2\cos 3\varphi$	$2\cos 4\varphi$	$\ldots$	-2		
$\ldots$	$\ldots$	$\ldots$	$\ldots$	$\ldots$	$\ldots$	$\ldots$	$\ldots$	$\ldots$	$\ldots$	$\ldots$	$\ldots$	$\ldots$		

Selection rules (forbidden vibrations) (See Meister et al. for selection rules for combinations and overtone bands)

Infrared

$$\overline{\Sigma_g^+,\ \Sigma_g^-,\ \Sigma_u^-,\ \Pi_g,\ \Delta_g,\ \Delta_u,\ \Phi_g,\ \Phi_u}$$

Raman

$$\overline{\Sigma_u^+,\ \Sigma_g^-,\ \Sigma_u^-,\ \Pi_u,\ \Delta_u,\ \Phi_g,\ \Phi_u}$$

TABLE 3.15 (*continued*)

Point Group T$_d$

Symmetry elements
 $I, 8C_3, 6\sigma_d, 6S_4, 3S_4^2 \equiv 3C_2$

Symmetry types and characters

T_d	I	$8C_3$	$6\sigma_d$	$6S_4$	$3S_4^2 \equiv 3C_2$		
A_1	1	1	1	1	1		$\alpha_{xx} + \alpha_{yy} + \alpha_{zz}$
A_2	1	1	-1	-1	1		
E	2	-1	0	0	2		$\alpha_{xx} + \alpha_{yy} - 2\alpha_{zz}, \alpha_{xx} - \alpha_{yy}$
F_1	3	0	-1	1	-1	R_x, R_y, R_z	
F_2	3	0	1	-1	-1	T_x, T_y, T_z	$\alpha_{xy}, \alpha_{yz}, \alpha_{xz}$

Selection rules (forbidden vibrations)

Infrared	Raman
$A_1, A_2, E, F_1, A_1 \times A_1, A_1 \times A_2,$ $A_1 \times E, A_1 \times F_1, A_2 \times A_2,$ $A_2 \times E, A_2 \times F_2, E \times E,$ A_1^n, A_2^n, E^n	$A_2, F_1, A_1 \times A_2, A_1 \times F_1,$ $A_2 \times F_2, A_2^n$ (n odd)

TABLE 3.15 (continued)

Point Group O_h

Symmetry elements

$I, 8C_3, 6C_2, 6C_4, 3C_4^2, i, 6S_4, 8S_6, 3\sigma_h, 6\sigma_d$

Symmetry types and characters

O_h	I	$8C_3$	$6C_2$	$6C_4$	$3C_4^2$	i	$6S_4$	$8S_6$	$3\sigma_h$	$6\sigma_d$		
A_{1g}	1	1	1	1	1	1	1	1	1	1		$\alpha_{xx}+\alpha_{yy}+\alpha_{zz}$
A_{1u}	1	1	1	1	1	-1	-1	-1	-1	-1		
A_{2g}	1	1	-1	-1	1	1	-1	1	1	-1		
A_{2u}	1	1	-1	-1	1	-1	1	-1	-1	1		
E_g	2	-1	0	0	2	2	0	-1	2	0		$(\alpha_{xx}+\alpha_{yy}-2\alpha_{zz}, \alpha_{xx}-\alpha_{yy})$
E_u	2	-1	0	0	2	-2	0	-1	-2	0		
F_{1g}	3	0	-1	1	-1	3	1	0	-1	-1	R_x, R_y, R_z	
F_{1u}	3	0	-1	1	-1	-3	-1	0	1	1	T_x, T_y, T_z	
F_{2g}	3	0	1	-1	-1	3	-1	0	-1	1		$(\alpha_{xy}, \alpha_{yz}, \alpha_{xz})$
F_{2u}	3	0	1	-1	-1	-3	1	0	1	-1		

Selection rules (forbidden vibrations)

Infrared

$A_{1g}, A_{1u}, A_{2g}, A_{2u}, E_g, E_u, F_{1g}, F_{2g}, F_{2u}, A_{1g} \times A_{1g},$
$A_{1g} \times A_{1u}, A_{1g} \times A_{2g}, A_{1g} \times A_{2g}, A_{1g} \times E_g, A_{1g} \times E_u, A_{1g} \times F_{1g},$
$A_{1g} \times F_{2g}, A_{1u} \times F_{1g}, A_{1u} \times F_{2g}, A_{1u} \times F_{2u}, A_{1u} \times E_g,$
$A_{1u} \times E_u, A_{2g} \times E_u, A_{2g} \times F_{1g}, A_{1u} \times F_{2u}, A_{2g} \times A_{2g},$
$A_{2g} \times E_g, E_g \times E_u, A_{2g} \times F_{1g}, E_g \times F_{2g}, E_g \times F_{1g},$
$E_g \times E_g, E_g \times E_u, F_{1g} \times F_{1g}, F_{1g} \times F_{2g}, F_{1u} \times F_{2g},$
$E_u \times F_{2u}, F_{1g} \times F_{1g}, F_{1g} \times F_{2g}, F_{1u} \times F_{1u}, F_{1u} \times F_{2u},$
$F_{2u} \times F_{2u}, A_{2u} \times A_{2u}, A_{2u} \times E_g, A_{2u} \times E_u, A_{2u} \times F_{1g},$
$A_{2u} \times F_{1u}, A_{2u} \times F_{2u}, A_{2u} \times A_{2g}, A_{2u} \times A_{2g}, A_{2u} \times A_{1u},$
$F_{1g} \times F_{2g}, F_{2g} \times F_{2g}, A_{1g}^n, A_{2g}^n, A_{2u}^n, E_g^n, E_u, F_{1g}^n,$
$F_{2g}^n, F_{1u}^n (n>2), F_{2u}^n (n>2)$

Raman

$A_{1u}, A_{2g}, A_{2u}, E_u, F_{1g}, F_{1u}, F_{2u}, A_{1g} \times A_{1u}, A_{1g} \times A_{2g},$
$A_{1g} \times E_u, A_{1g} \times F_{1g}, A_{1g} \times F_{1u}, A_{1g} \times F_{2u}, A_{1u} \times A_{2g},$
$A_{1u} \times E_g, A_{1u} \times F_{1u}, A_{1u} \times F_{1u}, A_{1u} \times F_{2g}, A_{2g} \times E_u,$
$A_{2g} \times F_{1u}, A_{2g} \times F_{2u}, A_{2u} \times A_{1g}, A_{2u} \times A_{2u}, A_{2g},$
$E_u \times F_{1g}, E_u \times E_u, E_u \times F_{1u}, E_g \times F_{1u}, F_{1g} \times F_{2g},$
$A_{2g} \times E_g, F_{2g} \times F_{2u}, F_{1u} \times F_{2g}, F_{1u} \times F_{2u},$
$A_{2g} \times E_g, F_{2g} \times F_{2u}, A_{2u} \times A_{1g}, A_{2u} \times A_{1u}, A_{2u} \times A_{2g},$
$A_{2u} \times E_g, A_{2u} \times F_{2u}, A_{2u} \times A_{1u} (n \text{ odd}),$
$A_{2g}^n (n \text{ odd}), A_{2u}^n (n \text{ odd}), E_u^n, F_{1u}^n, F_{2u}^n$

CHAPTER 4

THE VIBRATIONAL ORIGIN OF GROUP FREQUENCIES

4.1 Introduction

Given the masses, the molecular geometry, and the force constants, complete mathematical procedures exist for calculating the form and frequencies of all the normal modes of vibration of relatively simple molecules (see Chapter 14). However, as the molecules become more complex the difficulty of the mathematical treatment increases enormously, so empirical methods are frequently used. It is found empirically that certain submolecular groups of atoms consistantly produce bands in a characteristic frequency region of the vibrational spectrum. These bands are the characteristic *group frequencies*. For example, the vibrational spectra of *n*-heptane, *n*-octane, and *n*-nonane have a number of bands in common which are the group frequencies for normal alkanes. These spectra also have a number of bands *not* in common, which are the so-called fingerprint bands because they are characteristic of the individual chemical compound and are used to distinguish one compound from another.

The purpose of this chapter is to discuss the vibrational origin of group frequencies with an emphasis on mechanical effects. A characteristic of a group frequency vibration is that mechanical interaction effects which control the form of the vibration are relatively constant from molecule to molecule, making the frequency readily predictable. Mechanical interaction effects can only occur between vibrations belonging to the same symmetry species. For example, in a molecule with a plane of symmetry such as vinyl chloride, in-plane vibrations will not interact with out-of-plane vibrations. All group frequencies mentioned in this chapter will be discussed and referenced in later chapters.

4.2 Diatomic Oscillators

In this chapter the vibration of a diatomic molecule will be reviewed because of its importance. Consider a molecular model where the nuclei are represented by point masses and interatomic bonds are represented by massless springs which follow Hooke's law. The form and frequencies of the vibrations are determined by classical methods. During the vibration the internuclear distance changes sinusoidally but the center of gravity remains stationary. This means that at any time during the vibration the nuclear displacements are inversely proportional to the masses. The diatomic vibrational frequency, v (in sec^{-1}), is given by

$$v = \frac{1}{2\pi} \sqrt{F\left(\frac{1}{M_1} + \frac{1}{M_2}\right)} \qquad (4.1)$$

where M_1 and M_2 are the masses in grams and F is the force constant in dynes/cm. If the masses are expressed in unified atomic mass units (u) and the force constant is expressed in millidynes/angstrom, then $\bar{v}$, the frequency in cm^{-1} (v/c), is given by

$$\bar{v} = 1303 \sqrt{F\left(\frac{1}{M_1} + \frac{1}{M_2}\right)} \qquad (4.2)$$

where 1303 is equal to $(N_A \times 10^5)^{1/2}/(2\pi c)$. N_A is Avogadro's number, 6.02252×10^{23} mole^{-1}.

4.3 Coupled Oscillators

A complex molecule consists of a number of bonded atoms. Let us use the term "oscillator" for each bond between two atoms which can oscillate about some equilibrium bond length or for a bond angle between two bonds which can oscillate about some equilibrium bond angle. If two such oscillators are arranged so that they exert forces on each other when they oscillate, they are said to be *coupled*.

Let us consider the simplest possible type of coupled oscillators, the linear M—M—M model consisting of two M—M diatomic oscillators coupled together. Coupling occurs because the displacement of the central common mass M in a manner that distorts one M—M oscillator, unavoidably distorts the other also. The M—M diatomic frequency in cm^{-1} is $\bar{v} = 1303(F/M)^{1/2}(2)^{1/2}$. Another diatomic model M—$(M/2)$ where mass M is bonded to a mass equal to $(M/2)$ has a frequency in cm^{-1} which is $\bar{v} = 1303(F/M)^{1/2}(3)^{1/2}$. In Fig. 4.1(a) we have arranged two such M—$(M/2)$ models vibrating 180° out of phase side by side with equal amplitudes. The two separate oscillators vibrate with stationary centers of gravity and the two $(M/2)$ masses both vibrate sinusoidally and remain the same distance apart throughout the vibration. Since the spacing is constant, it can be made zero as in Fig. 4.1(b), where we have assembled a linear M—M—M model performing an out-of-phase stretching normal mode of vibration. This is verified by noting that the new center of gravity is stationary and all atoms move sinusoidally with

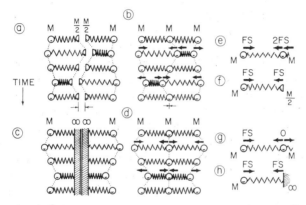

FIG. 4.1. Resolution of the stretching vibrations of the M—M—M linear triatomic ball and spring model into diatomic M—$(M/2)$ and M—(∞) components.

the same frequency and move through equilibrium simultaneously. The frequency and amplitudes are the same as those of the diatomic components in Fig. 4.1(a).

Next let us consider a diatomic model M—(∞), where mass M is bonded to an infinite mass, say a wall. The vibrational frequency in cm^{-1} is $\bar{v} = 1303(F/M)^{1/2}(1)^{1/2}$. Figure 4.1(c) shows two such oscillators vibrating in phase with equal amplitudes. The middle masses being infinite do not move so their spacing remains constant throughout the vibration. This spacing may be made zero which effectively combines the two infinite masses in one infinite mass. When we look at the result of the two forces exerted by the two symmetrically distorted springs on this middle mass, we see that this total force is zero at all times. This means that the middle mass can have any value from infinity to zero and still not move. In Fig. 4.1(d) we make the middle mass equal to M and in the process have assembled a linear M—M—M

model performing an in-phase stretching normal mode of vibration whose frequency and amplitudes are the same as those of the diatomic components in Fig. 4.1(c). These results are summarized below.

$$M—M \qquad\qquad \bar{v} = 1303\sqrt{F/M}\sqrt{2} \qquad (4.3)$$

$$\text{in-phase } M—M—M \qquad \bar{v} = 1303\sqrt{F/M}\sqrt{1} \qquad (4.4)$$

$$\text{out-of-phase } M—M—M \qquad \bar{v} = 1303\sqrt{F/M}\sqrt{3} \qquad (4.5)$$

If we ignore the hydrogen atoms, we can use these results to compare the $C{=}C$ stretch of ethylene $H_2C{=}CH_2$ at 1623 cm^{-1} with the in- and out-of-phase $C{=}C{=}C$ stretch of allene $H_2C{=}C{=}CH_2$ at 1071 and 1980 cm^{-1}, respectively, where the frequency ratios are not far from the ratio $2^{1/2} : 1^{1/2} : 3^{1/2}$ predicted by this very simple model.

From the above study we can see that the two $C{=}C$ bonds in allene do not vibrate separately. Because one $C{=}C$ bond is exactly like the other, both are vibrationally excited at the same time. If we start at equilibrium and one bond starts to stretch, the other bond must start to change in length also in one of two possible ways. It may stretch in-phase with the other bond or contract out-of-phase with the other bond. During the out-of-phase stretch in Fig. 4.1(b) the forces (represented by arrows) exerted by the two springs are exerted in the same direction on the middle mass so that the total force here is stronger. It can be said that the spring forces *cooperate* in restoring the distorted bond lengths to their equilibrium values. The "cooperation" condition tends to raise the oscillator combination frequency compared to the individual $M—M$ oscillator component frequency. During the in-phase stretch in Fig. 4.1(d) the two spring forces on the middle mass act in *opposition* in their efforts to restore distorted bond lengths to their equilibrium values. The "opposition" condition tends to *lower* the oscillator combination of frequency compared to the individual $M—M$ oscillator components. We will find these ideas useful when analyzing the form of group frequency type vibrations.

Since we now have the idea that two completely equivalent coupled oscillators will vibrate together, either in-phase or out-of-phase, we can use this as a starting point and go back to the diatomic oscillator components. In Fig. 4.1(b) and (d), each mass experiences a force (represented by an arrow) from each attached bond "spring." This force from Hooke's law is equal to FS, where F is the force constant and S is the amount the bond length differs from the equilibrium length. Figure 4.1(e) and (g) show one stretched bond from Fig. 4.1(b) and (d), respectively. Each differs from a conventional diatomic oscillator in that the forces on the two masses are unequal. These masses will vibrate with the same frequency and relative amplitudes as the masses of the conventional diatomic oscillators below them in Fig. 4.1(f) and (h) where equal forces (FS) are exerted on each mass as required, but the masses have been changed so the force/mass ratio is the same as that of the oscillator above them. The vibrational motion of a mass at a given location is specified by its acceleration (force/mass). If, at any location, the force/mass quotient is unchanged, the vibrational motion is unchanged. By this procedure we have resolved the in-phase and out-of-phase vibrations of the linear $M\!-\!M\!-\!M$ model into diatomic components whose frequencies in cm^{-1} can be obtained from Eq. (4.2).

Using the same sort of arguments as were used for the $M\!-\!M\!-\!M$ model, the stretching frequencies in cm^{-1} for the linear symmetrical $M_1\!-\!M_2\!-\!M_1$ model are derived as

$$\text{in-phase } M_1\!-\!M_2\!-\!M_1 \qquad \bar{\nu} = 1303\sqrt{F\left(\frac{1}{M_1}\right)} \qquad (4.6)$$

$$\text{out-of-phase } M_1\!-\!M_2\!-\!M_1 \qquad \bar{\nu} = 1303\sqrt{F\left(\frac{1}{M_1}+\frac{2}{M_2}\right)} \qquad (4.7)$$

4.4 Unsymmetrical Coupled Oscillators

The in-phase and out-of-phase stretching vibrations of the $M\!-\!M\!-\!M$ model are again illustrated in Fig. 4.2(a) and (d). The question is how are

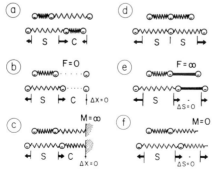

FIG. 4.2. The stretching vibrations of unsymmetrical triatomic linear models (bottom two rows) compared with those for a symmetrical model (top row). The atomic displacements are drawn as arrows and the bond length distortion is marked S for stretch and C for contract.

these vibrations modified if the two oscillator components become unequal. In Fig. 4.2(b) and (c) we have lowered the frequency of the right-hand oscillator component relative to that of the left-hand oscillator component. This is done in Fig. 4.2(b) by making the right-hand force constant zero (no spring) and in Fig. 4.2(c) by making the right-hand mass infinite. For these two extreme cases we can immediately illustrate the form of the out-of-phase vibration where we note that while both bond lengths change, the right-hand mass of the low frequency oscillator component is motionless ($\Delta X = 0$). In Fig. 4.2(b) this occurs because the forces on this mass are zero. In Fig. 4.2(e) and (f) we have raised the frequency of the right-hand oscillator component relative to that of the left-hand oscillator component. This is done in Fig. 4.2(e) by making the right-hand force constant infinite (rigid bond) and in Fig. 4.2(f) by making the right-hand mass zero. For these two extreme cases we can immediately illustrate the form of the low frequency vibration where we note that the right-hand high frequency oscillator component bond length does not change ($\Delta S = 0$). In Fig. 4.2(f) the right-hand spring cannot distort (except at an infinitely high frequency) because of the zero inertial mass on its end. Naturally if the changes made in one oscillator component are not as extreme as those discussed, the effects on the form of the vibration will be intermediate between these and those of the symmetrical coupled oscillators in Fig. 4.2(a) and (d).

From the above discussion we can write two approximations which, as we shall see, are very useful in the analysis of the form of group frequency vibrations of a wide variety of types.

1. *In the high frequency vibration of the unequal oscillator combination, the only atoms which move are those of the high frequency oscillator component.*
2. *In the low frequency vibration of the unequal oscillator combination, the only bond length (or angle) which is deformed is that of the low frequency oscillator component.*

These approximations improve as the frequencies of the oscillator components get further apart and break down when the component frequencies get closer together. When the oscillator components coupled together have the same or nearly the same frequencies, then the two resulting vibrations will involve both oscillator components more or less equally, vibrating in- or out-of-phase.

The first approximation above indicates that hydrogen stretching vibrations (X—H), triple bond stretching vibrations (C≡C, C≡N), and double bond stretching vibrations (C=C, C=N, C=O, P=O, S=O) should all give rise to good group frequencies when these groups are attached to the rest of the molecule by low frequency X—Y bonds where the attached atom is carbon or a heavier atom. Since the attached atom hardly moves, the vibration is mechanically insensitive to the nature of the attached group.

A group like $C=S$, having a frequency close to those of C—N, C—O, or C—C bonds which attach it to the rest of the molecule, should be a poor group frequency. The reason for this is that there may be strong $C=S$ interaction with the attached single bond oscillators which can, in turn, interact with other single bond oscillators attached to them. This makes a $C=S$ vibration mechanically sensitive to the nature of the attached groups.

4.5 X—H Stretching Frequencies

In general, X—H stretching frequencies are considerably higher than Y—X stretching frequencies where Y is a nonhydrogen atom directly attached to X. This means that hydrogen stretching vibrations are more or less mechanically independent of the rest of the molecule and tend to make good group frequencies. Table 4.1 lists the approximate X—H stretching frequencies and the force constants for various X—H groups arranged according to the periodic table.

Because the hydrogen stretching vibrations are relatively insensitive mechanically to the nonhydrogen part of the molecule, the different values in cm^{-1} in this table are mainly the result of effects within the X—H group, namely, a change in the mass of X or a change in the X—H force constant. If the

TABLE 4.1

APPROXIMATE X—H STRETCHING FREQUENCIES (cm^{-1}) AND FORCE CONSTANTS (mdynes/Å)

BH	CH	NH	OH	FH
2500	3000	3400	3600	3960 cm^{-1}
3.4	4.9	6.4	7.2	8.8 mdynes/Å
AlH	SiH	PH	SH	ClH
1820	2150	2350	2570	2890
1.9	2.6	3.2	3.8	4.8
	GeH	AsH	SeH	BrH
	2070	2150	2300	2560
	2.5	2.7	3.1	3.8
	SnH	SbH		IH
	1850	1890		2230
	2.0	2.1		2.9

C—H oscillator frequency is about 3000 cm^{-1}, substitution of $M_1 = 1$ and $M_2 = 12$ into the diatomic oscillator equation, Eq. (4.2) yields an approximate C—H force constant of 4.9 mdynes/Å. If we calculate the frequency of the oscillator where hydrogen is attached to an infinite mass and use the same force constant (4.9 mdynes/Å) we obtain 2880 cm^{-1}. This shows that changing the mass of X from twelve to infinity causes the frequency in cm^{-1} to become smaller by only 120 cm^{-1}. Inspection of Table 4.1 shows that the mass effect cannot account for the large variation in frequencies shown. Therefore, most of the shift is caused by changes in the X—H force constant. Approximate values for these force constants [calculated from Eq. (4.2)] are listed in Table 4.1. The force constants tend to increase as the electronegativity of X increases as it does when moving upwards or to the right in the periodic table. A plot of force constant versus electronegativity of X is a more or less a linear function for any one row or column of the periodic table.

A change in hybridization changes the spatial arrangement of electrons which can change force constants and affect the electronegativity. The C—H force constants in mdynes/Å are about 4.9 for sp^3 (alkanes), about 5.1 for sp^2 (aromatics, olefins), and about 5.9 for sp hybrids (acetylenes). The force field of the X—H bond can also be modified by hydrogen bonding. This effect, which will be discussed further in Section 4.18, is usually most important when X has a high electronegativity as in NH, OH, and FH bonds.

4.6 Triple Bond Vibrations

As an example of the use of the approximations presented earlier, consider a monosubstituted acetylene where the substituent X is carbon or a heavier atom. As illustrated in Fig. 4.3(a) the C—H oscillator is higher in frequency than the C≡C oscillator so in the C—H stretching vibration only the C—H atoms move appreciably while the center of gravity remains stationary (approximation 1). In Fig. 4.3(b) the C—H bond length does not change appreciably during the C≡C stretching vibration (approximation 2). The C≡C oscillator is low in frequency relative to the C—H oscillator but high in frequency relative to the X—C oscillator, so in Fig. 4.3(b) when the C≡C bond stretches the attached X atom is nearly stationary. In Fig. 4.3(c), when the X—C stretches, the C≡C and C—H bonds hardly change in length. The X—C≡C—H group can bend, but the resulting motions are perpendicular to the motions which occur during the stretching vibrations of this linear group, so bending interactions for these are zero. By using these approximations (which are accurate enough for group frequency analysis purposes) we have been able to deduce the approximate forms of the vibrations without calculation.

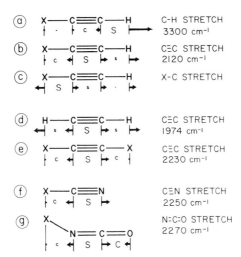

FIG. 4.3. Schematic representations of the forms of the vibrations of some molecules with triple bonds and with two cumulated double bonds.

The C—H and C≡C vibrations [Fig. 4.3(a) and (b)] scarcely involve any motion of the attached X atom. These two vibrations are good group frequencies since they are relatively independent mechanically of the mass of the X atom (carbon or heavier) or whether the carbon atom is a part of a methyl, ethyl, or phenyl group. The "C≡C stretching" vibration [Fig. 4.3(b)] to some extent also deforms the X—C bond. Therefore, the "C≡C stretch" will show a little sensitivity to any changes in the X—C force constant. If the chemical nature of the X group causes changes in the C—H or C≡C force constant then again shifts can occur. These shifts can be related directly to the chemical nature of the X group because the mechanical effects are relatively constant.

Using the same approximations we can illustrate the approximate form of the C≡C stretching vibration in acetylene [Fig. 4.3(d)] and a disubstituted acetylene [Fig. 4.3(e)]. In Fig. 4.3(e) it can be seen that the contracted X—C oscillators cooperate with the stretched C≡C oscillators in restoring bond lengths to equilibrium which raises the C≡C frequency in cm^{-1}. However, in Fig. 4.3(d) C—H the oscillators (which are actually slightly stretched when the C≡C bond is stretched) act in opposition and lower the C≡C stretching vibration in cm^{-1}. If the C—H bond (which does not change in length very much) is considered as effectively rigid for the C≡C stretching vibration, then the carbon atom has to drag its hydrogen with it. This gives the carbon an effective mass of 13 rather than 12 and is another way to picture the opposition to carbon atom motion and explain the lowering effect. Hence a hydrogen substituent on a C≡C bond has a different mechanical effect than a carbon or heavier atom substituent on the $C\equiv C$ stretching vibration. In the disubstituted acetylene (2230 cm^{-1}) two X—C bonds cooperate with the vibrating $C\equiv C$ bond in restoring equilibrium, whereas in a monosubstituted acetylene (2120 cm^{-1}) one X—C bond cooperates and one C—H bond acts in opposition, which accounts for the difference in cm^{-1}.

In Fig. 4.3(f) is illustrated a $C\equiv N$ vibration. Like the $C\equiv C$ vibration, the $C\equiv N$ is a good group frequency when the attached atom is carbon or a heavier atom.

4.7 Cumulated Double Bonds

Figure 4.3(g) illustrates the stretching vibration for an isocyanate ester X—N=C=O. While the C=O and C=N oscillators are not identical, they have nearly the same individual frequencies. When coupled together, they will both be involved more or less equally in isocyanate vibrations, meaning that the group should be considered as an entity. The isocyanate group is attached to the rest of the molecule by a low frequency X—N bond, so during the high frequency out-of-phase isocyanate vibration the X atom hardly moves, making this isocyanate vibration a good group frequency, mechanically independent of the nature of the X group. On the other hand, the in-phase isocyanate vibration has a frequency not very different from the X—N frequency, so more interaction will occur and the X atom will move, causing this vibration to be more mechanically sensitive to the nature of the X group.

The remarks made about the isocyanate group also apply to other X=Y=Z type cumulated double bond groups. In general the out-of-phase stretch of the cumulated double bond group X=Y=Z and the triple bond stretch of an $X\equiv Y$ group have about the same frequency. This can be explained roughly as follows. The out-of-phase stretch of the $M=M=M$ model was shown in Eq. (4.5) to have a frequency in cm^{-1} given by $\bar{v} = 1303(3F_d/M)^{1/2}$, where F_d is the double bond stretching force constant. The stretching frequency in cm^{-1} for the $M\equiv M$ model was shown in Eq. (4.3) to be given by $\bar{v} = 130 \times (2F_t/M)^{1/2}$, where F_t is the triple bond stretching force constant. If we specify that $F_d = \frac{2}{3}F_t$, the two frequencies are found to be equal. In a similar way the reader can verify that the in-phase stretching frequency in cm^{-1} of the $M=M=M$ model, $\bar{v} = 1303(F_d/M)^{1/2}$ from Eq. (4.4) is the same as that for the single bonded $M—M$ model, i.e., $\bar{v} = 1303(2F_s/M)^{1/2}$ from Eq. (4.3), where F_s is the single bond stretching force constant and equals $\frac{1}{2}F_d$. This accounts for the fact that the in-phase stretch of the X=Y=Z group may tend to interact with attached single bond vibrations.

4.8 The Linear M_1—M_2—M_3 Model

In this section we will present an unconventional derivation of the equations for the stretching frequencies of the linear M_1—M_2—M_3 model (assuming a simple valence force field). Figure 4.4(a) shows two unconnected diatomic models M_1—M_a and M_b—M_3 with force constants F_1 and F_3, respectively. These models, although dissimilar, have the same frequency and are vibrating in an out-of-phase manner with the amplitudes of M_a and M_b being made equal as in Fig. 4.1(a). Since the M_a, M_b spacing is constant

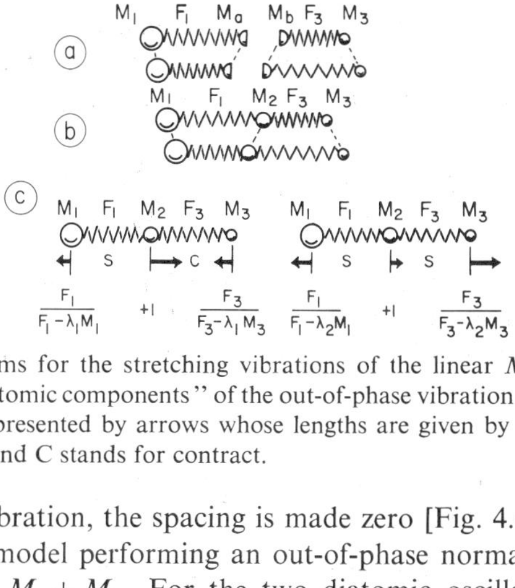

FIG. 4.4. Diagrams for the stretching vibrations of the linear M_1—M_2—M_3 model In (a) we see the "diatomic components" of the out-of-phase vibration (b). In (c) the atomic displacements are represented by arrows whose lengths are given by the formulas below. S stands for stretch and C stands for contract.

throughout the vibration, the spacing is made zero [Fig. 4.4(b)], resulting in a M_1—M_2—M_3 model performing an out-of-phase normal mode of vibration where $M_2 = M_a + M_b$. For the two diatomic oscillators which have the same frequency,

$$\bar{v} = 1303\sqrt{F_1\left(\frac{1}{M_1} + \frac{1}{M_a}\right)} \quad \text{and} \quad \bar{v} = 1303\sqrt{F_3\left(\frac{1}{M_3} + \frac{1}{M_b}\right)} \quad (4.8)$$

Solving for M_a and M_b yields

$$M_a = \frac{F_1 M_1}{\lambda M_1 - F_1} \quad \text{and} \quad M_b = \frac{F_3 M_3}{\lambda M_3 - F_3} \quad (4.9)$$

where

$$\lambda = \left(\frac{\bar{v}}{1303}\right)^2 \qquad M_2 = M_a + M_b$$

so

$$(4.10)$$

$$M_2 = \frac{F_1 M_1}{\lambda M_1 - F_1} + \frac{F_3 M_3}{\lambda M_3 + F_3}$$

Solving for λ yields

$$\lambda^2 - \lambda\left[F_1\left(\frac{1}{M_1} + \frac{1}{M_2}\right) + F_3\left(\frac{1}{M_2} + \frac{1}{M_3}\right)\right]$$

$$+ F_1 F_3\left(\frac{1}{M_1 M_2} + \frac{1}{M_2 M_3} + \frac{1}{M_1 M_3}\right) = 0 \qquad (4.11)$$

Given the masses and force constants we can calculate λ (using the quadratic equation formula) and then evaluate $\bar{v} = 1303\lambda^{1/2}$. Because Eq. (4.11) is quadratic we get two λ and $\bar{v}$ values.

The form of the vibration is found by remembering that in order to keep the center of gravity of a diatomic oscillator motionless the mass amplitudes X must be inversely proportional to the masses, thus

$$\frac{-X_1}{X_a} = \frac{M_a}{M_1} \quad \text{and} \quad \frac{-X_3}{X_b} = \frac{M_b}{M_3} \qquad (4.12)$$

We can substitute the values for M_a and M_b found in Eq. (4.9) into these equations. Since the amplitudes of M_2, M_a, and M_b are all necessarily equal, we can also substitute X_2 for X_a and X_b to give for the amplitude ratios

$$X_1 : X_2 : X_3 = \frac{F_1}{F_1 - \lambda M_1} : 1 : \frac{F_3}{F_3 - \lambda M_3} \qquad (4.13)$$

When the higher or lower of the two λ values from Eq. (4.11) is substituted into Eq. (4.13), the form of the vibration will be revealed [Fig. 4.4(c) and (d)] as either the out-of-phase or in-phase stretching vibration of the M_1—M_2—M_3 model, respectively. Because these two vibrations belong to the same symmetry species they are calculated from the same equations.

4.9 $X—C{\equiv}N$ Compounds

Using the M_1—M_2—M_3 equations, the stretching frequencies in cm^{-1} of some $X—C{\equiv}N$ molecules can be calculated. Listed in Table 4.2 are the experimentally observed stretching frequencies in cm^{-1}, $\bar{v}_1$, and $\bar{v}_2$, of $X—C{\equiv}N$ molecules.

When the force constants listed in Table 4.2 are substituted in Eq. (4.11), calculated stretching frequencies in cm^{-1} are obtained close to the observed

TABLE 4.2

STRETCHING VIBRATIONS OF X—C≡N MOLECULES

	$\bar{v}_1$	X_1/X_2	X_3/X_2	$\bar{v}_2$	X_1/X_2	X_3/X_2	F_1	F_3
^{1}HCN	3312	−8.8	−0.25	2089	−1.8	1.0	5.8	17.9
^{2}HCN	2629	−2.6	−0.48	1906	−4.1	1.4	5.8	17.9
^{3}HCN	2460	−1.3	−0.59	1724	−9.7	2.8	5.8	17.9
FCN	2290	−0.19	−0.61	1077	−2.4	2.4	9.2	16.3
ClCN	2201	−0.05	−0.72	729	−0.88	1.4	5.2	16.7
BrCN	2187	−0.02	−0.75	580	−0.36	1.2	4.2	16.9
ICN	2158	−0.01	−0.77	470	−0.22	1.1	3.0	16.7

values given in the table. The amplitude ratios X_1/X_2 and X_3/X_2 can be
calculated using Eq. (4.13). The results are listed in Table 4.2 and illustrated
in Fig. 4.5.

In every case the higher frequency in cm^{-1} is the out-of-phase stretch
mode (stretch—contract). The approximations outlined earlier can be
further verified here since $\bar{v}_1$ for HCN is chiefly CH stretch with the N nearly
stationary and $\bar{v}_1$ for ICN is chiefly CN stretch with the I nearly stationary.
For deuterium and tritium cyanide, however, the ^{2}H—C and ^{3}H—C frequencies are rather similar to the C≡N frequency and strong interaction
occurs. Only for FCN, ClCN, BrCN, and ICN is $\bar{v}_1$ really a typical C≡N
vibration where only the C and N atoms move appreciably. For these cases
there is nearly constant triple bond—single bond interaction, and good
group frequencies result.

In every case $\bar{v}_2$ is the in-phase stretch mode (stretch–stretch) and again the
approximations stated earlier can be verified. For HCN, $\bar{v}_2$ is chiefly C≡N
stretch with the CH bond changing only a little in length, and for ICN $\bar{v}_2$
is chiefly C—I stretch with the CN bond hardly changing in length. However,
for ^{2}HCN and ^{3}HCN strong interaction occurs as it did in $\bar{v}_1$. From this

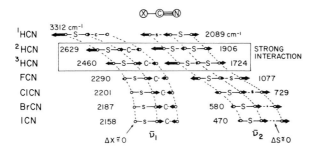

FIG. 4.5.   The stretching vibrations of various X—C≡N molecules.

result we see that the "nitrile" vibration in the halogen-substituted cyanides ($\bar{\nu}_1$) has the X—C bond cooperating with the C≡N bond in restoring equilibrium, while in the "nitrile" vibration in HCN ($\bar{\nu}_2$) the CH bond acts in opposition to the CN bond, which explains the lower frequency in cm^{-1}. In both deuterium and tritium cyanide, two vibrations strongly involve the nitrile group because of the strong interaction. As a result these two molecules do not have a vibrational frequency within the 2290–2089 cm^{-1} "nitrile" region shown by the other compounds.

4.10 Carbonyl and C=C Compounds

The stretching vibration of the C=O group is illustrated in Fig. 4.6(a). When the substituent is carbon or a heavier atom, it hardly moves during the carbonyl stretching vibration since the attached C—X oscillator has a low frequency compared to the C=O oscillator. This makes these vibrations good group frequencies. In a simple model, the slightly moving substituent can be replaced with a nonmoving infinite mass [Fig. 4.6(b)]. For a carbonyl with normal bond angles ($\sim 120°$) the C=O stretch frequency only decreases about 17 cm^{-1} when both substituent masses are changed from 14 (for CH_2) to ∞.

The contracted C—X attached bonds cooperate with the stretched C=O bond during the carbonyl stretch vibration and raise its frequency somewhat. The following treatment should help to visualize this effect. In Fig. 4.6(d) is illustrated a spring which is attached to an infinite mass (a wall) at one end. This spring is changed in length by displacing the free end a distance X at an

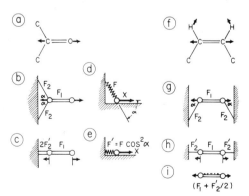

FIG. 4.6.   Stretching vibrations of C=O and C=C bonds, with diagrams used in text discussion.

angle α to the spring axis. For a small displacement in the $+X$ direction, the spring is lengthened by an amount $X \cos \alpha$, so the resorting force exerted by the spring in the spring axis direction is $FX \cos \alpha$, and the restoring force component in the $-X$ direction is $FX \cos^2 \alpha$. This is the same force that a spring oriented in the X direction exerts in the X direction if its force constant F' is equal to $F \cos^2 \alpha$ [Fig. 4.6(e)]. If we assume a bending force constant of zero we can use the above result to change the model in Fig. 4.6(b) into the linear "triatomic" model in Fig. 4.6(c) where the masses are ∞, 12, and 16 u and the force constants are $2F_{C-C} \cos^2 \alpha$ and $F_{C=O}$ and where α is one-half the X—C—X bond angle. Surprisingly, this simple model retains the most important characteristics of the carbonyl vibration (and other double bond stretch vibrations) which are as follows. The double bond stretching frequency in cm^{-1} is sensitive to changes in force constant of the double bond and, to a lesser extent, that of the attached C—X bonds. A change in double bond–single bond angle will shift the double bond stretching frequency. This occurs in cyclic double bond compounds. When the double bond–single bond angle is 90°, the frequency is lowest. As the angle becomes larger or smaller than 90°, the frequency increases.

The linear "triatomic" model in Fig. 4.6(c) can be treated using Eq. (4.11) where the highest λ and $\bar{v}$ values will be for the "carbonyl stretch" vibration. If F_{C-C} is 4.5 mdynes/Å and $F_{C=O}$ is 11.1 mdynes/Å, the effect of changing only the bond angle on the C=O stretching frequency can be calculated. The results along with the experimentally observed values are listed in Table 4.3.

In this model the "pure C=O" stretching frequency calculated with F_{C-C} equal to zero is 1660 cm^{-1}. This shows that for normal bond angles of 120° (cyclohexanone), the two attached C—C bonds together raise the "C=O" stretching frequency about 54 cm^{-1}. Although comparative data are not shown, the frequency shifts with ring size change calculated by this simplified model approach are in fairly good agreement with those that can be calculated by more sophisticated models. The relatively good agreement between calculated and observed frequencies indicates that most of the shifts are

TABLE 4.3

CALCULATED AND OBSERVED FREQUENCIES FOR CARBONYL STRETCHING VIBRATIONS

	Calculated (cm^{-1})	Experimental (cm^{-1})
Cyclohexanone	1714	1715
Cyclopentanone	1737	1740
Cyclobutanone	1774	1782
Cyclopropanone	1836	1822

caused by the mechanical effect of the geometry change and that no large change of double bond force constant is involved.

If both substituents on a C=O bond are carbon or heavier atoms and bond angles are normal ($\sim 120°$) then carbonyl frequency shifts can be related directly to mesomeric and inductive effects which account for the change in force constants of C=O bonds and substituent single bonds. In the inductive effect, electronegative substituents (like Cl) raise C=O frequencies by reducing the weight of carbonyl bond weakening resonance forms

$$(\overset{+}{C}-\overset{-}{O})$$

in the resonance hybrid by competing with the carbonyl oxygen's tendency to attract electrons, thus

$$
\begin{array}{ccc}
\overset{\displaystyle O}{\underset{\displaystyle \parallel}{}} & \overset{\displaystyle O}{\underset{\displaystyle \parallel}{}} & \overset{\displaystyle O^-}{\underset{\displaystyle |}{}} \\
CH_2-C-Cl & CH_2-C-CH_2 \longleftrightarrow & CH_2-C-CH_2 \\
1800\ cm^{-1} & 1715\ cm^{-1} & +
\end{array}
$$

Mesomeric effects involving conjugation or the donation of nonbonding electrons of attached heteroatoms also modify the C=O frequency.

$$
\begin{array}{lcl}
\overset{O}{\overset{\parallel}{}}\ \ H & & \overset{-O}{\overset{|}{}}\ \ H \\
CH_2-C-C=CH_2 \longleftrightarrow & CH_2-C=C-\overset{+}{C}H_2 & 1685\ cm^{-1} \\[2mm]
\overset{O}{\overset{\parallel}{}} & & \overset{-O}{\overset{|}{}} \\
CH_2-C-NH_2 \longleftrightarrow & CH_2-C=\overset{+}{N}H_2 & 1665\ cm^{-1} \\[2mm]
\overset{O}{\overset{\parallel}{}} & & \overset{-O}{\overset{|}{}} \\
CH_2-C-O-CH_3 \longleftrightarrow & CH_2-C=\overset{+}{O}-CH_3 & 1740\ cm^{-1}
\end{array}
$$

The greater the π electron delocalization, the weaker the C=O bond becomes.

When a carbon substituent on a carbonyl or double bond is replaced by a hydrogen, the mechanical interaction effects are somewhat different, but the overall mechanical effect is to lower the double bond stretching frequency slightly. In olefins the C=C stretching frequency is approximately 1680 cm^{-1} in tetralkyl ethylenes and 1623 cm^{-1} in ethylene with mono-, di-, and trialkyl ethylenes in between. During the double bond stretching vibration, the C—H bond hardly changes in length and the cooperative interaction effects provided by the replaced C—C bond is lost, but there is increased cooperative bending interaction with the =C—H in-plane bending vibration whose frequency is only about 300 cm^{-1} lower than the double bond stretch.

Interaction effects of double bond stretching and C—H vibrations are relatively constant for double bonds with the same hydrogen substitution. For a given substitution type, hydrogens can be left off simple models when investigating double bond frequency shifts as a result of changing force constants or bond angles. The group $R_2C=CH_2$ (minus its hydrogens) can be handled by the same treatment used for the $R_2C=O$ case where the terminal mass is changed from 16 to 12 and the C=C force constant is 8.9 mdynes/Å and only bond angles are varied. In Table 4.4 the frequencies calculated in this manner are compared with the experimentally observed values.

TABLE 4.4

CALCULATED AND OBSERVED FREQUENCIES FOR METHYLENE COMPOUNDS

	Calculated (cm^{-1})	Experimental (cm^{-1})
Methylene cyclohexane	1639	1651
Methylene cyclopentane	1661	1657
Methylene cyclobutane	1695	1678
Methylene cyclopropane	1756	1781

In cyclohexene the double bond in a *cis* configuration [Fig. 4.6(f)] forms one bond of the ring. If we leave off hydrogens and proceed as before, we can approximate the C=C vibration with the linear model in Fig. 4.6(h). Note that when the middle bond is stretched by an amount S the end bonds are each contracted by an amount $S/2$, so the restoring force on each mass is $F_1 S + F_2' S/2$ or $(F_1 + F_2'/2)S$. This means that the linear model, [Fig. 4.6(h)] can be replaced by the free diatomic model [Fig. 4.6(i)] which will have the same frequency if its force constant is $F_1 + F_2(\cos^2\alpha)/2$. Approximate frequencies for cycloalkenes can be calculated using Eq. (4.2) with both masses equal to 12 and with a force constant equal to $8.9 + 4.5\ (\cos^2\alpha)/2$ mdyne/Å. These results are shown in Table 4.5.

TABLE 4.5

CALCULATED AND OBSERVED FREQUENCIES FOR C=C
STRETCHING FREQUENCIES IN CYCLOAKENES

	Calculated (cm^{-1})	Experimental (cm^{-1})
Cyclohexene	1639	1646
Cyclopentene	1608	1611
Cyclobutene	1587	1566
Cyclopropene	1639	1641

Again these results are in fair agreement with those of more sophisticated calculations and indicate the frequency differences are mainly caused by geometry changes rather than force constant changes. In Fig. 4.7 similar results are illustrated using vibrating ball bearing and coil spring molecular models.[1]

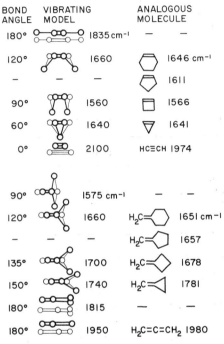

BOND ANGLE	VIBRATING MODEL		ANALOGOUS MOLECULE	
180°		1835 cm⁻¹	—	—
120°		1660		1646 cm⁻¹
—	—	—		1611
90°		1560		1566
60°		1640		1641
0°		2100	HC≡CH	1974
90°		1575 cm⁻¹	—	—
120°		1660	H₂C=	1651 cm⁻¹
—	—	—	H₂C=	1657
135°		1700	H₂C=	1678
150°		1740	H₂C=	1781
180°		1815	—	—
180°		1950	H₂C=C=CH₂	1980

FIG. 4.7. Vibrating mechanical model study showing the effect of geometry change on double bond stretching frequencies. In these models, masses and springs are unchanged. Only the angles are altered. The model wavenumbers are proportional to the model frequencies obtained. In these schematic representations of the vibrations it can be seen that in the 90° angle case each single bond remains the same length in the two extremes of the vibration illustrated. As the angles become increasingly larger or smaller than 90°, the single bond lengths are increasingly altered during these vibrations when the double bond is stretching. This added resistance to double bond stretching raises the frequency by an amount dependent on the angle.

4.11 Cyclic Compound Stretching Vibrations

Let us consider a model consisting of atoms of equal masses M bound in a regular polygon configuration where the M—M stretching force constant is F and the M—M—M angle is α. One type of ring vibration is the in-phase

[1] N. B. Colthup, *J. Chem. Educ.* **38**, 394 (1961).

stretching vibration (sometimes called ring breathing) seen in Fig. 4.8 where all the bondlengths change by the same amount in-phase and the atoms move radially. Such a vibration usually gives rise to a prominent band in the Raman spectrum. As shown in Fig. 4.8 this vibration does not change bond angles so a bending force constant is not involved. Each mass experiences a force from each attached bond directed along the attached bond axis. This force is equal to FS, where F is the bond stretching force constant and S is the amount the bond length differs from the equilibrium bond length. A force vector diagram is shown in Fig. 4.8. The total force experienced by each mass is the vector resultant of the attached bond forces. For the in-phase vibration, this force vector resultant has a force component directed along any bond equal to $FS(1 + \cos \alpha)$ and a force component directed perpendicular to the bond equal to $FS \sin \alpha$ (see Fig. 4.8). We have in essence an isolated diatomic model $M—M$ whose force constant is $F(1 + \cos \alpha)$, experiencing

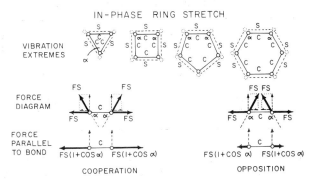

FIG. 4.8.   The in-phase ring stretching vibrations and force diagrams used in the analysis.

perpendicular forces which only translate the model. Since the same reasoning is true for every bond, this is a normal mode of vibration. Therefore, the frequency in cm^{-1} of the in-phase ring stretching vibration is

$$\bar{v}_{ip} = 1303 \sqrt{F(1 + \cos \alpha)\left(\frac{1}{M} + \frac{1}{M}\right)} \qquad (4.14)$$

Note that when $\alpha = 90°$ this equation is the same as that for the diatomic $M—M$ model. In Fig. 4.8 note that when α is less than 90° the forces cooperate and the frequency increases, whereas when α is greater than 90°, $\cos \alpha$ becomes negative and the forces act in opposition so the frequency is lower relative to that of the uncoupled $M\text{—}M$ oscillator. Using a half-angle formula from trigonometry, Eq. (4.14) becomes

$$\bar{\nu}_{ip} = 1303 \sqrt{\frac{F}{M}} \, 2 \cos \frac{\alpha}{2} \qquad (4.15)$$

which makes the in-phase ring stretching frequency directly proportional to $\cos \alpha/2$ for this model.

Experimental frequencies in cm^{-1} from Raman data[2] for the in-phase ring stretch vibrations for cyclohexane through cyclopropane and pentamethylene oxide through ethylene oxide are presented in Table 4.6. These can be seen to be roughly proportional to $\cos \alpha/2$.

TABLE 4.6

RING STRETCHING FREQUENCIES (IN cm^{-1})

Ring size	α (deg)	$1390 \times \cos \alpha/2$	In-phase $(CH_2)_n$	In-phase $O(CH_2)_{n-1}$	$1390 \times \sin \alpha/2$	Alternating $O(CH_2)_{n-1}$
$n = 6$	109.5	815	802	813	1155	1098
5	104	871	886	913	1114	1071
4	90	1000	1003	1028	1000	983
3	60	1224	1187	1270	707	877

Saturated six-membered rings are not planar and their effective $M\text{—}M\text{—}M$ angle is about 109.5°. Saturated five-membered rings are not quite planar so their ring angle is a little less than the pentagon angle of 108° (104° is used here).

For rings with an even number of masses, namely, six- and four-membered rings, there is an alternating stretching vibration where every other bond stretches while the intervening bonds contract (see Fig. 4.9). Using the same procedure and referring to Fig. 4.9 we see that the alternating stretching vibration in six- and four-membered rings is

$$\bar{\nu}_{alt} = 1303 \sqrt{F(1 - \cos \alpha)\left(\frac{1}{M} + \frac{1}{M}\right)}$$

$$\bar{\nu}_{alt} = 1303 \sqrt{\frac{F}{M}} \, 2 \sin \frac{\alpha}{2} \qquad (4.16)$$

[2] K. W. F. Kohlrausch, "Ramanspektren," p. 336. Akad. Verlagsges., Leipzig, 1943.

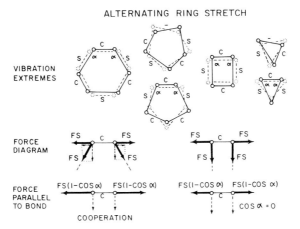

FIG. 4.9.   The "alternating" ring stretching vibrations and force diagrams used in the analysis.

These equations are exact for this model since the alternating stretching vibration does not change bond angles.

Since five- and three-membered rings do not have an even number of bonds they cannot have a full alternating stretching vibration. The closest they can come to this are the pseudo-alternating, doubly degenerate vibrations[3] shown in Fig. 4.9. In addition, since bond angles are changed for these, a bending force constant is involved, so Eqs. (4.16) are clearly only a crude first approximation, especially for the three-membered ring.

If one CH_2 group in a cyclic alkane is exchanged for an oxygen atom, then the polar C—O—C group motion in the alternating or pseudo-alternating stretch vibrations will usually give rise to a prominent absorption in the infrared spectrum. These are seen in the last column of Table 4.6 at 1098 cm^{-1} for the six-membered ring (pentamethylene oxide), 1071 cm^{-1} for the five-membered ring (tetrahydrofuran), 983 cm^{-1} for the four-membered ring (trimethylene oxide), and 877 cm^{-1} for the three-membered ring (ethylene oxide), showing at least a crude proportionality to sin $(\alpha/2)$. For the four-membered ring where the angle is $90°$, cooperative or opposition forces acting to restore bond lengths to equilibrium are effectively zero and both in-phase and alternating stretch vibrations are near 1000 cm^{-1}, which is close to the uncoupled C—O wavenumber in methanol (1033 cm^{-1}). When the forces cooperate or act in opposition the wavenumbers are higher or lower than this.

[3] G. Herzberg, "Infrared and Raman Spectra of Polyatomic Molecules." Van Nostrand-Reinhold, Princeton, New Jersey, 1945.

4.12   The Bent M_1—M_2—M_1 Stretching Frequencies

In Fig. 4.10 we illustrate the nonlinear M_1—M_2—M_1 model. Since the two bond oscillators are identical, we will have out-of-phase and in-phase bond stretching vibrations. On the left in Fig. 4.10 we illustrate the force vector diagram when one bond is slightly contracted and the other bond slightly stretched. The force resultant at M_2 has a force component directed along either bond axis equal to $FS(1 - \cos \alpha)$ which tries to change the length of the bond, and a force directed perpendicular to the bond which only serves to rotate the bond. We have in essence a "diatomic oscillator"

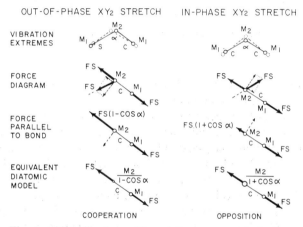

FIG. 4.10.   The stretching vibrations of the nonlinear symmetrical triatomic model and the force diagrams used in the analysis.

which differs from a conventional diatomic oscillator in that the forces on the two masses are unequal. The component of vibrational motion of M_2 in the bond-axis direction is specified by the component of acceleration in the bond-axis direction. The acceleration, which equals force/mass remains the same if both force and mass are divided by $(1 - \cos \alpha)$, in which case we get the diatomic model seen at the bottom left of Fig. 4.10 where the forces on the two masses are the same (FS) and the left-hand mass is $M_2/(1 - \cos \alpha)$. This mass will have the same bond-axis component of vibrational motion as M_2 which means that for the M_1—M_2—M_1 model the out-of-phase stretch frequency in cm^{-1} is

$$\bar{v}_{op} = 1303 \sqrt{F\left(\frac{1}{M_1} + \frac{1 - \cos \alpha}{M_2}\right)} \tag{4.17}$$

Since we get the same frequency in cm^{-1} for bond vibration for either bond, we have indeed been working on a normal mode of vibration.

For small amplitudes, the out-of-phase stretch vibration of the bent M_1—M_2—M_1 model does not change the bond angle so the bending force constant is not involved. This is not strictly true for the in-phase stretch vibration, however, and we can proceed for this vibration as before (Fig. 4.10, right side) *only* if the bending force constant is assumed to be zero (it is relatively small). In this case the in-phase stretching frequency in cm^{-1} for the M_1—M_2—M_1 model is

$$\bar{v}_{ip} = 1303\sqrt{F\left(\frac{1}{M_1} + \frac{1 + \cos \alpha}{M_2}\right)} \qquad (4.18)$$

The neglect of the bending force constant causes this equation to give slightly low results.

Let us consider such an XY_2 group as part of a molecule. It is clear that there is complete interaction between the two XY bonds during the stretching vibrations. However, if the XY_2 group, taken as a whole, is considered as an oscillator, and has stretching frequencies considerably higher than the frequencies of the bond oscillators connecting the XY_2 group to the rest of the molecule, then only the XY_2 group atoms move appreciably during the group stretching vibrations (approximation 1) which are good group frequencies. Examples of such XY_2 group stretching frequencies (approximate) are given in Table 4.7.

For the PH_2 group, the HPH angle is close to 90° ($\sim 93°$) and the middle mass is relatively heavy. From Eqs. (4.17) and (4.18) the PH_2 group should have stretching frequencies in cm^{-1} quite close together as is the case. These XY_2 equations also suggest that as long as the angle of a given XY_2 group is relatively constant, a change in force constant of the X—Y bond (caused by

TABLE 4.7

XY_2 GROUP STRETCHING FREQUENCIES (IN cm^{-1})

XY_2 Group	Out-of-phase	In-phase
R—NH_2	3370	3300
R_2—CH_2	2925	2850
R—PH_2	2300	2300
R_2—BH_2	2600	2500
R_2—SO_2	1310	1130
R—NO_2	1550	1375
R—$CO_2{}^-$	1600	1420

a substituent change) will cause both stretching frequencies in cm^{-1} to shift but that their *ratio* should remain relatively constant. This has been verified in two examples where for the amino NH_2 group $\bar{\nu}_{ip}/\bar{\nu}_{op} = 0.98$ and for the SO_2 group $\bar{\nu}_{ip}/\bar{\nu}_{op} = 0.86$. When the ratio is not constant, it is usually because of a change in the interaction of the in-phase XY_2 stretch with the stretching of the attaching bonds (an example is the NO_2 group), or because of unsymmetrical hydrogen bonding.

4.13 Noncyclic Single Bond Vibrations

In Fig. 4.11 we illustrate the alternating and in-phase stretching modes of an infinite chain model in a planar zig-zag form. Using the same procedure as before the reader can verify that the alternating stretch frequency in cm^{-1} is the same as Eq. (4.16) for rings.

$$\bar{\nu}_{alt} = 1303 \sqrt{F(1 - \cos \alpha)\left(\frac{1}{M} + \frac{1}{M}\right)} \qquad (4.19)$$

In linear alkane alcohols, ethers, and primary and secondary amines we have skeletal vibrations which involve the stretching of C—C, C—O, or C—N bonds which individually in CH_3—CH_3 (992 cm^{-1}), CH_3—OH (1033 cm^{-1}), and CH_3—NH_2 (1036 cm^{-1}) have nearly the same frequency in cm^{-1}. As a result we will have skeletal single bond interaction. When the C—C—O, C—O—C, C—C—N, or C—N—C bonds stretch in an alternating manner a prominant infrared band results. Such is the case during the

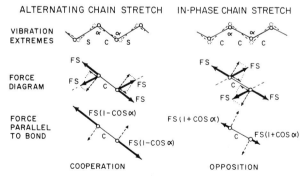

FIG. 4.11. The in-phase and "alternating" stretching vibrations in a zig-zag chain model along with force diagrams.

alternating stretch mode described in our long chain model [Eq. (4.19)] re-arranged to give

$$\bar{\nu}_{\text{alt}} = 1303\sqrt{\frac{F}{M}}\sqrt{2 - 2\cos\alpha} \qquad (4.20)$$

and the out-of-phase stretch frequency in cm^{-1} of the short chain M—M—M model [Eq. (4.17)]

$$\bar{\nu}_{\text{op}} = 1303\sqrt{\frac{F}{M}}\sqrt{2 - \cos\alpha} \qquad (4.21)$$

This suggests that the compounds discussed above should have alternating stretch group frequencies between the limits of the above two equations regardless of the chain length. We do indeed have such group frequencies for primary alcohols (~ 1050 cm^{-1}), ethers (~ 1125 cm^{-1}), primary amines (~ 1080 cm^{-1}), and secondary amines (~ 1135 cm^{-1}) in the single bond stretch region. We can also note that the C—O—C and C—NH—C groups in unstrained six-membered rings [see Eq. (4.16)] can be included with such correlations, since the model equations are the same as the chain model when the angles are the same.

In Fig. 4.12 we illustrate an actual study of ball bearing and coil spring vibrating models representing skeletal chain vibrations. Note that the highest frequency vibration is a "group frequency" when there are two or more bonds. Therefore the C—O—C or O—C—C group does not give rise to a group frequency because of zero interaction with the rest of the molecule, but because of relatively constant interaction, in particular with other connected C—C bonds. *This relatively constant mechanical interaction is what gives rise to a group frequency.*

The alternating stretch does not involve angle bending but the in-phase stretch vibration does. If the bending force constant is assumed to be zero, then, as seen in Fig. 4.11, the in-phase chain stretching frequency is the same as the in-phase ring stretching frequency [Eq. (4.14)] for comparable bond angles. If the bending force constant (Δ) is not zero, then the proper expression (which will not be derived here) for the in-phase chain stretching frequency is

$$\bar{\nu}_{\text{ip}} = 1303\sqrt{\left[F(1 + \cos\alpha) + \frac{4\Delta}{l^2}(1 - \cos\alpha)\right]\left(\frac{1}{M} + \frac{1}{M}\right)} \qquad (4.22)$$

where l is the bondlength and Δ/l^2 has the same units as F. For single bonds, Δ/l^2 is roughly an order of magnitude smaller than F. On the assumption that $\Delta/l^2 = 0.1F$ and $\alpha = 109.5°$ then Eq. (4.22) becomes $\bar{\nu}_{\text{ip}} = 1.55 \times [1303(F/M)^{1/2}]$ which can be compared with the comparable in-phase ring

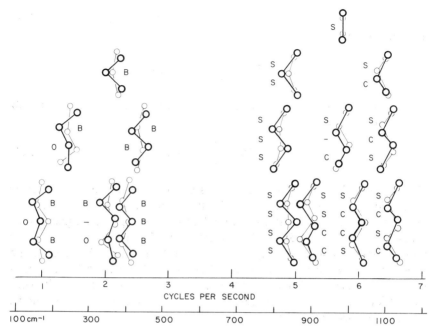

FIG. 4.12. Vibrations of ball and spring models representing the planar skeletal vibrations of some normal hydrocarbons. The extremes of the vibrations are illustrated. For one of these extremes, bonds which are stretched are labeled S, and those which are contracted, C. Those angles which become smaller are labeled B for "bend," and those which become larger, O for "open." A dash (—) means no change. Nonplanar skeletal vibrations are not shown. The butane model has one nonplanar vibration, the torsion of the middle bond, and the pentane model has two, the in- and out-of-phase torsion of the two middle bonds. Model construction: 7/16-in. diameter ball bearings, 16 mil helical spring, 9 coils 11/16 in. diameter, 2 in. between atoms, supporting threads 8 ft long.

frequency [Eq. (4.14)] where $\bar{v}_{ip} = 1.15[1303(F/M)^{1/2}]$. The frequency ratio is 1.35. The in-phase C—C stretch frequency for polyethylene, 1133 cm^{-1}, and cycloxexane, 803 cm^{-1}, can be compared with the above rough calculations, which show the effect of C—C—C bending interaction in the infinite chain.

4.14   Bend–Bend Interaction

(a) SKELETAL SINGLE BOND BENDING VIBRATIONS

The C—C—C bending oscillator in propane vibrates at a lower frequency than C—C stretching or CH bending or stretching vibrations, so according to our approximations we will have a vibration where the only oscillator appreciably deformed is the C—C—C bending oscillator. In butane we have two such oscillators coupled together (see Fig. 4.13) which will both be excited at the same time so the C—C—C—C group must be considered as a whole. During the in-phase vibration (421 cm^{-1}) (bend–bend) both C—C—C "hinge springs" in butane exert cooperating restoring forces on the two common middle masses. We can say that the "hinge springs" cooperate in restoring the two bond angles to equilibrium because both rotate the common middle bond in the same direction. During the out-of-phase vibration (271 cm^{-1}) (bend–open) both C—C—C "hinge springs" exert forces tending to oppose each other on the two common middle masses and act in opposition

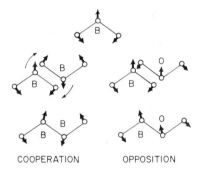

COOPERATION   OPPOSITION

FIG. 4.13. The top figure shows the bending vibration of the propane skeleton. The other figures show the bending vibrations of the n-butane skeleton where B stands for bend and O stands for open.

in rotating the middle bond which, as a result, does not rotate at all. These effects cause the butane bending frequencies (421 and 271 cm^{-1}) to diverge from the propane bending frequency at 375 cm^{-1}. See Fig. 4.12 for actual model bending vibrations.

(b) CH$_2$ DEFORMATION INTERACTION

Consider the group —CH$_2$—CH$_2$—. In Fig. 4.14(a) and (b) both CH$_2$ groups are performing the deformation vibration (deforming the shape of the CH$_2$ group by changing the HCH angle). Because they have similar individual frequencies both vibrate together, either in- or out-of-phase. However, the in-phase and out-of-phase vibrations have nearly the same frequency because the CH$_2$ deformation oscillators are weakly coupled. The deformation of one CH$_2$ group hardly exerts any force on the carbon of the next CH$_2$ group because of the intervening low frequency C—C bond, so the cooperative or opposition forces they exert on each other are quite

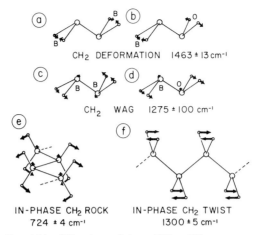

FIG. 4.14.   Vibrations of the $-CH_2-CH_2-$ group.

weak. As a result the restoring forces on one CH_2 group are nearly independent of what the next CH_2 group is doing. In a chain of CH_2 groups of the same type, all the CH_2 deformation vibrations can be considered together as one type of group frequency. The same arguments apply to CH_2 in- and out-of-phase stretching vibrations.

(c) CH_2 WAGGING INTERACTION

In the CH_2 wagging vibration [Fig. 4.14(c) and (d)] the CH_2 group rotates as a whole and the HCH angle does not change much. The vibration mainly changes the angle between the CH_2 plane and the adjacent C—C bonds. Therefore this vibration moves the carbon of the next CH_2 group and two adjacent CH_2 wag oscillators are more strongly coupled. As a result, the in- and out-of-phase CH_2—CH_2 wag vibrations have frequencies further apart than the deformations. In a CH_2 chain, the CH_2 deformation vibrations are found in a relatively narrow frequency region of the spectrum for a given type of CH_2 group, whereas the CH_2 wag vibrations are spread out over a wider frequency region.

(d) CH_2 ROCK and TWIST INTERACTION

In the CH_2 rock and twist vibrations illustrated in Fig. 4.14(e) and (f) the CH_2 group rotates as a whole and, as in the wagging vibration, this motion is compensated for by a counter-rotation of both attached C—C bonds. As in the case of the wag vibrations this means that there is coupling between two adjacent rocking CH_2 groups, which tends to spread CH_2 rock and twist vibrations over a region. In a CH_2 noncyclic chain, the *in-phase* rock [see in Fig. 4.14(e)] has the lowest frequency and is the most intense of the rock bands in the infrared spectrum since all the hydrogens move in the same direction while all the carbons move oppositely. This in-phase vibration has nearly the same frequency from $-(CH_2)_4-$ to $-(CH_2)_\infty-$ because in all these chains each CH_2 experiences virtually identical forces from its neighbors. Since the interaction is constant this makes the in-phase CH_2 rock vibration in a long CH_2 chain, a good group frequency. Similar arguments can be applied to the *in-phase* CH_2 twist vibration in CH_2 chains. This is the highest of the twist vibrational frequencies and is a good group frequency in the Raman spectra of CH_2 chains.

(e) ACETYLENES C—H BEND

In the group $X-C\equiv C-H$ the $C\equiv C-H$ bending oscillator has a higher frequency than the $X-C\equiv C$ bending oscillator when X is heavier than hydrogen. According to the approximations stated earlier we should have a good $C\equiv C-H$ bending group frequency (~ 650 cm^{-1}) because during this vibration the X atom is nearly stationary, making this vibration mechanically insensitive to the nature of X.

(f) OLEFINIC C—H OUT-OF-PLANE BEND OR WAG

Let us consider the vinylidene group $X_2C=CH_2$ [Fig. 4.15(a)]. There are two identical C—H bonds which can bend out of the molecular plane, either in-phase or out-of-phase. The only one which is infrared active is the in-phase

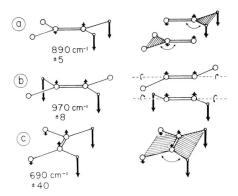

FIG. 4.15.   Out-of-plane C—H bending vibrations in some olefinic groups.

bend which is a change in the angle between the C=C bond and the CH_2 plane. This C=C—H_2 bending oscillator has a higher bending frequency than the X_2—C=C bending oscillator to which it is coupled so that during the CH_2 out-of-plane bending vibration the X atoms are nearly stationary. This makes this vibration a good group frequency in that it is mechanically relatively independent of X. All the wavenumber shifts for this vibration can be related to changes in the chemical nature of X (namely, whether it is an electron donor or acceptor) which can modify the bending force constants.

Consider the *trans*-disubstituted ethylene performing the in-phase, out-of-plane hydrogen bending vibration [Fig. 4.15(b)]. This vibration twists the C=C bond. The *trans* H—C=C—H group can bend in-phase with the X atoms hardly moving because the H—C=C—H twist oscillator is higher in frequency than the X—C=C—X twist oscillator to which it is coupled, making this a good group frequency.

Consider the *cis*-disubstituted ethylene performing the in-phase, out-of-plane hydrogen bending vibration. In Fig. 4.15(c) we can see that this vibration rotates the H—C=C—H plane relative to the X—C=C—X plane, so in reality this vibration is an X—C—H bending vibration and the X atom is mechanically involved. This makes this *cis* vibration mechanically sensitive to the nature of X and makes this vibration a poorer group frequency than the *trans* or vinylidine vibrations discussed above.

One may well wonder why the in-phase *cis*-CH wag near 690 cm^{-1} is so much lower in frequency than the in-phase *trans*-CH wag near 970 cm^{-1} in the infrared spectrum. In Fig. 4.16, the lone CH wag of the C_2C=CHC group has a frequency near 825 cm^{-1} in hydrocarbons. It can be seen that this vibration twists the C=C bond. Also in Fig. 4.16 are seen the in-phase and out-of-phase CH wag vibrations in *trans*- and *cis*-C—CH=CH—C groups. Only the in-phase wag is infrared active. In the *trans* isomer this in-phase vibration gives the C=C bond a double twist, and in the *cis* isomer the

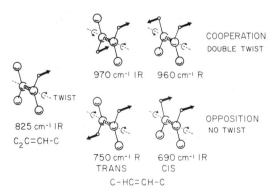

FIG. 4.16.   Out-of-plane C—H bending vibrations in *cis*-and *trans*-olefinic groups.

C=C bond is not twisted at all. This raises and lowers the *trans* and *cis* wag frequencies compared to that of the lone CH wag near 825 cm^{-1}. The out-of-phase CH wag vibrations in *trans* and *cis* isomers are Raman active.

(g) AROMATIC C—H OUT-OF-PLANE BEND OR WAG

Consider a pentasubstituted benzene ring (Fig. 4.17a). The isolated C—H out-of-plane bending vibration ($\sim$870 cm^{-1}) will be a group frequency because it is higher in frequency than other out-of-plane vibrations such as C—X bending with which it could interact. In this vibration part of the restoring force is a ring CC twisting restoring force. The CC bond is twisted because when a C—H bond bends out of the plane the angle between the HCC plane and the adjacent CCX plane is changed, where X is the substituent

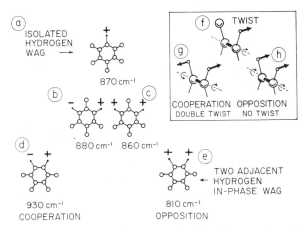

FIG. 4.17.   Out-of-plane C—H bending vibrations in some aromatic rings.

on the carbon adjacent to the C—H bond [see Fig. 4.17(f)]. Consider a tetra-substituted benzene ring in Figs. 4.17(b) and 4.17(c). The two C—H bonds being identical will bend together out of the plane either in-phase (infrared active) or out-of-phase (infrared inactive). (The substituents X are nearly motionless as before.) If the two hydrogens are not on adjacent carbons, both vibrations still involve changing the angle between the HCC plane and the adjacent CCX plane, twisting the adjacent CC bonds as before for the pentasubstituted case. Each bending C—H bond is between two nearly stationary C—X bonds and is only weakly coupled to the other C—H bond so the wavenumbers are still in the "isolated hydrogen" region. If the two hydrogens are on adjacent carbons as in Fig. 4.17(d) and (e), then the C—H bending oscillators are more strongly coupled. The CC bonds between the

C—H and adjacent C—X bonds are twisted as before. However, the CC bond between the two adjacent C—H bonds is given a double twist for the out-of-phase vibration [Fig. 4.17(g)] and no twist at all for the in-phase vibration [Fig. 4.17(h)]. Here we have twist forces cooperating or acting in opposition respectively in restoring equilibrium. Therefore the infrared-active in-phase out-of-plane vibration for two adjacent hydrogens (~ 810 cm^{-1}) is expected at a lower frequency than isolated C—H vibrations (~ 870 cm^{-1}). In general, it will be found that as the number of adjacent hydrogens increases, the frequency of the in-phase, out-of-plane C—H bend vibration decreases, giving rise to very useful group frequencies which can help distinguish *ortho-*, *meta-*, and *para*-substituted benzenes.

4.15   Bend Stretch Interaction

(a) XY$_2$ DEFORMATION

Imagine a bent XY$_2$ group attached to the rest of the molecule by one or two X—Z bonds. If the frequency of the XY$_2$ bending oscillator is considerably higher than the frequency of the X—Z stretching oscillator, then, during the XY$_2$ bending vibration only the XY$_2$ atoms move appreciably (approximation 1) and a good group frequency results. Examples of this are CH$_2$ deformation (~ 1460 cm^{-1}) and NH$_2$ deformation (~ 1600 cm^{-1}). We might imagine that if either of these groups were deuterated, the CD$_2$ and ND$_2$ oscillator deformation frequencies would be much closer to the attached C—C and N—C oscillator frequencies with greater possibilities for stronger interaction.

In the case of the C=CH$_2$ group the CH$_2$ oscillator deformation frequency is lower than the frequency of the attached C=C bond, so during this type of CH$_2$ deformation the attached C=C bond hardly changes in length. The relatively constant interaction results in a good group frequency (~ 1420 cm^{-1}).

Since the XY$_2$ deformation frequency is considerably lower than the XY stretching frequencies, a reasonably good approximation is that the XY bond hardly changes in length during the bending vibration (approximation 2). If the XY bonds are assumed to be rigid (which implies an infinitely large XY force constant) then the isolated XY$_2$ model deformation frequency in cm^{-1} is given (without derivation) as

$$\bar{v} = 1303 \sqrt{\frac{2\Delta}{l^2}\left(\frac{1}{M_Y} + \frac{1 - \cos \alpha}{M_X}\right)} \qquad (4.23)$$

which is the same as Eq. (4.17) for the out-of-phase XY_2 stretching frequency in cm^{-1}, but with F replaced by $2\Delta/l^2$ where Δ is the bending force constant and l is the X—Y bond length. On the average, Δ/l^2 is an order of magnitude smaller than F. This means that on the average the XY_2 deformation frequency should be a factor of very roughly $0.2^{1/2}$ lower than or somewhat less than half the out-of-phase XY_2 stretching frequency. This is roughly the case if the XY_2 bending vibration does not interact strongly with the attached X—Z stretching vibration. Some deformation and out-of-phase stretching frequencies are as follows: 1600 and 3370 cm^{-1} for NH_2, 1460 and 2925 cm^{-1} for CH_2, 1160 and 2600 cm^{-1} for BH_2, 1080 and 2300 cm^{-1} for PH_2, and 930 and 2140 cm^{-1} for SiH_2.

(b) THE H—X—Y GROUP

Let us consider a bent H—X—Y oscillator illustrated in Fig. 4.18(a) where X and Y are heavier than hydrogen, say nitrogen and carbon. The H—X stretching oscillator is much higher in frequency than the X—Y stretching oscillator, but it may happen that the H—X bending frequency may be nearly the same as that of the X—Y stretching oscillator. In such a case according to the principles stated earlier we should have two vibrations strongly involving both X—Y stretch and H—X—Y bend in two different phasings described as bend–stretch and open–stretch [see Figs. 4.18(a) and (b)]. A bending oscillator can be considered as two bonds connected at one end and rotating in opposite directions as in Fig. 4.18(e) and (d). The rotation of the X—Y

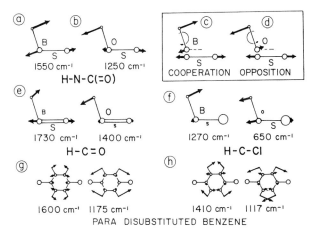

FIG. 4.18. Stretching–bending interaction in some H—X—Y groups. S stands for stretch, B for bend, and O for open.

bond has little effect on the X—Y stretching oscillation since the motions are essentially perpendicular to each other. However, the rotation of the H—X bond tends to move the X atom in either the opposite or the same direction that it moves during the X—Y stretching oscillation. This means that in the bend–stretch mode we have the cooperation condition and in the open–stretch we have the opposition condition, and these will have higher and lower frequencies respectively than a pure bend or a pure stretch mode would have. An example of such a group is the HNC group of a monosubstituted amide H—N—C=O where the bend–stretch mode absorbs at about 1550 cm^{-1} and the open–stretch mode absorbs at about 1250 cm^{-1}.

Let us see what happens to this interaction when the X—Y stretching oscillator is higher or lower in frequency than the hydrogen bending oscillator. When the X—Y oscillator is C=O or C=C, then its frequency is high relative to C—H bending. During the C=O or C=C stretching vibration the high frequency C—H oscillator hardly changes in length but there is a little bending motion of the hydrogen as indicated in Fig. 4.18(e), so that there is a little cooperative C—H bending interaction when the C=O or C=C stretches. For the C—H bending vibration the C=O or C=C oscillators hardly change in length.

When the X—Y oscillator is C—Cl [see Fig. 4.18(f)], then its frequency is low relative to C—H bending. During the C—H bending vibration the Cl is almost stationary but there is a little C—Cl stretching, and during the C—Cl stretching vibration the H—C—Cl angle hardly changes.

In Fig. 4.18(g) and (h) we illustrate vibrations of benzene, *para*-substituted with atoms such as chlorine. We have selected vibration pairs where the carbon displacements are nearly the same and each hydrogen move either oppositely (cooperation) or in about the same direction (opposition) as its attached carbon. In Fig. 4.18(g) the displacements are similar to the H—C=O system [Fig. 4.18(e)] so that we have a relatively high ring stretching frequency interacting a little with a low frequency C—H bending vibration. In Fig. 4.18(h) the displacements are similar to the H—N—C system [Fig. 4.18(a) and (b)], indicating relatively strong interaction between a ring stretching mode and a C—H bending mode which have frequencies relatively close together.

(c) THE C—Cl STRETCHING VIBRATION

In Fig. 4.19 we illustrate a " C—Cl stretching " vibration in a chloroalkane. These compounds frequently can exist as rotational isomers where either a hydrogen or a carbon is *trans* to the chlorine as shown. For both cases during the " C—Cl stretching " vibration, while the C—Cl bond stretches, the low

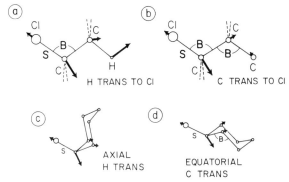

FIG. 4.19.   Some C—Cl stretching vibrations where S stands for stretch and B for bend. When a carbon is *trans* to the chlorine, the $C_1—C_2—C_3$ bond angle is sharply bent during the C—Cl stretching vibration.

frequency Cl—C—C bending oscillator yields by bending. However, when a hydrogen is *trans* to the chlorine [Fig. 4.19(a)] the high frequency C—C—H bending oscillator will hardly be changed in bond angle, but when a carbon is *trans* to the chlorine [Fig. 4.19(b)], the low frequency C—C—C bending oscillator yields and is sharply bent during the C—Cl stretching vibration. Here we have the C—C—C bending oscillator cooperating with the C—Cl stretching oscillator in restoring equilibrium which raises the frequency relative to the case where a hydrogen is *trans* to a chlorine.[4] For the group Cl—CH$_2$—CH$_2$—C the C—Cl stretching wavenumbers are about 725 cm^{-1} (carbon *trans*) and about 650 cm^{-1} (hydrogen *trans*), the major part of the difference being to the interaction effect described above. We can also note that equatorial chlorocycloalkanes (carbon *trans*) have a C—Cl stretch at about 750 cm^{-1} and axial chlorocycloalkanes (hydrogen *trans*) have a C—Cl stretch at about 680 cm^{-1} [see Fig. 4.19(c) and (d)].

4.16   Multiple Oscillator Groups

When more than two oscillators with similar frequencies are coupled together, all the oscillators must be considered together in picturing the normal modes of vibration which may be quite complex. In such cases it may be helpful to picture normal modes of vibrations as standing waves. In Fig. 4.20 we illustrate some of the stretching vibrations of a straight classical coil spring which has mass. The lowest frequency is whole chain stretch, the second lowest is half-chain stretch, the third lowest is one-third chain stretch,

[4] N. B. Colthup, *Spectrochim. Acta* **20**, 1843 (1964).

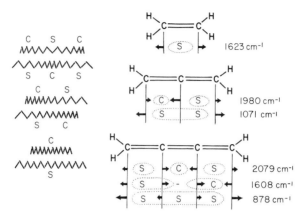

FIG. 4.20.   The CC stretching vibrations of ethylene, allene, and butatriene. The vibrations (S for stretch and C for contract) are compared with the standing waves of a helical spring with mass.

and so forth. These are compared with the C=C stretching vibrations of ethylene, allene, and butatriene in Fig. 4.20. The same approach can be used for sorting out both stretching and bending vibrations of the nonlinear chain models in Fig. 4.12. As we shall see, such an approach is quite useful in picturing the vibrations of aromatic rings and other groups.

4.17   Interaction Force Constant Effects

In the *keto* form of 2,4-pentanedione, CH_3—CO—CH_2—CO—CH_3, there will be C=O interaction in that we will have in-phase and out-of-phase C=O stretch vibrations. These appear at 1725 and 1707 cm^{-1} at nearly the same frequency because the C=O, C=O coupling is weak. If we replace this central CH_2 with an oxygen to make an anhydride, CH_3—CO—O—CO—CH_3, the carbonyl vibrations should be mechanically similar. However, the C=O frequencies at 1833 cm^{-1} (in-phase) and 1764 cm^{-1} (out-of-phase) are further apart than expected. This unexpectedly large separation comes about because of an effect not taken into account in a simple valence force field.

In a simple valence force field a bending or stretching oscillator can be thought of as a simple spring which exerts forces on its masses equal to a force constant times its distortion from equilibrium. The restoring force does *not* depend on what the other oscillators in the molecule are doing. However, in reality, if two or more oscillators in a molecule are distorted from equilibrium, they may be able to affect each others restoring forces to some extent because of a change in electronic structure. In this case, a more general valence force field is necessary where the restoring force on one oscillator is equal to a force constant times its own distortion *plus* a second force constant times the distortion of a second oscillator, and so forth. The second force constant, which can be positive or negative, is called an interaction force constant and is usually about an order of magnitude smaller than the main force constants. When interaction force constants are not zero they can cause coupling effects we have not been considering up to now.

In Fig. 4.21(a) and (b) we see the in-phase and out-of-phase C=O stretch vibrations of an anhydride. In the case of the out-of-phase stretch, at the extremes of the vibration seen in Fig. 4.21(b), one or the other of two charged

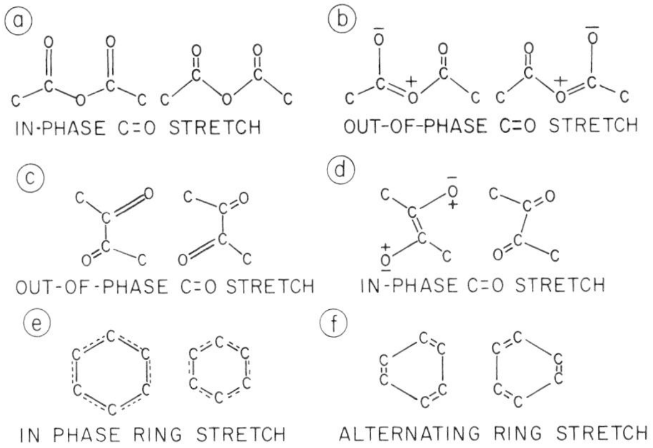

FIG. 4.21. Carbonyl vibrations in acetic anhydride and biacetyl and some ring vibrations in benzene. The vibrations on the right will be at lower frequencies than expected because, during the vibration, electrons can partially slip away from bonds becoming longer and slip into bonds becoming shorter. The vibrational extremes are illustrated.

resonance forms are favored because of the bondlength changes and have greater relative weights. The electrons in this case can partially slip away from the bonds which become longer and into bonds which become shorter. The result of this is that the restoring force for the out-of-phase vibration [Fig. 4.21(b)] is less than that for the in-phase vibration [Fig. 4.2(a)] where this kind of electron redistribution during the vibration cannot occur nearly as easily.[5]

The in- and out-of-phase C=O stretch vibrations of 2,3-butanedione, $CH_3-CO-CO-CH_3$, are illustrated in Fig. 4.21(c) and (d). During the in-phase vibration [Fig. 4.21(d)] the contracted central single bond cooperates with the stretched C=O bonds in restoring equilibrium. Therefore this

[5] N. B. Colthup and M. K. Orloff, *Spectrochim. Acta, Part A* **30**, 425 (1974).

vibration is expected to have a higher frequency than the out-of-phase vibration [Fig. 4.21(c)] where the central single bond is not deformed at all. A rough, simple valence force field calculation predicts a spacing of about 90 cm^{-1} but experimentally the frequencies are nearly the same, 1725 cm^{-1} for the in-phase stretch and 1713 cm^{-1} for the out-of-phase stretch. Again the unexpectedly small spacing is explained by a change in electronic structure that can occur during the in-phase vibration [Fig. 4.21(d)] to a much greater extent than in the out-of-in phase vibration [Fig. 4.21(c)]. For the in-phase stretch, electrons can partially slip away from bonds which become longer into bonds which become shorter, so the in-phase stretch restoring force is lower than that for the out-of-phase stretch.[5]

The alternating ring stretch of the benzene CC bonds is illustrated in Fig. 4.21(f). During this vibration the electronic structure can change and approach the Kekulé benzene structure form where the electrons have partially slipped away from the stretched bonds into contracted bonds. This causes the total restoring force to be lower than that for the in-phase stretch [Fig. 4.21(e)] at 992 cm^{-1} where such an electronic structure change during the vibration cannot occur. This causes the frequency of the alternating ring stretch vibration of benzene at 1311 cm^{-1} to be considerably lower than expected from a simple valence force field calculation (about 1700 cm^{-1}).

4.18 Hydrogen Bonding

Hydrogen bonding[6-10] has been recognized as an interaction between an X–H group of a molecule with a Y atom, usually of another molecule, so that a single hydrogen atom is associated with two atoms, X—H $\cdots$ Y. The atoms usually involved in strong hydrogen bond formation are those with high electronegativities such as F, O, N, and Cl. When X has a high electro-negativity the X—H bond is partially ionic in character. It has been observed that the higher the acidity of the X—H group, the stronger is the X—H $\cdots$ Y hydrogen bond. In this type of bond the Y atom has lone pair electrons in a nonspherical orbital such as sp hybrid orbitals which associate with the hydrogen atom through the hydrogen bond. It has been observed that the higher the electron donation ability (higher basicity) of Y, the stronger is the hydrogen bond. Typical hydrogen bonding acid groups include FH, OH, NH,

[6] G. C. Pimentel and A. L. McClellan, "The Hydrogen Bond." Freeman, San Francisco, California, 1960.
[7] C. G. Cannon, Spectrochim. Acta 10, 341 (1958].
[8] C. A. Coulson, Research 10, 149 (1957).
[9] A. S. N. Murthy and C. N. R. Rao, Appl. Spectrosc. Rev. 2, 1 (1968).
[10] P. A. Kollman and L. C. Allen, Chem. Rev. 72, 283 (1972).

and ClH groups. The CH in Cl_3CH and $\equiv C-H$ groups also show weak bonding properties. Typical hydrogen bonding base groups include the flourine in F^- and HF; the oxygen in H_2O, $C=O$, $H-O-C$, and $C-O-C$ groups; the nitrogen in NH_3, amines, and pyridines; and the chlorine ion Cl^-, for example. Aromatic compounds with their π electron clouds also have weak hydrogen bond base characteristics.

In boranes the $B \cdots H \cdots B$ bridge has similar characteristics to normal hydrogen bonds (where X and Y have high electronegativities) even though the BH bond and the electron deficient B atom do not fit in with the normal hydrogen bonding acids and bases. In a normal $X-H \cdots Y$ hydrogen bond there are four electron. Two electrons in the $X-H$ bond are somewhat delocalized to include Y in a bonding orbital, and two electrons on the Y atom are somewhat delocalized to include X in a nonbonding orbital. This second nonbonding orbital is stabilized if Y is a good electron donor and X is electronegative. In the electron deficient boranes there are only two electrons in the $B \cdots H \cdots B$ bridge which go in the bonding orbital so the second nonbonding orbital is unoccupied and the electronegativity requirements for X and Y are relaxed.[6]

Infrared and Raman spectroscopy provide two of the best means of detecting the hydrogen bond. The unbonded $X-H$ stretching band is usually relatively sharp when a compound such as an alcohol is run in a dilute solution in CCl_4, for example, or in the vapor state. In more concentrated solutions or in the condensed state, the hydrogen bond $X-H \cdots Y$ is formed. Then the hydrogen stretching band shifts to lower wavenumbers and becomes much broader in both the infrared and Raman spectrum.

In the infrared spectrum this band also becomes much more intense (see Fig. 4.22). The amount of the $X-H$ stretch wavenumber shift on formation of the $X-H \cdots Y$ hydrogen bond has been correlated with the bandwidth and band intensity for certain cases.[6] This shift has also been correlated with the $X \cdots Y$ distance, the smaller the distance the larger the shift.[11,12]

[11] R. C. Lord and R. E. Merrifield, *J. Chem. Phys.* **21**, 166 (1953).
[12] K. Nakamoto, M. Margoshes, and R. E. Rundle, *J. Amer. Chem. Soc.* **77**, 6480 (1955).

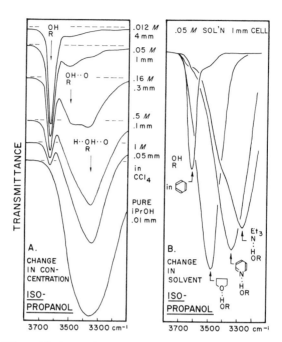

FIG. 4.22. Effect of hydrogen bonding on the OH stretching vibration of isopropanol. (A) Shows the effect of changing the isopropanol concentration in CCl_4 (M = molar) and compensating for dilution by increasing the cell thickness (given in mm). The number of isopropanol molecules in the spectrometer beam remains constant. The background level is adjusted to prevent overlap. (B) Shows the effect of changing the solvent (benzene, tetra-hydrofuran, pyridine, and triethylamine) while leaving the isopropanol concentration (0.05 M) and the cell thickness (1 mm) unchanged.

Bands involving X—H bending usually increase in wavenumber on the formation of the hydrogen bond, the shift being in the reverse direction from that in the stretching vibration, and the magnitude of the shift being notably smaller. The intensities and bandwidths of X—H bending bands do not change as much as the X—H stretching bands on the formation of the hydrogen bond.

After the formation of the hydrogen bond, new low frequency bands can be seen which involve the stretching and bending of the H ··· Y hydrogen bond. Also, vibrations involving the Y atom may be altered since the force field around Y is altered as a result of the formation of the hydrogen bond. An example is the shift in a C=O wavenumber in hydrogen bonding solvents.

Hydrogen bonding studies can sometimes be useful in structure determination where intramolecular hydrogen bonding may or may not occur, such as in the hydroxyacetophenones.

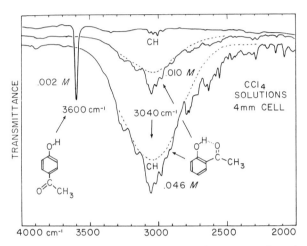

In the condensed state, both the *ortho* and *para* isomers are hydrogen bonded. In dilute CCl_4 solutions the hydrogen bonding in the *para* isomer is broken but it remains unbroken in the internally bonded *ortho* isomer (see Fig. 4.23).

Fig. 4.23.   Hydroxyacetophenones in CCl_4 solution in a 4-mm cell. In the *para* isomer the free OH stretch in dilute solution gives rise to a sharp band at 3600 cm^{-1} which is much shifted from the intermolecularly bonded OH $\cdots$ O stretch broad band in the solid state at 3130 cm^{-1} (not shown). In the *ortho* isomer the internally bonded OH $\cdots$ O group gives rise to a broad band centering at 3040 cm^{-1} with superimposed CH stretch bands and combination bands intensified by Fermi resonance. The OH $\cdots$ O group remains internally bonded even in dilute solution (shown here in two concentrations) in marked contrast to that for the *para* isomer.

CHAPTER 5

METHYL AND METHYLENE GROUPS

5.1 Introduction to Group Frequencies

In this chapter and following chapters, spectra-structure correlations will be discussed in detail. These will be infrared correlations unless they are specifically labeled as Raman correlations. Actually, molecular vibrational frequencies are the same in infrared and Raman spectra and for groups such as O—H, C—H, C≡N, C=O, C=C, etc., frequency correlations will be the same for both techniques. It is chiefly in the band intensities that the two types of spectra differ, sometimes markedly so. Because of this, a group frequency may only be useful in one technique because of its lack of intensity in the other. Indeed, when a center of symmetry is present, bands which are allowed in one technique will be forbidden in the other. Symmetrical groups such as symmetrically substituted C≡C or *trans* C=C or N=N or S—S groups will show only in the Raman spectra. Carbonyls and O—H groups show strongly in the infrared and weakly in the Raman, whereas S—H, amino N—H, and some nitriles show strongly in the Raman and weakly in the infrared, just to name a few examples. Raman spectra will occasionally be discussed for specific group frequencies but most of the discussion will apply to infrared spectra which are more commonly used for group frequency work. Examples of spectra of most of the group frequencies discussed can be found in Chapter 13. Refer to the Table of Contents of Group Frequency Spectra, Section 13.29(d).

5.2 Methyl Groups

The vibrations of the CH_3 group are shown in Fig. 5.1.[1,2] There are three CH bonds in a methyl group so there will be three CH stretching vibrations. In the symmetric, in-phase stretching vibration, the whole CH_3 group is

[1] N. Sheppard and D. M. Simpson, *Quart. Rev., Chem. Soc.* **7**, 19 (1953).
[2] N. Sheppard, *Trans. Faraday Soc.* **51**, 1465 (1955).

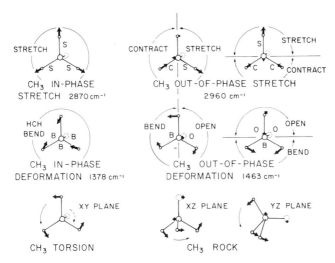

Fig. 5.1. The vibrations of the CH_3 group. In this figure: S, stretch; C, contract; B, bend; and O, open.

stretching in-phase. The two out-of-phase stretching vibrations are anti-symmetrical with respect to the CH_3 axis of symmetry and can be most simply characterized as "half-methyl" stretch. As seen in Fig. 5.1, a horizontal or vertical nodal plane is drawn containing the CH_3 group axis of symmetry. The CH bonds on one side of this plane stretch while those on the other side contract. If the nodal plane is drawn in any other orientation, the resulting vibration can be described as a linear combination of the two fundamental components described above. The reader may recognize these as standing waves in the CH_3 group. If the attached group has a threefold axis of symmetry, then the two "half-methyl" out-of-phase stretch vibrations are doubly degenerate and so absorb at the same frequency. If the symmetry of the attached group is lower, the two components will have slightly different frequencies.

In group theory descriptions, an out-of-phase vibration is described as being antisymmetrical with respect to one (or more) specific symmetry elements. In group frequency work, such a vibration is frequently referred to simply as "asymmetric" which means without symmetry. Strictly speaking this is not always correct since a vibration which is antisymmetrical with respect to one symmetry element may remain symmetrical with respect to another symmetry element. However, the term "asymmetric" with the implied meaning "not totally symmetric" is in rather common usage in the group frequency literature and no real confusion results.

There are three HCH angles and three HCH bending vibrations in the

TABLE 5.1

SPECTRAL BANDS FOR THE CH_3 GROUP (IN cm^{-1})

(aliphatic)—CH_3	2972–2952 strong (asymmetric stretch) 2882–2862 weaker than 2960 (symmetric stretch) 1475–1450 medium (asymmetric deformation) 1383–1377 medium (symmetric deformation)
(aromatic)—CH_3	2930–2920 medium 2870–2860 medium
CH_3 | —CH_2—CH—CH_2—	1159–1151 medium
(R)CH⟨CH_3 / CH_3 (R = hydrocarbon group)	1389–1381 medium 1372–1368 equal to the 1385 intensity 1171–1168 medium 1150–1130 variable
CH_3 | (R)—C—(R) | CH_3	1391–1381 medium 1368–1366 intensity 5/4 1385 1221–1206 weak 1191–1185 medium
CH_3 | (R)—C—CH_3 | CH_3	1401–1393 medium 1374–1366 about twice the 1390 intensity 1253–1235 medium 1208–1163 medium
CH_3—(C=O)	3000–2900 much weaker than hydrocarbons 1440–1405 medium 1375–1350 stronger than hydrocarbons
(aliphatic)—N—CH_3 (amine) (aromatic)—N—CH_3 (aliphatic)—N—$(CH_3)_2$ (aromatic)—N—$(CH_3)_2$	2805–2780 strong 2820–2810 2825–2810 2775–2765 2810–2790
(RO)CH_3	2992–2955 strong (asymmetric stretch) 2897–2867 strong (symmetric stretch) 2840–2815 variable (deformation overtone) 1470–1440 medium (asymmetric and symmetric deformation)
(RS)CH_3 (R—S—S)CH_3	2992–2955 medium (asymmetric stretch) 2897–2867 medium (symmetric stretch) 1440–1415 medium (asymmetric deformation) 1330–1290 weaker (symmetric deformation)
Si—CH_3	1440–1410 weak (asymmetric deformation) 1270–1255 strong (symmetric deformation)
P—CH_3	1330–1280 weak (symmetric deformation)

methyl group illustrated in Fig. 5.1. The bending vibrations are described as a symmetric, in-phase CH_3 deformation and two degenerate or nearly degenerate out-of-phase "half-methyl" deformations where a nodal plane separates the bending half from the opening half as in the stretching case.

The stretching and deformation vibrations are all more or less localized in the CH_3 group and give rise to good group frequencies. If the CH_3 group is compared to NH_3, these vibrations correspond to NH_3 vibrations. The CH_3 rocking and torsion vibrations (see Fig. 5.1) correspond to NH_3 rotations about three different axes. The rotation of the CH_3 group as a whole is compensated for by a counter-rotation of the rest of the molecule, so the rest of the molecule is involved to some extent in these vibrations. Furthermore, the frequency of the CH_3 rock vibration is not very different from the frequencies of the $C-C(H_3)$, $N-C(H_3)$, and $O-C(H_3)$ single bond stretching vibrations. In these cases, when they belong to the same symmetry species, $X-C(H_3)$ stretching can interact with CH_3 rocking and this interaction frequently complicates the spectra. The spectral regions for the CH_3 group are listed in Table 5.1.

5.3 CH₃ Stretching Vibrations

The infrared spectra of some alkanes are illustrated in Figs. 5.2 and 5.3. The out-of-phase or asymmetric CH_3 stretching vibration in an aliphatic compound absorbs[3] near 2960 ± 10 cm^{-1} and can be differentiated from the comparable CH_2 absorption near 2925 ± 10 cm^{-1} which has two or three times less intensity per group.

The in-phase or symmetric CH_3 stretching vibration in an aliphatic compound absorbs[3] near 2870 ± 10 cm^{-1} and can be distinguished from the comparable CH_2 absorption near 2855 ± 10 cm^{-1}. Additional CH_3 bands are seen in some compounds[3] near 2934 and 2912 cm^{-1}.

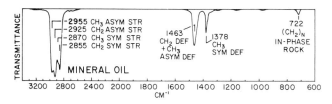

FIG. 5.2.   The infrared spectrum of mineral oil which contains CH_3 groups and $(CH_2)_n$ groups.

[3] J. J. Fox and A. E. Martin, *Proc. Roy. Soc., Ser. A* **167**, 257 (1938).

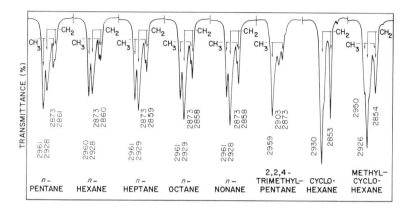

FIG. 5.3. The infrared spectra of some alkanes in the CH stretch region. These were run on a grating spectrometer in CCl_4 solution.

In Fig. 5.4 we can see an example of a methyl group on an aromatic ring. This type of CH_3 group has prominent bands near 2925 and 2865 cm^{-1}[4] both assigned to the in-phase symmetric CH_3 stretch in Fermi resonance with a CH_3 deformation overtone.[5] The two out-of-phase CH_3 stretching vibrations

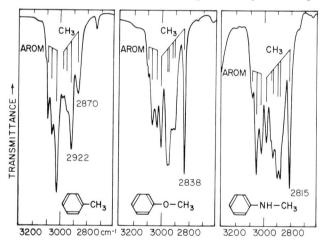

FIG. 5.4. The infrared spectra in the CH stretching region are illustrated for the aryl-methyl, the aryl-methoxy, and the aryl-methylamino groups. A triplet of bands at 3100–3000 cm^{-1} are caused by aromatic CH stretching vibrations. The rest of the bands are caused by the CH_3 group.

[4] G. M. Badger and A. G. Moritz, *Spectrochim. Acta* **15**, 672 (1959).
[5] A. B. Dempster, D. B. Powell, and N. Sheppard, *Spectrochim. Acta, Part A* **28**, 373 (1972).

are seen near 2975 and 2945 cm^{-1},[4] where the dipole moment change (which is perpendicular to the C—CH$_3$ bond) is in the ring plane and perpendicular to it, respectively.[5] These out-of-phase vibrations are broadened when CH$_3$ internal rotation (torsion) rotates the changing dipole moment with the result that the peak height intensities are lower than the 2925–2865 cm^{-1} bands. If free CH$_3$ rotation is hindered by bulky *ortho*-substituents, or is decreased by lowering the temperature to depopulate elevated CH$_3$ torsional states, then the out-of-phase CH$_3$ stretching bands at 2975–2945 cm^{-1} become sharper and their peak heights increase.[5] The symmetrical CH$_3$ stretching vibration has a dipole moment change along with the C—CH$_3$ band which is not changed in orientation by CH$_3$ torsional rotation, so the 2925–2865 cm^{-1} bands are not broadened and their peak heights are consistently prominent in methyl-substituted aromatics.

Methyl groups[6-8] on oxygen, sulfur,[7] and on nitrogens in amines[9,10] have been studied. Methoxy groups attached to an aromatic ring (Fig. 5.4) usually have a sharp isolated band near 2835 cm^{-1}. The methylamino groups in amines and anilines (in the free base form) absorb near 2800 cm^{-1} at the low end of the CH$_3$ region. Methyl deformation overtones which belong to the same symmetry species as the in-phase symmetrical CH$_3$ stretch probably are involved in Fermi resonance interaction with the in-phase CH$_3$ stretch vibration in these cases.

5.4 CH$_3$ Deformation Vibrations

The infrared spectra of some alkane groups in the CH$_3$ deformation region are illustrated in Fig. 5.5. In aliphatic compounds the asymmetric CH$_3$ deformation vibration absorbs near 1465 cm^{-1} and the symmetric deformation absorbs near 1375 cm^{-1}.[11] When there are two methyls on one saturated carbon atom, two bands with nearly equal intensity appear in the symmetric deformation region near 1385 and 1368 cm^{-1} because of interaction between the two CH$_3$ groups[11] (see Fig. 5.5). A *tert*-butyl group gives rise to a strong band near 1368 cm^{-1} and a weaker one near 1390 cm^{-1} from the same cause.[11] When a methyl is next to a carbonyl, the symmetric CH$_3$ deformation band

---

[6] S. E. Wiberley, S. C. Bunce, and W. H. Bauer, *Anal. Chem.* **32**, 217 (1960).

[7] A. Pozefsky and N. D. Coggeshall, *Anal. Chem.* **23**, 1611 (1951).

[8] H. B. Henbest, G. D. Meakins, B. Nicholls, and A. A. Wagland, *J. Chem. Soc., London* p. 1462 (1957).

[9] R. D. Hill and G. D. Meakins, *J. Chem. Soc., London* p. 760 (1958).

[10] W. B. Wright, Jr., *J. Org. Chem.* **24**, 1362 (1959).

[11] H. L. McMurry and V. Thornton, *Anal. Chem.* **24**, 318 (1952).

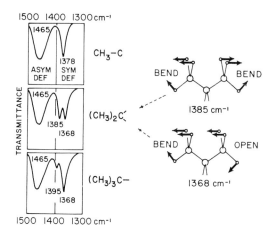

FIG. 5.5.   The infrared spectra of alkanes in the CH_3 deformation region. The in-phase deformation band is split when there are two or three CH_3 groups on one saturated carbon atom.

appears at 1375–1350 cm^{-1} and is intensified,[12] which makes it stand out in the spectra of larger molecules.

The methyl frequencies show some sensitivity to the electronegativity of the attached atoms,[2,13,14] the symmetric deformation being particularly sensitive. See Table 5.2 which is arranged according to the periodic chart of elements. These symmetric CH_3 deformation frequency variations are largely the result of changes in the HCX bending force constant[2] which is strongly

TABLE 5.2

Symmetrical CH_3 Deformation Frequencies (in cm^{-1}) (± 20 cm^{-1})

BCH_3	CCH_3	NCH_3	OCH_3	FCH_3
1310	1380	1410	1445	1475
	$SiCH_3$	PCH_3	SCH_3	$ClCH_3$
	1265	1295	1310	1355
	$GeCH_3$	$AsCH_3$	$SeCH_3$	$BrCH_3$
	1235	1250	1282	1305
	$SnCH_3$	$SbCH_3$		ICH_3
	1190	1200		1252
	$PbCH_3$			
	1165			

[12] N. B. Colthup, *J. Opt. Soc. Amer.* **40**, 397 (1950).
[13] L. J. Bellamy and R. L. Williams, *J. Chem. Soc., London* p. 2753 (1956).
[14] J. K. Wilmshurst, *J. Chem. Phys.* **26**, 426 (1957).

involved in this vibration. For the symmetric CH$_3$ deformation there is little overlap in position for different kinds of methyls. For example, a band at 1378 cm^{-1} can only arise if a methyl is on a carbon atom.

The asymmetric CH$_3$ deformation is less variable, ranging from 1471 to 1410 cm^{-1}. The asymmetric X—CH$_3$ deformation and the X—CH$_2$ scissors deformation are usually found in about the same region.

5.5 *CH$_3$ Rock Vibrations*

The X—CH$_3$ rock vibration deforms the HCX bond angle and, like the CH$_3$ symmetric deformation, is sensitive to the electronegativity of the X substituent.[13,14] For example, approximate CH$_3$ rock frequencies in cm^{-1} are as follows: Si—CH$_3$ near 800 cm^{-1}, P—CH$_3$ near 880 cm^{-1}, S—CH$_3$ near 960 cm^{-1}, and Cl—CH$_3$ at 1016 cm^{-1} (see sections on Si, P, and S compounds). The frequency increases roughly as the electronegativity increases.

In the groups C—CH$_3$, N—CH$_3$, and O—CH$_3$ there is usually interaction between the CH$_3$ rock and X—C(H$_3$) stretch, unless such interaction is forbidden by symmetry. In normal alkanes, vibrations involving in-phase C—C—C stretch mixed with in-plane CH$_3$ rock absorb weakly at 1150–1120 cm^{-1} and 900–880 cm^{-1}.[15–19] In these compounds the alternating C—C—C stretch vibration (mixed with a little CH$_2$ wag) absorbs weakly at 1090–1065 cm^{-1}.[15–19] The in-phase and alternating C—C stretching vibrations show in the Raman spectrum of polyethylene at 1133 and 1061 cm^{-1}.[17–19]

In isopropyl and *tert*-butyl groups, CH$_3$ rock vibrations mix with C—C stretch vibrations to give rise to bands near 1250–1130 and 950–810 cm^{-1}. In alkanes with isopropyl groups, bands are usually found near 1170, 1140, and 920 cm^{-1}, and in alkanes with *tert*-butyl groups, bands are usually found near 1245, 1200, and 930 cm^{-1}.[1,11,19] For normal coordinates of alkanes, see Snyder and Schachtschneider,[17–19] and for hydrocarbon intensity measurements, see McMurry and Thornton,[11] Jones,[16] Francis,[20] and Egorov *et al.*[21]

[15] J. K. Brown, N. Sheppard, and D. M. Simpson, *Phil. Trans Roy. Soc. London, Ser. A* **247**, 35 (1954).

[16] R. N. Jones, *Spectrochim. Acta* **9**, 235 (1957).

[17] R. G. Snyder and J. H. Schachtschneider, *Spectrochim. Acta* **19**, 85 (1963); **21**, 169 (1965).

[18] J. H. Schachtschneider and R. G. Snyder, *Spectrochim. Acta* **19**, 117 (1963).

[19] R. G. Snyder, *J. Chem. Phys.* **47**, 1316 (1967).

[20] S. A. Francis, *Anal. Chem.* **25**, 1466 (1953).

[21] Y. P. Egorov, V. V. Shlyapochnikov, and A. D. Petrov, *J. Anal. Chem. USSR* **14**, 617 (1959).

TABLE 5.3

SPECTRAL BANDS FOR THE CH_2 GROUP (IN cm^{-1})

(R)—CH_2—(R) (R = hydrocarbon group)	2936–2916 strong (asymmetric stretch) 2863–2843 weaker than 2930 (symmetric stretch) 1475–1450 medium (deformation)
—$(CH_2)_6$—(CH_3)	724– 722 weak (CH_2 rock)
—$(CH_2)_5$—(CH_3)	724– 723
—$(CH_2)_4$—(CH_3)	726– 724
—$(CH_2)_3$—(CH_3)	729– 726
—$(CH_2)_2$—(CH_3) (4th carbon branched)	743– 734
—CH_2—(CH_3) (3rd carbon branched)	785– 770
(R)—CH_2—$(CH=CH_2)$ (R)—CH_2—$(C≡CH)$	2936–2916 strong 2863–2843 weaker than 2930 1455–1435 medium
(R)—CH_2—$(C=O)$ (R)—CH_2—$(C≡N)$ (R)—CH_2—(NO_2)	3000–2900 medium (stretch) 1445–1405 stronger than in hydrocarbons (deformation)
(R)—CH_2—$(O—R)$ (R)—CH_2—(OH) (R)—CH_2—(NH_2)	2955–2922 strong 2878–2835 as strong as 2930 1475–1445 medium
(R)—CH_2—(NHR) (R)—CH_2—(NR_2)	2960–2920 strong 2820–2760 as strong as 2930 1475–1445 medium
(R)—CH_2—(SH) (R)—CH_2—$(S—C)$ (R)—CH_2—$(S—S)$	2948–2922 strong 2878–2846 weaker than 2930 1440–1415 medium (deformation) 1270–1220 strong (CH_2 wag)
(R)—CH_2(Cl)	3000–2950 weak 1460–1430 medium (deformation) 1300–1240 strong (CH_2 wag)
(R)—CH_2(P)	1445–1405 medium
CH_2 in cyclopropanes	3100–3072 (asymmetric stretch) 3033–2995 (symmetric stretch) 1050–1000
CH_2 in cyclobutanes	2999–2977 (asymmetric stretch) 2924–2875 (symmetric stretch)
CH_2 in cyclopentanes	2959–2952 (asymmetric stretch) 2866–2853 (symmetric stretch)
CH_2 in cyclohexanes	ca. 2927 (asymmetric stretch) ca. 2854 (symmetric stretch)
Epoxy	3058–3029 3004–2990 (usually one band only)

5.6 Methylene Groups

The vibrations of the CH$_2$ group[1] are illustrated in Fig. 5.6. The out-of-phase or asymmetric stretch, the in-phase or symmetric stretch, and the scissors deformation vibrations appear at 2925 ± 10, 2855 ± 10, and 1463 ± 13 cm^{-1}, respectively, in hydrocarbons. If the CH$_2$ group is compared to H$_2$O these vibrations correspond to H$_2$O vibrations. They are more or less localized in the CH$_2$ group and give rise to good group frequencies. The CH$_2$ wag, twist, and rock vibrations correspond to H$_2$O rotations. In the CH$_2$

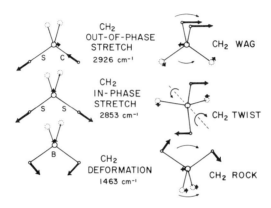

FIG. 5.6.   The vibrations of the CH$_2$ group.

group these rotations are opposed by motions of the adjacent carbon atoms attached to the CH$_2$ group, and these attached atoms are definitely involved in the vibrations. This causes a coupling between adjacent CH$_2$ groups and, as a result, in a chain of CH$_2$ groups, the wag (1382–1170 cm^{-1}), twist (1295–1063 cm^{-1}), and rock (1174–724 cm^{-1}) frequencies are each spread over a region. In the stretching and deformation vibrations, the coupling between adjacent CH$_2$ groups is weak, so each of these fall in a narrow correlatable region of the spectrum. The CH$_2$ spectral regions are listed in Table 5.3.[6,11]

5.7 CH$_2$ Stretching Vibrations

In alkanes the asymmetric CH$_2$ stretch band at 2925 ± 10 cm^{-1} is somewhat more intense than the symmetric CH$_2$ stretch band at 2855 ± 10 cm^{-1}. In ethers, alcohols, and amines the symmetric stretch band may be somewhat intensified. In secondary and tertiary amines (but not in amides) a band involving the symmetric stretch of the CH_2 next to the nitrogen atom shifts to about 2800 cm^{-1} below the alkane stretching frequencies.[10] This effect in amines is eliminated when the nitrogen acquires a positive charge in salts.

Overtones and combination bands of CH bending vibrations intensified by Fermi resonance appear sometimes in the CH stretching region, usually on the low frequency side of the main CH stretching bands. A particularly distinct band is usually seen in cyclohexanes near 2700 cm^{-1}.

5.8 CH_2 Deformation Vibrations

The CH_2 deformation band which comes near 1463 cm^{-1} in alkanes[11] is lowered to about 1440 cm^{-1} when the CH_2 group is next to a double or triple bond. A carbonyl, nitrile, or nitro group each lowers the frequency of the adjacent CH_2 group to about 1425 cm^{-1} and intensifies it.[12]

5.9 CH_2 Wag Vibrations

The CH_2 wagging bands in CH_2 chains are spread over a region. They can be most clearly seen in the solid phase spectra of long straight chain compounds such as fatty acids and soaps.[22] In Fig. 5.7 a series of sharp regularly spaced bands can be seen between 1347 and 1182 cm^{-1}. The number of bands in this region increases as the length of the CH_2 chain increases (see Section 9.32). The CH_2 wag bands are intensified in CH_2—Cl (1275 $\pm$ 25 cm^{-1}), CH_2—S (about 1250 cm^{-1}), CH_2—Br (about 1230 cm^{-1}), and CH_2—I (about 1170 cm^{-1}). The CH_2 wag vibration changes the HCX bond angles like the

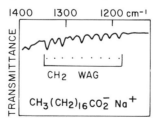

FIG. 5.7. CH_2 wag band progression seen in the solid state infrared spectrum of sodium stearate.

[22] R. A. Meiklejohn, R. J. Meyer, S. M. Aronovic, H. A. Schuette, and V. W. Meloche, *Anal. Chem.* **29**, 329 (1957).

CH_3 symmetrical deformation and the CH_3 rock vibrations and shows the same sensitivity to substituent electronegativity. The CH_2 wagging frequencies in $C-CH_2-X$ compounds are roughly 70 cm^{-1} lower than the $X-CH_3$ symmetrical deformation frequencies (Table 5.2) in comparable $X-CH_3$ compounds. For example, the $O=C-O-CH_3$ symmetrical deformation absorbs near 1450 cm^{-1}. The $O=C-O-CH_2$ wag band falls near 1380 cm^{-1} close to the symmetrical deformation band of the $C-CH_3$ group, and interaction between these probably occurs in ethyl and n-propyl esters. In Fig. 5.8

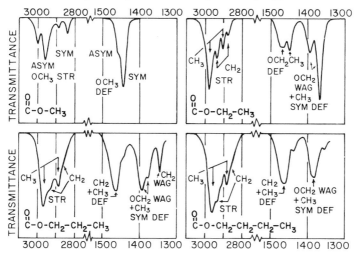

FIG. 5.8. The CH stretch and deformation region of the infrared spectra of alkane group substituents in esters.

the correlatable bands for methyl, ethyl, n-propyl, and n-butyl groups on an ester oxygen[23] are shown where an extra band from the $O-CH_2$ wag can be seen in the $C-CH_3$ symmetrical deformation region.

5.10 CH₂ Rock Vibrations

Like the CH_2 wag vibrations, the CH_2 rock vibrations are spread over a region. At the lower limit of this region is the *in-phase* CH_2 chain rock band. It is the most intense of the rock bands in the infrared spectrum and appears at 724 ± 4 cm^{-1} in noncyclic CH_2 chains.[11] If the material containing the CH_2 chain is in the crystalline state, then the CH_2 in-phase rock band is

[23] A. R. Katritzky, J. M. Lagowski, and J. A. T. Beard, *Spectrochim. Acta* **16**, 954 (1960).

replaced by two bands at 730 and 720 cm^{-1},[24] seen in Fig. 5.9. This splitting is caused by the crystallinity.

A single crystal can be considered as a large molecule with a tremendous number of normal modes of vibration ($3N - 6$, where N is very large). However, the only modes which can be infrared or Raman active are those where each unit cell vibrates in-phase with every other unit cell.[25] For any other phase relationship, the change in dipole moment or polarizability summed over the whole crystal is zero. It is therefore necessary only to examine the normal modes of one unit cell with the above restriction in mind. (The unit cell is the smallest volume within the crystal from which, by translation alone, the whole crystal can be generated.) In a crystalline straight chain alkane there

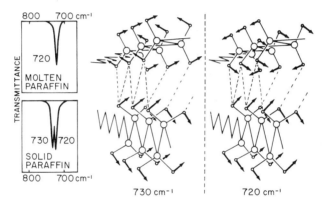

FIG. 5.9. Crystal splitting of the 730–720 cm^{-1} band caused by the in-phase (CH$_2$)$_n$ rock vibration. Two adjacent differently oriented long CH$_2$ chains in a crystal are illustrated, each performing the in-phase CH$_2$ rock vibration as shown. In the crystalline state, similarly oriented pairs of chains lie parallel to, and on all sides of, those illustrated, all pairs vibrating in the same manner as those shown for a given mode.

are sections of two differently oriented chains in every unit cell. These are seen in Fig. 5.9. Each chain is performing the in-phase CH$_2$ rock vibration but the phase relation between the two differently oriented chain vibrations are different for the 730 and 720 cm^{-1} bands.[25] In principle this crystal splitting effect occurs for every vibration when there is more than one molecule per unit cell but frequently the amount of splitting is too small to be noticed because of the weak coupling between nonbonded atoms in different molecules in the unit cell.

[24] S. Krimm, C. Y. Liang, and G. B. B. M. Sutherland, *J. Chem. Phys.* **25**, 549 (1956).
[25] S. Krimm, *in* "Infrared Spectroscopy and Molecular Structure" (N. Davies, ed.), Chapter 8, p. 273. Amer. Elsivier, New York, 1963.

5.11 CH$_2$ Twisting Vibrations

The CH$_2$ twisting vibrations in CH$_2$ chains have frequencies which are spread over a region like the wag and rock and have little intensity in the infrared. However, the in-phase CH$_2$ twist vibration in CH$_2$ chains is a good group frequency in the Raman spectrum at 1300 ± 5 cm^{-1} which is the high frequency limit for the CH$_2$ chain twist vibrational frequencies.[17, 18]

5.12 CH$_2$ in Cyclic Compounds

In cyclic rings there is a steady increase in the CH$_2$ asymmetric stretching frequency from the normal six-membered ring near 2930 cm^{-1} to the strained three-membered ring near 3080 cm^{-1}.[6, 26–29] The CH$_2$ and CH groups in a cyclopropane ring ring absorbs at 3100–2990 cm^{-1}.[6,27,29] A band at 1050–1000 cm^1 is characteristic for cyclopropyl rings which are not highly substituted.[29, 30] Bands for monoalkyl-substituted cyclopropanes appear near 3080, 1020–1000, and 820–810 cm^{-1}.[31] Bands for 1,1-dialkyl cyclopropanes appear near 3080, 1020–1000 and 880–840 cm^{-1}.[31] Highly substituted cyclopropanes may be difficult to characterize. Monoalkyl and 1,1-dialkylcyclobutanes absorb at 925–910 cm^{-1}.[30, 32] Monoalkyl and 1,1-dialkylcyclopentanes absorb at 945–925 and 900–880 cm^{-1}.[33] Monoalkylcyclohexanes absorb at 1005–950, 900–880, and 850–830 cm^{-1}.[33]

Raman spectroscopy is frequently useful in characterizing ring compounds.[34, 35] Most cyclohexane and cyclopentane derivatives have an in-phase ring stretching vibration which gives rise to a prominent band in the Raman spectrum, usually in the 800–700 and 900–800 cm^{-1} regions, respectively, even when there is a double bond in or on the ring.[34]

---

[26] S. H. Hastings, A. T. Watson, R. B. Williams, and J. A. Anderson, Jr., *Anal. Chem.* **24**, 612 (1952).

[27] E. K. Plyler and N. Acquista, *J. Res. Natl. Bur. Stand.* **43**, 37 (1949).

[28] H. B. Henbest, G. D. Meakins, B. Nicholls, and K. J. Taylor, *J. Chem. Soc., London* p. 1459 (1957).

[29] S. A. Liebman and B. J. Gudzinowicz, *Anal. Chem.* **33**, 931 (1961).

[30] J. M. Derfer, E. E. Pickett, and C. E. Boord, *J. Amer. Chem. Soc.* **71**, 2482 (1949).

[31] L. M. Sverdlov and E. P. Krainov, *Opt. Spektrosk.* **7**, 460 (1959).

[32] H. E. Ulery and J. R. McClenon, *Tetrahedron* **19**, 749 (1963).

[33] H. A. Szymanski, "Interpreted Infrared Spectra," Vol. 1. Plenum, New York, 1964.

[34] S. K. Freeman, "Applications of Laser Raman Spectroscopy." Wiley, New York, 1974.

[35] F. R. Dollish, W. G. Fateley, and F. F. Bently, "Characteristic Raman Frequencies of Organic Compounds." Wiley, New York, 1974.

5.13   Carbon Hydrogen Group

In Table 5.4 are listed correlations for the CH group.

The hydrocarbon CH stretch occurs[3] near 2900 cm^{-1} and is usually lost among other aliphatic absorptions. The CH deformation absorbs weakly in hydrocarbons at 1350–1315 cm^{-1} in the infrared. This vibration is more distinctive in the Raman effect. In nonhydrocarbons, such as *ortho*-formates, acetals, secondary peroxides, or α-substituted amines, the CH deformation at 1350–1315 cm^{-1} is intensified in the infrared.[12] In secondary alcohols, the CH and OH deformation interact to give two bands sensitive to hydrogen bonding at 1440–1350 and 1350–1200 cm^{-1} (see alcohols). A chlorine next to the CH will raise the stretching (ca. 3000 cm^{-1}) and lower and intensify the bending frequency (1250–1200 cm^{-1}).

TABLE 5.4

SPECTRAL REGIONS FOR THE CH GROUP (IN cm^{-1})

—O—CH	*ortho*-formates				
	Acetals, *Sec*-peroxides	1350–1315 medium			
—N—CH	Substituted amine				
HO—CH	*Sec*-alcohol free	1410–1350	1300–1200		
	Sec-alcohol bonded	1440–1400	1350–1285		
—CHO	Aldehyde	2900–2800	2775–2695	1420–1370	

When the CH is attached to a carbonyl, as in the formates and formamides, the stretching and bending frequencies are about 2930–2900 and 1400 cm^{-1}, respectively, but are not very useful for correlation purposes.

In aldehydes, the CH in-plane deformation absorbs at 1410–1370 cm^{-1}. Two aldehyde bands are usually observed in the CH stretch region[6,36] at 2900–2800 and 2775–2695 cm^{-1} which are probably the result of an interaction between the CH stretch and the overtone of the CH deformation. The few aldehydes which do not show this doubling (e.g., chloral) have a larger discrepancy between the overtone and the fundamental, so that interaction does not occur (see aldehydes).

[36] S. Pinchas, *Anal. Chem.* **27**, 2 (1955).

CHAPTER 6

TRIPLE BONDS AND CUMULATED DOUBLE BONDS

6.1 Introduction

Triple bonds (X≡Y—Z) and cumulated double bonds (X=Y=Z) absorb in roughly the same region, 2300–1900 cm^{-1}. The vibrations involved are the stretching of the triple bond and the out-of-phase stretching of the cumulated double bonds. These are illustrated in Fig. 6.1. Many of the compounds of

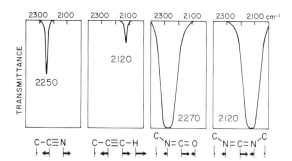

FIG. 6.1. Bands expected for various triple bond and cumulated double bond stretching vibrations.

these types are actually resonance hybrids of triple–single and double–double bond forms, so that nomenclature distinction is difficult in some cases. Usually, however, compounds which are predominantly X=Y=Z absorb in the infrared considerably more strongly than compounds which are predominantly X≡Y—Z. The reverse intensity relationship tends to hold for Raman spectra. Since this region of the spectrum is relatively free of bands, even weak bands here can be distinctive and reliable. Sometimes multiple bands appear in this region where a single band was expected. When this occurs it is usually the result of Fermi resonance type interaction. Table 6.1 lists the spectral regions for these types of groups.

235

TABLE 6.1

THE 2300–1900 cm^{-1} REGION

—C≡C—H	2140–2100 cm^{-1}	Weak–variable
—C≡C—	2260–2190	Very weak-variable
C=C=CH$_2$	2000–1900	Strong
CH$_2$—C≡N	2260–2240	Medium weak
C=C—C≡N	2235–2215	Medium
aryl-C≡N	2240–2220	Variable
—C≡N→O	2304–2288	Strong
—N=C=O	2275–2263	Very strong
—S—C≡N	2170–2135	Medium strong
—N=C=S	2150–2050	Very strong
>N—C≡N	2225–2175	Strong
—N=C=N—	2150–2100	Very strong
>C=C=N—	2050–2000	Very strong
—CH=$\overset{+}{N}$=$\overset{-}{N}$	2132–2012	Very strong
—N=$\overset{+}{N}$=$\overset{-}{N}$	2170–2080	Very strong
aryl-$\overset{+}{N}$≡N	2309–2136	Medium
—$\overset{+}{N}$≡$\overset{-}{C}$	2165–2110	Strong
>C=C=O	2200–2100	Very strong
[C≡N]$^-$	2200–2070	Medium
[Fe(C≡N)$_6$]$^{4-}$	2010	Medium
[Fe(C≡N)$_6$]$^{3-}$	2100	Medium
[N=C=O]$^-$	2220–2130	Strong
[N=C=S]$^-$	2090–2020	Strong
metal(CO)	2170–1900	Strong

6.2 Monosubstituted Acetylenes

The ≡CH stretching band near 3300 cm^{-1} has a narrower half-intensity bandwidth than the usual bonded NH or OH bands which appear in this region. The C≡C stretching frequency absorbs weakly but clearly in alkyl acetylenes near 2120 cm^{-1}. Conjugation with a carbonyl increases the intensity of this band. The broad strong band due to ≡CH wag is usually prominent at 700–610 cm^{-1} and has a broad weak overtone at 1375–1225 cm^{-1}. The regions for the C≡CH group are summarized in Table 6.2.[1–3]

[1] R. A. Nyquist and W. J. Potts, *Spectrochim. Acta* **16**, 419 (1960).
[2] J. H. Wotiz and F. A. Miller, *J. Amer. Chem. Soc.* **71**, 3441 (1949).
[3] N. Sheppard and D. M. Simpson, *Quart. Rev., Chem. Soc.* **6**, 1 (1952).

TABLE 6.2

GROUP FREQUENCIES FOR THE C≡CH GROUP

≡C—H stretch	3340–3267 cm⁻¹	Strong
C≡C stretch	2140–2100	Weak] in hydrocarbons
Overtone CH wag	1375–1225	Weak–broad
≡C—H wag	700–610	Strong–broad

6.3 Disubstituted Acetylenes

The $C-C≡C-C$ stretching frequency appears near 2260–2190 cm⁻¹ in hydrocarbons,[2-4] but because of symmetry, when the substituents are similar in mass and inductive and mesomeric properties, the intensity may be very weak or zero in the infrared. This vibration is far better studied in the Raman effect where it shows up strongly. In the Raman spectra of dialkyl acetylenes, two bands appear near 2300 and 2235 cm⁻¹, probably due to Fermi-resonance doubling.[2-4] When the two substituents are sufficiently different in their properties, the bond is made more polar and a strong band may result in the infrared.

A CH_2 wagging band in the narrow range 1336–1325 cm⁻¹ is characteristic for most molecules containing the group $-C≡C-CH_2-CH_2-$.[5]

6.4 Allenes

Allenes $(C=C=CH_2)$ absorb strongly at 2000–1900 cm⁻¹ as a result of the out-of-phase CCC stretch. The band is sometimes double. The terminal $=CH_2$ wagging vibration gives rise to a strong band near 850 cm⁻¹ with its overtone near 1700 cm⁻¹.[6] In the Raman spectra of mono n-alkyl-substituted allenes a strong doublet appears near 1130 and 1100 cm⁻¹ presumably involving the in-phase stretch of the allene double bands.[7]

6.5 Nitriles

Nitriles are characterized by the C≡N stretching frequency which occurs[8] at 2260–2240 cm⁻¹ in aliphatic nitriles. This medium intensity band is weakened in intensity when an electron attracting atom such as oxygen or chlorine is placed on the α carbon, whereas the same group on the β carbon

[4] M. J. Murray and F. F. Cleveland, *J. Amer. Chem. Soc.* **63**, 1718 (1941).

[5] J. J. Mannion and T. S. Wang, *Spectrochim. Acta* **17**, 990 (1961).

[6] J. H. Wotiz and D. E. Mancuso, *J. Org. Chem.* **22**, 207 (1957).

[7] L. Piaux, M. Gaudemar, and L. Henry, *Bull. Soc. Chim. Fr.* **23**, 794 (1956).

[8] R. E. Kitson and N. E. Griffith, *Anal. Chem.* **24**, 334 (1952).

has less effect.[8,9] A CH_2 group next to the nitrile absorbs near 1425 cm^{-1}. Conjugation lowers the nitrile frequency, due to resonance, to 2235–2215 cm^{-1} in most cases.[8]

The β-aminoacrylonitriles appear lower still and absorb strongly at 2210–2185 cm^{-1} due to resonance.[10]

$$N-CH=CH-C\equiv N \quad \longleftrightarrow \quad \overset{+}{N}=CH-CH=C=\overset{-}{N}$$

Benzonitriles absorb at 2240–2220 cm^{-1}. The intensity of this band is quite variable and depends on the nature of substituents. Electron attracting groups such as nitro groups decrease the band intensity and increase the frequency, whereas electron donating groups such as amino groups increase the intensity and decrease the frequency.[11] The frequencies and the log of the intensity A have been correlated with Hammett σ values.[11]

6.6   Nitrile N-Oxides and Complexes

Aryl nitrile N-oxides (aryl$-C\equiv N\rightarrow O$) absorb strongly at 2304–2288 cm^{-1} ($C\equiv N$ stretch) and strongly at 1393–1365 cm^{-1} ($N-O$ stretch).[12]

When the nitrile group is coordinated to a metal ion, the $R-C\equiv N\rightarrow$ (metal) group has a $C\equiv N$ stretching band which has a higher frequency, 2325–2265 cm^{-1}, and a greater intensity than the uncomplexed $C\equiv N$ band.[13,14] This positive frequency shift in nitrile complexes is in contrast to the general negative frequency shifts observed in $C=O$, $P=O$, and $S=O$ frequencies when these groups are coordinated to metal ions.[13]

6.7   Cyanates

Alkyl and aryl cyanates ($C-O-C\equiv N$) have a $C\equiv N$ stretch absorption at 2256–2245 cm^{-1}.[15-17] In addition, alkyl cyanates have a prominent band involving $C-O$ stretch at 1123–1084 cm^{-1}.[15] The comparable band in aryl cyanates appears at 1190–1110 cm^{-1}.[16,17]

[9] J. P. Jesson and H. W. Thompson, *Spectrochim. Acta* **13**, 217 (1958).

[10] S. Baldwin, *J. Org. Chem.* **26**, 3288 (1961).

[11] H. W. Thompson and G. Steel, *Trans. Faraday Soc.* **52**, 1451 (1956).

[12] R. H. Wiley and B. J. Wakefield, *J. Org. Chem.* **25**, 546 (1960).

[13] J. Reedijk, A. P. Zuur, and W. L. Groeneveld, *Rec. Trav. Chim. Pays-Bas* **86**, 1127 (1967).

[14] K. F. Purcell and R. S. Drago, *J. Amer. Chem. Soc.* **88**, 919 (1966).

[15] N. Groving and A. Holm, *Acta Chem. Scand.* **19**, 443 (1965).

[16] H. Hoyer, *Chem. Ber.* **94**, 1042 (1961).

[17] D. Martin, *Chem. Ber.* **97**, 2689 (1964).

6.8 Isocyanates

Compounds containing the organic isocyanate group $(C-N=C=O)$ absorb very strongly at 2275–2263 cm^{-1}. This absorption involves the out-of-phase stretching of the $N=C=O$ bonds[18-21] and is relatively unaffected by conjugation. In alkyl isocyanates the in-phase $N=C=O$ stretch vibration gives rise to a Raman band at 1450–1400 cm^{-1}.[22-24]

6.9 Thiocyanates

Organic thiocyanates $(-S-C\equiv N)$ show a medium strong sharp peak at 2170–2135 cm^{-1} as a result of the $C\equiv N$ stretching vibration.[18,25]

6.10 Isothiocyanates

Aliphatic and aromatic isothiocyanates $(-N=C=S)$ give rise to a very strong band at 2150–2050 cm^{-1} as a result of the out-of-phase stretching of the NCS bonds.[21,25-29] A shoulder usually appears at 2221–2150 cm^{-1}. In compounds with an adjacent CH_2 or CH group $(R-CH_2-N=C=S)$ the CH_2 or CH wagging gives rise to a strong band at 1347–1318 cm^{-1}.[27] The in-phase NCS vibration absorbs at 945–925 cm^{-1} in aryl isothiocyanates and 700–650 cm^{-1} in akyl isothiocyanates.[26]

[18] H. Hoyer, Chem. Ber. **89**, 2677 (1956).

[19] W. H. T. Davison, J. Chem. Soc., London p. 3712 (1953).

[20] N. Bortnick, L. S. Luskin, M. D. Hurowitz, and A. W. Rytina, J. Amer. Chem. Soc. **78**, 4358 (1956).

[21] G. L. Caldow and H. W. Thompson, Spectrochim. Acta **13**, 212 (1958).

[22] R. P. Hirschmann, R. N. Kniseley, and V. A. Fassel, Spectrochim. Acta **21**, 2125 (1965).

[23] D. F. Koster, Spectrochim. Acta, Part A **24**, 395 (1968).

[24] C. V. Stevenson, W. G. Coburn, Jr., and W. S. Wilcox, Spectrochim. Acta **17**, 933 (1961).

[25] E. Lieber, C. N. R. Rao, and J. Ramachandran, Spectrochim. Acta **13**, 296 (1959).

[26] N. S. Ham and J. B. Willis, Spectrochim. Acta **16**, 279 (1960).

[27] E. Svatek, R. Zahradnik, and A. Kjaer, Acta Chem. Scand. **13**, 442 (1959).

[28] L. S. Luskin, G. E. Gantert, and W. E. Craig, J. Amer. Chem. Soc. **78**, 4965 (1956).

[29] A. J. Coustoulas and R. L. Werner, J. Aust. Chem. **12**, 601 (1959).

6.11 Nitriles on a Nitrogen Atom

Cyanamides absorb strongly at 2225–2210 cm^{-1} [19] This low wavenumber and high intensity is due to resonance which weakens the C≡N bond.

$$>N-C\equiv N \quad \longleftrightarrow \quad >\overset{+}{N}=C=\overset{-}{N}$$

Cyanoguanidines [(N$_2$)—C=N—C≡N] absorb strongly at 2210–2175 cm^{-1} due also to resonance.

The band in cyanoguanidines is frequently multiple. Other nitrogen–nitrile compounds also absorb near here.

6.12 Carbodiimides

Disubstituted carbodiimides (R—N=C=N—R) have strong absorption at 2150–2100 cm^{-1} involving the out-of-phase stretching of the NCN bonds.[30] The in-phase stretching of the NCN bonds has been reported in the Raman spectra near 1460 cm^{-1}.[31]

6.13 Ketene Imines

Ketene imines

$$(>C=C=N-)$$

trisubstituted with aliphatic or aromatic groups absorb strongly at 2050–2000 cm^{-1}.[32] The unusual compound (CH$_3$—SO$_2$)$_2$—C=C=N—CH$_3$ absorbs at 2170 cm^{-1}.[33]

[30] G. D. Meakins and R. J. Moss, J. Chem. Soc., London p. 993 (1957).
[31] P. H. Mogul, Nucl. Sci. Abstr. 21, 47014 (1967).
[32] C. L. Stevens and J. C. French, J. Amer. Chem. Soc. 76, 4398 (1954).
[33] R. Dijkstra and H. J. Backer, Rec. Trav. Chim. Pays-Bas 73, 575 (1954).

6.14 Diazo Compounds

Diazo compounds are resonance hybrides of resonance forms 1 and 2.

$$[1] \quad R-CH=\overset{+}{N}=\overset{-}{N} \quad \longleftrightarrow \quad [2] \quad R-\overset{-}{C}H-\overset{+}{N}\equiv N$$

They are characterized by a strong band[34] at 2132–2012 cm^{-1} best described as out-of-phase CNN stretching. The compounds

$$R-CH=\overset{+}{N}=\overset{-}{N}$$

absorb at 2049–2036 cm^{-1}, and the compounds

$$RR'-C=\overset{+}{N}=\overset{-}{N}$$

absorb at 2032–2012 cm^{-1} where the R groups are either aliphatic or aromatic. No other correlatable bands are detected for the diazo group in these compounds.

Diazocarbonyl compounds are found at somewhat higher wavenumbers due to resonance contribution No.[3].[34]

$$[3] \quad \overset{-}{O}-C=CH-\overset{+}{N}\equiv N$$

Diazo ketones such as

$$\overset{\quad R \quad H}{O=C-C}=\overset{+}{N}=\overset{-}{N}$$

absorb at 2100–2087 cm^{-1}, and the compounds

$$\overset{\quad R \quad R'}{O=C-C}=\overset{+}{N}=\overset{-}{N}$$

absorb at 2074–2057 cm^{-1}. The carbonyls are lowered around 60 cm^{-1} by the same resonance. When R is aliphatic the carbonyl absorbs at 1647–1644 cm^{-1}, and when R is aromatic the carbonyl absorbs at 1628–1605 cm^{-1}.[34]

In addition to the band at 2100–2057 cm^{-1}, diazo carbonyl compounds

[34] P. Yates, B. L. Shapiro, N. Yoda, and J. Fugger, *J. Amer. Chem. Soc.* **79**, 5756 (1957).

have a second strong band at .1388–13333 cm^{-1} which is probably the in-phase CNN stretch. This does not appear in the aliphatic or aromatic diazo compounds.[34]

The quinone diazides or diazo oxides seem to be resonance hybrids also.

60976-6-3 $O={\large\langle}\ {\large\rangle}{=}\overset{+}{N}{=}\overset{-}{N} \longleftrightarrow \overset{-}{O}{-}{\large\langle}\ {\large\rangle}{-}\overset{+}{N}{\equiv}N$

Compounds have been studied having 1,2- and 1,4-benzo- and naphtho-quinone diazide groups.[35,36] They absorb at 2173–2014 and 1642–1562 cm^{-1}, due to the asymmetric CNN stretch and the C=O stretch, respectively.

6.15 Azides

Organic azides

$$-N{=}\overset{+}{N}{=}\overset{-}{N} \longleftrightarrow -\overset{-}{N}{-}\overset{+}{N}{\equiv}N$$

are characterized by a strong band [37-40] at 2170–2080 cm^{-1} as a result of the vibration best described as an out-of-phase NNN stretch. The band is relatively insensitive to conjugation effects or to changes in electronegativities of the substituent.

The in-phase NNN stretch gives rise to a weaker band at 1343–1177 cm^{-1} not so useful for identification.

In acid azides, the out-of-phase N_3 stretch which is usually a singlet appears as a doublet in $C_6H_5C(O)-N_3$ and compounds where the aromatic ring is substituted with nitro groups. For these compounds the following strong bands are observed.[41]

2237–2179 and 2155–2141 cm^{-1}	N_3 out-of-phase stretch
1258–1238	N_3 in-phase stretch
1709–1692	C=O stretch

[35] R. J. W. Le Fèvre, J. B. Sousa, and R. L. Werner, J. Chem. Soc., London p. 4686 (1954).

[36] K. B. Whetsel, G. F. Hawkins, and F. E. Johnson, J. Amer. Chem. Soc. 78, 3360 (1956).

[37] E. Lieber, C. N. R. Rao, T. S. Chao, and C. W. W. Hoffman, Anal. Chem. 29, 916 (1957).

[38] Y. N. Sheinker and Y. K. Syrkin, Izv. Akad. Nauk SSSR, Ser. Fiz. 14, 478 (1950).

[39] J. H. Boyer, J. Amer. Chem. Soc. 73, 5248 (1951).

[40] E. Lieber, D. R. Levering, and L. J. Patterson, Anal. Chem. 23, 1594 (1951).

[41] E. Lieber and E. Oftedahl, J. Org. Chem. 24, 1014 (1959).

6.16 Aryl Diazonium Salts

The diazonium group

$$(aryl-\overset{+}{N}\equiv NX^-)$$

is characterized by a medium intensity band at 2309–2136 cm^{-1}. The position of the band depends mainly on the identity of the diazonium cation. A shift of no more than 40 cm^{-1} occurs when the anion is changed.[36,42,43]

6.17 Isocyanides

Aliphatic isocyanides

$$(Aliphatic-\overset{+}{N}\equiv\overset{-}{C})$$

absorb strongly at 2146–2134 cm^{-1}. Aromatic isocyanides

$$(Aromatic-\overset{+}{N}\equiv\overset{-}{C})$$

absorb at 2125–2109 cm^{-1}.[44]

6.18 Ketenes

The out-of-phase stretching vibration in ketenes $>C=C=O$ gives rise to a strong band at 2200–2100 cm^{-1}. Specific values are ketene, 2153 cm^{-1};[45] diphenyl ketene 2105 cm^{-1};[46] dimethyl ketene, 2122 cm^{-1};[47] and $(CF_3)_2C=C=O$ at 2197 cm^{-1}.[46,48]

[42] M. Aroney, R. J. W. Le Fèvre, and R. L. Werner, *J. Chem. Soc., London* p. 276 (1955).

[43] R. H. Nuttall, E. R. Roberts, and D. W. A. Sharp, *Spectroccim. Acta* 17, 946 (1961).

[44] I. Ugi and R. Meyr, *Chem. Ber.* 93, 239 (1960).

[45] F. Halverson and V. Z. Williams, *J. Chem. Phys.* 15, 552 (1947).

[46] S. Nadzhimutdinov, N. A. Slovokhotova, and V. A. Kargin, *Russ, J. Phys, Chem.* 40, 479 (1966).

[47] W. H. Fletcher and W. B. Borish, *Spectrochim. Acta* 21, 1647 (1965).

[48] F. A. Miller and F. E. Kiviat, *Spectrochim. Acta, Part A* 25, 1577 (1969).

6.19   Cyanide Ions

The cyanide ion in sodium and potassium and related simple cyanides absorbs at 2080–2070 cm^{-1}. Silver cyanide absorbs at 2178 cm^{-1}, reflecting the more covalent character of the metal—carbon bond in this case. The ferro-cyanide ion [Fe (C≡N)$_6$]$^{4-}$ absorbs near 2010 cm^{-1}, and the ferri-cyanide ion [Fe (C≡N)$_6$]$^{3-}$ near 2100 cm^{-1}.[49]

6.20   Cyanate Ions

Inorganic cyanate ions absorb at 2220–2130 cm^{-1} (out-of-phase stretch), 1334–1292 cm^{-1} (in-phase stretch), 1293–1203 cm^{-1} (bending-overtone-Fermi-resonance with in-phase stretch), and 640–605 cm^{-1} (bending).[49,50]

6.21   Thiocyanate Ions

Inorganic thiocyanate ions absorb strongly at 2090–2020 cm^{-1} (out-of-phase stretch) and weakly near 950 cm^{-1} (bending overtone) and 750 cm^{-1} (in-phase stretch). The bending frequency is near 470 cm^{-1}.[49,51] In thiocyanate ion coordination a distinction can be made between M←NCS, 860–780 cm^{-1} and M←SCN, 720–690 cm^{-1}.[52]

6.22   Metal Carbonyls

Metal carbonyls absorb strongly at 2170–1700 cm^{-1}.[53] When the carbon of the carbon monoxide is associated with only one metal, absorption usually occurs at 2170–1900 cm^{-1}. The CO frequency shows variation with the number and availability of electrons in the rest of the molecule. The CO frequency

[49] F. A. Miller and C. H. Wilkins, *Anal. Chem.* **24**, 1253 (1952).

[50] T. C. Waddington, *J. Chem. Soc., London* p. 2499 (1959).

[51] P. Kinell and B. Strandberg, *Acta Chem. Scand.* **13**, 1607 (1959).

[52] A. Turco and C. Pecile, *Nature (London)* **191**, 66 (1961).

[53] J. Chatt, P. L. Pauson, and L. M. Venanzi, in "Organometallic Chemistry," (H. Zeiss, ed.), p. 477. Van Nostrand-Reinhold, Princeton, New Jersey, 1960.

and bond order are lowered when the ligand is an electron donor. In complexes with strong electron donors the frequency may fall below 1900 cm^{-1}.

Bridging carbonyl compounds, where the carbon is associated with two metal atoms, usually absorb at 1900–1700 cm^{-1}.[53]

6.23   Three and Four Cumulated Double Bonds

Butatriene $H_2C{=}C{=}C{=}CH_2$ has three double bonds and three double bond stretching vibrations. These are at 2079 cm^{-1} Raman active, 1608 cm^{-1} infrared active, and 878 cm^{-1} Raman active.[54] The vibrations involved are stretch–contract–stretch, stretch–no change–contract, and stretch–stretch–stretch, respectively, of the $C{=}C{=}C{=}C$ chain.

Carbon suboxide $O{=}C{=}C{=}C{=}O$ has two infrared active stretching bands at 2258 and 1573 cm^{-1} and two Raman active stretching bands at 2200 and 830 cm^{-1}.[55] Carbon subsulfide $S{=}C{=}C{=}C{=}S$ has two infrared active stretching bands at 2065 and 1019 cm^{-1} and two Raman active stretching bands at 1663 and 485 cm^{-1}.[56] Using S for stretch and C for contract, the relative phasings of the four bonds in these vibrations in order of decreasing vibrational frequency are SCSC (1R), SCCS (R), SSCC (1R), and SSSS (R) for both compounds.

[54] F. A. Miller and I. Matsubara, *Spectrochim. Acta* 22, 173 (1966).
[55] F. A. Miller and W. G. Fately, *Spectrochim. Acta* 20, 253 (1964).
[56] W. H. Smith and G. E. Leroi, *J. Chem. Phys.* 45, 1778 (1966).

CHAPTER 7

OLEFIN GROUPS

7.1 Noncyclic Olefins

In addition to the C=C bond, a vinyl group has three carbon—hydrogen bonds and so will have three CH stretching vibrations, three in-plane CH bending and three out-of-plane CH bending vibrations. These vibrations each interact to give the modes illustrated in Fig. 7.1.[1-3] Also illustrated are the in-phase, out-of-plane CH wagging vibrations for *trans*- and *cis*-disubstituted olefins, 1,1-disubstituted ethylene, and trisubstituted ethylene. The spectral regions for the alkyl-substituted olefins are given in Table 7.1.[1-6] Figure 7.2 shows the general spectrum expected for a vinyl group with an alkane substituent.

Olefinic carbon—hydrogen stretching frequencies occur at 3100–3000 cm^{-1}. There is a tendency for the asymmetric =CH$_2$ stretch of the vinyl and vinylidine groups (3092–3077 cm^{-1})[4] to absorb at a higher frequency than =CH vibrations (3050–3000 cm^{-1}) in hydrocarbons.[4]

The C=C stretching frequency near 1640 cm^{-1} in vinyl hydrocarbons is a medium intensity band which becomes inactive in the infrared region in a symmetrical *trans*- or symmetrical tetrasubstituted double bond compound, both of which have centers of symmetry. Even when the substituents are not exactly alike in *trans*-and tetrasubstituted olefins, the infrared absorption may be quite weak.[5] These double bond vibrations all appear strongly in the Raman effect, however. *The trans-*, tri-, and tetraalkyl-substituted olefins have somewhat higher C=C stretching frequencies than *cis*, vinylidine, or vinyl groups.[5] Conjugation, which weakens the C=C force constant, lowers the frequency 10–50 cm^{-1}. In 1,3-dienes the two double bonds interact to give

[1] N. Sheppard and D. M. Simpson, *Quart. Rev. Chem. Soc.* **6**, 1 (1952).

[2] W. J. Potts and R. A. Nyquist, *Spectrochim. Acta* **15**, 679 (1959).

[3] J. R. Scherer and W. J. Potts, *J. Chem. Phys.* **30**, 1527 (1959).

[4] S. E. Wiberley, S. C. Bunce, and W. H. Bauer, *Anal. Chem.* **32**, 217 (1960).

[5] H. L. McMurry and V. Thornton, *Anal. Chem.* **24**, 318 (1952).

[6] W. H. Tallent and I. J. Siewers, *Anal. Chem.* **28**, 953 (1956).

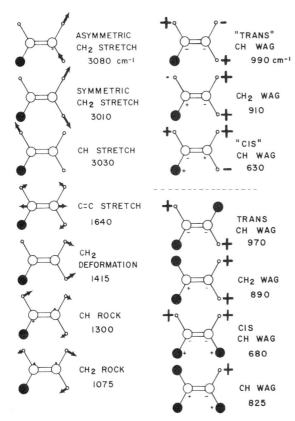

FIG. 7.1. Olefinic vibrations. The left-hand column illustrates the in-plane vibrations of the vinyl group. The right-hand column illustrates the out-of-plane vibrations ($+$ and $-$) of the vinyl group and for comparison the in-phase, out-of-plane hydrogen wagging vibrations of *trans*-, *cis*-, and 1,1-disubstituted olefins and trisubstituted olefins. The approximate frequencies are given for hydrocarbon-substituted olefins.

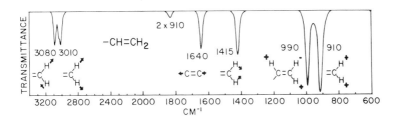

FIG. 7.2. General spectrum expected for a vinyl group with an alkane substituent.

TABLE 7.1

SPECTRAL REGIONS FOR THE ALKYL-SUBSTITUTED OLEFINS (IN cm^{-1})

Vinyl	3092–3077	Medium CH_2 asymmetric stretch
	3025–3012	Medium CH stretch, CH_2 symmetric stretch
	1840–1805	Medium $2 \times CH_2$ wag
	1648–1638	Medium C=C stretch
	1420–1412	Medium CH_2 deformation
	995– 985	Strong *trans* CH wag
	910– 905	Strong CH_2 wag
trans	3050–3000	Medium CH stretch
	1678–1668	Weak or absent (strong Raman band) C=C stretch
	980–965	Strong *trans* CH wag
cis	3050–3000	Medium CH stretch
	1662–1631	Medium C=C stretch
	1429–1397	Medium CH rock
	730–650	Medium to strong *cis* CH wag
Vinylidine	3100–3077	Medium CH_2 asymmetric stretch
	1792–1775	Medium $2 \times CH_2$ wag
	1661–1639	Medium C=C stretch
	895–885	Strong CH_2 wag
Trisubstituted	3050–2990	Weak CH stretch
	1692–1667	Weak C=C stretch
	840–790	Medium to strong CH wag
Tetrasubstituted	1680–1665	Weak or absent (strong Raman band) C=C stretch

two stretching bands. In both *cis*- and *trans*-1,3-pentadiene, in-phase stretching bands occur at 1658 cm^{-1} and out-of-phase at 1605 cm^{-1}.[7] In 1,3-butadiene only the out-of-phase (asymmetric) stretch is infrared active, absorbing at 1603 cm^{-1}.[7] Fluorinated olefins have an unusually high C=C frequency[8] at 1755–1735 cm^{-1} for the group C=CF$_2$ and 1800–1780 cm^{-1} for the group CF=CF$_2$. Other noncyclic, nonhydrocarbon double bond bands are usually found at 1680–1570 cm^{-1}.

The C=C frequency is usually lowered if an aliphatic group substituent is replaced by a heavy atom such as chlorine or sulfur. The C=C vibration interacts to some extent with =CH$_2$ deformation vibration and to some extent with the attached single bond stretching vibration, since nonhydrogen substituent atoms remain nearly motionless for this mode. The C=C stretching frequency will be affected by changes in these interactions and by mesomeric and inductive effects which alter the strength of the C=C and attached C—X bonds.

The CH$_2$ scissors deformation of the vinyl and vinylidine groups give rise to a medium intensity band in the infrared and Raman spectra near 1415 cm^{-1}[5] In *cis*-1,2-dialkyl ethylenes, the in-plane CH rock where both CH bonds rotate in the same direction (clockwise or counterclockwise) appears near 1405 cm^{-1} in the infrared, and the rock where the CH bonds rotate oppositely appears near 1265 cm^{-1} in the Raman spectrum.[1] In *trans*-1,2-dialkyl ethylenes, the in-plane rock where both CH bonds rotate in the same direction appears in the Raman spectrum near 1305 cm^{-1}, and the rock where the CH bonds rotate oppositely appears near 1295 cm^{-1} in the infrared spectrum.[1] In monoalkyl ethylenes the CH rock shows up in the Raman spectrum near 1295 cm^{-1} where all the ethylene CH bonds rotate in the same direction to some extent.[1]

7.2 Olefinic Hydrogen Wagging Vibrations

The strongest bands in olefins arise from the hydrogen wag vibrations (out-of-plane)[1,2,5,7,9,10] (see Fig. 7.1).

The lone CH of a trisubstituted olefin has its wagging (out-of-plane) frequency near 825 cm^{-1} in hydrocarbons, but this medium intensity band is not always easy to pick out among other bands.[2]

[7] R. S. Rasmussen and R. R. Brattain, *J. Chem. Phys.* **15**, 120 and 131 (1947).

[8] L. J. Bellamy, *Spectrochim. Acta* **13**, 60 (1958).

[9] J. K. Brown and N. Sheppard, *Trans. Faraday Soc.* **51**, 1611 (1955).

[10] H. W. Thompson and P. Torkington, *Trans. Faraday Soc.* **41**, 246 (1945).

The CH_2 in-phase, out-of-plane wagging vibration in a hydrocarbon vinylidine compound absorbs strongly near 890 cm^{-1} and is closely related to the 910 cm^{-1} vibration in the hydrocarbon vinyl group,[2,3,5,9] which we shall also call CH_2 wag. Both have an unusually strong first overtone, useful as a check.

The in-phase, out-of-plane CH wagging vibration of *trans*-disubstituted hydrocarbon olefins absorbs strongly near 970 cm^{-1} and is closely related to the 990 cm^{-1} vibrations in the hydrocarbon vinyl group,[2,3,5,9] which we shall call in both groups *trans*-CH wag. This band has no prominent overtone.

The in-phase, out-of-plane CH wagging of the *cis*-disubstituted hydrocarbon olefins gives rise to a more variable, less reliable, weaker band near 730–650 cm^{-1} in hydrocarbons.[2,5] This band is related to the variable vinyl vibration somewhere near 630 cm^{-1} in hydrocarbons,[3,9] which we shall also call *cis* CH wag.

In the hydrogen wagging vibrations of *trans*- and vinylidine-disubstituted olefins and in the comparable vibrations of the vinyl olefins, the motions of the hydrogens are balanced by motions of the carbons within the group. The rest of the molecule is relatively uninvolved mechanically in these vibrations. In the hydrogen wagging vibration of the *cis*-disubstituted olefin and the lowest vinyl olefin wagging vibration, the motions of the hydrogens are partially balanced by motions of the attached groups. These groups are thus more mechanically involved in the vibrations which makes the vibrational frequency of the *cis*-disubstituted olefin more variable than the *trans* or vinylidine vibrations or the comparable vinyl vibrations.

The out-of-phase, out-of-plane CH vibrations of the *cis*, *trans*, or vinylidine hydrocarbon olefins do not usually result in useful infrared bands.[2,9] They are weakly Raman active at about 960, 740, and 700 cm^{-1} respectively.[9]

The most reliable of these bands mentioned above are the CH_2 wag of vinyl and vinylidine groups and the *trans*-CH wag of vinyl- and *trans*-disubstituted olefins. When nonhydrocarbon olefins are considered, the CH_2 wagging frequency is not mass sensitive but is quite sensitive to mesomeric effects.[2] Those groups which mesomerically withdraw electrons from the CH_2 group raise the frequency, and those groups which mesomerically

$$\left(\overset{\overset{\text{O}}{\|}}{C-O-C}-CH=CH_2 \quad \longleftrightarrow \quad \overset{\overset{\text{O}^-}{|}}{C-O-C}=CH-CH_2^+ \right)$$

donate electrons

$$\left(C-O-CH=CH_2 \quad \longleftrightarrow \quad C-\overset{+}{O}=CH-\overset{-}{CH_2} \right)$$

lower the frequency relative to hydrocarbon olefins. In vinylidines, both groups shift the CH_2 frequency additively. For example, $CH_2=CH-CN$

absorbs 50 cm^{-1} higher than $CH_2{=}CH{-}R$, and $CH_2{=}C(CN)_2$ absorbs about 100 cm^{-1} higher than $CH_2{=}CR_2$. The CH_2 wagging frequency has been correlated with the *ortho-para* directing ability of the group,[2] and with the charge on the $={CH_2}$ carbon as calculated from molecular orbital theory.[11] See Table 7.2 for representative compounds.

TABLE 7.2

CH_2 WAGGING FREQUENCIES (IN cm^{-1})

$R{-}O{-}CO{=}CH{=}CH_2$	961
$N{\equiv}C{-}CH{=}CH_2$	960
$R{-}CH{=}CH_2$	910
$Cl{-}CH{=}CH_2$	894
$R{-}CO{-}O{-}CH{=}CH_2$	870
$R{-}CO{-}N{-}CH{=}CH_2$	840
$R{-}O{-}CH{=}CH_2$	813
$(N{\equiv}C)_2C{=}CH_2$	985
$(R)(R{-}O{-}CO)C{=}CH_2$	939
$(N{\equiv}C)(Cl)C{=}CH_2$	916
$R_2C{=}CH_2$	890
$Cl_2C{=}CH_2$	867
$(R)(R{-}O)C{=}CH_2$	795
$(R{-}O)_2C{=}CH_2$	711

[11] N. B. Colthup and M. K. Orloff, *Spectrochim. Acta, Part A* **27**, 1299 (1971).

The *trans*-CH wagging vibration of the vinyl- and *trans*-disubstituted olefins is relatively insensitive to mass or mesomeric effects but shows some sensitivity to the inductive power of the group. Electronegative groups tend to lower this frequency relative to hydrocarbons.[2] In *trans*-disubstitution, both groups shift the frequency. See Table 7.3.[2]

TABLE 7.3

Trans-CH WAGGING FREQUENCIES (IN cm^{-1})

$(CH_3)_3Si-CH=CH_2$	1009
$R-CH=CH_2$	990
$R-O-CO-CH=CH_2$	982
$R-CO-N-CH=CH_2$	972
$R-O-CH=CH_2$	960
$R-CO-O-CH=CH_2$	950
$Cl-CH=CH_2$	938
$R-O-CO-CH=CH-CO-O-R$ *trans*	976
$CH_3-CH=CH-CO-O-R$ *trans*	968
$R-CH=CH-R$ *trans*	964
$CH_3-CH=CH-C\equiv N$ *trans*	953
$CH_3-CH=CH-Cl$ *trans*	926
$Cl-CH=CH-C\equiv N$ *trans*	920
$Cl-CH=CH-Cl$ *trans*	892

In conjugated polyenes containing conjugated *trans*- and *cis*-substituted double bonds, mechanical interaction can affect the 965 cm^{-1} region.[12] The in-phase *trans*-CH wag vibration and the infrared inactive out-of-phase *cis* CH wag vibration have about the same frequency, 965 cm^{-1}.[9] This means that both kinds of groups will be excited at the same time, but with various phase relationships. The expected frequencies and relative intensities of a few examples are listed in Table 7.4 where, for example, the last entry indicates that a *cis-trans-cis* conjugated triene has bands within 10 cm^{-1} of 994 and 936 cm^{-1}, each about one-third as intense as an isolated *trans* group band at 965 cm^{-1}.[12]

TABLE 7.4

APPROXIMATE CH WAG FREQUENCIES AND RELATIVE INTENSITIES FOR *cis* (C), *trans* (T) CONJUGATED POLYENES

Diene	986 cm^{-1}	944 cm^{-1}	Triene	994 cm^{-1}	965 cm^{-1}	936 cm^{-1}
TT	2	0	TTT	3	0	$\frac{1}{3}$
CT	$\frac{1}{2}$	$\frac{1}{2}$	CTT	$\frac{4}{3}$	$\frac{1}{2}$	0
CC	0	0	CTC	$\frac{1}{3}$	0	$\frac{1}{3}$

[12] N. B. Colthup, *Appl. Spectrosc.* **25**, 368 (1971).

7.3 Cyclic C=C

An external $C=CH_2$ on an otherwise saturated six-membered ring is not noticeable different from its noncyclic vinylidine counterparts. As the ring gets smaller, the double bond–single bond angle increases due to strain. There is a steady increase in the $C=C$ stretching frequency[13] due predominantly to an increasing interaction with the $C-C$ bonds directly attached[14] (see Fig. 4.7). As the double bond stretches, the attached single bonds must be contracted, the amount of this interaction varying with the double bond–single bond angle.[14] See Table 7.5.

TABLE 7.5

$H_2C=C$ EXTERNAL CYCLIC DOUBLE BOND STRETCHING FREQUENCIES (IN cm^{-1})

6-Membered ring	Methylene cyclohexane	1651[13]
5-Membered ring	Methylene cyclopentane	1657[13]
4-Membered ring	Methylene cyclobutane	1678[13]
3-Membered ring	Methylene cyclopropane	1781[15]
(2-Membered ring)	(Allene)	1980[16]

Substitution of the hydrogens of the $C=CH_2$ group with methyl groups increases the frequency,[17] due in part to out of phase interaction with the attached $C-C$ bonds.

1657 cm^{-1} 1687 cm^{-1} 1651 cm^{-1} 1668 cm^{-1}

An unsubstituted $C=CH_2$ group on a bridged five-membered ring is under more strain than methylene cyclopentane and resembles methylene cyclobutane.[17]

1672 cm^{-1} 1678 cm^{-1}

[13] R. C. Lord and F. A. Miller, *Appl. Spectrosc.* **10**, 115 (1956).

[14] N. B. Colthup, *J. Chem. Educ.* **38**, 394 (1961).

[15] E. J. Blau, *Diss. Abstr.* **18**, 1628 (1958).

[16] G. Herzberg, "Infrared and Raman Spectra of Polyatomic Molecules," Van Nostrand-Reinhold, Princeton, New Jersey, 1945.

[17] G. Chiurdoglu, J. Laune, and M. Poelmans, *Bull. Soc. Chim. Belg.* **65**, 257 (1956).

An internal —CH=CH— in an otherwise saturated six-membered ring is not noticeable different from its noncyclic *cis* counterpart. As the ring gets smaller, the C=C stretching frequency decreases from the six- to the four-membered ring case as the double bond–single bond angle decreases to 90°.[13,18] This is due predominantly to changes in the interaction with the C—C bonds directly attached, which are altered in length as the double bond vibrates (see Fig. 5.12). This interaction is at a minimum at 90° and increases as the angle gets larger or smaller than 90°.[14] The =CH stretching frequency is also sensitive to ring size.[14] See Table 7.6.

TABLE 7.6

CH=CH INTERNAL CYCLIC DOUBLE BOND AND CH BOND STRETCHING FREQUENCIES (IN cm^{-1})

6-Membered ring	Cyclohexene	1646	3017[13]
5-Membered ring	Cyclopentene	1611	3045[13]
4-Membered ring	Cyclobutene	1566	3060[13]
3-Membered ring	Cyclopropene	1641	3076[19]
(2-Membered ring)	(Acetylene)	1974	3374[16]

Substitution of the remaining hydrogens with carbons will change the interaction.

1566 cm^{-1} [13]   1641 cm^{-1} [20]   1675 cm^{-1} [21]

All the above cyclobutene double bonds are equally strained, but the added out-of-phase interaction with the noncyclic C—C bonds increases the frequency.[14]

If the carbons in the ring which are not part of the C=C group are substituted, there is little mechanical effect and frequency shifts are small. A band at 1684 cm^{-1} is reported for 1,2-dimethylcyclobutenes where the

[18] R. C. Lord and R. W. Walker, *J. Amer. Chem. Soc.* **76**, 2518 (1954).
[19] K. B. Wiberg, B. J. Nist, and D. F. Eggers, Jr., private communication (1961).
[20] F. F. Cleveland, M. J. Murray, and W. S. Gallaway, *J. Chem. Phys.* **15**, 742 (1947).
[21] R. Srinivasan, *J. Amer. Chem. Soc.* **84**, 4141 (1962).

3 and 4 positions are substituted.[22] The same thing occurs in alkyl-substituted cyclopentenes[23] and cyclohexenes.[24]

1617–1614 cm^{-1} 1657–1650 cm^{-1} 1686–1671 cm^{-1}

1655–1645 cm^{-1} 1682–1668 cm^{-1} 1685–1677 cm^{-1}
                                              (Raman)

As seen above the frequencies for the 1,2-dialkyl cyclo-enes are nearly the same from the six- to the four-membered ring (1690–1670 cm^{-1}). This is markedly different from the unsubstituted case (1646→1611→1566 cm^{-1}). The decrease in the cyclic single bond–double bond angle as the ring gets smaller which reduces interaction and frequency is compensated for by an increase in the noncyclic single bond–double bond angle which increases interaction and frequency. The effect of interaction is to leave the frequencies nearly unchanged even though the strain is altered.[14] The 1,2-dialkylcyclo-propenes have a ring angle much less than 90° where interaction again increases the frequency, and this combined with an increased noncyclic single bond–double bond angle[25] of about 150° markedly increases the frequency which occurs at 1900–1865 cm^{-1}.[26-28] The compound 1-methylcyclo-propene absorbs at 1788 cm^{-1} [29] somewhat lower than the 1,2-disubstituted case as expected.

Table 7.7 lists double bond stretching frequencies (some from Raman data) of the 1,2-disubstituted cyclo-enes. Other positions as well as 1 and 2 may be substituted.

[22] R. Criegee and G. Louis, *Chem. Ber.* **90**, 417 (1957).

[23] L. M. Sverdlov and E. P. Krainov, *Opt. Spectrosc. (USSR)* **6**, 214 (1959).

[24] "Landolt-Börnstein Tables," 6th ed., Vol. 1, Part 2, pp. 480-510. Springer-Verlag, Berlin and New York, 1951.

[25] P. H. Kasai, R. J. Myers, D. F. Eggers, Jr., and K. B. Wiberg, *J. Chem. Phys.* **30**, 512 (1959).

[26] W. E. Doering and T. Mole, *Tetrahedron* **10**, 65 (1960).

[27] R. Breslow and H. Höver, *J. Amer. Chem. Soc.* **82**, 2644 (1960).

[28] K. Faure and J. C. Smith, *J. Chem. Soc., London* p. 1818 (1956).

[29] R. W. Mitchell and J. A. Merritt, *Spectrochim. Acta, Part A* **25**, 1881 (1969).

TABLE 7.7

INTERNAL CYCLIC DOUBLE BOND, 1,2-DISUBSTITUTED (IN cm^{-1})

6-Membered ring	Cyclohexenes, 1,2-dialkyl	1685–1677[24]
5-Membered ring	Cyclopentenes, 1,2-dialkyl	1686–1671[23]
4-Membered ring	Cyclobutenes, 1,2-dimethyl	1685–1675[21,22]
3-Membered ring	Cyclopropenes, 1,2-dialkyl	1900–1865[26–28]
(2-Membered ring)	(Acetylene, 1,2-dimethyl)	2313[16]

Bridged rings with unsubstituted internal double bonds such as in dicyclopentadiene have extra strain resulting in smaller bond angles than in unbridged rings of the same size, resulting in absorption at lower frequencies for these double bonds (ca. 1570 cm^{-1} for a bridged five-membered ring and ca. 1615 cm^{-1} for a bridged six-membered ring[13,18]).

1568 cm^{-1} 1566 cm^{-1} 1614 cm^{-1} 1611 cm^{-1}

CHAPTER 8

AROMATIC AND HETEROAROMATIC RINGS

8.1 Benzene Rings

In Fig. 8.1 all the modes of vibration of a monosubstituted benzene ring are illustrated.[1-10] The bottom row contains substituent sensitive vibrations which involve some substituent bond stretching or bending. All the rest are mechanically relatively insensitive to the nature of the nearly stationary substituent and are all group frequencies in the sense that they are found within a relatively small frequency range. They are not all *useful* group frequencies however, because some are quite weak in the infrared or Raman spectrum. However, it is clear that the substituted benzene ring is one of the most complicated groups we shall consider in group frequency studies.

Figure 8.2 shows the general appearance of the monosubstituted benzene group in the infrared spectrum. The aromatic C—H stretching vibrations give rise to multiple bands in the region 3100–3000 cm^{-1}.[11-13] In all the other vibrations, the C—H bond length hardly changes so that any *radially* moving carbon and its hydrogen move in the same direction. In substituted benzene

[1] A. R. Katritzky, *Quart. Rev., Chem. Soc.* **13**, 353 (1959).

[2] R. R. Randle and D. H. Whiffen, *Rep. Conf. Mol. Spectrosc., 1954* Inst. Petrol. Pap. No. 12.

[3] K. S. Pitzer and D. W. Scott, *J. Amer. Chem. Soc.* **65**, 803 (1943).

[4] A. M. Bogomolov, *Opt. Spectrosc. (USSR)* **9**, 162 (1960); **10**, 162 (1961).

[5] E. W. Schmid, J. Brandmuller, and G. Nonnenmacher, *Z. Elektrochem.* **64**, 726 and 940 (1960).

[6] J. R. Scherer, "Planar Vibrations of Chlorinated Benzenes." Dow Chemical Company, Midland, Michigan, 1963.

[7] G. Varsányi, "Vibrational Spectra of Benzene Derivatives." Academic Press, New York, 1969.

[8] D. H. Whiffen, *J. Chem. Soc., London* p. 1350 (1956).

[9] J. R. Scherer, *Spectrochim. Acta* **21**, 321 (1965).

[10] J. R. Scherer, *Spectrochim. Acta, Part A* **24**, 747 (1968).

[11] M. L. Josien and J. M. Lebas, *Bull. Soc. Chim. Fr.* pp. 53, 57, and 62 (1956).

[12] J. M. Lebas, C. Carrigou-Lagrange, and M. L. Josien, *Spectrochim. Acta* **15**, 225 (1959).

[13] S. E. Wiberley, S. C. Bunce, and W. H. Bauer, *Anal. Chem.* **32**, 217 (1960).

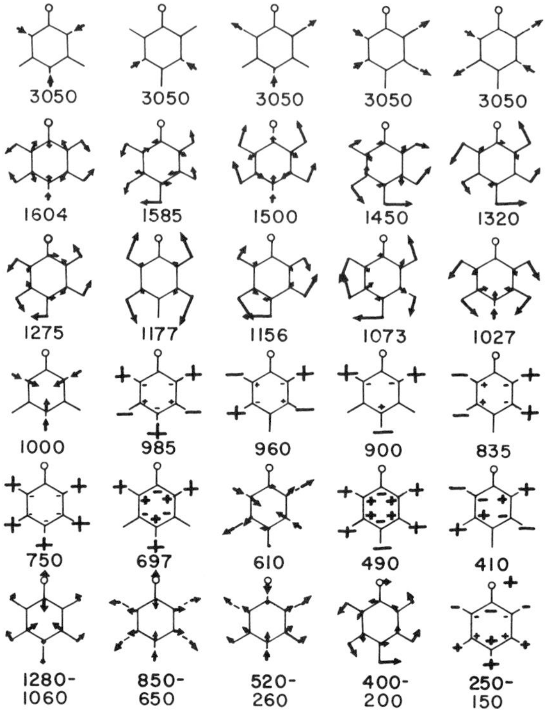

3050
3050
3050
3050
3050
1604
1585
1500
1450
1320
1275
1177
1156
1073
1027
1000
985
960
900
835
750
697
610
490
410
1280-1060
850-650
520-260
400-200
250-150

FIG. 8.1. The approximate normal modes of vibration of the monosubstituted benzene ring. These have frequencies in cm^{-1} usually within 30 cm^{-1} or better of the values given, except for the substituent sensitive modes which are collected in the bottom row.

ring compounds the C—H out-of-plane bending vibrations give rise to bands in the region 1000–700 cm^{-1}. In these vibrations a carbon and its hydrogen move in opposite directions while for out-of-plane vibrations below about 700 cm^{-1} they usually move in the same direction. Vibrations involving some C—H in-plane bending are found throughout the 1600–1000 cm^{-1} region. These in-plane CH bending vibrations interact, sometimes strongly, with vibrations involving ring CC stretching. When there is in-

TRANSMITTANCE
3100-3000
CH STRETCH
2000-1700
SUMMATION
BANDS
1585
1604
RING
STRETCH 1500 1450
1156
1177
1000
900
1073 1027
5 ADJ.
H WAG
750
697
3200 2800 2400 2000 1800 1600 1400 1200 1000 800 600
CM-1

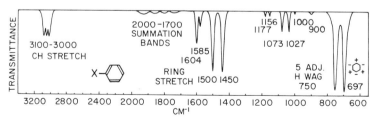

FIG. 8.2. The general appearance of the monosubstituted benzene group in the infrared spectrum. Some intensity variations may be expected with different substituents.

plane interaction above about 1200 cm^{-1}, a *tangentially* moving carbon and its hydrogen usually move in opposite directions, while below about 1200 cm^{-1}, they usually move in the same direction.[9] Except when strong mixing occurs, we will separate vibrations of benzene rings into modes which are predominantly CH vibrations (small carbon amplitudes, large hydrogen amplitudes) and modes which predominantly involve CC vibrations (large carbon amplitudes).

8.2 The Carbon—Carbon Vibrations

Ring modes of benzene[1-10] which involve C≕C bonds are illustrated (without hydrogens) in Fig. 8.3. In a simple C$_6$ ring there are six equal

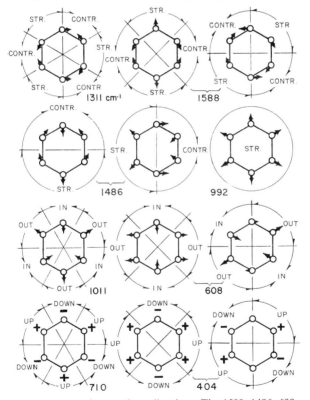

FIG. 8.3.   Benzene ring carbon–carbon vibrations. The 1588, 1486, 608, and 404 cm^{-1} modes are doubly degenerate in benzene. For descriptive purposes the stretching vibrations are described as whole ring stretching, semicircle stretching, quadrant stretching, and sextant stretching. The in-plane and out-of-plane bendings are described as quadrant and sextant bending.

C═C bonds and so there will be six C═C stretching vibrations. If for descriptive purposes the ring is thought of as a continuous helical spring in the form of a doughnut, the two components of the 1588 cm^{-1} vibration may each be described as opposite quadrants of the ring stretching while the intervening quadrants contract. The two components of the 1486 cm^{-1} vibration may be described as one semicircle of the ring stretching while the other contracts. The 710 cm^{-1} vibration is described as every other sextant going up out of the plane while the intervening sextants go down. The two components of the 404 cm^{-1} vibration may be described as opposite quadrants going up out of the plane while the intervening quadrants go down. The form of these above-mentioned vibrations remain nearly the same in substituted benzenes, and good group frequencies are found for these.

When a threefold axis of symmetry is present as in 1,3,5 symmetrically substituted benzene, symmetrically hexasubstituted benzene, and benzene itself, then the vibrations with two components (1588, 1486, 608, and 404 cm^{-1}) are doubly degenerate. For less symmetrical substitutions the degeneracy is broken, the two separate components can be found at slightly different frequencies. See Section 4.17 for a discussion of the 1311 cm^{-1} sextant stretching vibration in benzene. This frequency is lower than expected from the simple standing wave approach.

8.3   The 1600 cm^{-1} Region

The 1600 cm^{-1} vibrations mainly involve "quadrant stretching" of the ring C═C bonds but there is a little interaction with CH in plane bending. When a carbon moves tangentially clockwise, its hydrogen moves counterclockwise to some extent, but the substituents are nearly motionless. There are two quadrant stretching components which can sometimes be resolved in substituted benzenes[1,2] in the regions 1620–1585 and 1590–1565 cm^{-1}. The second band is not always seen clearly. The 1600 cm^{-1} doublet is not very frequency sensitive to changes in *ortho*, *meta*, or *para* substitution.

There is a center of symmetry in benzene ring compounds which have identical groups on all the *para* pairs of ring carbons. This includes symmetrical *para*-, symmetrical tetra-, hexasubstituted benzenes, and benzene itself. For these types of substitution, the quadrant stretching vibrations are infrared inactive because all the ring atoms *para* to each other move in exactly opposite directions. When the groups on the *para* pairs are different, the symmetry is destroyed and the vibrations are allowed in the infrared, the intensity being dependent on the nature of the group. In *para* compounds where one group is *ortho–para* directing and the other is *meta* directing, the 1600–1585 cm^{-1} bands can be quite intense in the infrared spectrum because of the dipole moment change provided by the different groups.

The infrared intensities of the 1600 cm^{-1} band and its relatively weaker companion near 1585 cm^{-1} in monosubstituted benzenes are high for electron donor or acceptor groups and low for weakly interacting groups.[1] *Meta*-disubstituted rings have 1600–1585 cm^{-1} infrared band intensities which vary as the algebraic sum of the electronic effects of the substituents, and *para*-disubstituted benzenes, as the algebraic difference.[1] *Ortho*-disubstituted aromatics are intermediate.[1] In the Raman effect[14] most aromatics show bands at 1620–1565 cm^{-1} whether there is symmetry or not.

8.4 The 1500 cm^{-1} Region

In hexachlorobenzene the vibration involving semicircle stretching of the carbon ring absorbs in the infrared at 1350 cm^{-1}.[6] In benzene this semicircle stretching vibration mixes strongly with C—H in plane bending and absorbs in the infrared at 1486 and 1037 cm^{-1}.[9] When a carbon moves *tangentially* clockwise, its hydrogen moves counterclockwise for the 1486 cm^{-1} band and clockwise with its carbon for the 1037 cm^{-1} band. In substituted benzenes without a threefold axis of symmetry each of these doubly degenerate vibrations are split. In monosubstituted benzenes (see Fig. 8.1) these vibrations have frequencies of about 1500, 1450, 1073, and 1027 cm^{-1}, all involving semicircle stretching and C—H bending.

The first component involving semicircle stretching mixed with CH bending usually appears at 1510–1470 cm^{-1} for mono, *ortho*, and *meta* substitution, and at 1525–1480 cm^{-1} for *para* substitution.[1,2] For *para* isomers this band is usually 10–20 cm^{-1} higher than that for the other substitutions, and so stands out in a mixture of isomers. For some rings conjugated with carbonyls, the infrared intensity may be quite weak. The infrared intensity is strong for electron donor groups and weak or absent when these are not present.[1] The Raman intensity is usually weak.[14]

The second component involving semicircle stretching mixed with C—H bending appears at 1465–1430 cm^{-1} for mono, *ortho*, and *meta* substitution and at 1420–1400 cm^{-1} for *para*.[1] The infrared intensity of this component is relatively independent of the nature of the substituent.[1] The Raman intensity is usually weak.[14] It should be noted that CH$_2$ and CH$_3$ deformation vibrations frequently obscure this component.

Table 8.1 gives spectral regions for the vibrations above 1400 cm^{-1} involving ring stretching. The symbol 1510→1480 cm^{-1} indicates electron

[14] K. W. F. Kohlrausch, "Ramanspektren." Akad. Verlagsges., Leipzig, 1943.

donor substituents cause absorption near 1510 cm^{-1} while electron acceptors cause absorption near 1480 cm^{-1}. Arithmetic means and standard deviations of band position in cm^{-1} are indicated otherwise.[1,2]

TABLE 8.1

BENZENE RING STRETCHING FREQUENCIES (IN cm^{-1})

Monosubstituted benzene	1604 ± 3	1585 ± 3	$1510 \rightarrow 1480$	1452 ± 4
o-Disubstituted benzene	1607 ± 9	1577 ± 4	$1510 \rightarrow 1460$	1447 ± 10
m-Disubstituted benzene	$1600 \rightarrow 1620$	1586 ± 5	$1495 \rightarrow 1470$	$1465 \rightarrow 1430$
p-Disubstituted benzene	1606 ± 6	1579 ± 6	$1520 \rightarrow 1480$	1409 ± 8
1,2,3-Trisubstituted benzenes	1616 ± 8	1577 ± 8	1510 ± 8	1456 ± 1

8.5 The 700 cm^{-1} Region

The 700 cm^{-1} band involves out-of-plane ring bending by sextants. In mono, *meta*, and 1,3,5 trisubstitution a prominent IR band appears at 710–665 cm^{-1} [1,2] which is absent or weak in the Raman spectrum. For these substitutions the hydrogens in the 2, 4, and 6 positions move in the same direction as their carbons but the substituents and hydrogens on the 1, 3, and 5 carbons are nearly motionless.[4–6,10]

In 1,2,3- and 1,2,4-trisubstitution the band is usually weaker in the infrared and somewhat higher in frequency, usually 730–685 cm^{-1}. *Ortho* and *para* isomers can also absorb here in the infrared when the two substituents differ. When the substituents are identical in *ortho* and *para* isomers, this band is infrared inactive.

8.6 The 900–700 cm^{-1} CH Wag Bands

During the out-of-plane C—H bending vibrations of substituted aromatics, the substituents tend to stay nearly motionless, so these vibrations tend to be substituent insensitive. They are illustrated in Fig. 8.4 along with their approximate frequencies.[1,10] It is most useful to designate aromatic hydrogens into sets of adjacent hydrogens.[15] For example, monosubstituted benzenes have five adjacent hydrogens, *ortho* has four adjacent hydrogens,

[15] N. B. Colthup, *J. Opt. Soc. Amer.* **40**, 397 (1950).

meta has three adjacent hydrogens and one isolated hydrogen, and *para* has two pairs of two adjacent hydrogens. For a given set of adjacent hydrogens, the modes can be described using the standing wave approach as in Fig. 8.4, where the more nodes in the vibration of a set of adjacent hydrogens, the higher the frequency. Although all these vibrations are group frequencies in the sense that they are found within a relatively small frequency range, they are not all useful group frequencies because many are weak or absent in the spectrum. The most useful of the modes illustrated in Fig. 8.4 are the modes marked (a) where all the hydrogens move *in-phase* and thus give rise to a large dipole moment change and a strong infrared band. Another useful mode is the (b) mode in *meta*- and 1,2,4-trisubstituted benzenes which involves the isolated hydrogen wag, usually seen in the infrared near 870 cm^{-1}. The (c) mode in monosubstituted benzenes usually absorbs weakly near 900 cm^{-1} in the infrared.

The in-phase, out-of-plane C—H bending vibrations give rise to strong infrared bands as follows[1,2,15]; six adjacent hydrogens in benzene at 671

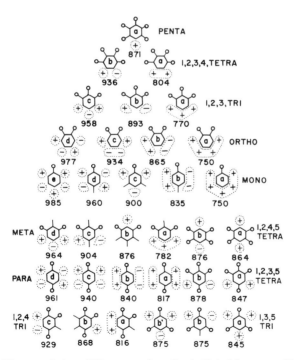

FIG. 8.4. The out-of-plane CH wag modes of substituted benzenes. Plus and minus stand for out-of-plane hydrogen displacement. Standing wave patterns are indicated by enclosing adjacent hydrogens which are wagging in-phase within an antinodal region (dotted region). Frequencies in cm^{-1} are usually within 30 cm^{-1} of those given.

cm^{-1}; five or four adjacent hydrogens in mono and *ortho* at 780–730 cm^{-1}; three adjacent hydrogens in *meta* and 1,2,3-trisubstitution at 820–760 cm^{-1}; two adjacent hydrogens in *para*-, 1,2,4-trisubstitution, and 1,2,3,4-tetra-substitution at 850–795 cm^{-1}; and an isolated hydrogen as in *meta*-, 1,2,4-trisubstitution, and 1,3,5-trisubstitution at 920–820 cm^{-1}. The general usefulness of the adjacent hydrogen wag correlation is illustrated by the fact that it can be extended to include naphthalenes[15-19] and pyridines.[15,20-22] The α-substituted naphthalenes have a set of four adjacent hydrogens and a set of three adjacent hydrogens. The β-substituted naphthalenes have four adjacent hydrogens, two adjacent hydrogens, and an isolated hydrogen. In pyridines the nitrogen (which has no hydrogen) is counted as a substituted carbon atom and the correlations used for benzene rings can then be applied. The adjacent hydrogen correlations can be used as a guide for many poly-cyclic benzenoid compounds and heteroaromatics although the actual values may deviate somewhat from those for substituted benzenes.[23,24]

In Table 8.2 the intense infrared bands in the 850–675 cm^{-1} region are summarized for mono- and disubstituted benzenes where arithmetic means in cm^{-1} and standard deviations are given.[1,2]

Examples of these infrared bands are seen in Fig. 8.5. The first two rows of spectra illustrate benzene rings with alkane substituents. For these, the adjacent hydrogen wag bands tend to increase in frequency somewhat as the α carbon of the substituent is more highly substituted.[25,26] We can note

TABLE 8.2

IN-PHASE, OUT-OF-PLANE HYDROGEN WAGGING AND OUT-OF-PLANE SEXTANT
RING BENDING FREQUENCIES (IN cm^{-1})

Monosubstituted benzene	751 ± 15	5 Adjacent H wag	697 ± 11 Ring bend
ortho-Disubstituted benzene	751 ± 7	4 Adjacent H wag	
meta-Disubstituted benzene	782 ± 9	3 Adjacent H wag	690 ± 15 Ring bend
para-Disubstituted benzene	817 ± 13	2 Adjacent H wag	

[16] J. G. Hawkins, E. R. Ward, and D. H. Whiffen, *Spectrochim. Acta* **10**, 105 (1957).

[17] R. L. Werner, W. Kennard, and D. Rayson, *Aust. J. Chem.* **8**, 346 (1955).

[18] T. S. Wang and J. M. Sanders, *Spectrochim. Acta* **15**, 1118 (1959).

[19] L. Cencelj and D. Hadži, *Spectrochim. Acta* **7**, 274 (1955).

[20] A. R. Katritzky, ed., "Physical Methods in Heterocyclic Chemisty," Vol. 2. Academic Press, New York, 1963.

[21] H. E. Podall, *Anal. Chem.* **19**, 1423 (1957).

[22] G. L. Cook and F. M. Church, *J. Phys. Chem.* **61**, 458 (1957).

[23] C. G. Cannon and G. B. B. M. Sutherland, *Spectrochim. Acta* **4**, 373 (1951).

[24] M. P. Groenewege, *Spectrochim. Acta* **11**, 579 (1958).

[25] H. L. McMurry and V. Thornton, *Anal. Chem.* **24**, 318 (1952).

[26] D. A. McCaulay, A. P. Lien, and P. A. Launer, *J. Amer. Chem. Soc.* **76**, 2354 (1954).

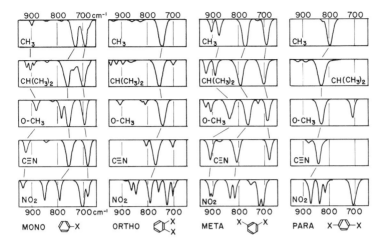

FIG. 8.5. The 900–700 cm^{-1} region in the infrared spectra of some benzene ring derivatives. The X substituents are indicated on each spectrum. In the disubstituted compounds both X substituents are the same.

absorptions for toluene at 728 cm^{-1}, ethylbenzene at 745 cm^{-1}, isopropylbenzene at 759 cm^{-1}, and *tert*-butylbenzene at 763 cm^{-1}.

There is some shift caused by the electron donating or with drawing characteristics of substituents.[27] An example of this is illustrated in the third and fourth row of Fig. 8.5 where the effect of O—CH$_3$ (electron donating) substitution can be compared with that for C≡N (electron withdrawing) substitution. This suggests that the C—H wag force constant tends to be reduced when electron donor substituents increase the electron density on ring carbons *ortho* and *para* to the substituent, similar to the case for the olefinic C=CH$_2$ wag vibration.

If the substituent group has internal vibrations with frequencies in the hydrogen wag region (for example, the CH$_2$—Cl group) the interpretation of this region becomes more complex. This is particularly true if the internal vibration of the group is an out-of-plane vibration which can interact with the CH wag vibrations, causing frequency shifts and intensity changes. Such is thought to be the case for nitro benzenes and derivatives of aromatic carboxylic acids. In benzene compounds with NO$_2$ substituents[28] a prominent extra band near 700 cm^{-1} in many of these is thought to involve NO$_2$ out-of-plane bending, interacting somewhat with the out-of-plane aromatic CH wagging vibrations (see Fig. 8.5, bottom row). The band

[27] L. J. Bellamy, *J. Chem. Soc., London* p. 2818 (1955).
[28] J. H. S. Green and D. J. Harrison, *Spectrochim. Acta, Part A* **26**, 1925 (1970).

normally associated with CH wag is shifted, usually to somewhat higher frequencies, because of NO_2 wag interaction, and the relative intensity is sometimes lower, making it harder to recognize. In addition there is frequently an in-plane NO_2 vibration near 850 cm^{-1} involving NO_2 scissors bending and C—N stretch which complicates the interpretation.[28] The COO group of aromatic carboxylic acids, esters, and salts affects the spectra in almost the same way as the aromatic NO_2 group does. For these particular substituents (NO_2, CO_2X) this region is not always as helpful as it is with most other types of substituents and should be used with discretion, preferably by comparison with closely related compounds.

8.7   The 2000–1650 cm^{-1} Summation Bands

The out-of-plane aromatic C—H bending vibrations in the 1000–700 cm^{-1} region give rise to relatively prominent summation bands in the infrared 2000–1650 cm^{-1} region,[29] seen in Fig. 8.6. Since all the CH out-of-plane bending vibrations in Fig. 8.4 have relatively well-defined fundamental frequency ranges, they give rise to summation bands with relatively well-defined frequencies even though the intensities of some of the fundamentals

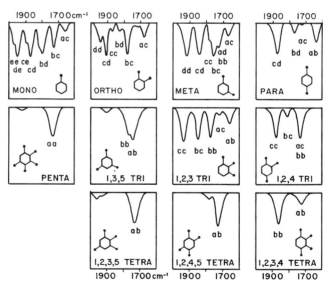

FIG. 8.6.   Characteristic patterns of summation bands for various substituted benzene rings. The double letters (ee) are the binary combinations of modes seen in Fig. 8.4.

[29] C. W. Young, R. B. DuVall, and N. Wright, *Anal. Chem.* **23**, 709 (1951).

are weak or even zero. The summation band patterns have been found to be relatively constant for a given ring substitution although the pattern as a whole can shift somewhat.[28] This means this band pattern can be used to help determine the ring substitution. The band assignments[30] in Fig. 8.6 (such as bc) refer to binary combinations of appropriate vibrations in Fig. 8.4. To be seen clearly, the sample should be run somewhat thicker than normal.

8.8   The 1300–1000 cm^{-1} Region

In Fig. 8.3 the 992 cm^{-1} ring stretching vibration and the 1011 cm^{-1} ring bending vibration have rather similar frequencies in benzene. These modes do not mix in benzene or *ortho-* or *para*-substituted benzenes with identical substituents because they belong to different symmetry species. In mono- and *meta* isomers, however, these modes belong to the same symmetry species and, since the frequencies are similar, they interact and mix. The result is a good group frequency[14] appearing at 1000 ± 5 cm^{-1} in mono-, *meta*-, and 1,3,5-trisubstitution *only*, where the 2, 4, and 6 carbons and hydrogens move radially in phase while the 1, 3, and 5 carbons (with one or more substituents) are nearly stationary.[1–19] This 1000 cm^{-1} band is very strong in the Raman spectrum[14] and is sometimes seen as a weak sharp band in the infrared for mono and *meta* isomers only.[1]

The vibration just discussed is one linear combination of the 992, 1011 cm^{-1} modes in Fig. 8.3. In the other linear combination, the unsubstituted 2, 4, and 6 carbons are nearly stationary and the 1, 3, and 5 carbons move radially in-phase in mono-, *meta*-, and 1,3,5-trisubstitution. For these isomers this vibration interacts with the stretching of the bands attaching substituents to the ring and gives rise to substituent sensitive bands.[1–9] For monosubstituted benzenes this vibration involving substituent bond stretching has the following frequencies for the given substituents: NH_2, 1280 cm^{-1}; OH, 1230 cm^{-1}; CH_3, 1211 cm^{-1} (weak in IR); F, 1219 cm^{-1}; Cl, 1086 cm^{-1}; Br, 1068 cm^{-1}; I, 1063 cm^{-1}; and Si, P, or S, 1130–1090 cm^{-1}.

In the infrared spectrum there are usually a number of medium-to-weak sharp bands in the 1300–1000 cm^{-1} region many of which involve in-plane C—H bending vibrations[1] sometimes strongly mixed with C—C vibrations.[9] For example, in Fig. 8.1 it can be seen that the 1027 and 1073 cm^{-1} vibrations in mono-substituted benzene rings involve some semicircle ring stretch like the 1500 and 1450 cm^{-1} vibrations. Infrared correlations for this region are listed in Table 8.3 where a vibrational description such as (2,3 vs. 5,6) means

[30] D. H. Whiffen, *Spectrochim. Acta* 7, 253 (1955).

TABLE 8.3

SUBSTITUTED BENZENE RING FREQUENCIES IN THE 1200–1000 cm^{-1} REGION

Mono	Ortho	Meta	Para
1000 ± 5	—	1000 ± 5	—
1027 ± 3 (2,3 vs. 5,6)	1033 ± 11 (3,4 vs. 5,6)	1076 ± 7 (2 vs. 5)	1013 ± 5 (2,3 vs. 5,6)
1073 ± 4 (2,6 vs. 3,4,5)	1125 ± 14 (3,6 vs. 4,5)	1096 ± 7 (4 vs. 6)	1117 ± 7 (2,6 vs. 3,5)
1156 ± 5 (3,5 vs. 4)	1160 ± 4 (3,5 vs. 4,6)	1157 ± 5 (2,5 vs. 4,6)	1175 ± 6 (2,5 vs. 3,6)
1177 ± 7 (2,5 vs. 3,6)	—	—	—

that the 2 and 3 hydrogens tend to move clockwise while the 5 and 6 hydrogens tend to move counterclockwise. Below about 1200 cm^{-1} for in-plane vibrations a carbon tends to move in roughly the same direction as its hydrogen. These medium-weak bands are of variable intensity, some being quite weak. Other vibrations occur in this region such as substituent single band stretching modes, methyl rocking, and substituent sensitive aryl vibrations, so this region should be used with discretion.

In *para*-substituted benzenes with identical substituents the 1175 cm^{-1} vibration (2,5 vs. 3,6) is centro-symmetric and is infrared inactive but is Raman active,[14] whereas the 1013 and 1117 cm^{-1} centro-antisymmetric vibrations are Raman inactive. For the other isomers in Table 8.3 the vibrations are active (but not necessarily strong) in the Raman[14] as well as the infrared spectrum. Two especially prominent Raman bands are the 1027 cm^{-1} band for mono and the 1033 cm^{-1} band for *ortho* which show up at 1015–1030 cm^{-1} and 1030–1045 cm^{-1} respectively.[14] As was mentioned before, mono-, *meta*-, and 1,3,5-trisubstituted benzenes have a very strong Raman band at 995–1005 cm^{-1}.

8.9 The 850–600 cm^{-1} Region

In Fig. 8.2 one component of the in-plane quadrant bend has the 2,5 carbons moving out and the 3,6 carbons in and the 1,4 carbons moving tangentially. In mono- and *para*- (1,4-) substitution this mode cannot interact with substituent stretching and gives rise to good group frequencies, 625–605 cm^{-1} for mono and 660–615 cm^{-1} for *para* isomers.[14,30] These bands are medium-strong in the Raman spectrum and are depolarized.[14] They are medium-weak in the infrared.[31] For other isomers this mode becomes substituent sensitive.

In the other quadrant in-plane bending mode the 1 and 4 carbons move radially. In mono- and *para* (1,4)-substituted benzenes this quadrant bend interacts with the stretching of the bond connecting the substituent to the ring. At 850–650 cm^{-1} a band occurs where the C—X distance increases as the distance between carbons 1 and 4 decreases, and at 520–260 cm^{-1} a band occurs where the C—X and the carbon 1,4 spacing increase together. The 850–650 cm^{-1} band in mono- and *para*-substituted benzenes is usually prominent in the Raman spectrum[14] and is polarized. This band is medium-weak in the infrared. The out-of-plane vibrations which occur in this region and are strong in the infrared tend to be weak in the Raman spectrum. In *ortho* and *meta* isomers a comparable strong band involving quadrant in-plane bend appears in the Raman spectrum[14] at somewhat lower frequencies than for mono and *para* isomers, at 750–650 cm^{-1}. In *ortho*- (1,2-) substitution the 1,2,4, and 5 ring carbons move out while the 3 and 6 carbons move in.[6] In *meta*- (1,3-) substitution the 1,3,4, and 6 ring carbons move out when the 2 and 5 carbons move in.[6] This frequency depends to some extent on the substituent mass and the C—X force constant.

8.10   The 550–400 cm^{-1} Region

There are two benzene ring vibrations in this region which involve out-of-plane ring bending by quadrants. The hydrogens move in the same direction as their carbons. In mono- and disubstituted benzenes the less symmetrical component is always infrared active. In the disubstituted case, the two ring carbons which have substituents move in the same direction out of the plane for this component. It gives rise to a medium-strong bond in the infrared at 560–418 cm^{-1} for mono, 470–418 cm^{-1} for *ortho*, 490–415 cm^{-1} for *meta*, 552–446 cm^{-1} for *para*, and 476–428 cm^{-1} for 1,2,4-trisubstitution.[31] This vibration is somewhat variable in frequency because sometimes there is interaction with out-of-plane substituent bending vibrations.

The other component usually retains a twofold axis of symmetry going through the substituents in *para* and mono and between identical substituents

[31] F. F. Bently, L. D. Smithson, and A. L. Rozek, "Infrared Spectra of Characteristic Frequencies 700-300 cm^{-1}." Wiley (Interscience), New York, 1968.

in *ortho* and *meta* isomers. This quadrant out-of-plane bending component is infrared inactive but becomes infrared active (but not necessarily a strong band) when the twofold axis is destroyed by unsimilar *ortho* or *meta* substituents or by nonaxially symmetrical mono and *para* substituents. It is observed as a weak band in the Raman spectra of monosubstituted benzenes at 415–400 cm^{-1}.[14]

8.11   Condensed Ring Aromatic Compounds

In naphthalenes[15–18] and other condensed ring compounds[23] the classification of substitutions in terms of adjacent hydrogens[12] proves useful in interpreting the 900–700 cm^{-1} region. Most α-naphthalenes substituted only in the 1 position have infrared bands at 810–785 cm^{-1} (3 adjacent hydrogens) and infrared bands at 780–760 cm^{-1} (4 adjacent hydrogens).[17,18] Most β-naphthalenes substituted only in the 2 position absorb at 862–835 cm^{-1} (isolated hydrogen), 835–805 cm^{-1} (2 adjacent hydrogens), and 760–735 cm^{-1} (4 adjacent hydrogens).[17,18] These and other out-of-plane vibrations for more highly substituted naphthalenes are summarized in Table 8.4.[16]

In this table the hydrogen pattern of each side of the naphthalene ring is considered separately. Hydrogen positions are given the lowest numbering

TABLE 8.4

CHARACTERISTIC BANDS FOR SUBSTITUTED NAPHTHALENES (IN cm^{-1})

Hydrogen pattern on one side	Isolated H	2 Adjacent H	3 Adjacent H	4 Adjacent H
1 2 3 4				800–761 s 770–726 s-vs
1 2 3			820–776 s 774–730 s	
1 2 4	894–835 m-s	847–805 vs		
1 2		835–799 s		
2 3		834–812 s-vs		
1 3	875–843 s 905–867 m-s			
1 4	889–870 s			
1 or 2	896–858 m			

which is not the case in conventional nomenclature. Regions are given with average intensities indicated by m—medium, s—strong, vs—very strong. In the 3 and 4 adjacent hydrogen region, two bands are usually seen but not always with the same intensity. One is probably a ring bending mode.[16] Other naphthalene bands may be found in regions similar to those of benzene derivatives.[16,23] Bands for alkyl-substituted naphthalenes have been found near 166 cm^{-1} (a doublet), 1520–1505 cm^{-1}, and 1400–1390 cm^{-1}.[23] There are many bands at 1400–1000 cm^{-1}.

In alkyl-substituted anthracenes,[23] bands appear at 1640–1620 cm^{-1}. A band is present near 1550 cm^{-1} which disappears with 9,10-substitution. There is no 1500 cm^{-1} band. There are one or two strong bands at 900–650 cm^{-1}. The band near 900 cm^{-1} is associated with the 9,10 isolated hydrogens and disappears when they are substituted.[23]

In alkyl-substituted phenanthrenes,[23] bands are found near 1600 cm^{-1} (often double on the high frequency side) and 1500 cm^{-1} which distinguishes it from anthracenes.

Tetralins[32] with 4 adjacent hydrogens on the aromatic ring absorb at 770–740 cm^{-1} (H wag) with 3 adjacent hydrogens at 815–785 cm^{-1} (H wag) and 760–730 cm^{-1} (ring vibration), with two adjacent hydrogens at 850–810 cm^{-1} (H wag) and with isolated hydrogens at 835–773 cm^{-1} (ring vibration).

In the ring stretching region, quinolines[1,33–35] show three bands near 1600 cm^{-1} and 5 bands in the 1500–1300 cm^{-1} region. The 900–700 cm^{-1} region has adjacent hydrogen wagging bands similar to the corresponding naphthalenes.

Indoles[36] absorb near 1460, 1420, and 1350 cm^{-1}.

Quinazolines[37] absorb at 1628–1612, 1581–1566, and 1517–1478 cm^{-1}.

8.12 Pyridines

The characteristic bands for pyridine compounds[21–24] are listed in Tables 8.5 and 8.6. Table 8.5 is similar to the corresponding Table 8.1.

[32] T. Momose, Y. Ueda, and H. Yano, *Talanta* **3**, 65 (1959).

[33] H. Shindo and S. Tamura, *Pharm. Bull.* **4**, 292 (1956).

[34] C. Karr, P. A. Estep, and A. J. Papa, *J. Amer. Chem. Soc.* **81**, 152 (1959).

[35] J. T. Braunholtz and F. G. Mann, *J. Chem. Soc., London* p. 3368 (1958).

[36] J. B. Brown, H. B. Henbest, and E. R. H. Jones, *J. Chem. Soc., London* p. 3172 (1952).

[37] H. Culbertson, J. C. Decius, and B. E. Christensen, *J. Amer. Chem. Soc.* **74**, 4834 (1952).

TABLE 8.5

PYRIDINE RING STRETCHING FREQUENCIES[1] (IN cm^{-1})

2-Substituted pyridine	1615 → 1585	1572 ± 4	1471 ± 6	1433 ± 5
3-Substituted pyridine	1595 ± 5	1577 ± 5	1485 → 1465	1421 ± 4
4-Substituted pyridine	1603 ± 5	1561 ± 8	1520 → 1480	1415 ± 4
Polysubstituted pyridine	1610 — 1597	1588 — 1564	1555 — 1490	
2-Substituted pyridine N-oxides	1640 → 1600	1567 ± 10	1540 → 1480	1435 ± 10
3-Substituted pyridine N-oxides	1605 ± 4	1563 ± 3	1480 ± 6	1434 ± 5
4-Substituted pyridine N-oxides	1645 → 1610	—	1483 ± 6	1443 ± 7

TABLE 8.6

OUT-OF-PLANE VIBRATIONS[1] (IN cm^{-1})

2-Substituted pyridine	780–740 4 adjacent H wag	
3-Substituted pyridine	820–770 3 adjacent H wag	730–690 ring bend
4-Substituted pyridine	850–790 2 adjacent H	
2-Substituted pyridine N-oxides	790–750 4 adjacent H	
3-Substituted pyridine N-oxides	820–760 3 adjacent H wag	680–660 ring bend
4-Substituted pyridine N-oxides	855–820 2 adjacent H	

8.13   Pyridine N-Oxides

Pyridine N-oxides[1,38] and the N-oxides of pyrimidine[38] and pyrazine[39,40] absorb strongly at 1320–1200 cm^{-1} due to the NO stretching frequency. A second band is found at 880–845 cm^{-1}.

8.14   Pyrimidines

Nontautomeric derivatives of pyrimidine[41] absorb strongly at 1580–1520 cm^{-1} and absorb near 990 and 810 cm^{-1}. Amino-substituted pyrimidines show NH_2 absorption bands at 3500–3100 and 1680–1635 cm^{-1}, in addition

[38] R. H. Wiley and S. C. Slaymaker, *J. Amer. Chem. Soc.* **79**, 2233 (1957).

[39] B. Klein and J. Berkowitz, *J. Amer. Chem. Soc.* **81**, 5160 (1959).

[40] C. F. Koelsch and W. H. Grumprecht, *J. Org. Chem.* **23**, 1602 (1958).

[41] L. N. Short and H. W. Thompson, *J. Chem. Soc., London* p. 168 (1952).

to the strong absorption at 1600–1500 cm^{-1}. Pyrimidines substituted with hydroxyl groups are generally in the "keto" form with C=O absorption near 1700 cm^{-1}.[41]

8.15 Triazines

The spectra of 1,3,5-s-triazines show absorption in three main regions, 1550 cm^{-1} ("quadrant stretching"), 1420 cm^{-1} ("semicircle stretching"), and 800 cm^{-1} (out-of-plane ring bending by "sextants").[42-47]

8.16 Alkyl- or Aryl-Substituted Triazines

Triazine rings, mono-, di-, or trisubstituted with an alkyl or aryl carbon directly attached to the ring have at least one strong band in the region 1580–1525 cm^{-1} (which may be double) and at least one weak band in the region 860–775 cm^{-1}. It does not seem to be possible to distinguish mono-, di-, or trisubstitution except possibly for ring CH stretch absorption for mono or di which can sometimes be seen in the 3100–3000 cm^{-1} region. There is usually at least one band in the 1450–1350 cm^{-1} region.

8.17 Melamines and Guanamines

There is complicated absorption at 3500–3100 cm^{-1} (NH$_2$ stretch) and absorption at 1680–1640 cm^{-1} (NH$_2$ deformation) when there is an NH$_2$ group on the triazine ring. If the NH$_2$ groups are all mono- or disubstituted, no absorption occurs at 1680–1640 cm^{-1}.

There is strong absorption in the 1600–1500 cm^{-1} region usually centering near 1550 cm^{-1} and multiple absorption in the 1450–1350 cm^{-1} region. In the

[42] A. Roosens, *Bull. Soc. Chim. Belg.* **59**, 377 (1950).

[43] J. Goubean, E. L. Jahn, A. Kreutzberger, and C. Grundmann, *J. Phys. Chem.* **58**, 1078 (1954).

[44] W. M. Padgett, II and W. F. Hammer, *J. Amer. Chem. Soc.* **80**, 803 (1958).

[45] J. E. Lancaster, R. F. Stamm, and N. B. Colthup, *Spectrochim. Acta* **17**, 155 (1961).

[46] H. Schroeder, *J. Amer. Chem. Soc.* **81**, 5658 (1959).

[47] Over 200 triazine spectra run by the author (1962).

800 cm^{-1} region, a sharp medium intensity band occurs at 825–800 cm^{-1} which in most melamines is found in the narrow range 812 ± 5 cm^{-1}. This band falls in frequency to 795–750 cm^{-1} when the triazine is in the "*iso*" form with less than three double bonds in the ring and at least one double bond external to the ring. The ring *N*-alkyl *iso*-melamines and ammeline, both in the *iso* form,[48] exhibit a band here as does the HCl salt of melamine indicating protonation of a ring nitrogen.

8.18   Chloro-, Oxy-, and Thio-substituted Triazines

Triazines with "hydroxyl" groups on the ring all exhibit strong C=O absorptions at 1775–1675 cm^{-1}, indicating the presence of the "keto" form. Ammeline and ammelide exhibit broad absorption near 2650 cm^{-1} (not in melamine) due to the ring NH bonded to C=O oxygen, possible in a dimer configuration. Absorption occurs at 795–750 cm^{-1} due to the ring in the *iso* form. The sodium salts of these compounds go back to the enol C—O$^-$Na$^+$ form,[48] as in tri sodium cyanurate which has no carbonyl and "normal" absorption at 820 cm^{-1}.

Trialkyl cyanurates absorb at 1600–1540, 1380–1320, 1160–1110 (O—CH$_2$ stretch), and at 820–805 cm^{-1} (normal ring).

Thioammeline exhibits a broad band at 2900–2800 cm^{-1} (ring NH··S) and a band at 1200 cm^{-1} (C=S) not in *S*-alkyl thioammeline or the sodium salt of thioammeline. In addition, thioammeline absorbs at 775 cm^{-1} (*iso* ring) where the *S*-alkyl and salt compounds absorb at 812 cm^{-1} (normal ring), all of which are consistent with the *iso* form for thioammeline.

In all these cases of hydroxy- and mercaptan-substituted heterocyclic rings, the tautomerization is best determined by comparison with alkylated compounds where the form is fixed. For correlation purposes, the ring vibrations of "*iso*" or "keto" forms should be considered separately from "normal" or "enol" forms. The band near 800 cm^{-1} is usually quite reliable in telling a normal triazine ring (825–795 cm^{-1}) from an *iso* ring (795–750 cm^{-1}) in rings with any combination of nitrogen or oxygen substituents, or with two nitrogen substituents plus any third substituent. It fails to hold the same position in the trialkyl- or aryl-substituted triazines, indicating some sensitivity to mechanical interaction here.

Dichlorotriazines absorb near 850 cm^{-1}.[49]

---

[48] R. C. Hirt and R. G. Schmitt, *Spectrochim. Acta* **12**, 127 (1958).
[49] W. A. Heckle, H. A. Ory, and J. M. Talbert, *Spectrochim. Acta* **17**, 600 (1961).

8.19   Tetrazines

Symmetrically substituted s-tetrazines have a center of symmetry, so the "quadrant" ring stretching band at 1600–1500 cm^{-1} is forbidden in the infrared. When the two substituents are electronically different, a strong band may result in this region. The "semicircle" ring stretching vibration (two components) remains active regardless of symmetry and results in absorption at 1495–1320 cm^{-1}. Another tetrazine band is found at 970–880 cm^{-1}.

8.20   Heteroaromatic Five-Membered Ring Compounds

Five-membered ring heteroaromatic compounds with two double bonds in the ring generally show three ring stretching bands near 1590, 1490 and 1400 cm^{-1}.[1] Assignments for the skeletal vibrations in Fig. 8.7 are based on

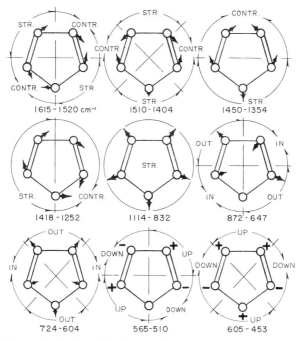

FIG. 8.7.   Schematic skeletal vibrations of five-membered ring compounds with two double bonds. For descriptive purposes the stretching vibrations may be described as whole ring, half-ring, and quarter-ring stretching. As the bonds are not all identical, the two highest frequencies consist predominantly of double bond stretching vibrations.

those made for pyrrole[50,51] and furan[52,53] Table 8.7 gives the regions for various types of rings.[1]

Most of these five-membered heteroaromatics with a $CH=CH$ group unsubstituted have strong hydrogen wag absorption at 800–700 cm^{-1}.

Compounds with hydroxyl or mercapto substituents can and usually do exist in the "keto" form. This is best determined by comparison with alkylated products where the structures are fixed.

TABLE 8.7

HETEROAROMATIC FIVE-MEMBERED RINGS[1] (IN cm^{-1})

2-Substituted furans	1605–1570	1510–1475	1400–1380
Polysubstituted furans	ca. 1560	ca. 1510	
2-Substituted thiophenes	1534–1514	1454–1430	1361–1347
3-Substituted thiophenes	ca. 1530	ca. 1410	ca. 1370
Substituted pyrroles	1565	1500	
Substituted thiazoles	1610	1500	1380
Furoxans	1610	1460	1420
Substituted isoxazoles	1600	1460	1380
Substituted furazans	1570	1425	1380
1,2,4-Oxadiazoles[54]	1590–1560	1470–1430	1390–1360

The NH stretch in pyrroles and indoles causes absorption at 3450–3400 cm^{-1} in dilute solution and at 3400–3100 cm^{-1} for hydrogen bonded solids. The bonded NH··N band in the solid state in most heteroaromatics with more than one nitrogen in the five-membered ring results in broad absorption at 2800–2600 cm^{-1}.

8.21 Cyclopentadienyl Ring–Metal Complexes

Metal complexes of the cyclopentadienyl ring with a large number of different metals have been studied.[55] The bands in Table 8.8 are observed for the cyclopentadienyl ring.

[50] R. C. Lord and F. A. Miller, *J. Chem. Phys.* **10**, 328 (1942).
[51] P. Mirone, *Gazz. Chim. Ital.* **86**, 165 (1956).
[52] H. W. Thompson and R. B. Temple, *Trans. Faraday Soc.* **41**, 27 (1945).
[53] B. Bak, S. Brodersen, and L. Hansen, *Acta Chem. Scand.* **9**, 749 (1955).
[54] J. Barrans, *C.R. Acad Sci.* **249**, 1096 (1959).
[55] H. P. Fritz, *Chem. Ber.* **92**, 780 (1959).

TABLE 8.8

CYCLOPENTADIENYL RING

3108–3027 cm^{-1}	CH stretch	Medium
1443–1440 cm^{-1}	CC stretch	Medium
1112–1090 cm^{-1}	CC stretch	Medium-strong
1009–990 cm^{-1}	CH deformation in plane	Strong
830–701 cm^{-1}	CH deformation out of plane	Strong

CHAPTER 9

CARBONYL COMPOUNDS

9.1 Introduction

Carbonyl compounds give rise to a strong band at 1900–1550 cm^{-1} caused by the stretching of the C=O bond. Some carbonyl vibrations are illustrated in Fig. 9.1. The carbonyl spectral regions are summarized in Table 9.1.

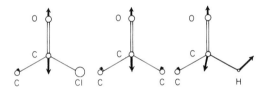

FIG. 9.1. The carbonyl stretching vibration. This vibration changes the C=O bond-length and changes bond angles around the carbon. Non-hydrogen substituent atoms do not move much so the attached C—X bonds are changed in length somewhat. Attached C—H bonds are not changed in length much but there is more bending involved.

TABLE 9.1

C=O SPECTRAL REGIONS (IN cm^{-1})

Dialkyl ketones	1725–1705
Singly conjugated ketones	1700–1670
Doubly conjugated ketones	1680–1640
α-Chloro ketone (Cl near O) (cyclic equatorial)	ca. 1745
α-Chloro ketone (Cl not near O) (cyclic axial)	ca. 1725
Ketone in five membered ring (unconjugated)	1750–1740
o-Hydroxy aryl ketones	1670–1630
1, 3-Diketones, enol form	1640–1580
Aliphatic aldehydes	1740–1720
Aromatic aldehydes	1710–1685

TABLE 9.1 (*continued*)

Formate esters	1725–1720
Other saturated esters	1750–1735
Conjugated esters (electron attracting groups on the oxygen raise the C=O frequency)	1735–1715
Lactones six-membered ring	1750–1715
Lactones five-membered ring	1795–1740
Carbonates noncyclic organic	1780–1740
Carbonates in five-membered ring	1850–1790
Carboxylic acid dimer	1720–1680
Carboxylic acir monomer	1800–1740
Carboxylic acid salt	$\begin{cases} 1650\text{–}1540 \\ 1450\text{–}1360 \end{cases}$
Amides and ureas (electron attracting groups on the nitrogen raise the C=O frequency)	1695–1630
Lactams five-membered ring	1750–1700
Lactams four-membered ring	1780–1730
Carbamates	1740–1683
Anhydrides, noncyclic, unconjugated	$\begin{cases} 1825\text{–}1815 \\ 1755\text{–}1745 \end{cases}$ higher band stronger
Anhydrides, noncyclic, conjugated	$\begin{cases} 1780\text{–}1770 \\ 1725\text{–}1715 \end{cases}$ higher band stronger
Anhydrides, cyclic, unconjugated	$\begin{cases} 1870\text{–}1845 \\ 1800\text{–}1775 \end{cases}$ lower band stronger
Anhydrides, cyclic, conjugated	$\begin{cases} 1860\text{–}1850 \\ 1780\text{–}1760 \end{cases}$ lower band stronger
Acid chlorides, aliphatic	1810–1795
Acid chlorides, aromatic	$\begin{cases} 1785\text{–}1765 \\ 1750\text{–}1735 \end{cases}$
Chloroformates, aliphatic	1785–1775
Thiol esters, unconjugated	1710–1680
Thiol esters, conjugated	1700–1640

Factors which cause shifts in the carbonyl frequencies are discussed in the following sections.

9.2 Mass Effects

The effect of replacing a carbon atom attached to a carbonyl group with a heavier atom is very small due to the mass effect alone.[1-5] The effect of replacing a carbon atom with a hydrogen atom has been calculated and would result in an aldehyde absorbing about 17 cm^{-1} lower than a ketone from the mass effect alone.[1]

The fact that aldehydes (1730 cm^{-1}) absorb about 15 cm^{-1} higher than ketones (1715 cm^{-1}), and the fact that acid chlorides (1800 cm^{-1}), esters (1740 cm^{-1}), and amides (1670 cm^{-1}) are different from ketones indicate that there are changes in the force constant of the $C=O$ bond.

9.3 Bond Angle Effects

During a carbonyl vibration, the atoms attached to the carbonyl do not move much when they have a mass equal to or greater than a carbon atom, hence the relatively small mass effect. However, since the attached atoms are nearly motionless, their single bonds must be contracted a little due to the motion of the carbonyl carbon atom as it moves away from the oxygen. This additional resistance to the motion of the carbon atom raises the $C=O$ vibrational frequency and is a function of bond angle. In most carbonyl compounds, the bond angle is not a variable, but in compounds where the angles are different as in strained ring compounds, this effect should be considered. Here, the double bond–single bond interaction is altered, the interaction increasing as the double bond–single bond angle increases or as the $C-C-C$ angle decreases. Studies, indicate that most of the frequency shifts in strained ring carbonyls are due to the geometry change rather than to a large change in force constant.[1-3, 6]

The following equation which gives the carbonyl frequency in cm^{-1} has been derived for saturated ketones[4]:

$$\bar{v} = 1278 + 68F - 2.2\phi$$

[1] S. Bratoz and S. Besnainou, *C.R. Acad. Sci.* **248**, 546 (1959).
[2] K. Frei and H. H. Gunthard, *J. Mol. Spectrosc.* **5**, 218 (1960).
[3] J. O. Halford, *J. Chem. Phys.* **24**, 830 (1956).
[4] T. Miyazawa, *J. Chem. Soc. Jap., Pure Chem. Sect.* **77**, 366 (1956).
[5] J. K. Wilmshurst, *Can. J. Chem.* **37**, 1896 (1959).
[6] N. B. Colthup and M. K. Orloff, *Chem. Phys. Lett.* **19**, 298 (1973).

The carbonyl frequency in cm^{-1} is dependent on F, the force constant in mdyne/Å, and ϕ, the $C-C-C$ angle in degrees. For unconjugated ketones, $F = 10.2 \pm 0.3$ mdyne/Å.

The average difference between the frequencies of various types of cyclic carbonyl compounds and their acyclic analogs is given in cm^{-1} as follows[7]: seven-membered ring, -8 ± 3; six-membered ring, $+7 \pm 14$; five-membered ring, $+37 \pm 11$; and four-membered ring, $+76 \pm 7$. Approximately the same relationships apply for carbonyls in bicyclic compounds when the smallest ring containing the C=O is chosen.

9.4 Inductive Effects

The tendency of the carbonyl oxygen to attract electrons

$$(\overset{+}{C}-\overset{-}{O})$$

results in a weakening of the C=O force constant and a lowering of the C=O frequency. Electron attracting groups attached to the carbon atom compete with the oxygen for electrons, resulting in less contribution from the polar form

$$(\overset{+}{C}-\overset{-}{O})$$

and a higher frequency. As an example of this, ketones absorb near $1715\ cm^{-1}$ whereas acid chlorides absorb near $1800\ cm^{-1}$. The carbonyl frequency has been correlated with the sum of the electronegatives of the attached atoms,[8] for cases where mesomerism is unimportant. A change in the carbonyl force constant due to a change in substituent electronegativity has the largest effect on the carbonyl stretching frequency, but other force constants involved in the carbonyl stretching vibration such as $C(O)-X$ stretching and $X-C(O)-Y$ bending also change with substituent electronegativities and have some effect on the carbonyl stretching frequency. Acid fluorides have carbonyl frequencies about $50\ cm^{-1}$ higher than acid chlorides or bromides largely because of changes in these other force constants rather than in the carbonyl force constant.[9]

In α-chlorocarbonyl compounds, a difference is observed between rotational isomers due to a field effect. When the chlorine is near the carbonyl oxygen in space, a higher frequency results relative to the isomer in which the

[7] H. K. Hall, Jr. and R. Zbinden, *J. Amer. Chem. Soc.* **80**, 6428 (1958); **82**, 1215 (1960).
[8] R. E. Kagarise, *J. Amer. Chem. Soc.* **77**, 1377 (1955).
[9] J. Overend and J. R. Scherer, *Spectrochim. Acta* **16**, 773 (1960).

chlorine is rotated away from the oxygen. The electron cloud of the chlorine atom near the oxygen electrostatically restrains the tendency of oxygen to attract electrons

$$(\overset{+}{C}-\overset{-}{O})$$

resulting in an increase in the C=O frequency of about 20 cm^{-1}.[10,11]

9.5 Mesomeric Effects

Acid chlorides (1800 cm^{-1}), esters (1740 cm^{-1}), and amides (1670 cm^{-1}) have progressively decreasing carbonyl frequencies. This shift cannot be explained by inductive effects alone, particularly in amides where the nitrogen is more electronegative than a carbon atom, but an amide carbonyl (1670 cm^{-1}) has a lower frequency than a ketone (1717 cm^{-1}). The dominant effect in amides is the mesomeric effect ($^{-}$O—C=N^{+}), where the oxygen atom draws electrons and the nitrogen atom donates electrons, resulting in a weaker C=O bond but a stronger C—N bond. When an electron attracting group is placed on the nitrogen atom, it competes with the oxygen for electrons and a stiffer C=O bond and a higher frequency result.[12] This also happens in esters when electron attracting groups are put on the single bonded oxygen atom.

Conjugation of carbonyls with double bonds or aromatic rings usually results in a band shift to lower wavenumbers of about 20–30 cm^{-1}, due again to a redistribution of electrons which weakens the C=O bond. Steric effects may remove the coplanarity of the conjugated system and thus reduce the effect of conjugation.

Hydrogen bonding affects carbonyl frequencies, but the largest effects occur when hydrogen bonding is combined with mesomeric effects. When a carbonyl is hydrogen bonded and resonance can occur which puts a partial negative charge on the oxygen atom accepting the hydrogen bond and a positive charge on the atom donating the hydrogen, the partial " transfer of allegiance " of the proton enhances resonance

$$(C=O\cdots H-X- \quad \longleftrightarrow \quad C-\overset{-}{O}\cdots H-\overset{+}{X}=)$$

and lowers the C=O frequency. This effect can be seen in o-hydroxy acetophenones, the enol form of acetylacetone types, carboxylic acid dimers, etc.

[10] R. N. Jones, D. A. Ramsay, F. Herling, and K. Dobriner, J. Amer. Chem. Soc. 74, 2828 (1952).

[11] L. J. Bellamy and R. L. Williams, J. Chem. Soc., London p. 4294 (1957).

[12] L. J. Bellamy, Spectrochim. Acta 13, 60 (1958).

9.6   Ketones

Ketones are best characterized by the strong C=O stretching frequency absorption near 1715 cm^{-1}. Ketone carbonyls between saturated hydrocarbon groups absorb strongly at 1725–1705 cm^{-1},[13] in most cases. A CH$_3$ group next to the carbonyl has a strong band due to symmetric CH$_3$ deformation at 1370–1350 cm^{-1},[14] which is lower and more intense than the alkane CH$_3$ group absorption. A CH$_2$ group next to the carbonyl has a strong band due to CH$_2$ deformation at 1440–1405 cm^{-1}.[14] In CH$_2$—CO—CH$_2$ compounds, the asymmetric C—C—C stretching vibration gives rise to a medium intensity band at 1230–1100 cm^{-1}. Methyl alkyl ketones have this band at 1170 cm^{-1} and a carbonyl deformation band at 595 cm^{-1}.[15]

9.7   α-Chloro Ketones

Ketones with a chlorine on the α-carbon will absorb at higher frequencies when the chlorine is rotated near the oxygen than when the chlorine is not near the oxygen due to a field effect.[10,11]

Examples of this may be seen in the difference between the rotational isomers of α-chloroacetone at 1745 and 1725 cm^{-1} or in equatorial (ca. 1745 cm^{-1}) and axial (ca. 1725 cm^{-1}) α-halogenated cyclic ketones.[10,11]

9.8   Conjugated Ketones

Conjugation without hydrogen bonding results in a lowering of the carbonyl stretching frequency to 1700–1640 cm^{-1}. Singly conjugated ketones such as alkyl phenones and carbonyl conjugated olefins usually absorb near 1700–1670 cm^{-1} though the frequency may be lower in certain cases.[16,17] The intensities of aryl ketone carbonyls are considerably lower than the intensities of amide carbonyls which absorb in the same region.[18] Doubly conjugated ketones, quinones, and benzophenones usually absorb near 1680–1640 cm^{-1}. The aromatic ketones have a medium band at 1300–1230 cm^{-1} [14]

[13] C. N. R. Rao, G. K. Goldman, and C. Lurie, *J. Phys. Chem.* **63**, 1311 (1959).
[14] N. B. Colthup, *J. Opt. Soc. Amer.* **40**, 397 (1950).
[15] H. W. Thompson and P. Torkington, *J. Chem. Soc., London* p. 640 (1945).
[16] R. N. Jones, W. F. Forbes, and W. A. Mueller, *Can. J. Chem.* **35**, 504 (1957).
[17] R. N. Jones and E. Spinner, *Can. J. Chem.* **36**, 1020 (1958).
[18] H. W. Thompson and D. A. Jameson, *Spectrochim. Acta* **13**, 236 (1958).

due to the phenyl-carbonyl C—C stretch. The 1600 cm^{-1} aromatic band is usually a doublet, but the 1500 cm^{-1} aromatic band may be very weak.

The carbonyl frequencies in *para*- and *meta*-substituted acetophenones have been correlated with Hammett σ constants.[16] Amino substitution lowers and nitro substitution raises the carbonyl frequency.

In α,β-unsaturated ketones, a difference has been noted between compounds where the C=O and C=C have *trans* or *cis* orientation.[19, 20] In *s-cis*-ketones the C=O and C=C bands are further apart than in *s-trans*. The infrared intensities of the *s-cis* C=C are much higher and the C=O somewhat lower than those of *s-trans*-ketones.

s-trans	C=O 1690–1675 cm^{-1}	C=C 1644–1618 cm^{-1}
s-cis	C=O 1700–1687 cm^{-1}	C=C 1624–1617 cm^{-1}

Carbonyls conjugated with other carbonyls in α-diketones, such as diacetyl[21] and benzil, are hardly different in frequency from comparable ketones without the carbonyl-carbonyl conjugation (1726 cm^{-1} in 11:12 diketone steroids).[22, 23]

9.9   Conjugated Hydrogen Bonded Ketones

When ketones are conjugated and hydrogen bonded in such a way that resonance puts a negative charge on the carbonyl oxygen and a positive charge on the atom which carries the bonding proton, a further change occurs. Because of hydrogen bonding, resonance is increased, resulting in weakening of the C=O and C=C bonds. An example of this is seen in *ortho*-hydroxyaryl carbonyl compounds which absorb at 1670–1630 cm^{-1}.[24] Another example is the enol form of compounds where two carbonyls are separated by one CH$_2$ group (β-diketones).

$$
\begin{array}{ccc}
\underset{\text{—C—CH}_2\text{—C—}}{\overset{\text{O}\quad\ \ \text{O}}{\|\quad\ \ \|}} & \underset{\text{—C—CH=C—}}{\overset{\text{O}\cdots\text{H—O}}{\|\qquad\ |}} \longleftrightarrow & \underset{\text{—C=CH—C—}}{\overset{-\text{O}\cdots\text{H—O}^+}{|\qquad\quad\|}}
\end{array}
$$

keto (1727, 1707 cm^{-1})   enol (1600 cm^{-1})

[19] R. Mecke and K. Noack, *Spectrochim. Acta* **12**, 391 (1958).

[20] R. Barlet, M. Montagne, and P. Arnaud, *Spectrochim. Acta. Part A* **25**, 1081 (1969).

[21] N. B. Colthup and M. K. Orloff, *Spectrochim. Acta, Part A* **30**, 425 (1974).

[22] R. S. Rasmussen, R. D. Tunnicliff, and R. R. Brattain, *J. Amer. Chem. Soc.* **71**, 1068 (1949).

[23] R. N. Jones, P. Humphries, and K. Dobriner, *J. Amer. Chem. Soc.* **72**, 956 (1950).

[24] I. M. Hunsberger, *J. Amer. Chem. Soc.* **72**, 5626 (1950).

A vibration best described as asymmetric OCC stretching in the enol tautomer absorbs broadly and strongly at 1640–1580 cm^{-1},[22, 25–27] the low frequency implying a large amount of resonance. As there are four nearly equivalent bonds in the enol nucleus (two CO and two CC bonds), four stretching frequencies are expected involving all the bonds. These usually occur near 1600, 1500, 1450, and 1260 cm^{-1}.[25] The bonded OH group in these compounds absorbs weakly and broadly around 3000–2700 cm^{-1}. Certain 1,3-diketones with bulky substituents (t-Bu-CO-CH$_2$-CO-t-Bu)[27] have no 1700 cm^{-1} absorption at all, indicating the absence of the keto form. This is due to steric effects which hinder the formation of rotational isomers other than that form required for the chelated enol ring.[27] Unsubstituted and 2-monosubstituted 1,3-cyclohexanedione derivatives enolize strongly in CCl$_4$ solution.[28] The enol form of the monosubstituted ones exists as a hydrogen bonded dimer

which absorbs at 1630–1607 cm^{-1}, and as a monomer, absorbing at 1649–1646 cm^{-1} more like an ordinary conjugated ketone. The 2-disubstituted derivatives have no enol form of course. The keto form of the 1,3-cyclohexane diones has two bands, asymmetric C=O stretch, 1707–1694 cm^{-1} (strong) and symmetric C=O stretch, 1739–1725 cm^{-1}, less than half as intense as the 1700 cm^{-1} band.[28] Metal chelates or salts[29–32] of the 1,3-diketones absorb near 1580 cm^{-1} since here the two C$\doteq$O and C$\doteq$C bonds are nearly equivalent (all approximately a " bond and a half"). The regions for the vibrations of the metal chelates, each involving all four " bond and a half" bonds are 1600–1560, 1530–1500, ca. 1450, and ca. 1250 cm^{-1}. There is some relation between the 1580 cm^{-1} band frequency and the stability of the complex.[30, 31] Detailed assignments have been made for acetylacetonates. Calculations

[25] R. Mecke and E. Funk, *Z. Elektrochem.* **60**, 1124 (1956).

[26] D. Hadzi and N. Sheppard, *Trans. Faraday Soc.* **50**, 911 (1954).

[27] G. S. Hammond, W. G. Bordium, and G. A. Guter, *J. Amer. Chem. Soc.* **81**, 4682 (1959).

[28] S. N. Anachenko, I. V. Berezin, and I. V. Torgov, *Izv. Akad. Nauk SSSR* **9**, 1644 (1960).

[29] L. Bellamy, G. S. Spicer, and J. D. H. Strickland, *J. Chem. Soc., London* p. 4653 (1952).

[30] H. F. Holtzclaw, Jr. and J. P. Collman, *J. Amer. Chem. Soc.* **79**, 3318 (1957).

[31] J. Charette and P. Teyssié, *Spectrochim. Acta* **16**, 689 (1960).

[32] K. Nakamoto, P. J. McCarthy, and A. E. Martell, *J. Amer. Chem. Soc.* **83**, 1272 (1961).

indicate that the highest frequency band in the 1600 cm^{-1} region involves more C≎C stretching than C≎O.[32]

Another example of conjugated hydrogen bonded systems is found in 6 hydroxyfulvenes with a carbonyl in the 1-position.

The OH stretch is very weak like acetylacetone, and bands involving C≎O and C≎C stretching occur at 1635 and 1545 cm^{-1}.

Tropolones, in solution, absorb near 3200–3100 cm^{-1} (OH), near 1620 cm^{-1} (C=O), and 1550 cm^{-1} (C=C).[33]

9.10 Bond Angle Effects in Ketones

Ketones in strained rings have C=O frequencies which are progressively higher than normal as the C—C—C bond angle decreases,[3] caused predominantly by increased interaction with the C—C stretching vibration.[1-3, 6] Ketone carbonyls in six-membered rings are about the same as noncyclic ketone carbonyls. The C=O frequency steadily increases as the ring gets smaller. The carbonyl frequencies in cm^{-1} measured in CCl$_4$ solution are given below.

Acetone[34]	1719 cm^{-1}
Cyclohexanone[34]	1719 cm^{-1}
Cyclopentanone[35]	1746 cm^{-1}
Cyclobutanone[35]	1788 cm^{-1}
Cyclopropanone[36]	1822 cm^{-1}

[33] W. E. Doering and L. H. Knox, *J. Amer. Chem. Soc.* **75**, 297 (1953).
[34] L. J. Bellamy and R. L. Williams, *Trans. Faraday Soc.* **55**, 15 (1959).
[35] K. E. Wiberg and B. J. Nist, *J. Amer. Chem. Soc.* **83**, 1226 (1961).
[36] A. Ohno, D. J. Grosse, and R. E. Davis, *Tetrahedron Lett.* p. 959 (1968).

The carbonyl absorption in cyclopentanone is a Fermi resonance doublet.[37] Unconjugated ketones in five-membered rings absorb at 1750–1740 cm^{-1} in sterols, conjugation lowering the band to near 1716 cm^{-1}.[23]

The compound di-*tert*-butyl ketone absorbs at 1686 cm^{-1}, lower than normal ketone C=O bands, primarily due to a steric increase in the C—C—C angle.[3]

9.11 Aldehyde CH Vibrations

In most aldehydes with alkyl groups or an aromatic ring next to the carbonyl, the aldehyde CH give rise to two bands at 2900–2800 and 2775–2695 cm^{-1} and to a band at 1410–1380 cm^{-1}.[38–40] The 1390 cm^{-1} band is assigned to the in plane hydrogen rock vibrations.[40]

Aliphatic aldehydes have bands at 2830–2810 and 2720–2695 cm^{-1}.[38] Some *ortho*-substituted benzaldehydes with substituents such as halogen, nitro, or methoxy absorb at 2900–2860 and 2765–2747 cm^{-1}, these regions differing somewhat from all other benzaldehydes which absorb at 2832–2812 and 2745–2720 cm^{-1},[39] due possibly to a steric effect.[41]

The two bands in the CH stretch region where only one fundamental is expected is most satisfactorily explained as an interaction of the CH stretch fundamental with the overtone of the CH bending vibration near 1390 cm^{-1}.[40] Since aldehydes with a large variety of side groups give rise to this same doublet, it is unconvincing to assign this to any interactions with overtones or combinations of the side groups. Since the intensities of both bands in the stretching region are comparable in intensity to the 1390 cm^{-1} band, both are too intense for either to be a simple overtone. This must involve Fermi resonance of the overtone with the fundamental, which means that both bands near 2830 and 2740 cm^{-1} involve aldehyde CH stretch, and both bands involve the overtone of the 1390 cm^{-1} CH rock.

A few aldehydes do not have the doublet, having only one band near 2870–2830 cm^{-1}. The CH rock is usually shifted in these cases so that there is no longer close coincidence between the frequency of the overtone and fundamental. Chloral is an example with bands at 2870 and 1360 cm^{-1}.

[37] C. L. Angell, P. J. Krueger, R. Lauzon, L. C. Leitch, K. Noack, R. D. Smith, and R. N. Jones, *Spectrochim. Acta* **15**, 926 (1959).

[38] S. Pinchas, *Anal. Chem.* **27**, 2 (1955).

[39] S. Pinchas, *Anal. Chem.* **29**, 334 (1957).

[40] D. F. Eggers, Jr. and W. E. Lingren, *Anal. Chem.* **28**, 1328 (1956).

[41] R. West and L. S. Whatley, *Chem. Ind. (London)* p. 333 (1959).

The CH out-of-plane wag (about 900–700 cm^{-1} in some simple aldehydes from the literature) is disappointingly weak and is therefore useless for correlation purposes.

9.12   Aldehyde Carbonyl Vibrations

The most intense band in the aldehyde spectrum is usually due to the carbonyl stretching vibration. Most aliphatic aldehydes absorb strongly at 1740–1720 cm^{-1}. Most aromatic aldehydes absorb at 1710–1685 cm^{-1}.[39] Salicylaldehyde absorbs at 1666 cm^{-1} due to conjugated internal hydrogen bonding as in the *ortho*-hydroxyphenones. As with other carbonyl compounds, conjugation lowers the frequency while electronegative groups next to the carbonyl raise the frequency. The chloro acetaldehydes in CCl_4 absorb as follows: trichloro, 1768 cm^{-1}; dichloro, 1748 cm^{-1}; monochloro, 1730 cm^{-1}; and acetaldehyde, 1730 cm^{-1}. No doubling is observed as in the chloro ketones.[42]

9.13   Aldehyde C—C

Aromatic aldehydes have a band[14,40] usually at 1210–1160 cm^{-1} which probably involves stretching of the phenyl-C bond. An aromatic ring band usually absorbs at 1310–1260 cm^{-1}.

9.14   Ester C=O

Esters are characterized by the strong absorption due to the C=O stretching frequency[43] near 1740 cm^{-1} and by the strong absorption involving the stretching of the C—O near 1200 cm^{-1}.[44]

The C=O stretching frequency for normal formate esters is at 1725–1720 cm^{-1}.[15] For most other alkyl-substituted saturated esters, the C=O frequency is at 1750–1735 cm^{-1}.[15] Ester C=O groups conjugated with C=C groups or aromatic rings absorb at 1740–1715 cm^{-1}.[45–47] When an electron withdraw-

[42] L. J. Bellamy and R. L. Williams, *J. Chem. Soc., London* p. 3465 (1958).

[43] E. J. Hartwell, R. E. Richards, and H. W. Thompson, *J. Chem. Soc., London* p. 1436 (1948).

[44] R. G. Fowler and R. M. Smith, *J. Opt. Soc. Amer.* **43**, 1054 (1953).

[45] W. L. Walton and R. B. Hughes, *Anal. Chem.* **28**, 1388 (1956).

[46] W. L. Walton and R. B. Hughes, *J. Amer. Chem. Soc.* **79**, 3985 (1957).

[47] J. L. Mateos, R. Cetina, E. Olivera, and S. Mesa, *J. Org. Chem.* **26**, 2494 (1961).

ing group such as a vinyl or phenyl group is placed on the single bonded oxygen, the carbonyl frequency will be raised. Vinyl acetate, for example, absorbs at 1770 cm^{-1} whereas ethyl acetate absorbs at 1740 cm^{-1}. The tendency for the carbonyl oxygen to draw electrons from the other oxygen and thus weaken the C=O bond [1] is reduced by the action of the group [2].[12]

$$
\text{[1]} \quad CH_3-\overset{\overset{\textstyle O-}{|}}{C}=\overset{+}{O}-CH=CH_2 \quad \longleftrightarrow \quad \text{[2]} \quad CH_3-\overset{\overset{\textstyle O}{\|}}{C}-\overset{+}{O}=CH-\overset{-}{C}H_2
$$

This same group which stiffens the C=O bond will also weaken the C—O bond at the same time, resulting in a higher C=O frequency but a lower C—O frequency.

9.15 Ester C—O

The C—O stretching frequency, so-called for convenience, actually involves some interaction with all the C—C bonds in the molecule. The most intense and therefore the most useful of these single bond vibrations is the highest asymmetric stretching frequency near 1200 cm^{-1}. Actually the C—O next to the carbonyl is stiffer than the other single bonds due to resonance which tends to localize the high vibration in the C—O bond. All the normal saturated esters except acetates absorb strongly at 1210–1160 cm^{-1}.[15] Normal acetates absorb at 1260–1230 cm^{-1}.[15] Acetates which were made from primary alcohols, where there is a CH_2 or CH_3 group on the single bonded oxygen, have a second correlatable single bond vibration at 1060–1035 cm^{-1}.[15] When the acetate is made from other than primary alcohols, a band appears at higher frequencies (1100 cm^{-1} for *sec*-alcohol). We can ascribe the band near 1250 cm^{-1} to asymmetric C—C—O stretching and the band near 1050 cm^{-1} to O—CH_2—C asymmetric stretching. This is a convenient nomenclature, since the lower band is more sensitive to branching changes in the alcohol part of the ester. There is a tendency for this lower band near 1050 cm^{-1} to appear in other saturated esters made from primary alcohols, but it is usually not as clear as in the acetates.

Esters conjugated with double bonds, such as acrylates, methacrylates, fumarates, etc., usually have multiple bands at 1300–1160 cm^{-1}.[45, 46] Esters conjugated with an aromatic ring, such as phthalates and benzoates, usually have a strong band near 1280 cm^{-1} and a second band for esters of primary alcohols near 1120 cm^{-1}.

Tables 9.2–9.4 summarize the absorption regions for various esters.

TABLE 9.2

BANDS FOR *O*-ALKYL SATURATED ESTERS (IN cm^{-1})[5]

Formates	Acetates	Propionates	*n*-Butyrates	Isobutyrates
1724–1722 (C=O)	1740 (C=O)	1740–1735 (C=O)	1735 (C=O)	1735 (C=O)
1214–1185 strong (C—O)	1370 (CH$_3$)	1275	1300	1260
1160	1245 strong (C—O)	1200–1190 strong (C—O)	1255	1200 strong (C—O)
	1040 (OCH$_2$)	1080 (OCH$_2$)	1190 strong (C—O)	1160
	640	1020 (OCH$_2$)	1100	1080
	612	810	750	755
		610	590	
		590		

TABLE 9.3

Bands for O-Alkyl α,β-Unsaturated Esters[45,46] (in cm^{-1})

Acrylates	Methacrylates	Crotonates (*trans*)	Maleates	Fumarates
1725 (C=O)	1725 (C=O)	1720 (C=O)	1725 (C=O)	1725 (C=O)
1640 (C=C)	1640 (C=C)	1660 (C=C)	1645 (C=C)	1650 (C=C)
1625	1410 (=CH$_2$ deformation)	—	1410 (*cis* CH rock)	—
1410 (=CH$_2$ deformation)	1325	1280	1290	1300
1280 (=CH rock)	1300	1190	1250	1265
1200 (C—O)	1180	—	1220 strong	1225 weak
1065 ± 5 (=CH$_2$ rock)	—	975 ± 5 (*trans* CH wag)	1162	1175
985 ± 5 (*trans* CH wag)	—	838 ± 4	—	1165
965 ± 5 (=CH$_2$ wag)	939 ± 4 (=CH$_2$ wag)	—	850 ± 50 (*cis* CH wag)	979 ± 3 (*trans* CH wag)
811 ± 3 (=CH$_2$ twist)	815 ± 2 (=CH$_2$ twist)	—	—	775 ± 3
657 ± 12 (C=O wag)	652 ± 6 (C=O wag)	685 ± 10 (C=O wag)	—	667 ± 3

TABLE 9.4

BANDS FOR O-ALKYL AROMATIC ESTERS (IN cm^{-1})

Benzoates	Phthalates	Terephthalates	Isophthalates
1725 (C=O)	1725 (C=O)	1725 (C=O)	1725 (C=O)
1605 (ring)	1605 (ring)	(no 1600 aromatic bands)	1613 (ring)
1585 (ring)	1585 (ring)	—	1587 (ring)
1280 strong (C—O)	1280 strong (C—O)	1280 strong (C—O)	1250 (C—O)
1110 (OCH$_2$)	1120 (OCH$_2$)	1110 (OCH$_2$)	1110 (OCH$_2$)
1070 (ring)	1080	—	—
1030 (ring)	1040 (ring)	—	—
710 (ring)	745 (ring)	730 (ring)	730 (ring)

9.16 Out-of-Plane Hydrogen Vibrations in Unsaturated Esters

Polarized Raman spectra of α,β-unsaturated esters with a plane of symmetry help in distinguishing out-of-plane vibrations (weak depolarized Raman lines) from in-plane vibrations (medium-strong polarized Raman lines) (see Table 9.5).

TABLE 9.5

POLARIZED RAMAN SPECTRA[a]

	CH$_2$ wag	C—C—O, in-phase stretch	CH$_2$ twist	C=O wag, out-of-plane
Methyl acrylate	975 DP weak	855 P medium	812 DP weak	665 DP weak
Methyl methacrylate	937 DP weak	836 P medium	817 DP weak	660 DP weak
Methyl crotonate (*trans*)		840 P medium		690 DP weak

[a] P, polarized; DP, depolarized.

Thus the Raman spectrum helps to identify the polarized 840 cm^{-1} band in methyl crotonate, which is the only band between 900 and 725 cm^{-1}, as an in plane vibration (probably in-phase C—C—O—C stretch) and confirms the 815 cm^{-1} assignment in acrylates, methacrylates, acrylic acid, and acrylamide as CH$_2$ twist with the carbonyl wag out-of-plane band at 700–640 cm^{-1} in these compounds. All the above depolarized bands have "C" type contours in the infrared spectra of the gas phase, verifying their out-of-plane character.

The lowest hydrogen wag of the vinyl group in hydrocarbons[48]

$$\left(\begin{array}{c} {}^{+}H \qquad H^{+} \\ C{=}C \\ H^{-} \end{array} \qquad \text{ca. } 630 \text{ cm}^{-1} \right)$$

involving *cis* wag and CH_2 twist seems to be the parent of the vinylidene twist

$$\left(\begin{array}{c} H^{+} \\ C{=}C \\ H^{-} \end{array} \qquad \text{ca. } 700 \text{ cm}^{-1} \right)$$

and the in-phase *cis* wag

$$\left(\begin{array}{c} {}^{+}H \qquad H^{+} \\ C{=}C \end{array} \qquad \text{ca. } 700 \text{ cm}^{-1} \right)$$

all three at about the same frequency ($+$ and $-$ here are motions out-of-plane). This suggests comparable assignments for acrylates (811 cm^{-1}), methacrylates (815 cm^{-1}), and maleates (850 cm^{-1}). These particular hydrogen bending modes all require a small out-of-plane counter-motion of the attached group so it is likely that some interaction with the C=O wag occurs. In acrylates and methacrylates this may account in part for the unusually high CH_2 twist frequencies as well as the C=O wagging frequencies which are somewhat lower than they are in crotonates. The CH_2 wagging frequencies are insensitive to interaction.[48]

9.17 Groups Next to the Carbonyl in Esters

There are alterations of the group next to the carbonyl in esters whicy are useful for identification.[49] In acetates, the methyl next to the C=O absorbs strongly near 1374 cm^{-1} due to the symmetric CH_3 deformation; the asymmetric CH_3 deformation absorbs weakly near 1430 cm^{-1}; and the CH_3 stretching absorbs weakly near 2990 cm^{-1}. The CH_2 next to the carbonyl in other saturated esters has a deformation frequency near 1420 cm^{-1}.

Acrylates have a doublet in the double bond region at 1640 and 1625 cm^{-1} where the intensity ratio is roughly 10:7 in the infrared and roughly 10:1 in the Raman spectrum. Normal benzoates with an unsubstituted ring have ring vibrations near 1070 and 1030 cm^{-1} and a strong band near 710 cm^{-1}. The aromatic esters are exceptions to the normal *ortho*, *meta*, and *para* distribution of bands in the 850–700 cm^{-1} region, the complications possibly due to

[48] J. K. Brown and N. Sheppard, *Trans. Faraday Soc.* **51**, 1611 (1955).

[49] A. R. Katritzky, J. M. Lagowski, and J. A. T. Beard, *Spectrochim. Acta* **16**, 964 (1960).

interaction with the out-of-plane carbonyl wag. Normal phthalates have aromatic bands at 1040 and 745 cm^{-1}. Terephthalates and isophthalates have an aromatic band near 730 cm^{-1}. Terephthalates have no 1600 cm^{-1} aromatic bands because of the center of symmetry.

9.18 Groups on the Oxygen Atoms in Esters

The alcohol from which the ester is made may be characterized in some cases.[50] A methyl ester has an asymmetric stretch near 2960 cm^{-1} and a deformation near 1440 cm^{-1}. An ethyl ester has an asymmetric CH$_3$ stretch near 2980 cm^{-1} with weaker bands near 2950 and 2900 cm^{-1}. The deformation of the O—CH$_2$ group absorbs near 1475 cm^{-1}, the asymmetric CH$_3$ deformation near 1455 cm^{-1}, the O—CH$_2$ wag near 1400 cm^{-1}, and the symmetric CH$_3$ deformation near 1375 cm^{-1}. When the CH$_2$ of the O—CH$_2$—CH$_3$ group is substituted as in O-sec-butyl, O—CH(C$_2$H$_5$)—CH$_3$, the 1400 cm^{-1} band disappears and the CH$_3$ absorbs near 1380 cm^{-1}. In the O-ethyl case there is probably interaction between the CH$_3$ symmetric deformation and the CH$_2$ wag so that both bands involve both vibrations. The O—CH$_2$ wag is still visible in n-propyl esters near 1390 cm^{-1} (CH$_3$ deformation near 1380 cm^{-1}). As with any other correlations, the usefulness of these bands depends on how much interference there is from the rest of the molecule. See Fig. 5.8.

9.19 Lactones

Lactones with unstrained six-membered rings are similar to noncyclic esters. As the ring becomes smaller the C—C—O angle decreases and the frequency of the carbonyl stretching vibration increases. Saturated δ-lactones (six-membered ring) have a carbonyl band at 1750–1735 cm^{-1}.[51] Saturated γ-lactones (five-membered ring) have carbonyl bands at 1795–1760 cm^{-1}.[52]

Unsaturated lactones of certain types in solution exhibit two bands in the carbonyl region due probably to a Fermi resonance effect.[52]

(I)   (II)

[50] A. R. Katritzky, J. M. Lagowski, and J. A. T. Beard, Spectrochim. Acta 16, 954 (1960).
[51] R. N. Jones and B. S. Gallagher, J. Amer. Chem. Soc. 81, 5242 (1959).
[52] R. N. Jones, C. L. Angell, T. Ito, and R. J. D. Smith, Can J. Chem. 37, 2007 (1959).

Lactones of type I have bands at 1790–1777 and 1765–1740 cm^{-1}. Lactones of type II have bands at 1775–1740 and 1740–1715 cm^{-1}.[52]

9.20   Thiol Esters and Related Compounds

Compounds which contain the group —S—C=O are characterized by carbonyl bands lower than normal esters[53] due to increased resonance

$$(-\overset{+}{S}=C-\overset{-}{O})$$

As with normal esters, S-aryl substitution raises the C=O frequency relative to S-aliphatic substitution. Bands can also be assigned to vibrations involving the C—S and C—C stretch. See Table 9.6.[53]

TABLE 9.6

THIOL CARBONYL COMPOUNDS

	C=O (cm^{-1})	C—C (cm^{-1})	C—S (cm^{-1})
Aliph—CO—S—aliphatic	1700–1680	1140–1070	1030–930
Aliph—CO—S—aryl	1710–1690	1110–1060	1020–920
Aryl—CO—S—aliphatic	1680–1640	1210–1190	940–905
Aryl—CO—S—aryl	1700–1650	1205–1190	920–895
Aliph—S—CO—CO—S—aliphatic	1680	—	790
Aryl—S—CO—CO—S—aryl	1698	—	770
H—CO—S—aliphatic	1675	—	755
H—CO—S—aryl	1693	—	730
CH$_3$—CO—SH	1712	1122	988
Aryl—CO—SH	1700–1690	1210–1205	950–945

The thiol formates have CH vibrations which absorb at 2835–2825 cm^{-1} (CH stretch), 1345–1340 cm^{-1} (CH deformation), and 2680–2660 cm^{-1} (deformation overtone). The thiol acids in solution absorb at 2585–2565 cm^{-1} (SH stretch) and at 837–828 cm^{-1} (SH in-plane deformation)[53].

9.21   Organic Carbonate Derivatives and Related Compounds

Table 9.7 lists frequencies for a number of compounds related to organic carbonates run in CCl$_4$ above 1333 cm^{-1} and CS$_2$ below.[54] The list clearly demonstrates that both substituents affect the carbonyl frequency. An alkoxy

[53] R. A. Nyquist and W. J. Potts, *Spectrochim. Acta* **15**, 514 (1959).
[54] R. A. Nyquist and W. J. Potts, *Spectrochim. Acta* **17**, 679 (1961).

group or nitrogen next to the carbonyl lowers its frequency. Changing the aliphatic group on the sulfur or oxygen to an aromatic group raises the frequency. Carbonyls in five-membered rings are somewhat higher in frequency than in noncyclic compounds. Note ethylene carbonate at 1831 cm^{-1}, ethylene monothiol carbonate, 1757 cm^{-1}, and ethylene dithiol carbonate, 1678 cm^{-1}.[54] In carbonic ester derivatives most of the noncyclic carbonates absorb at 1780–1740 cm^{-1} while the five-membered ring cyclic carbonates absorb at 1850–1790 cm^{-1}.[55, 56]

TABLE 9.7

ORGANIC CARBONATE DERIVATIVES AND RELATED COMPOUNDS[54]

	C=O Stretching region (cm^{-1})	Other bands (cm^{-1})
R—O—CO—O—R	1741–1739	1280–1240 (O—C—O asymmetric)
R—O—CO—O—ϕ^a	1787–1754	1248–1211
ϕ—O—CO—O—ϕ	1819–1775	1221–1205
R—S—CO—S—R	1655–1640	880–870 (S—C—S asymmetric)
R—S—CO—S—ϕ	1649	839
ϕ—S—CO—S—ϕ	1718–1714	833–827
R—O—CO—S—R	1710–1702	1162–1142 (C—O)
R—O—CO—S—ϕ	1731–1719	1141–1125
ϕ—O—CO—S—R	1739–1730	1102–1056
Cl—CO—O—R	1780–1775	1169–1139
Cl—CO—O—ϕ	1784	1113
Cl—CO—S—R	1772–1766	—
Cl—CO—S—ϕ	1775–1769	—
Cl—CO—NR$_2$	1739	—
Cl—CO—Nϕ_2	1742	—
R—O—CO—NH—R	1738–1732	1250–1210 (C—N)
R—S—CO—NH$_2$	1699	—
R—S—CO—NH—R	1695–1690	1230–1170
R—S—CO—NH—ϕ	1662–1649	1165–1152
ϕ—S—CO—NH—ϕ	1659–1652	1160–1152
R—S—CO—NR$_2$	1666	1248
R—S—CO—Nϕ_2	1670	1275

^{a} ϕ = Phenyl.

[55] L. Hough, J. E. Priddle, R. S. Theobald, G. R. Barker, T. Douglas, and J. W. Spoors, *Chem. Ind. (London)* p. 148 (1960).

[56] J. L. Hales, J. I. Jones, and W. Kynaston, *J. Chem. Soc., London* p. 618 (1957).

9.22   Oxalates

Oxalate esters ($-O-CO-CO-O-$) have two carbonyl groups. These can vibrate in-phase, symmetrically, or out-of-phase, asymmetrically, to give rise to bands near 1765 and 1740 cm^{-1}, respectively, for alkyl esters, and near 1795 and 1770 cm^{-1}, respectively, for aryl esters. When the carbonyls are *cis* to each, other both in-phase and out-of-phase vibrations are infrared active, but when the carbonyls are *trans* to each other (as in the solid state) only the out-of-phase lower frequency vibration is infrared active.

9.23   Anhydrides

Anhydrides are characterized by two bands in the carbonyl region due to the asymmetric and symmetric carbonyl stretching vibrations.[57,58] Noncyclic saturated anhydrides absorb near 1820 and 1750 cm^{-1}, while noncyclic conjugated anhydrides absorb near 1775 and 1720 cm^{-1}. The band at higher wavenumbers is the more intense in both cases.

Saturated anhydrides in a five-membered ring absorb near 1860 and 1780 cm^{-1} while conjugated anhydrides in a five-membered ring absorb near 1850 and 1760 cm^{-1}. The band at lower wavenumbers is the more intense in both cases. The relative intensities of the two bands is a good indication of whether the anhydride is cyclic or not. See Table 9.8.

Absorbance ratios of the low frequency band to the high $A_L/A_H = 0.81$–0.93 for saturated noncyclic anhydrides and 6.3–11.1 for five-membered ring cyclic anhydrides. Glutaric anhydride (six-membered ring) is intermediate where $A_L/A_H = 2.7$.[57] This is explained in terms of the relative orientation of the two

TABLE 9.8

ANHYDRIDE C=O REGIONS (IN cm^{-1})

Noncyclic, unconjugated	1825–1815 stronger	1755–1745 weaker
Noncyclic, conjugated	1780–1770 stronger	1725–1715 weaker
Cyclic, unconjugated (five-membered ring)	1870–1845 weaker	1800–1775 stronger
Cyclic, conjugated (five-membered ring)	1860–1850 weaker	1780–1760 stronger

[57] W. G. Dauben and W. W. Epstein, *J. Org. Chem.* **24**, 1595 (1959).
[58] L. J. Bellamy, B. R. Connelly, A. R. Philpotts, and R. L. Williams, *Z. Elektrochem.* **64** 563 (1960).

carbonyl groups, which in five-membered rings are nearly linearly and oppositely oriented. As a result, in the symmetric stretching vibration, the dipoles almost cancel each other and the intensity is weak in five-membered rings. The high frequency band is assigned to the symmetric stretching vibration.[57,58] In the six-membered ring, the orientation is about 120° which gives rise to a smaller A_L/A_H ratio. Noncyclics have several possible orientations and may orient with the two carbonyls pointing in the same direction, where the symmetric stretch would give rise to a stronger band than the asymmetric stretching vibration.[57,58]

Unconjugated straight chain anhydrides (except acetic, 1125 cm^{-1}) absorb strongly at 1050–1040 cm^{-1}. The cyclic anhydrides absorb strongly at 955–895 and 1300–1000 cm^{-1}. The nonconjugated cyclic anhydrides also absorb strongly at 1130–1000 cm^{-1}. These bands involve stretching of the C—C—O—C—C bands.

Mixed carboxyl carbonic anhydrides of the type R—CO—O—CO—O—C$_2$H$_5$, absorb at 1827–1810 and 1755–1750 cm^{-1} when R is saturated. When R is unsaturated or aromatic, the absorption occurs at 1812–1792 and 1735–1727 cm^{-1}.[59]

9.24 Peroxides, Acyl or Aroyl

The peroxide group O=C—O—O—C=O gives rise to two bands due to the carbonyl vibrations found at 1816–1787 and 1790–1765 cm^{-1}.[58]

9.25 Halogen-Substituted Carbonyls

Most acid chlorides with aliphatic groups attached to the carbonyl absorb strongly near 1810–1795 cm^{-1}. The presence of the electronegative chlorine atom next to the carbonyl greatly reduces the tendency for oxygen to draw electrons. Thus the contribution from the

$$[1] \quad R-\overset{\overset{\textstyle O}{\|}}{C}-Cl \quad \longleftrightarrow \quad [2] \quad R-\overset{\overset{\textstyle O^-}{|}}{\underset{+}{C}}-Cl$$

carbonyl bond weakening resonance form [2] is greatly reduced relative to ketones, resulting in absorption at higher wavenumbers. Another band is found at 965–920 cm^{-1} (probably involving C—C= stretch).

[59] D. S. Tarbell and N. A. Leister, *J. Org. Chem.* **23**, 1149 (1958).

In the gas phase, acetyl bromide has a carbonyl absorption at 1821 cm^{-1}; acetyl chloride, 1822 cm^{-1}; and acetyl fluoride, 1869 cm^{-1}.[9]

Aromatic acid chlorides absorb strongly at 1785–1765 cm^{-1}. A second weaker band at 1750–1735 cm^{-1} involves an overtone of a strong band near 875 cm^{-1} intensified by Fermi resonance with the carbonyl.[52] The aryl acid chlorides have absorption near 1200 and at 890–850 cm^{-1} (C—C stretch). Benzoyl fluoride has strong bands at 1812 1253–1237, and 1008 cm^{-1}.[60]

In the gas phase, COF_2 absorbs at 1928 cm^{-1}, $COCl_2$ absorbs at 1827 cm^{-1}, and $COBr_2$ absorbs at 1828 cm^{-1}.[9]

Chloroformates (also called chlorocarbonates) R—O—CO—Cl, absorb at 1800–1760 cm^{-1} (C=O), 1172–1134 cm^{-1} (C—O), 694–689 cm^{-1} (C—Cl), and 487–471 cm^{-1} (deformation).[54, 56, 61]

9.26 Carboxylic Acid OH Stretch

Carboxylic acids are usually characterized in the condensed state by a strongly bonded, very broad OH stretching band centering near 3000 cm^{-1}. While this is superimposed on the CH stretching bands (3100–2800 cm^{-1}) the broad wings of the OH stretch can be seen on either side of the narrow CH bands. Distinctive shoulders betwen 2700 and 2500 cm^{-1} appear regularly in carboxyl dimers and are due to overtones and combinations of the 1300 and 1420 cm^{-1} bands, due to interacting C—O stretch and OH deformation vibrations.[62] Monomeric acids absorb weakly and sharply at 3580–3500 cm^{-1}.[63]

9.27 Carboxyl—Carbonyl Stretch

Carboxylic acids with no other polar groups in the molecule usually exist predominantly as the hydrogen-bonded dimer, even in CCl_4 solution, although some acids exist at least partially in the hydrogen-bonded polymeric form.

[60] F. Seel and J. Langer, *Chem. Ber.* **91**, 2553 (1958).
[61] H. A. Ory, *Spectrochim. Acta* **16**, 1488 (1960).
[62] S. Bratoz, D. Hadzi, and N. Sheppard, *Spectrochim. Acta* **8**, 249 (1956).
[63] J. D. S. Goulden, *Spectrochim. Acta* **6**, 129 (1954).

When the dimer is considered as a whole, there will be two carbonyl stretching frequencies, symmetric and asymmetric. The dimer molecule has a center of symmetry, so the symmetric carbonyl stretch will be Raman active only, and the asymmetric stretch will be infrared active only. Most carboxyl dimers have a band in the infrared at 1720–1680 cm^{-1} (asymmetric C=O stretch). In the Raman spectrum, a band appears at 1680–1640 cm^{-1} (symmetric C=O stretch).[64]

Solid trichloroacetic acid is dimeric and has a Raman band[65] at 1687 cm^{-1} and an infrared band at 1742 cm^{-1}.[62] This is an example of the effect of strong electron withdrawing groups on the carbon of the acid, but studies of this kind are complicated due to hydrogen bonding permutations.

Resonance causes a positive charge to appear on the proton donor atom and a negative charge to appear on the receiver atom which tends to encourage the hydrogen bond. The increased association of the proton with the acceptor atom and the decreased association with the donor atom tends to encourage the resonance. Thus, hydrogen bonding and resonance are mutually enhancing. Resonance weakens the carbonyl bond and lowers the frequency.

When a hydroxyl or an ether group is in the same molecule with the carboxyl group, or when an acid is dissolved in a solvent containing hydroxyl or ether groups, there exists the opportunity for other types of hydrogen bonding to occur involving hydroxyl–carboxyl bonds or carboxyl–ether bonds as an alternate to carboxyl dimer bonding. When this happens, symmetry is destroyed and the carbonyl vibrations appear in both the infrared and Raman. When the carbonyl is not hydrogen bonded as in carboxyl–ether bonding, a band appears at 1760–1735 cm^{-1} in both infrared and Raman. When the carbonyl is hydrogen bonded but not dimerized, as in alcohol–carbonyl bonds, a band appears at 1730–1705 cm^{-1} in both infrared and Raman. When the carboxyl group is dimerized, as we have seen above, a band appears at 1680–1640 cm^{-1} in the Raman effect and at 1720–1690 cm^{-1} in the infrared. From this it can be seen that the most unambiguous band in this region for the carboxyl dimer is the Raman band at 1680–1640 cm^{-1}.[64]

Lactic acid, for example, has no Raman bands between 1700 and 1600 cm^{-1}. Its carbonyl in the Raman effect resembles the infrared carbonyl at 1723 cm^{-1}. Therefore, no dimer with a center of symmetry is present. Methoxyacetic acid has a very weak Raman band at 1666 cm^{-1} and a relatively strong Raman band at 1741 cm^{-1} like the infrared band.[64] This indicates only a small amount of dimer is present.

[64] K. W. F. Kohlrausch, "Ramanspektren." Akad. Verlagsges., Leipzig, 1943.
[65] I. D. Poliakova and S. S. Raskin, *Opt Spectrosc.* (*USSR*) **6**, 220 (1959).

9.28 Carboxyl OH Bending and C—O Stretching

The best infrared band for the carboxyl dimer is a broad, medium intensity band at 960–875 cm^{-1} due to out of plane OH $\cdots$ O hydrogen deformation.[66] This band is present here only for the dimer and is usually noticeably broader than other bands in this region. The absence of a band here would be fairly good evidence for the lack of the dimer form.

Carboxylic acids have a strong band in the region 1315–1200 cm^{-1}. Dimers usually absorb in the narrower region 1315–1280 cm^{-1}.[66] Another somewhat weaker band is found at 1440–1395 cm^{-1}.[66] These bands involve stretching of the C—O bond and in-plane deformation of the C—O—H angle which interact somewhat[66] so that both bands to some extent involve both OH deformation and C—O stretch, in- and out-of-phase. Due to the similarity of the 1315–1200 cm^{-1} band to the ester C—O stretching bands, it is a nomenclature convenience to call this band in acids predominantly C—O stretching. In the same region as the "OH deformation" band at 1440–1395 cm^{-1} is found the deformation of the CH$_2$ group next to the carbonyl when it is present. The carboxyl dimer infrared bands are summarized in Table 9.9.[62,66,67]

TABLE 9.9

CARBOXYL DIMER SPECTRAL REGIONS (IN cm^{-1})

OH stretch	ca. 3000 very broad
Overtones and combinations	2700–2500
C=O stretch	1740–1680
OH deformation in plane	1440–1395
C—O stretch	1315–1280
OH deformation out of plane	960–875

9.29 Monomeric Acids

When carboxylic acids are run in the vapor state at about 150°C, carboxylic acids are monomeric in form.[68] Some monomer is present in the vapor state at room temperature or in solutions.[62] Bands at 1380–1280 cm^{-1} (medium)

[66] D. Hadzi and N. Sheppard, *Proc. Roy. Soc., Ser. A* **216**, 247 (1953).

[67] F. Gonzalez-Sanchez, *Spectrochim. Acta* **12**, 17 (1958).

[68] D. Hadzi and M. Pintar, *Spectrochim. Acta* **12**, 162 (1958).

and 1190–1075 cm^{-1} (strong) involve interacting OH deformation in-plane and C—O stretching.[68]

The OH stretching vibration absorbs sharply and weakly at 3550–3500 cm^{-1} in CCl$_4$ solution[63] and somewhat higher in the vapor phase.[62] The C=O stretching vibration absorbs at 1800–1740 cm^{-1}.[62] Table 9.10 lists carboxyl monomer bands.

TABLE 9.10

CARBOXYL MONOMER SPECTRAL REGIONS (IN cm^{-1})

OH stretch	3580–3500
C=O stretch	1800–1740
OH deformation	1380–1280
C—O stretch	1190–1075

9.30 Aliphatic Peroxy Acids

Peroxy acids in the vapor state have the bands listed in Table 9.11.[69,70] The peroxy acid band at 3280 cm^{-1} is distinctly different from bands of either the monomer or dimer of the normal acid vapor.

TABLE 9.11

SPECTRAL BANDS FOR ALIPHATIC PEROXY ACIDS IN THE VAPOR STATE (IN cm^{-1})

OH stretch	ca. 3280
C=O stretch	1760
OH bend	1450
C—O stretch	1175
O—O stretch	865

9.31 Aromatic Acids

Derivatives of benzoic and toluic acids[71] have C=O bands due to dimers at 1715–1680 cm^{-1} and to monomers in dilute solution about 40–45 cm^{-1} higher at 1760–1730 cm^{-1}. A hydroxyl group in the *ortho* position causes

[69] E. R. Stephens, P. L. Hanst, and R. C. Doerr, *Anal. Chem.* 29, 776 (1957).

[70] P. A. Giguère and A. W. Olmos, *Can. J. Chem.* 30, 821 (1952).

[71] D. Peltier, A. Pichevin, P. Dizabo, and M. L. Josien, *C.R. Acad. Sci.* 248, 1148 (1959).

about a 55 cm^{-1} lowering of the band and an *ortho*-amino group about a 30 cm^{-1} lowering due to internal hydrogen bonding. The band position has been related to the pK of the acid.[71]

The methyl esters absorb about 13 cm^{-1} lower than the comparable acid monomers.[71] The aromatic bands at 900–700 cm^{-1} in acids and derivatives with a carbonyl group on the ring do not resemble those of unconjugated aromatics and should be considered separately.

9.32 Aliphatic Bands in Long Chain n-Aliphatic Carboxylic Acids, Esters, and Soaps

In the solid state spectra of fatty acids, esters, and soaps a regularly spaced progression of fine bands appears between 1345 and 1180 cm^{-1},[72–74] due to the various CH$_2$ waggings.[75] See Fig. 5.7.

A correlation has been made for the number of bands in the " band progression " with the number of carbon atoms.[73]

For acids with even numbers of carbons, the number of bands in the progression series equals the number of carbons in the acid divided by two. For the acids with odd numbers of carbons, the number of bands equals the number of carbons in the acid plus one, divided by two.[73]

This works for either acids or soaps and is more definite in the latter, particularly above C$_{26}$, due to lack of interference from the C—O in the acid.

9.33 Carboxyl Salts

When a salt is made from a carboxylic acid, the C=O and C—O are replaced by two equivalent carbon–oxygen bonds which are intermediate in force constant between the C=O and C—O.

[72] R. N. Jones, A. F. McKay, and R. G. Sinclair, *J. Amer. Chem. Soc.* **74**, 2575 (1952).
[73] R. A. Meiklejohn, R. J. Meyer, S. M. Aronovic, H. A. Schuette, and V. W. Meloche, *Anal. Chem.* **29**, 329 (1957).
[74] P. J. Corish and D. Chapman, *J. Chem. Soc. London* p. 1746 (1957).
[75] J. K. Brown, N. Sheppard, and D. M. Simpson, *Phil. Trans. Roy. Soc. London, Ser. A* **247**, 35 (1954).

These two "bond and a half" oscillators are strongly coupled, resulting in a strong asymmetric CO_2 stretching vibration at 1650–1540 cm^{-1} and a somewhat weaker symmetric CO_2 stretching vibration at 1450–1360 cm^{-1}.[14]

Acetate salts absorb strongly at 1600–1550 and 1450–1400 cm^{-1} and weakly near 1050, 1020, and 925 cm^{-1}.[76] Formate salts absorb near 2930, 1600, 1360, and 775 cm^{-1}. Oxalate salts, $(O_2C-CO_2)^{2-}$ with interacting carboxylate groups, absorb near 1620, 1320, 770, and 515 cm^{-1}.

9.34   Amino Acids

Amino acids usually exist as the zwitterion.

They have, therefore, the absorption of an ionized carboxyl group and an amine salt.

The NH_3^+ stretching frequencies are found between 3100 and 2600 cm^{-1} in the form of a broad strong band with multiple peaks on the low frequency wing which continue until about 2200 cm^{-1}. Between 2200 and 2000 cm^{-1} is found a relatively prominent combination band which resembles the similar band in primary amine hydrochlorides. This is assigned as a combination band of NH_3^+ asymmetric deformation and NH_3^+ hindered rotation (see amine salts).

The spectra of D- and α-L-amino acids are identical[77] but may differ from that of the DL form. They have the following characteristic bands[77–79]: 1665–1585 cm^{-1} (asymmetric NH_3 deformation), not resolved in most normal chain amino acids; 1605–1555 cm^{-1} (asymmetric CO_2 stretch); 1530–1490 cm^{-1} (symmetric NH_3 deformation); 1425–1393 cm^{-1} (symmetric CO_2 stretch); and 1340–1315 cm^{-1} (CH deformation). More variable bands are found at 3100–2850, 2650–2500, and 2120–2010 cm^{-1}.

[76] F. Vratny, C. N. R. Rao, and M. Dilling, *Anal. Chem.* **33**, 1455 (1961).

[77] R. J. Koegel, J. P. Greenstein, M. Winitz, S. M. Birnbaum, and R. A. McCallum, *J. Amer. Chem. Soc.* **77**, 5708 (1955)

[78] K. Fukushima, T. Onishi, T. Shimanouchi, and S. Mizushima, *Spectrochim. Acta* **15**, 236 (1959).

[79] H. M. Randall, R. G. Fowler, N. Fuson, and R. Dangl, "Infrared Determination of Organic Structures." Van Nostrand-Reinhold, Princeton, New Jersey, 1949.

9.35 Amido Acids

Bands for α-amido acids are found in the solid state at 3390–3260 cm^{-1} (NH), 1724–1695 cm^{-1} (acid C=O), 1621–1600 cm^{-1} (amide C=O), and 1565–1508 cm^{-1} (CNH).[79–81] The amide C=O at or below 1620 cm^{-1} is characteristic for the α-amido acids, other amido acids absorbing on the high frequency side of 1620 cm^{-1} [79] in the solid state.

9.36 Unsubstituted Amides

Normal amides are characterized by a strong absorption at 1695–1630 cm^{-1} due to C=O stretching. Unsubstituted amides in the solid state are characterized by asymmetric and symmetric NH_2 stretching frequencies near 3350 and 3180 cm^{-1}, and a strong band at 1670–1620 cm^{-1}, which is usually a doublet near 1655 and 1630 cm^{-1} involving C=O stretch and NH_2 deformation.[82] In solution these bands are replaced by bands near 3520, 3400, 1690–1670 (C=O), and 1620–1614 cm^{-1} (NH_2 deformation weak). These differences result from hydrogen bonding.[82,83] The amide group has a large contribution from resonance structure [2] which weakens the C=O bond and stiffens the C—N bond.

$$[1] \quad -\!\overset{\overset{\displaystyle O}{\|}}{C}\!-\!NH_2 \quad \longleftrightarrow \quad [2] \quad -\!\overset{\overset{\displaystyle O^-}{|}}{C}\!=\!\overset{+}{N}H_2$$

An electron withdrawing group on the nitrogen will reduce the contribution form 2, and a higher frequency carbonyl band will result.[12]

In unsubstituted amides, $C-CO-NH_2$, a vibration involving the C—N stretch absorbs somewhere near 1400 cm^{-1}. A weaker band somewhere near 1150 cm^{-1} can sometimes be seen which involves the NH_2 rock (in plane). A broad band at 750–600 cm^{-1} is due to NH_2 wag (out of plane).

[80] N. Fuson, M. L. Josien, and R. L. Powell, J. Amer. Chem. Soc. **74**, 1 (1952).

[81] H. H. Freedman, J. Amer. Chem. Soc. **77**, 6003 (1955).

[82] R. E. Richards and H. W. Thompson, J. Chem. Soc., London p. 1248 (1947).

[83] T. L. Brown, J. F. Regan, R. D. Schuetz, and J. Sternberg, J. Phys. Chem. **63**, 1324 (1959).

9.37   N-Substituted Amides (trans)

N-Monosubstituted amides exist mainly with the NH and C=O in the *trans* configuration. In the solid state, the NH stretch gives rise to a strong band near 3300 cm^{-1}. A weaker band appears near 3100 cm^{-1} due to an overtone of the 1550 cm^{-1} band.[84] The carbonyl vibration absorbs strongly at 1680–1630 cm^{-1}.[82] The in plane NH bending frequency and the resonance stiffened C—N bond stretching frequency fall close together and therefore interact. The CNH vibration where the nitrogen and hydrogen move in opposite directions relative to the carbon involves both NH bending and C—N stretching and absorbs strongly near 1550 cm^{-1}.[85–89] This band is very characteristic for monosubstituted amides. The CNH vibration where the N and H atoms move in the same direction relative to the carbon gives rise to a weaker band near 1250 cm^{-1}. The out-of-plane NH wag absorbs broadly near 700 cm^{-1}.[85–89]

In solution comparable bands appear near 3470–3400, 1700–1670, and 1540–1510 cm^{-1}.[82,90] Electron withdrawing groups on the nitrogen raise the carbonyl frequency.

9.38   N-Substituted Amides (cis) (Lactams)

Monosubstituted amides which are forced into the *cis* configuration in cyclic structures such as lactams have a strong NH stretching absorption in the solid state near 3200 cm^{-1} and a weaker band near 3100 cm^{-1} due to a combination band of the C=O stretching and NH bending bands.[84,91] The carbonyl vibration absorbs near 1650 cm^{-1} in six- or seven-membered rings as in the noncyclic *trans* case. Lactams in five-membered rings absorb at 1750–1700 cm^{-1}. Unfused lactams in four membered rings absorb at 1760–1730 cm^{-1}, while β-lactams fused to unoxidized thiazolidine rings absorb at

[84] T. Miyazawa, *J. Mol. Spectrosc.* **4**, 168 (1960).

[85] T. Miyazawa, T. Shimanouchi, and S. Mizushima, *J. Chem. Phys.* **24**, 408 (1956).

[86] T. Miyazawa, T. Shimanouchi, and S. Mizushima, *J. Chem. Phys.* **29**, 611 (1958).

[87] R. D. B. Fraser and W. C. Price, *Nature* (London) **170**, 490 (1952).

[88] M. Beer, H. B. Kessler, and G. B. B. M. Sutherland, *J. Chem. Phys.* **29**, 1097 (1958).

[89] C. G. Cannon, *Spectrochim. Acta* **16**, 302 (1960).

[90] R. L. Jones, *Spectrochim. Acta, Part A* **23**, 1745 (1967).

[91] T. Miyazawa, *J. Mol. Spectrosc.* **4**, 155 (1960).

$1780-1770$ cm^{-1}.[92] There is, in the cyclic monosubstituted amide, no band in the $1600-1500$ cm^{-1} region comparable to the 1550 cm^{-1} CNH band in the *trans* case. The *cis*-NH bending vibration absorbs at $1490-1440$ cm^{-1} and the C—N stretching vibration at $1350-1310$ cm^{-1}.[91,93] There is much less interaction between these modes than in the *trans* case.[91] The NH wag absorbs broadly near 800 cm^{-1}.

Trans- and *cis*-monosubstituted amides are best distinguished by the NH vibrations.[94] Hydrogen-bonded *trans*-monosubstituted amides (polymers) absorb near 3300, 3080, and 1550 cm^{-1}, and hydrogen-bonded *cis*-monosubstituted amides (dimers) absorb at 3200 and 3080 cm^{-1}. In dilute solution, the *trans* form absorbs at $3440-3400$ cm^{-1}, while the *cis* form absorbs $20-40$ cm^{-1} lower. In solution, the association NH band of the *trans* form shifts to higher frequencies $3370-3300$ cm^{-1} with dilution (shorter polymers), while the associated *cis*-NH band does not shift (dimers). The *cis* form remains associated at lower concentration than the *trans* form. At a concentration of 0.003 mole/liter in CCl$_4$, the *trans* form is mostly unassociated, while the *cis* is still mostly associated.[94]

9.39   Disubstituted Amides

N-Disubstituted amides are characterized by the strong C=O stretching at $1680-1630$ cm^{-1}.[82] Strong electron withdrawing groups substituted on the nitrogen atom will reduce the tendency for the carbonyl oxygen to draw electrons (N=C—O) and thus raise the carbonyl frequency.[12]

9.40   Ureas

Ureas are not too different from amides. A single substituent on the nitrogen will give the same CNH band as monosubstituted amides. As with amides, a strong electron withdrawing group on the nitrogen atom will raise the C=O stretching frequency.

[92] H. T. Clarke, ed., "The Chemistry of Penicillin," p. 390. Princeton Univ. Press, Princeton, New Jersey, 1949.

[93] H. Brockmann and H. Musso, *Chem. Ber.* **89**, 241 (1956).

[94] I. Suzuki, M. Tsuboi, T. Shimanouchi, and S. Mizushima, *Spectrochim. Acta* **16**, 471 (1960).

9.41   Carbamates

The carbonyl frequencies in carbamates are somewhat higher than in amides and lower than in esters, being found in the region 1740–1683 cm^{-1}.[95-97] In CHCl$_3$ solution, most primary carbamates absorb at 1728–1722 cm^{-1}, secondary carbamates at 1772–1705 cm^{-1}, and tertiary at 1691–1683 cm^{-1}.[96] In the solid state they are much the same,[96] except that some primary carbamates give very broad bands which may absorb as low as 1690 cm^{-1}. The C=O frequency for linear carbamates with a cyclic N atom depends on the electronic effects in the ring.[96]

In N-monosubstituted carbamates, the NH group absorbs at 3340–3250 cm^{-1}, and the CNH group absorbs near 1540–1530 cm^{-1} in the condensed state as in amides. In solution these bands appear at 3480–3390 and 1530–1510 cm^{-1}. The NH$_2$ group when present also resembles the NH$_2$ in amides absorbing at 3450–3200 cm^{-1} and near 1620 cm^{-1}.

The 2-oxazolidones absorb at 1810–1746 cm^{-1} and also at 1059–1029 cm^{-1}.[96]

9.42   Hydroxamic Acids

Hydroxamic acids (R—CO—NH—OH) are characterized in the solid state by three bands between 3300 and 2800 cm^{-1}, a band near 1640 cm^{-1} (C=O), and a band near 1550 cm^{-1}, a variable intensity band at 1440–1360 cm^{-1}, and a strong band near 900 cm^{-1}.[98]

9.43   Imides

In the solid state, imides (R—CO—NH—CO—R) have a strong band at 1740–1670 cm^{-1} and a bonded NH band near 3250 cm^{-1}.[79,99-101] In noncyclic imides the CNH group gives rise to bands at 1507–1503 and 1236–1167

[95] D. A. Barr and R. N. Haszeldine, *J. Chem. Soc., London* p. 3428 (1956).
[96] S. Pinchas and D. Ben Ishai, *J. Amer. Chem. Soc.* **79**, 4099 (1957).
[97] A. R. Katritzky and R. A. Jones, *J. Chem. Soc., London* p. 676 (1960).
[98] D. Hadzi and D. Prevorsek, *Spectrichim. Acta* **10**, 38 (1957).
[99] J. Uno and K. Machida, *Bull. Chem. Soc. Jap.* **34**, 545 and 551 (1961).
[100] R. A. Abramovitch, *J. Chem. Soc., London* p. 1413 (1957).
[101] N. A. Borisevich and N N. Khovratovich, *Opt Spectrosc. (USSR)* **10**, 309 (1961).

cm^{-1} like the monosubstituted amides. Most of the dialkyl imides exist in the "B" form where the carbonyls are parallel and *trans-trans* relative to the NH.[99] These are characterized by bands at 3280, 3200, 1737–1733, 1505–1503, 1236–1167, and 739–732 cm^{-1} (NH wag).[99] Sometimes a weak band is visible at 1695–1690 cm^{-1}. Diacetamide can be in either "A" or "B" forms but is usually found in the "A" form (*trans-cis*) which is distinguished from the "B" form by a NH band at 3245 with weaker companions at 3270 and 3190, C=O bands at 1700 with weaker companions at 1734 and 1650 cm^{-1}, and NH wag bands at 836–816 cm^{-1}.[99]

Imides which are part of five-membered rings such as phthalimides[101] have C=O bands at 1790–1735 and 1745–1680 cm^{-1}. The lower frequency band is the more intense. The cyclic imides do not have the 1505 cm^{-1} CNH band.

9.44 Isocyanurates

Aliphatic isocyanurates (isocyanate trimers) have a strong C=O band at 1700–1680 cm^{-1} with a weak shoulder near 1755 cm^{-1}. Aromatic isocyanurates[102] have a higher C—O frequency at 1715–1710 cm^{-1} with a weak shoulder near 1780 cm^{-1} because of the electron withdrawing group on the nitrogen.

Aromatic isocyanate dimers.[102]

$$aryl-N-C=O$$
$$\,\,\,\,\,\,\,\,\,\,|\,\,\,\,\,\,|$$
$$O=C-N-aryl$$

have a carbonyl band at 1785–1775 cm^{-1}.

9.45 Acid Hydrazides

Monoacid hydrazides[103] (—CO—NH—NH$_2$) have NH or NH$_2$ bands at 3320–3180 cm^{-1}, a C=O band at 1700–1640 cm^{-1}, an NH$_2$ deformation band at 1633–1602 cm^{-1}, a CNH band at 1542–1522 cm^{-1}, and an NH$_2$ band at 1150–1050 cm^{-1}.

[102] B. Taub and C. E. McGinn, *Dyestuffs* **42**, 263 (1958).
[103] M. Mashima, *Bull. Chem. Soc. Jap.* **35**, 1882 and 2020 (1962).

Diacid hydrazides[104] (—CO—NH—NH—CO) have bands in acetonitrile solution at 3330–3280 cm^{-1} (NH stretch), 1742–1700 cm^{-1} (C=O), and 1707–1683 cm^{-1}, (C=O). In the solid state, aliphatic diacid hydrazides absorb at 3210–3100 cm^{-1}, 3060–3020 cm^{-1}, 1623–1580 cm^{-1} (C=O), 1506–1480 cm^{-1} (CNH), and 1260–1200 cm^{-1} (CNH); and aromatic diacid hydrazides absorb at 3280–2980 cm^{-1}, 1730–1669 cm^{-1} (C=O), 1658–1635 cm^{-1} (C=O), 1535–1524 cm^{-1} (CNH), and 1285–1248 cm^{-1} (CNH).

Phthalhydrazides in the solid state have a very broad NH band centering near 3000 cm^{-1}, and a C=O band at 1670–1635 cm^{-1}.

[104] M. Mashima, *Bull. Chem. Soc. Jap.* **35**, 332 and 423 (1962).

CHAPTER 10

ETHERS, ALCOHOLS, AND PHENOLS

10.1 Aliphatic Ethers

When an oxygen atom is substituted for a carbon atom in a normal aliphatic chain, the positions of the skeletal stretching frequencies (ca. 1150–800 cm^{-1}) are altered somewhat. Far more spectacular, however, is the change in intensity of the highest skeletal stretching frequency. The group CH_2-O-CH_2 gives rise to a strong band at 1140–1085 cm^{-1}. Most of the simpler aliphatic ethers absorb near 1125 cm^{-1}. While we call this band the asymmetric $C-O-C$ stretching band, it is understood that during this vibration in a normal ether, other skeletal bonds are involved.[1]

Branching of the carbon atoms next to the oxygen causes complications, but usually there remains a band near 1100 cm^{-1} involving the stretching of the $C-O$ bond and, in the case of a *tert*-butoxy ether type, absorption near 1200 cm^{-1} involving the asymmetric motion of the branched carbon atom and methyl rocking.

The aliphatic groups next to the oxygen have bands listed in Table 10.1.[2]

The intensity of the symmetric stretching band of the OCH_2 group is enhanced, so that it is more nearly comparable to the intensity of the

TABLE 10.1

Aliphatic Groups in Ethers (in cm^{-1})

$(R-O-)CH_3$	2992–2955 Asymmetric stretch
	2897–2867 Symmetric stretch
	2840–2815 Deformation overtone
	1470–1440 Asymmetric and symmetric deformation
$(R-O-)CH_2-$	2955–2922 Asymmetric stretch
	2878–2835 Symmetric stretch
	1475–1445 Deformation

[1] R. G. Snyder and G. Zerbi, *Spectrochim. Acta, Part A* **23**, 391 (1967).
[2] S. E. Wiberley, S. C. Bunce, and W. H. Bauer, *Anal. Chem.* **32**, 217 (1960).

311

asymmetric stretch in contrast to the hydrocarbon case where the symmetric stretch is always weaker.

The largest frequency shift is in the symmetric CH_3 deformation which shifts from 1375 cm^{-1} for $C-CH_3$ to about 1450 cm^{-1} for $O-CH_3$. This is due to the electronegativity of the oxygen (see CH_3 groups).

Table 10.2 gives the regions for the strong bands due to the asymmetric stretching of the $C-O-C$ bonds.

TABLE 10.2

ASYMMETRIC $C-O-C$ SPECTRAL REGIONS (IN cm^{-1})

Aliphatic ethers	1140–1085
Aromatic ethers	1310–1210 and 1050–1010 ($O-CH_2$)
Vinyl ethers	1225–1200
Oxirane ring (monosubstituted)	880– 805

10.2 Aromatic Ethers

A methoxy group attached to an aromatic ring usually gives rise to a sharp isolated band near 2835 cm^{-1}. An alkoxy on an aromatic ring usually gives rise to two correlatable bands, 1310–1210 and 1050–1010 cm^{-1}.[3–5] The band near 1250 cm^{-1} may be looked on as an aromatic carbon–oxygen stretching frequency (like phenols), and the band near 1040 cm^{-1}, as the highest aliphatic carbon–oxygen stretching frequency (like primary alcohols). Diphenyl ether does not have a 1040 cm^{-1} "OCH_2" band, only the 1240 cm^{-1} "aryl-O" band (asymmetric $C-O-C$ stretch). Since there is undoubtedly interaction, the convenient nomenclature of aryl-O and OCH_2 stretch could be described as asymmetric and symmetric $C-O-C$ stretching in anisole. In phenetoles we have three skeletal single bonds and three stretching frequencies absorbing near 1240, 1040, and 920 cm^{-1} (weaker), the first two of which are correlatable with other alkoxy aromatics. Methylene-1,2-dioxybenzenes ($O-CH_2-O$) have aromatic ether bands at 1266–1227 and 1047–1025 cm^{-1}. Other bands for this group appear at 1376–1350 (variable intensity) and 938–919 cm^{-1} (strong).[4]

---

[3] A. R. Katritzky and N. A. Coats, J. Chem. Soc., London p. 2062 (1959).

[4] L. H. Briggs, L. D. Colebrook, H. M. Fales, and W. C. Wildman, Anal. Chem. **29**, 904 (1957).

[5] K. J. Sax, W. S. Saari, C. L. Mahoney, and J. M. Gordon, J. Org. Chem. **25**, 1590 (1960).

The aromatic carbon–oxygen bond has a higher force constant than the aliphatic carbon–oxygen bond due to resonance, which is one of the reasons for the higher vibrational frequency.

The contributions of the

$$=\overset{+}{\text{O}}-$$

forms stiffen that bond and concentrate negative charges in the *ortho* and *para* position. When electron repelling groups such as methyls are put on the ring in the *ortho* or *para* positions, the resonance is reduced relative to *meta* substitution. The aryl C—O band is found at higher frequencies in *meta* than in *ortho* or *para* as a result. Conversely, electron attracting groups such as chlorine enhance the resonance in *ortho* and *para* substitution relative to *meta*, and here the aryl C—O band is lower in *meta*.

In 1,2,3-trimethoxybenzene the strongest band in the C—O region is at 1115 cm^{-1}. Some other highly substituted methoxybenzenes are similar. This reflects the fact that the aryl C—O vibration interacts with ring frequencies and is therefore not completely mechanically independent of other single bonds attached to the ring.

10.3 Vinyl Ethers

The asymmetric C—O—C stretch of alkyl vinyl ethers gives rise to a strong band at 1225–1200 cm^{-1},[6] very near 1203 cm^{-1} in most cases. The vinyl carbon–oxygen bond is stiffened by resonance as was the aromatic carbon–oxygen bond.

$$\text{CH}_2{=}\text{CH}{-}\text{O}{-}\text{C} \qquad \overset{-}{\text{C}}\text{H}_2{-}\text{CH}{=}\overset{+}{\text{O}}{-}\text{C}$$

Frequencies of the vinyl group are altered by the oxygen, the most notable of which is a doubling and intensifying of the strong C=C stretching frequency. One band is found near 1640 cm^{-1} and a slightly stronger one near 1620 cm^{-1}. These bands are temperature dependent, with the 1640 cm^{-1}

[6] Y. Mikawa, *Bull. Chem. Soc. Jap.* **29**, 110 (1956).

band increasing in intensity with increasing temperature. In 2-ethylhexyl vinyl ether the ratio of absorbancies is A 1615 cm^{-1}/A 1640 cm^{-1} = 1.91 at 25°C and 1.21 at 125°C. This shows that the doubling is caused by rotational isomers, resulting from rotation of the vinyl carbon–oxygen bond (see Fig. 10.1). The 1615 cm^{-1} band is the result of the more stable form, which is the resonance stabilized planar *cis* configuration with a resonance weakened C=C bond.[7] The 1640 cm^{-1} band is the result of a nonplanar *gauche* form,

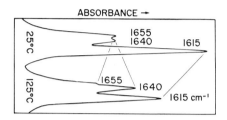

FIG. 10.1.   Infrared spectra of 2-ethylhexyl vinyl ether at two temperatures. The 1640 cm^{-1} C=C band is assigned to the less stable isomer (*gauche*) and the 1615 and 1655 cm^{-1} bands are from the more stable isomer (*cis*) and are assigned to C=C stretch and an overtone of the 815 cm^{-1} band.

where resonance (which requires coplanarity) is reduced. Sometimes an additional band near 1655 cm^{-1} is seen. This has been assigned to an overtone of the =CH$_2$ wag band near 815 cm^{-1} in vinyl ethers, interacting in Fermi resonance with the C=C stretch band at 1615 cm^{-1}.[7] Phenyl vinyl ether absorbs at 1653 and 1635 cm^{-1},[6] where vinyl ether resonance is reduced by the electron withdrawing characteristics of the phenyl group.

The bands for alkyl vinyl ethers are summarized in Table 10.3.[6]

TABLE 10.3

SPECTRAL REGIONS FOR ALKYL VINYL ETHERS (IN cm^{-1})

Vinyl CH stretch	3125–3098, 3078–3060, 3050–3000
C=C *gauche*	1660–1635
C=C *cis*	1620–1610
=CH rock	1323–1320
C—O—C asymmetric stretch	1225–1200
trans =CH wag	970– 960
=CH$_2$ wag	820– 810

[7] N. L. Owen and N. Sheppard, *Trans. Faraday Soc.* **60**, 634 (1964).

10.4 Cyclic Ether Linkages

The CH_2-O-CH_2 linkage in six-membered rings absorbs in about the same region that the noncyclic ethers do, owing to an asymmetric $C-O-C$ stretching vibration. As the ring becomes smaller, the frequency of a band involving the asymmetric stretch of the $C-O-C$ bonds decreases, and the symmetric $C-O-C$ (ring breathing) frequency increases. This is mainly the result of the geometry change rather than of force constant changes. The asymmetric $C-O-C$ stretch is usually a strong infrared band and a weaker depolarized Raman band. The symmetric $C-O-C$ is usually a strong polarized Raman band[8] and a weaker infrared band (see Table 10.4).

TABLE 10.4

CYCLIC ETHER COMPOUNDS

Ring size	Compound	Asymmetric C—O—C (cm^{-1})	Symmetric C—O—C (cm^{-1})
Six-membered	Pentamethylene oxide	1098	813
Five-membered	Tetramethylene oxide	1071	913
Four-membered	Trimethylene oxide	983	1028
Three-membered	Ethylene oxide[9]	892	1270
(Two-membered)	(Formaldehyde)	—	(1740)

10.5 Oxirane Ring Compounds

Epoxy ring compounds[9-15] absorb at 1280–1230 cm^{-1} as a result of the ring breathing vibration ($C-C$, $C-O$, and $C-O$ bonds all stretching in phase). There are two other ring vibrations approximately described as

[8] K. W. F. Kohlrausch, "Ramanspektren." Akad. Verlagsges., Leipzig, 1943.

[9] R. C. Lord and B. Nolin, *J. Chem. Phys.* **24**, 656 (1956).

[10] W. J. Potts, *Spectrochim. Acta* **21**, 511 (1965).

[11] J. Bomstein, *Anal. Chem.* **30**, 544 (1958).

[12] W. A. Patterson, *Anal. Chem.* **26**, 823 (1954).

[13] O. D. Shreve, M. R. Heether, H. B. Knight, and D. Swern, *Anal. Chem.* **23**, 277 (1951).

[14] J. E. Field, J. O. Cole, and D. E. Woodford, *J. Chem. Phys.* **18**, 1298 (1950).

[15] H. B. Henbest, G. D. Meakins, B. Nicholls, and K. J. Taylor, *J. Chem. Soc., London* p. 1459 (1957).

follows: (1) The C—C bond contracts while both C—O bonds stretch and (2) the C—C bond does not change in length while one C—O bond stretches and the other contracts. In ethylene oxide (1) is observed at 877 cm^{-1} and (2) is calculated to be at 892 cm^{-1}[9,10] (see Fig. 10.2). These two probably give rise to observed bands at 880–750 and 950–815 cm^{-1}. The lower frequency band has been used to distinguish monosubstituted oxirane rings at 880–805 cm^{-1}, disubstituted rings (1,1 and 1,2) at 850–775 cm^{-1}, and trisubstituted rings at 770–750 cm^{-1}.

The CH and CH$_2$ groups in the rings absorb either at 3050–3029 or 3004–2990 cm^{-1} or in a few cases in both regions.[15]

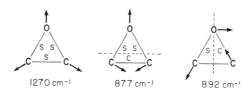

FIG. 10.2. The ring vibrations of ethylene oxide, a whole-ring stretch, and two nearly degenerate half-ring stretching vibrations.

10.6 OH Stretch in Alcohols and Phenols

Alcohols are characterized by several bands, all of which are sensitive to the environment. In the pure liquid or solid state, alcohols and phenols usually exist as hydrogen bonded polymers. In dilute solution in nonpolar solvents, the alcohols are essentially free. In concentrated solutions or mixtures, both free and bonded forms may be present.

Alcohols in the liquid or solid state have very broad strong absorption near 3300 cm^{-1} owing to the stretching of the O—H··O bonds.[16] In CCl$_4$ solution they absorb near 3640 cm^{-1} as a sharp weak band owing to free OH stretch.[16] In the triterpenoids and similar compounds in CCl$_4$ solution, primary alcohols absorb at 3641–3640 cm^{-1}, axial secondary at 3638–3635 cm^{-1}, equatorial secondary at 3630–3628 cm^{-1}, and tertiary at 3618–3613 cm^{-1} (axial 3617 cm^{-1}—equatorial 3613 cm^{-1}).[17,18]

[16] A. V. Stuart and G. B. B. M. Sutherland, *J. Chem. Phys.* **24**, 559 (1956).

[17] A. R. H. Cole, G. T. A. Muller, D. W. Thornton, and R. L. S. Willix, *J. Chem. Soc., London* p. 1218 (1959).

[18] A. R. H. Cole, P. R. Jeffries, and G. T. A. Muller, *J. Chem. Soc., London* p. 1222 (1959).

Phenols absorb at 3250–3200 cm^{-1} in the bonded state and at 3617–3593 cm^{-1} in solutions.[19,20] The vibrational frequency diminishes as the electron attracting power of the substituent on the ring increases.[20]

In some cases such as 2,6-ditertbutyl phenol, steric hindrance prevents bonding even in the pure state. In some cases internal bonding takes place, as in salicyladehyde where bonding remains even in the dilute state.

10.7 C—O Stretch

The vibrations involving the stretching of the C—O bond give rise to strong infrared bands (see Table 10.5).[21]

TABLE 10.5

C—O SPECTRAL REGIONS (IN cm^{-1})

Primary alcohols CH$_2$—OH	1075–1000
Aliphatic secondary alcohols (alkyl—CHOH—alkyl)	1150–1075
Aromatic secondary alcohols (phenyl—CHOH—)	1075–1000
Cyclic equatorial secondary alcohols	1065–1037
Cyclic axial secondary alcohols	1036– 970
Tertiary alcohols	1210–1100
Phenols	1260–1180

In primary alcohols, this band involves asymmetric C—C—O stretching. In secondary and tertiary alcohols, the asymmetric motion of the branched carbon atom against its neighbors gives rise to bands which might be described as asymmetric C—C—O when the branched carbon moves along the C—O bond and asymmetric C—C—C when it moves at right angles to the C—O bond. Both of these can interact with methyl rocking vibrations of the same symmetry type. The vibration involving C—O usually shifts a little when hydrogen bonding is eliminated by dilution. As seen in Table 10.5 an aromatic ring next to the secondary alcohol group lowers its frequency.[21] Equatorial cyclic secondary alcohols absorb near 1050 cm^{-1} and axial cyclic secondary alcohols absorb near 1000 cm^{-1}.[22–24]

[19] R. E. Richards and H. W. Thompson, *J. Chem. Soc., London* p. 1260 (1947).

[20] P. J. Stone and H. W. Thompson, *Spectrochim. Acta* **10**, 17 (1957).

[21] H. Zeiss and M. Tsutsui, *J. Amer. Chem. Soc.* **75**, 897 (1953).

[22] R. N. Jones and F. Herling, *J. Amer. Chem. Soc.* **78**, 1152 (1956).

[23] I. N. Nazarov, A. F. Vasil'ev, and I. A. Gurvich, *J. Gen. Chem. USSR* **29**, 749 (1959).

[24] W. Hückel and Y. Riad, *Justus Liebigs Ann. Chem.* **637**, 33 (1960).

Additional bands for secondary alcohols[25] are $-CHOH-CH_3$, 950–910 cm^{-1} (weaker than 1100 cm^{-1}); $-CHOH-C_2H_5$, 975–950 cm^{-1} (stronger than 1100 cm^{-1}); and $-CHOH-C_3H_7$ (and larger), 1025–975 cm^{-1} (stronger than 1100 cm^{-1}). Tertiary alcohols frequently absorb at 960–900 cm^{-1}.

10.8 OH Deformation

The out-of-plane C—OH deformation in the bonded state absorbs very broadly and diffusely near 650 cm^{-1}.[16] The in-plane C—OH deformation is complicated by interaction with hydrogen wagging vibrations in primary and secondary alcohols.[16, 26, 27] The hydrogens on the oxygen and on the carbon move more or less parallel to the C—O bond in the same direction in one case and the opposite direction in the other case, giving rise to more than one band that involves OH deformation.

Primary alcohols in the bonded state have diffuse association bands near 1420 and 1330 cm^{-1} (OH deformation and CH$_2$ wag), which disappear on dilution, the lower band shifting to 1300–1200 cm^{-1}.[16]

Secondary alcohols in the bonded state have bands near 1420 and 1330 cm^{-1} (OH deformation + CH wag) which disappear in solution, the lower band shifting on dilution to about 1225 cm^{-1}.[16,26] In both the primary and secondary alcohols, the 1420 cm^{-1} band shifts to about 1385 cm^{-1} where it is overlapped by CH$_3$ absorption.

Tertiary alcohols absorb near 1410 cm^{-1} (OH deformation) in the bonded state and near 1320 cm^{-1} in solution. There is no second band at 1330–1325 cm^{-1} as in the alcohols with α-hydrogens.[16]

10.9 Phenols

Phenols[19] in the solid state absorb at 1390–1330 cm^{-1} (medium) and 1260–1180 cm^{-1} (strong) which represent OH deformation and C—O stretch which interact somewhat.[28] In solution these bands are shifted to lower frequencies. Detailed correlations have been made for the alkyl phenols in solution[29] from

[25] C. J. Pouchert, "The Aldrich Library of Infrared Spectra." Aldrich Chemical Co., Inc., Milwaukee, Wis., 1970.

[26] S. Krimm, C. Y. Liang, and G. B. B. M. Sutherland, *J. Chem. Phys.* **25**, 778 (1956).

[27] D. M. W. Anderson, L. J. Bellamy, and R. L. Williams, *Spectrochim. Acta* **12**, 233 (1958).

[28] R. Mecke and G. Rossmy, *Z. Elektrochem.* **59**, 866 (1955).

[29] D. D. Shrewsbury, *Spectrochim. Acta* **16**, 1294 (1960).

1400 to 650 cm^{-1}, only some of which will be given here. Phenols substituted with alkane groups in the 2 position absorb strongly at 1321–1319, 1256–1242, and 1171–1160 cm^{-1}; 3 alkyl phenols absorb strongly at 1285–1269, 1188–1180, and 1160–1149 cm^{-1}; and 4 alkyl phenols absorb strongly at 1260–1248 and 1174–1166 cm^{-1}.[29] Some of these bands are aromatic ring bands.

10.10   Noncyclic Acetals and Related Compounds

Acetals and *ortho*-formates and compounds such as dialkoxyacetonitriles have strong multiple bands in the region 1160–1040 cm^{-1} involving stretching of the C—O bonds. The O—CH—O group gives rise to a band at 1350–1325 cm^{-1} (CH deformation) which disappears when the hydrogen is missing as in *ortho*-acetates. When the carbon between the oxygens is tetrasubstituted, additional absorption usually occurs at 1235–1175 cm^{-1}, probably involving the asymmetric motion of the branched carbon atom.

10.11   Cyclic Acetals

In substituted dioxolane rings, ring vibrations occur at 1181–1153 and 1093–1070 cm^{-1}.[30,31] Bands are also found at 1126–1104 and 1055–1029 cm^{-1}.[31]

10.12   Carbohydrates

Carbohydrates[32,33] have many bands in the region 1125–1000 cm^{-1} involving stretching of the C—O bond, and broad absorption near 3300 cm^{-1} due to bonded OH stretch.

10.13   Peroxides

In primary and secondary alkyl hydroperoxides the bonded OH stretching band is usually at a little higher frequency (30 cm^{-1}) than the corresponding alcohols[34] but this is not necessarily true for all types of peroxides.[35]

[30] S. A. Barker, E. J. Bourne, R. M. Pinkard, and D. H. Whiffen, *J. Chem. Soc., London* p. 807 (1959).
[31] R. S. Tipson, H. S. Isbell, and J. E. Stewart, *J. Res. Nat. Bur. Stand.* **62**, 257 (1959).
[32] L. P. Kuhn, *Anal. Chem.* **22**, 276 (1950).
[33] R. L. Whistler and L. R. House, *Anal Chem.* **25**, 1463 (1953).
[34] H. R. Williams and H. S. Mosher, *Anal. Chem.* **27**, 517 (1955).
[35] N. A. Milas and O. L. Mageli, *J. Amer. Chem. Soc.* **75**, 5970 (1953).

Primary hydroperoxides (CH_2—O—OH) show distinct satellite bands at 1488 and 1435 cm^{-1} on either side of the 1465 cm^{-1} band while most corresponding alcohols have only the 1465 cm^{-1} band.[34] Secondary hydroperoxides (CH—O—OH) have a band at 1352–1334 cm^{-1}.[34] (CH wag). The comparable band in secondary alcohols is shifted from this position owing to interaction with the deformation of the adjacent OH group.

The bands between 1150 and 1030 cm^{-1} (C—O) follow the pattern of the alcohols.[34] The primary alcohol band near 1050 cm^{-1} shifts only about 10 cm^{-1} to lower frequencies in the corresponding hydroperoxides.

A weak band at 880–845 cm^{-1} has been proposed characteristic for the O—O linkage[36] but its lack of intensity in this well-populated spectral region[34,37] limits its usefulness. In the normal alkyl hydroperoxides a second characteristic region is found at 835–800 cm^{-1}. When there are an even number of carbon atoms, two bands are found at 827 and 807 cm^{-1}. When the carbon atoms are odd in number, one band appears at 814 cm^{-1}.[34]

A vibration involving some stretching of the O—O bond gives rise to a prominant band in the Raman spectrum of CH_3—O—O—CH_3 at 779 cm^{-1} and $(CH_3)_3$C—O—O—C$(CH_3)_3$ at 769 cm^{-1}.[38]

[36] O. D. Shreve, M. R. Heether, H. B. Knight, and D. Swern, *Anal. Chem.* **23**, 282 (1951).
[37] A. R. Philpotts and W. Thain, *Anal. Chem.* **24**, 638 (1952).
[38] K. O. Christie, *Spectrochim. Acta, Part A* **27**, 463 (1971).

CHAPTER 11

AMINES, C=N, AND N=O COMPOUNDS

11.1 NH₂ Stretch in Amines

The NH_2 group gives rise to absorption at 3550–3330 cm^{-1} (asymmetric stretch) and at 3450–3250 cm^{-1} (symmetric stretch). A relationship has been developed relating the two bands,[1] namely, $v_{sym} = 345.5 + 0.876\, v_{asym}$, this relationship holding for NH_2 groups in which the two NH bonds are equivalent. Cases of asymmetric bonding are revealed by a breakdown of the relationship. Using equations related to those for the triatomic case [Eqs. (4.17 and (4.18)] this relationship has been reduced[2] to $v_{sym} = 0.98\, v_{asym}$. The relationship holds since the NH_2 angle is not a variable, while variations in the force constant for the two equal NH bonds changes both frequencies proportionately.

In liquid amines there is usually a shoulder near 3200 cm^{-1} which is probably an overtone of the 1600 cm^{-1} NH_2 deformation intensified by Fermi resonance due to its proximity to the in-phase stretching fundamental.

Most liquid aliphatic primary amines absorb at 3400–3350 and 3330–3270 cm^{-1} with a shoulder near 3200 cm^{-1}. Most liquid aromatic amines absorb at 3500–3390 and 3420–3330 cm^{-1} with a weaker band at 3250–3170 cm^{-1}. Aromatic primary amines in solution absorb at 3520–3420 and 3420–3325 cm^{-1}.

11.2 NH₂ Deformation in Amines

Most primary amines have an NH_2 deformation band at 1650–1590 cm^{-1}.[2,3]

In liquid aliphatic amines the NH_2 wagging and twisting vibrations give rise to broad, strong, usually multiple absorption bands at 850–750 cm^{-1}.[2]

[1] L. J. Bellamy and R. L. Williams, *Spectrochim. Acta* 9, 341 (1957).
[2] J. E. Stewart, *J. Chem. Phys.* 30, 1259 (1959).
[3] A. R. Katritzky and R. A. Jones, *J. Chem. Soc., London* p. 3674 (1959).

11.3 NH

The NH stretching vibration gives rise to a weak band at 3500–3300 cm^{-1}. A stronger band at 750–700 cm^{-1} in secondary aliphatic amines is due to NH wagging.[2] In solution *N*-methylanilines absorb at 3460–3420 cm^{-1}.[3–5]

Secondary aromatic amines have a CNH bending absorption near 1510 cm^{-1} near the 1500 cm^{-1} aromatic band.[6] Most aliphatic secondary amines have no visible NH bending band above the 1470 cm^{-1} aliphatic bands.

11.4 C—N in Aliphatic Amines

In amines, as in alcohols, single bond frequencies are affected by branching at the α-carbon atom. In the case of amines, they are also affected by amine substitution, primary and secondary amines being comparable to alcohols and ethers respectively.

Aliphatic primary amines with a primary α-carbon (CH_2—NH_2) have a medium intensity band at 1090–1068 cm^{-1}.[2] Aliphatic secondary amines with primary α-carbons (CH_2—NH—CH_2) have a medium-strong band at 1146–1132 cm^{-1}.[2] These bands involve asymmetric C—C—N and C—N—C stretching, respectively.

Secondary branching at the α-carbon in noncyclic secondary amines (CH—NH—C) gives rise to a band at 1191–1171 cm^{-1}.[2]

As in the alcohols, several vibrations result from the asymmetric motion of the branched carbon atom against its neighbors. Motion along the C—N bond can be described as asymmetric C—C—N, where motion at right angles to the C—N bond can be described as asymmetric C—C—C.

Primary amines with secondary α-carbon atoms (CH—NH_2) absorb weakly[2] at 1043–1037 cm^{-1} and more strongly at 1140–1080 cm^{-1}.

Primary amines with tertiary α-carbon atoms (C—NH_2) absorb weakly[2] at 1038–1022 cm^{-1} and more strongly at 1240–1170 cm^{-1}.

Tertiary aliphatic amines are difficult to characterize in the 1300–1000 cm^{-1} region. If there is no absorption in the NH stretch region and a tertiary amine is suspected, a spot check may be made by mixing 2 drops of the suspected amine and 1 drop of a 50% alcohol–50% concentrated HCl

[4] J. J. Elliott and S. F. Mason, *J. Chem. Soc., London* p. 1275 (1959).

[5] P. J. Krueger and H. W. Thompson, *Proc. Roy. Soc., Ser. A* **243**, 143 (1957).

[6] D. Hadzi and M. Skrbljak, *J. Chem. Soc., London* p. 843 (1957).

mixture. Absorption near 2600 cm^{-1} indicates the presence of a tertiary amine hydrochloride. Tertiary amines may also be detected by changes in the aliphatic groups next to the nitrogen.

11.5 C—N in Aromatic Amines

Primary aromatic amines with the nitrogen directly on the ring absorb strongly at 1330–1260 cm^{-1},[3] due to stretching of the phenyl carbon—nitrogen bond.

Secondary aromatic amines absorb strongly at 1342–1320 and 1315–1250 cm^{-1}.[3,6]

Dimethyl anilines absorb strongly at 1380–1332 cm^{-1}.[3] For all tertiary anilines the region is 1380–1265 cm^{-1}.

The C—N in anilines is higher in frequency than aliphatic amines due to resonance which stiffens the C—N bond.

Thus electron withdrawing and donating groups on the *ortho* and *para* position can affect this resonance contribution relative to *meta* substitution and shift the frequency.

The intensity of this band is remarkably reduced in aniline salts where this type of resonance is suppressed.

11.6 Aliphatic Bands in Amines

The CH$_2$ or CH$_3$ groups next to the nitrogen atom in amines are shifted somewhat. The symmetric stretch at 2830–2770 cm^{-1} is lowered in frequency and intensified and so stands out among other aliphatic bands. This only applies for amines and not for amides or amine salts[7,8] (see Table 11.1).

[7] R. D. Hill and G. D. Meakins, *J. Chem. Soc., London* p. 760 (1958).
[8] W. B. Wright, Jr., *J. Org. Chem.* **24**, 1362 (1959).

TABLE 11.1

ALIPHATIC GROUPS IN AMINES (IN cm^{-1})

Aliphatic $-N-CH_3$	2805–2780	
Aromatic $-N-CH_3$	2820–2810	
Aliphatic $-N-(CH_3)_2$	2825–2810	2775–2765
Aromatic $-N-(CH_3)_2$	2810–2790	
$N-CH_2$ (aliphatic secondary and tertiary amines)	2820–2760	

11.7 The Ammonium Ion

The ammonium ion[9,10] has bands at 3332–3100 cm^{-1} (NH_4 stretch) and 1484–1390 cm^{-1} (NH_4 deformation). There is also a variable combination band assigned as NH_4 deformation plus NH_4 torsional oscillation[9] appearing at 1712 cm^{-1} in NH_4Br, 1762 cm^{-1} in NH_4Cl, 2007 cm^{-1} in NH_4F, and 1990 cm^{-1} in $NH_4(CH_3CO_2)$, these changes being due mainly to changes in the strength of the hydrogen bond.

11.8 Amine Salts

Primary amine salts are characterized by strong absorption between 3200 and 2800 cm^{-1} due to the asymmetric and symmetric NH_3^+ stretch.[11,12] Aromatic amine salts absorb at somewhat lower frequencies than aliphatic amine salts. In addition, between 2800 and 2000 cm^{-1} there are a number of weaker bands, the most interesting of which is an isolated band usually near 2000 cm^{-1}, the intensity of which increases when the symmetry is reduced.[11] It is strong in aromatic primary amine salts. This band is sensitive to changes in hydrogen bonding, having a tendency to increase in frequency in salts made from weaker acids where stronger hydrogen bonding can occur. Tert-butylamine salts absorb as follows: bromide, 2030 cm^{-1}; acid sulfate, 2065 cm^{-1}; chloride, 2080 cm^{-1}; nitrate, 2080 cm^{-1}; sulfate, 2085 cm^{-1}; and acetate 2220 cm^{-1}. This band is assigned to a combination band of the NH_3^+ torsional oscillation (about 480 cm^{-1}) and the asymmetric NH_3^+ deformation (about 1580 cm^{-1}).[11] Sharp bands at 2800–2400 cm^{-1} are due to combination bands of NH_3^+ bending vibrations.[11] Aromatic primary amine salts have two main bands, about 2800 cm^{-1} (NH stretching vibrations) and 2600 cm^{-1} (deformation combination bands).[11] The complicated fine structure associated with the NH stretching bands is a characteristic of amine salts.

[9] T. C. Waddington, *J. Chem. Soc., London* p. 4340 (1958).
[10] E. L. Wagner and D. F. Hornig, *J. Chem. Phys.* **18**, 296 (1950).
[11] C. Brissette and C. Sandorfy, *Can. J. Chem.* **38**, 34 (1960).
[12] B. Chenon and C. Sandorfy, *Can. J. Chem.* **36**, 1181 (1958).

The asymmetric NH_3^+ deformation absorbs at 1625–1560 cm^{-1}, and the symmetric NH_3^+ deformation absorbs at 1550–1505 cm^{-1}.[12]

As with other hydrogen bonded compounds, stronger hydrogen bonding in amine salts ($I^- < Br^- < Cl^-$) lowers hydrogen stretching frequencies but raises deformation frequencies (and their overtones).

Secondary amine hydrochlorides have strong multiple absorption bands between 3000 and 2700 cm^{-1} [11–15] involving the asymmetric and symmetric NH_2^+ stretch, weaker combination bands at 2700–2300 cm^{-1}, and a medium absorption band[12–15] at 1620–1560 cm^{-1} due to NH_2^+ deformation.

Tert-amine hydrochlorides have multiple absorption bands[11–13,15,16] between 2700 and 2330 cm^{-1} involving the NH_3^+ stretching vibration. This absorption in a given tertiary amine moves toward higher frequencies in the order Cl < Br < I.

Trimmethyl ammonium quaternary groups $C-N(CH_3)_3^+$ usually give rise to several bands in the 980–900 cm^{-1} region.[17] These groups also usually have bands near 3020, 1485, and 1410 cm^{-1}.

11.9 C=N Groups

Aliphatic Schiff ($R-C-N-R$) bases absorb near 1670 cm^{-1}. Aromatic Schiff bases absorb near 1630 cm^{-1}.[18] The C=N intensities are intermediate between C=O and C=C bands on the whole.

Oximes (C=N—OH), including aliphatic,[19–21] aromatic, and amide oximes, absorb broadly at 3300–3150 cm^{-1} due to bonded OH stretch, at 1690–1620 cm^{-1} due to C=N stretch,[18] and near 930 cm^{-1} due to N—O stretching.

Imino carbonates [$(RO)_2C=NH$] and imidates[18,22] absorb near 3300 cm^{-1} due to bonded NH stretch, and at 1690–1645 cm^{-1} due to C=N stretch. Bands appear near 1325 and 1100 cm^{-1} which involve C—O stretching like the esters. Imidate hydrochlorides absorb strongly and broadly near 3000 cm^{-1} due to NH_2^+ stretch, 1685–1635 cm^{-1} due to C=N stretch, and 1590–1540 cm^{-1} due to NH_2^+ deformation.

[13] E. A. V. Ebsworth and N. Sheppard, *Spectrochim. Acta* **13**, 261 (1959).
[14] R. A. Heacock and L. Marion, *Can. J. Chem.* **34**, 1782 (1956).
[15] P. J. Stone, J. C. Craig, and H. W. Thompson, *J. Chem. Soc., London* p. 52 (1958).
[16] R. C. Lord and R. E. Merrifield, *J. Chem. Phys.* **21**, 166 (1953).
[17] A. S. Hume, W. C. Holland, and F. Fry, *Spectrochim. Acta, Part A* **24**, 768 (1968).
[18] J. Fabian, M. Legrand, and P. Poirier, *Bull. Soc. Chim. Fr.* **23**, 1499 (1956).
[19] J. F. Brown, Jr., *J. Amer. Chem. Soc.* **77**, 6341 (1955).
[20] A Palm and H. Werbin, *Can. J. Chem.* **32**, 858 (1954).
[21] D. Hadzi, *J. Chem. Soc., London* p. 2725 (1956).
[22] D. Hadzi and D. Prevorsek, *Spectrochim. Acta* **10**, 38 (1957).

Guanidines and their salts (N_2C=N) usually have absorption[18,23] at 1670–1500 cm^{-1} due to NH deformation and CN stretching vibrations. Biguanides are quite similar. The C=N is weakened by resonance, so its vibrations more appropriately should be described as asymmetric NCN stretch, particularly in the salts where the three CN bonds should be identical as the CO bonds are in ionized carboxyls. The NH_2 groups absorb at 3400–3200 and 1670–1620 cm^{-1}. If there are no NH_2 groups, there is usually no absorption above 1635 cm^{-1}. On the whole, there is considerably less change in the NH_2 frequencies when going from the free base to the hydrochloride than there is in amines, due to the fact that the charge is distributed over several nitrogen atoms. Other compounds involving the N_2—C=N group are in general similar to the guanidines.

Trialkyl isoureas

$$\begin{array}{c} O-R' \\ | \\ R-NH-C=N-R'' \end{array}$$

absorb near 1672–1655 cm^{-1} (C=N).[18] Conjugation lowers the C=N frequency (N-cyano 1582 cm^{-1}).

Substituted amidines[24–27] (N—C=N) are characterized by strong CN absorption, 1685–1580 cm^{-1}. Amidines of the type

$$\begin{array}{c} N- \\ | \ \| \\ -C-C-NH-R' \\ | \end{array}$$

show a band near 1540 cm^{-1} which shifts to about 1515 cm^{-1} in solution. This is assigned to the CNH group as in monosubstituted amides (NH bend and C—N stretch interacting). This band is absent in N-dialkyl amidines. The free NH groups absorb at 3470–3380 cm^{-1} in solution and 3300–3100 cm^{-1} in the condensed state.

Unsubstituted amidine hydrochlorides or bromides R—$C(NH_2)_2{}^+$ absorb strongly at 1710–1675 cm^{-1} and weakly at 1530–1500 cm^{-1} with a third weak band appearing at 1575–1540 cm^{-1} in about a third of the samples. Substituted amidine salts absorb strongly at 1700–1600 cm^{-1}.[27,28] In N-disubstituted amidine hydrochlorides the $NH_2{}^+$ deformation absorbs at 1590–1530 cm^{-1}.[27,28]

In all compounds with the group C=$NH_2{}^+$, there is undoubtedly much interaction between the C=N stretching vibration and the NH_2 deformation.

[23] E. Lieber, D. R. Levering, and L. J. Patterson, *Anal. Chem.* **23**, 1594 (1951).
[24] J. Fabian, V. Delaroff, and M. Legrand, *Bull. Soc. Chim. Fr.* **23**, 287 (1956).
[25] D. Prevorsek, *C.R. Acad. Sci.* **244**, 2599 (1957).
[26] D. Prevorsek, *Bull. Soc. Chim. Fr.* **25**, 788 (1958).
[27] J. C. Grivas and A. Taurins, *Can. J. Chem.* **37**, 795 (1959).
[28] J. C. Grivas and A. Taurins, *Can. J. Chem.* **37**, 1260 (1959).

The cyclic C=N group in pyrrolines and related compounds absorbs at 1655–1560 cm^{-1} depending on the substituents.[29]

The nitrone group

$$\begin{array}{c}\diagdown \\ \diagup\end{array}C=\overset{+}{N}\begin{array}{c}\diagup \\ \diagdown \\ \underset{\bar{O}}{}\end{array}$$

absorbs at 1620–1540 and 1280–1088 cm^{-1}.[30,31] These bands involve the stretching of the

$$C=N \quad \text{and} \quad \overset{+}{N}-\bar{O} \quad \text{bonds}$$

The spectral regions for the C=N group are summarized in Table 11.2.[18]

TABLE 11.2

SPECTRAL REGIONS FOR C=N GROUPS (IN cm^{-1})

R—CH=N—R	1674–1665
C_6H_5—CH=N—R	1656–1629
C_6H_5—CH=N—C_6H_5	1637–1626
R_2C=N—R	1662–1649
$\begin{array}{c}C_6H_5 \\ \diagdown \\ R \diagup\end{array}$C=N—R	1650–1640
$\begin{array}{c}C_6H_5 \\ \diagdown \\ CH_3 \diagup\end{array}$C=N—$C_6H_5$	1640–1630
R_2C=NH	1646–1640
$\begin{array}{c}C_6H_5 \\ \diagdown \\ R \diagup\end{array}$C=NH	1633–1620
R—CH=NOH	1673–1652
C_6H_5—CH=NOH	1645–1614
R_2C=NOH	1684–1652
$\begin{array}{c}C_6H_5 \\ \diagdown \\ R \diagup\end{array}$C=NOH	1640–1620
—S—C=N—	1640–1607
—O—C=N—	1690–1645
—N—C=N—	1685–1582

[29] A. I. Meyers, *J. Org. Chem.* **24**, 1233 (1959).
[30] J. Hamer and A. Macaluso, *Chem. Rev.* **64**, 474 (1964).
[31] P. A. S. Smith and J. E. Robertson, *J. Amer. Chem. Soc.* **84**, 1197 (1962).

11.10 Nitro Group

The nitro group has two identical NO bonds

$$-\overset{+}{N}\overset{\displaystyle O}{\underset{\displaystyle O^-}{\big\backslash\!\!\big/}} \longleftrightarrow -\overset{+}{N}\overset{\displaystyle O^-}{\underset{\displaystyle O}{\big/\!\!\big\backslash}} \quad \text{or} \quad -\overset{+}{N}\overset{\displaystyle O}{\underset{\displaystyle O}{\big\backslash\!\!\big/}}^-$$

which vibrate asymmetrically, causing strong absorption at 1556–1545 cm^{-1} in aliphatic nitro compounds and symmetrically causing somewhat weaker absorption at 1390–1355 cm^{-1}.[19,32-34] Electronegative substituents such as halogens on the α-carbon cause the nitro stretching frequencies to diverge. When a CH_3 group is on the α-carbon, interaction occurs between the symmetric NO_2 stretch and the symmetric CH_3 deformation. The group CH_3-C-NO_2 gives rise to two bands near 1390 and 1360 cm^{-1}.[19]

Nitro alkenes usually absorb at 1550–1500 and 1360–1290 cm^{-1},[19] conjugation lowering both frequencies.

Aromatic nitro groups absorb strongly at 1530–1500 cm^{-1} and somewhat more weakly at 1370–1330 cm^{-1}.[19,35,36] In addition, aromatic nitro compounds usually have strong aromatic ring absorption at 760–705 cm^{-1}. The usual o-m-p bands at 900–700 cm^{-1} are upset in the nitro aromatics and not very reliable, probably due to interaction with the out-of-plane NO_2 bending frequency. Steric effects which destroy the nitro-ring coplanarity and thus reduce conjugation make aromatic nitro groups resemble aliphatic nitro groups.[37,38] The frequency of the asymmetric NO_2 stretch in p-substituted nitrobenzenes has been correlated with the electron donating or withdrawing characteristics of the substituent.[36] The symmetric stretch does not correlate due to interaction with the C—N bond.[19,36] In the nitroanilines, for example, the electron withdrawing properties of the nitro group and the electron donating properties of the amino group cause a large amount of resonance to occur in the ortho and para isomers which weaken the N—O

$$\underset{H}{\overset{H}{}}\overset{+}{N}=\!\!\!\left\langle\!\!\!\bigcirc\!\!\!\right\rangle\!\!\!=\overset{+}{N}\overset{O^-}{\underset{O^-}{\big/}}$$

[32] R. N. Hazeldine, J. Chem. Soc. London p. 2525 (1953).

[33] W. H. Lunn, Spectrochim, Acta 16, 1088 (1961).

[34] Z. Eckstein, P. Gluzinsky, W. Sobotka, and T. Urbanski, J. Chem. Soc., London p. 1370 (1961).

[35] R. R. Randle and D. H. Whiffen, J. Chem. Soc., London p. 4153 (1952).

[36] R. D. Kross and V. A. Fassel, J. Amer. Chem. Soc. 78, 4225 (1956).

[37] C. P. Conduit, J. Chem. Soc., London p. 3273 (1959).

[38] R. J. Francel, J. Amer. Chem. Soc. 74, 1265 (1952).

bonds relative to the *meta* isomers where this mesomeric effect does not occur. The strongest band in the spectra of nitro anilines probably involves the NO_2 symmetric stretch, since it shifts as expected from mesomeric effects, and the $C-N(H_2)$ band which is an alternate assignment would be expected to shift the other way. It occurs in the pure materials at 1350 cm^{-1} in *meta*, 1305 cm^{-1} in *para*, and 1245 cm^{-1} in *ortho*. The usual strong nitro aromatic band near 740 cm^{-1} is also altered in the *ortho* and *para* isomers.

A number of 3- and 5-nitro-2-aminopyridines with various other substituents have been studied and show a very strong band at 1310–1270 cm^{-1} for the 5 nitro-2-amino and at 1250–1210 cm^{-1} for the 3-nitro-2 amino pyridines. As in the nitroanilines, when the nitro and amino groups are *ortho* to each other, there is an extra shift due to the proximity of the two groups.

The NO_2 stretching frequencies are summarized in Table 11.3.

TABLE 11.3

NO_2 STRETCHING FREQUENCIES

	Asymmetric stretch (cm^{-1})	Symmetric stretch (cm^{-1})
$C-CH_2-NO_2$	1556–1545	1388–1368
$C-CH(CH_3)-NO_2$	1549–1545	1364–1357
$C(CH_3)_2-NO_2$	1553–1530	1359–1342
$CHX-NO_2(X=Cl, Br)$	1580–1556	1368–1340
CX_2-NO_2	1597–1569	1339–1323
CCl_3-NO_2	1610	1307
$C(NO_2)_2$	1590–1570	1340–1325
$N-NO_2$	1630–1530	1315–1260
$O-NO_2$	1660–1625	1285–1270
Nitro alkene	1550–1500	1360–1290
Aromatic nitro	1530–1500	1370–1330
p-Aminonitro aromatic	(Hydrogen bonded)	1330–1270
o-Aminonitro aromatic	(Hydrogen bonded)	1260–1210

In aromatic nitro compounds some of the nitro group bending bands can be confused with aromatic bands. In aromatic nitro compounds, bands are usually seen at 890–835 cm^{-1} (NO_2 scissors deformation), 785–300 cm^{-1} (NO_2 out-of-plane wag), and 580–520 cm^{-1} (NO_2 in-plane rock).[39,40]

[39] J. H. S. Green and D. J. Harrison, *Spectrochim. Acta, Part A* **26**, 1925 (1970).
[40] J. H. S. Green and H. A. Lauwers, *Spectrochim. Acta, Part A* **27**, 817 (1971).

11.11 Organic Nitrates

Organic nitrates $R—O—NO_2$ are characterized by strong bands near 1660–1625 cm^{-1} (NO_2 asymmetric stretch), 1285–1270 cm^{-1} (NO_2 symmetric stretch), and 870–840 cm^{-1} (N—O stretch), 760–745 cm^{-1} (out of plane deformation), and 710–690 cm^{-1} (NO_2 deformation).[19,41-43]

11.12 Nitramines

Nitramines ($N—NO_2$), nitroguanidines, and related compounds have an asymmetric NO_2 stretch at 1630–1530 cm^{-1} and a symmetric stretch at 1315–1260 cm^{-1}.[23,44,45]

11.13 Organic Nitrites

Organic nitrites $R—O—N=O$ are characterized by two bands in the N=O region due to rotational isomers.[46] A strong band occurs at 1681–1648 cm^{-1} (*trans*) and a somewhat weaker band at 1625–1605 cm^{-1} (*cis*).[46,47] A strong N—O absorption occurs at 814–751 cm^{-1}.[46]

Alkyl thionitrites $R—S—N=O$ absorb near 1534 cm^{-1} due to N=O stretch.[48]

11.14 Inorganic Nitrates and Nitrites

Nitrate salts absorb strongly at 1380–1350 cm^{-1} (asymmetric NO_3 stretch), 835–815 cm^{-1} (medium-sharp), and 740–725 cm^{-1} (weak).[49]

Nitrite salts absorb strongly at 1275–1235 cm^{-1} (asymmetric NO_2 stretch) and have a medium sharp band at 835–820 cm^{-1}.[49]

[41] F. Pristera, *Anal, Chem.* **25**, 844 (1953).
[42] R. A. G. Carrington, *Spectrochim. Acta* **16**, 1279 (1960).
[43] R. D. Guthrie and H. Spedding, *J. Chem. Soc., London* p. 953 (1960).
[44] W. D. Kumler, *J. Amer. Chem. Soc.* **76**, 814 (1954).
[45] L. J. Bellamy, "The Infrared Spectra of Complex Molecules." Wiley, New York, 1954.
[46] P. Tarte, *J. Chem. Phys.* **20**, 1570 (1952).
[47] R. N. Hazeldine and B. J. H. Mattinson, *J. Chem. Soc., London* p. 4172 (1955).
[48] R. J. Philippe and H. Moore, *Spectrochim. Acta* **17**, 1004 (1961).
[49] F. A. Miller and C. H. Wilkins, *Anal. Chem.* **24**, 1253 (1952).

11.15 N=N Azo

The N=N stretching vibration of the *trans* symmetrically substituted azo group is forbidden in the infrared spectrum as this is a symmetrical vibration in a molecule having a center of symmetry. It is active in the Raman effect, however.

The Raman spectrum of azomethane has an N=N stretching band at 1576 cm^{-1}.[50] The Raman spectrum of azobis(isobutyronitrile) has an N=N band at 1580 cm^{-1} (see Fig. 11.1). These frequencies are probably

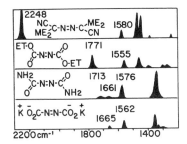

FIG. 11.1. Raman spectra of some nonaromatic azo compounds, The N=N stretching band is seen at 1580–1555 cm^{-1}.

representative of the N=N stretching frequency in unconjugated cases of the type C—N=N—C. The Raman spectrum of *trans*-azobenzene has a N=N stretching band at 1442 cm^{-1}.[51,52] Other azo aromatics have a prominent N=N band in the Raman spectrum at 1463–1380 cm^{-1}.[51-53] The Raman spectra of some azo aromatics, run by one of the authors, are illustrated in Fig. 11.2 where the N=N group give rise to a strong band. The N=N stretching frequencies range from 1463 cm^{-1} for the electron withdrawing *p*-nitrophenyl group to 1391 cm^{-1} for the electron donating *p*-dimethyl-aminophenyl group. Also seen is a prominent Raman band near 1140 cm^{-1} probably involving aryl-N stretch.[51-53]

In unsymmetrical aromatic azo compounds of the type

[50] K. W. F. Kohlrausch, "Ramanspektren." Akad Verlagsges., Leipzig, 1943.
[51] J. Brandmüller, H. Hacker, and H. W. Schrötter, *Chem Ber.* **99**, 765 (1966).
[52] H. Hacker, *Spectrochim Acta* **21**, 1989 (1965).
[53] J. Brandmüller and Hacker, *Z. Phys.* **184**, 14 (1965).

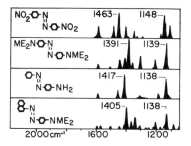

FIG. 11.2. Raman spectra of some aromatic azo compounds, A band involving N=N stretch is seen at 1463–1391 cm^{-1} and a band involving aryl-N stretch is seen near 1140 cm^{-1}.

where X=OH, NH$_2$, and N(CH$_3$)$_2$, the N=N stretching frequency occurs at 1416, 1418, and 1410 cm^{-1}, respectively, in the infrared spectrum.[54] The assignment was proved by the shifts due to ^{15}N substitution. In the same manner, the N=N stretching frequency in *cis*-azobenzene was located at 1511 cm^{-1} in the infrared.[54] Here, steric effects prevent coplanarity so conjugation is reduced relative to the *trans* isomer. Infrared correlations indicate the region 1450–1400 cm^{-1} for the N=N stretching frequency in azoaryls and other compounds studied[55–58] when it is expected to be infrared active. Since this region is well populated with both aryl and aliphatic bands, this infrared correlation should be used with discretion.

The infrared spectrum of ethyl and *n*-propyl azodiformates have high frequency carbonyls near 1775 cm^{-1} due to the election withdrawing character of the N=N bond.[58] Strong bands involving the C—O bonds appear near 1260–1220 and 1060–1020 cm^{-1}. As seen in Fig. 11.1, the N=N stretching frequency is 1555 cm^{-1} in the Raman effect.[59]

In diazoaminobenzene (ϕ—N=N—NH—ϕ) ^{15}N substitution studies indicated the following assignments: NH deformation, 1522, N=N stretch, 1416; phenyl-N, 1248; N—N stretch, 1202; phenyl-N, 1178; and N—N—N bend, 637 cm^{-1}.[54]

[54] R. Kubler, W. Lüttke, and S. Weckherlin, *Z. Elektrochem.* **64**, 650 (1960).
[55] R. J. W. Le Fèvre, M. F. O'Dwer, and R. L. Werner, *Chem. Ind. (London)* p. 378 (1953).
[56] R. J. W. Le Fèvre, M. F. O'Dwyer, and R. L. Werner, *Aust. J. Chem.* **6**, 341 (1953).
[57] R. J. W. Le Fèvre and R. L. Werner, *Aust. J. Chem.* **10**, 26 (1957).
[58] R. J. W. Le Fèvre, W. T. Oh, I. H. Reece, R. Roper, and R. L. Werner, *Aust. J. Chem.* **11**, 92 (1958).
[59] A. Simon and H. Wagner, *Naturwissenschaften* **47**, 540 (1960).

11.16 Azoxy and Azothio Groups

When azoxy compounds are compared with the analogous azo compounds, the N=N bond is now unsymmetrically substituted. The azoxy group has two bands associated with the N=N→O group, a band at 1480–1450 cm^{-1} (near the 1500 cm^{-1} aromatic band) described as asymmetric NNO (or mainly N=N) stretch and at 1335–1315 cm^{-1} due to symmetric NNO (or mainly N→O stretch).[60-62] Aliphatic azoxy compounds absorb at 1530–1495 cm^{-1}.[63]

The azothio group has two bands due to the N=N→S group at 1465–1445 cm^{-1} (mainly N=N stretch) and at 1071–1058 cm^{-1} (mainly N→S stretch).[64]

11.17 C-Nitroso Compounds

Aliphatic and aromatic C-nitroso compounds usually exist in the solid state as the dimer in a *cis* or *trans* configuration,[65]

$$
\begin{array}{cc}
\underset{/}{\overset{O}{\nwarrow}}N=N\underset{\searrow}{\overset{O}{\nearrow}} & \overset{\diagdown}{\underset{O^{\nearrow}}{}}N=N\overset{O}{\underset{\searrow}{\nearrow}}
\end{array}
$$

These compounds have strong infrared absorption in the following regions.[19,65-69]

Aliphatic *trans* dimer	1290–1176 cm^{-1}
Aliphatic *cis* dimer	1420–1330 and 1344–1323 cm^{-1}
Aromatic *trans* dimer	1299–1253 cm^{-1}
Aromatic *cis* dimer	ca. 1409 and 1397–1389 cm^{-1}

[60] W. Maier and G. Englert, *Z. Elektrochem.* **62**, 1020 (1958).
[61] B. Witkop and H. M. Kissman, *J. Amer. Chem. Soc.* **75**, 1975 (1953).
[62] J. Jander and R. N. Hazeldine, *J. Chem. Soc., London* p. 919 (1954).
[63] B. W. Langley, B. Lythgoe, and N. V. Riggs, *J. Chem. Soc., London* p. 2309 (1951).
[64] A. Foffani, G. Leandri, I. Zanon, and C. Carpanelli, *Tetrahedron Lett.* No. 11, p. 21 (1959).
[65] B. G. Gowenlock and W. Lüttke, *Quart. Rev. Chem. Soc.* **12**, 321 (1958).
[66] W. Lüttke, *Angew. Chem.* **68**, 417 (1956); **69**, 99 (1957).
[67] W. Lüttke, *Z. Elektrochem.* **61**, 976 (1957).
[68] P. Tarte, *Bull. Soc. Chim. Belg.* **63**, 525 (1954).
[69] B. G. Gowenlock, H. Spedding, L. Trotman, and D. H. Whiffen, *J. Chem. Soc., London* p. 3927 (1957).

Monomeric C-nitroso compounds give rise to absorption in the following regions due to the N=O stretching frequency.[65,68,70-72]

| Aliphatic nitroso monomer | $1621-1539$ cm^{-1} |
| Aromatic nitroso monomer | $1513-1488$ cm^{-1} |

The C–N frequency couples with vibrations of the rest of the molecule, generally resulting in a band about 1100 and at $860-750$ cm^{-1}.[65]

Nitrosophenols are in tautomeric equilibrium with the quinone oximes. Nitrosoanilines are resonance hybrids.

11.18   Nitrosamines

Aliphatic nitrosamines (R_2N—N=O) absorb at $1460-1425$ cm^{-1}.[47,73,74] This band is assigned to the N=O stretching vibration, and another band at $1150-0030$ cm^{-1} is assigned to the N—N stretching vibration. Aromatic nitrosamines have bands at $1500-1450$ cm^{-1} (N=O), $1200-1160$ cm^{-1} (C–N), and $1025-925$ cm^{-1} (N—N).[74] The N=O stretching vibration can interact somewhat with CH_2 and CH_3 deformation vibrations and with the $1500-1450$ cm^{-1} aromatic vibrations.[74]

[70] W. Lüttke, J. Phys. Radium [8] 15, 633 (1954).

[71] W. Lüttke, Z. Elektrochem. 61, 302 (1957).

[72] J. Mason and J. Dunderdale, J. Chem. Soc., London p. 754 (1956).

[73] P. Tarte, J. Chem. Phys. 23, 979 (1955).

[74] R. L. Williams, R. J. Pace, and G. J. Jeacocke, Spectrochim. Acta 20, 225 (1964).

CHAPTER 12

COMPOUNDS CONTAINING BORON, SILICON, PHOSPHORUS, SULFUR, OR HALOGEN

12.1 Boron Compounds

The spectral regions for boron compounds are listed in Table 12.1.

TABLE 12.1

BORON CORRELATIONS (IN cm^{-1})

BH	2640–2350
BH (boron octet complete)	2400–2200
B··H··B bridge	2220–1540
B—O	1380–1310
B—N	1465–1330
B—O—H	3300–3200

12.2 B—O

Compounds containing the B—O linkage, such as borates, boronates boronites, boronic anhydrides, borinic acids, and boronic acids, are characterized by intense absorption at 1380–1310 cm^{-1},[1–9] involving the stretching

[1] L. J. Bellamy, W. Gerrard, M. F. Lappert, and R. L. Williams, *J. Chem. Soc., London* p. 2412 (1958).

[2] R. L. Werner and K. G. O'Brien, *Aust. J. Chem.* **8**, 355 (1955).

[3] R. L. Werner and K. G. O'Brien, *Aust. J. Chem.* **9**, 137 (1956).

[4] S. H. Dandegaonker, W. Gerrard, and M. F. Lappert, *J. Chem. Soc., London* p. 2872 (1957).

[5] E. W. Abel, W. Gerrard, and M. F. Lappert, *J. Chem. Soc., London* p. 3833 (1957).

[6] W. J. Lehmann, T. O. Onak, and J. Shapiro, *J. Chem. Phys.* **30**, 1215 (1959).

[7] W. J. Lehmann, T. O. Onak, and J. Shapiro, *J. Chem. Phys.* **30**, 1219 (1959).

[8] W. J. Lehmann, H. G. Weiss, and J. Shapiro, *J. Chem. Phys.* **30**, 1222 (1959).

[9] W. J. Lehmann, H. G. Weiss, and J. Shapiro, *J. Chem. Phys.* **30**, 1226 (1959).

of the B—O bond. The high frequency and intensity is due to the fact that the B—O has some polar double bond character

$$(\bar{B}{=}\overset{+}{O}{-})$$

However, in B—O compounds where a nitrogen atom is coordinated to the boron atom, no strong band is usually found near this region. It is difficult in this case to find any B—O band, the intensity of which is markedly reduced when the boron octet is complete[1] due to a reduction of bond polarity. In alkoxyboranes with CH_2 or CH groups next to the oxygen, there is possibly interaction[8,9] between the B—O stretch near 1340 cm^{-1} and the OCH_2 wag, which in carboxyl and phosphorus esters absorbs near 1400 cm^{-1}.

12.3 B—OH

The OH groups in boronic acids (and boric acid) in the solid state absorb broadly near 3300–3200 cm^{-1} due to bonded OH stretch. In the aromatic boronic acids, there are usually broad, medium intensity bands near 1000 cm^{-1} and in the 800–700 cm^{-1} region, not present in the comparable anhydrides.

12.4 B—N

Compounds have a B—N linkage, such as borazines and amino boranes,[10,11] have strong absorption at 1465–1330 cm^{-1} involving the stretching of the B—N bond. This bond has some polar double bond character

$$(\bar{B}{=}\overset{+}{N})$$

like the B—O bond. Aliphatic groups adjacent to the nitrogen atom in B—N compounds have intensified deformation frequencies. In aminoboranes, the dimethylamino group has strong absorption near 1515 cm^{-1}. The CH_2 deformation in diethylaminoboranes absorbs strongly near 1490 cm^{-1}.

The boron—nitrogen dative bond in amine–borane complexes (B←N) gives rise to a band in the region near 800–650 cm^{-1}.[12] This band may be

[10] K. Neidenzu and J. W. Dawson, *J. Amer. Chem. Soc.* **81**, 5553 (1959).

[11] H. Watanabe, M. Narisada, T. Nakagawa, and M. Kubo, *Spectrochim. Acta* **16**, 78 (1960).

[12] R. C. Taylor and C. L. Cluff, *Nature (London)* **182**, 390 (1958).

shifted by interaction and is not necessarily a strong infrared band. Its analytical value is limited.

12.5   B—H

Normal BH and BH_2 groups absorb at 2640–2350 cm^{-1} due to BH stretch.[1] The BH_2 absorption is usually a doublet due to symmetric and asymmetric vibrations. The BH_2 deformation vibration absorbs at 1205–1140 cm^{-1}, and the BH_2 wagging vibration absorbs at 975–920 cm^{-1}.[1] The B··H··B bridge bonds cause a series of absorptions at 2220–1540 cm^{-1} with usually at least one band at 1900–1800 cm^{-1}.[1]

In alkyl diboranes[13] the terminal BH_2 absorbs at 2532–2488 cm^{-1}, strong (symmetric stretch); 2640–2571 cm^{-1}, strong (asymmetric stretch); 1170–1140 cm^{-1}, medium-strong (deformation); and 940–920 cm^{-1}, medium (wag). The single terminal BH absorbs at 2565–2481 cm^{-1}, strong (stretch); 1180–1110 cm^{-1}, strong (in plane bend); and 920–900 cm^{-1}, medium weak (out-of-plane bend). The diborane

$$
\begin{array}{c}
\text{H} \\
\text{B} \quad \quad \text{B} \\
\text{H}
\end{array}
$$

bridge absorbs very strongly at 1610–1540 cm^{-1} due to asymmetric, in-phase hydrogen motions (⇉). The symmetric, out-of-phase hydrogen motions (↑↑) give rise to weak bands at 1990–1850 cm^{-1}. The other two motions give rise to weak-to-medium intensity bands depending on the molecular symmetry, 1800–1710 cm^{-1}, asymmetric, out of phase (⇄), and 2140–2080 cm^{-1}, symmetric, in phase (↑↓).

The BH stretch in borazines absorbs at 2580–2450 cm^{-1}. Compounds where the boron octet is complete such as in borohydride salts or amine-borane coordination complexes absorb at 2400–2200 cm^{-1} due to BH stretch.

12.6   B—Cl

The B—Cl stretching vibration is variable in position (1037–579 cm^{-1})[7] and not too characteristic. However, in alkyl phenyl chloroboronites the range is 909–893 cm^{-1}.[1]

[13] W. H. Lehmann and I. Shapiro, *Spectrochim. Acta* **17**, 396 (1961).

12.7 B—CH₃

The B—CH$_3$ group has asymmetric and symmetric CH$_3$ deformation frequencies which absorb at 1460–1405 and 1330–1280 cm^{-1}.[1,13]

12.8 B-Phenyl

Compounds which have the B-phenyl group have a strong sharp band at 1440–1430 cm^{-1} due to the ring vibration[1] normally found in this region.

The diphenyl boron group has strong absorption[1] at 1280–1250 cm^{-1} which is probably due to C—B—C asymmetric stretch. Alkyl phenyl chloroboronites absorb strongly at 1220–1198 cm^{-1}. When there is more than one phenyl on a boron atom, the 760 cm^{-1} phenyl CH wag frequency becomes a doublet with about a 20 cm^{-1} separation.[4]

12.9 Silicon Compounds

The more common silicon correlations are summarized in Table 12.2.[14]

TABLE 12.2

SILICON CORRELATIONS (IN cm^{-1})

SiH	2250–2100 and 950–800
SiCH$_3$	1280–1255 and 860–760
SiC$_2$H$_5$	1250–1220, 1020–1000, and 970–945
SiC$_6$H$_5$	1430, 1125–1100, 730, and 700–690
Si—O—CH$_3$	2860 and 1190
Si—O—CH$_2$—R	1190–1140, 1100–1075, and 990–945
Si—O—aryl	970–920
Si—Cl	625–420
Si—OH	3700–3200, and 955–835
Si—O—Si	1125–1010
Si—CH=CH$_2$	1615–1590, 1020–1000, and 980–950

[14] A. L. Smith, *Spectrochim. Acta* **16**, 87 (1960).

12.10 Si—H

The Si—H gives rise to absorption at 2250–2100 cm^{-1}.[15,16] Si—H bending frequencies absorb in the region 950–800 cm^{-1}.[14,16] A summary of the approximate positions of Si—H absorption bands for alkyl- and aryl-substituted Si—H compounds is presented in Table 12.3[16]

TABLE 12.3

ALKYL AND ARYL Si—H SPECTRAL REGIONS

Compound	SiH Stretch (cm^{-1})	Deformation (cm^{-1})		Wag (cm^{-1})
Aryl SiH$_3$	2157–2152	945–930	930–910	—
Alkyl SiH$_3$	2153–2142	945–930	930–910	—
Diaryl SiH$_2$	2147–2130	940–928		870–843
Aryl-alkyl SiH$_2$	2142–2128	938–923		870–843
Dialkyl SiH$_2$	2138–2117	942–933		895–885
Triaryl SiH	2132–2112	—		842–800
Diaryl-alkyl SiH	2125–2115	—		842–800
Aryl-dialkyl SiH	2115–2103	—		842–800
Trialkyl SiH	2100–2094	—		842–800

SiH$_3$ groups have two bands in the 950–900 cm^{-1} region due to asymmetric and symmetric deformation, where the SiH$_2$ group has one band due to SiH$_2$ deformation.[16] The SiH$_2$ wag vibration absorbs at 900–845 cm^{-1}, and the SiH wag vibration absorbs at 845–800 cm^{-1}.[16] The SiH group is best distinguished by the lack of strong absorption at 945–910 cm^{-1}.[16] The SiH stretching frequencies are relatively insensitive to mesomeric or mass effects. They are sensitive to inductive effects, however,[15] the SiH stretching frequencies being increased by the substitution of increasingly electronegative groups. The contributions of various substituents to the SiH stretching frequency in CCl$_4$ solution have been evaluated, indicating that each group except for oxygen substitution has a strictly additive effect on the SiH frequency which can be predicted closely.[15]

[15] A. L. Smith and N. C. Angelotti, *Spectrochim. Acta* **15**, 412 (1959).
[16] R. N. Kniseley, V. A. Fassel, and E. E. Conrad, *Spectrochim. Acta* **15**, 651 (1959).

12.11   Si—CH₃

The Si—CH₃ group is characterized by a very strong sharp band[14,17] at 1280–1255 cm⁻¹ due to CH₃ symmetric deformation and an absorption at 860–760 cm⁻¹ due to methyl rocking and Si—C stretching.[14] One methyl on a silicon usually absorbs near 765 cm⁻¹, two methyls near 855 and 800 cm⁻¹, and three methyls near 840 and 765 cm⁻¹.[17,18] The asymmetric CH₃ deformation absorbs weakly near 1410 cm⁻¹.

12.12   Si—CH₂—R

A medium intensity band due to the Si—CH₂—R group absorbs[14] at 1250–1200 cm⁻¹ with longer aliphatic chains absorbing at the low frequency end of the region. In straight chain compounds, absorption at 760–670 cm⁻¹ is due to CH₂ rock.[14] In line with CH₂ assignments on halogens, phosphorus, and sulfur, a band near 1410 cm⁻¹ is probably due to CH₂ deformation and the band at 1250–1200 cm⁻¹ due to CH₂ wag.

The Si—C₂H₅ group absorbs at 1250–1220, 1020–1000, and 970–945 cm⁻¹.[14]

12.13   Si—C₆H₅

The Si—C₆H₅ group gives rise to a band[16] at 1125–1100 cm⁻¹ which is due to a planar ring vibration[19] having some Si—C stretching character. Phenyl bands are found near 1430, 730, and 695 cm⁻¹.[14]

12.14   Si—CH=CH₂

The Si—CH=CH₂ group is characterized by vinyl vibrations at 1615–1590 cm⁻¹ (C=C stretch), 1410–1390 cm⁻¹ (CH₂ deformation), 1020–1000 cm⁻¹ (*trans* CH wag), and 980–950 cm⁻¹ (CH₂ wag).[14]

[17] N. Wright and M. J. Hunter, *J. Amer. Chem. Soc.* **69**, 803 (1947).
[18] R. E. Richards and H. W. Thompson, *J. Chem. Soc., London* p. 124 (1949).
[19] D. H. Whiffen, *J. Chem. Soc., London* p. 1350 (1956).

12.15 Si—O—R

The Si—O—R group (R = saturated aliphatic group) has at least one strong band[14] at 1110–1000 cm^{-1} due to an asymmetric Si—O—C stretching vibration. This absorbs at the same place as the Si—O—Si group.

The Si—O—CH$_3$ group absorbs near 2820 cm^{-1} (CH$_3$ symmetric stretch) and 1190 cm^{-1} (CH$_3$ rock). The asymmetric and symmetric stretchings of the Si—O—C bonds give rise to bands near 1100 and 850–800 cm^{-1}. The Si—O—C$_2$H$_5$ group absorbs at 1175–1165 cm^{-1}, 1100–1075 cm^{-1} (double band), and 965–940 cm^{-1}.[14]

12.16 Si—O—C$_6$H$_5$

The Si—O—C$_6$H$_5$ group absorbs[14,20] at 970–920 cm^{-1} due to stretching of the Si—O bond.

12.17 Si—O—Si

Siloxanes are characterized by at least one strong band[14,17,18] at 1100–1000 cm^{-1} due to asymmetric Si—O—Si stretching. In infinite siloxane chains, absorption maxima occur near 1085 and 1020 cm^{-1}.[17,18]

Cyclotrisiloxane rings absorb near 1020 cm^{-1}.[17,21] Cyclic tetramers and pentamers have absorption near 1090 cm^{-1}.[21]

12.18 Si—OH

The Si—OH group absorbs, like the alcohols, at 3700–3200 cm^{-1} due to monomers (in solution) and hydrogen bonded polymers. A strong band due to Si—O stretching vibration occurs at 910–830 cm^{-1}.[14] In the condensed

[20] G. R. Wilson, A. G. Smith, and F. C. Ferris, *J. Org. Chem.* **24**, 1717 (1959).

[21] C. W. Young, P. C. Servais, C. C. Currie, and M. J. Hunter, *J. Amer. Chem. Soc.* **70**, 3758 (1948).

state a broad medium-weak intensity band occurs near 1030 cm^{-1} which shifts appropriately on deuteration[22] and may be described as Si—OH deformation.

12.19   Si—Halogen

The SiF group absorbs at 1000–800 cm^{-1}. The SiF$_3$ group absorbs at 980–945 cm^{-1} (strong) and 910–860 cm^{-1} (medium). The SiF$_2$ group absorbs at 945–915 cm^{-1} (strong) and 910–870 cm^{-1} (medium), and the SiF vibration absorbs at 920–820 cm^{-1}.[14] The two bands in SiF$_2$ and SiF$_3$ probably represent asymmetric and symmetric stretching vibrations.[14]

The SiCl group absorbs at 625–420 cm^{-1}.[14] The SiCl$_3$ group absorbs at 620–570 cm^{-1} (strong) and 535–450 cm^{-1} (medium). The SiCl$_2$ group absorbs at 600–535 cm^{-1} (strong) and 540–460 cm^{-1} (medium), and the SiCl vibration absorbs at 550–470 cm^{-1}.[14,23]

Halosilanes react so readily with moisture to form siloxanes that precautions should be observed.[24]

12.20   Si—N

The Si—NH$_2$ group has two bands between 3570 and 3390 cm^{-1} in solution. The NH$_2$ deformation band occurs near 1540 cm^{-1}.[14] The Si—NH—Si group absorbs near 3400 cm^{-1}, 1175 cm^{-1}, and 935 cm^{-1}.[25]

12.21   Phosphorus Compounds

Correlations for phosphorus groups are listed in Table 12.4.

[22] Y. I. Ryskin, M. G. Voronkov, and Z. I. Shabarova, *Izv. Akad. Nauk SSSR* p. 1019 (1959).

[23] A. L. Smith, *J. Chem. Phys.* **21**, 1997 (1953).

[24] A. L. Smith and J. A. McHard, *Anal. Chem.* **31**, 1174 (1959).

[25] R. Fessenden, *J. Org. Chem.* **25**, 2191 (1960).

TABLE 12.4

Phosphorus Correlations (in cm^{-1})

PH	2440–2275
P=O	1300–1140
P—OH	2700–2550, 2300–2100, 1040–910
Phosphinic acids	ca. 1660 broad
P—O—P	1000–870
PO_3^{2-}	1030–970
PO_4^{3-}	1100–1000
P—O—C (aliphatic)	1050–970
P—O—CH$_3$	ca. 2960, 1460, 1190–1170
P—O—C$_2$H$_5$	ca. 2990, 1485, 1450, 1395, 1375, 1165–1155
P—O—C$_6$H$_5$	1240–1160, 994–855
P—CH$_2$	1440–1405
P—CH$_3$	1310–1280, 960–860
P—C$_6$H$_5$	1450–1425, 1130–1090, 1010–990
P=S	750–580
P—N	1110–930
P=N (cyclic)	1320–1100
P—Cl	580–440

12.22 P—H

The PH stretching vibrations give rise to absorption at 2440–2275 cm^{-1}.[26-29] This band is usually sharp and of medium intensity. In aliphatic and aromatic phosphines, absorption occurs at the low frequency end of the region at 2320–2275 cm^{-1}. With rocksalt optics, the PH$_2$ in monoalkyl phosphines is not resolved but gives rise to a single band in the PH stretch region. The PH$_2$ deformation gives rise to a medium intensity band near 1090–1080 cm^{-1} in aliphatic and aromatic phosphines. The PH$_2$ wagging vibration gives rise to a band at 940–910 cm^{-1}.

The PH wagging vibration is probably involved in medium-to-strong bands in R$_2$—P(=O)—H at 990–965 cm^{-1} and R$_2$—P(=S)—H at 950–910 cm^{-1} where R = alkyl or phenyl groups. Phosphorus acid esters (RO)$_2$—P(=O)—H have a very strong band at 980–960 cm^{-1} which is some-

[26] L. W. Daasch and D. C. Smith, *Anal. Chem.* **23**, 853 (1951).

[27] L. J. Bellamy and L. Beecher, *J. Chem. Soc.*, pp. 475 and 1701 (1952).

[28] L. J. Bellamy and L. Beecher, *J. Chem. Soc., London* p. 728 (1953).

[29] D. E. C. Corbridge, *J. Appl. Chem.* **6**, 456 (1956).

what lower than the normal position for the P—O—C asymmetric stretch band when the hydrogen is not present probably due to P—O—C interaction with the PH wagging vibration.

12.23 P=O

The stretching of the P=O bond gives rise to a strong band at 1300–1140 cm^{-1}.[26-30] The exact position of the band varies with the sum of the electronegativities of the attached groups.[31] Aliphatic phosphine oxides usually absorb near 1150 cm^{-1} and aromatic phosphine oxides near 1190 cm^{-1}. Most phosphorus esters with two or three alkoxy groups on the P=O absorb at 1300–1250 cm^{-1}. Attached electronegative groups tend to pull electrons from the phosphorus thus competing with the bond weakening tendency for the double bonded oxygen to attract electrons

$$(\overset{+}{P}—\overset{-}{O})$$

resulting in a stiffer P=O bond and a higher frequency. Hydrogen bonding will lower the P=O frequency as seen in the hydrogen bonded phosphorus acids, where the P=O absorption is broadened and shifted to lower frequencies and may be hard to identify.

12.24 P—OH

Most organic phosphorus acids which have a P—OH group have a strong band at 1040–910 cm^{-1}, probably involving the stretching of the P—O(H) bonds.[32,33] This includes phosphinic acids, phosphonic acids, phosphonous acids, half esters of phosphonic acids, and acid salts of phosphonic acids.

The OH stretching of the POH group gives rise to broad medium intensity bands at 2700–2550 and 2350–2100 cm^{-1}. These have been assigned to the OH stretching vibration interacting with the overtone of the in-plane POH bending vibration which absorbs weakly near 1230 cm^{-1} and is usually

[30] R. C. Gore, *Discuss. Faraday Soc.* **9**, 138 (1950).
[31] J. V. Bell, J. Heisler, H. Tannenbaum, and J. Goldenson, *J. Amer. Chem. Soc.* **76**, 5185 (1954).
[32] R. A. McIvor and C. E. Hubley, *Can. J. Chem.* **37**, 869 (1959).
[33] L. C. Thomas and R. A. Chittenden, *Chem. Ind. (London)* p. 1913 (1961).

somewhat obscured by P=O bands.[34,35] When there is one P—OH with one P=O as in phosphinic acids, phosphonous acids, dialkyl phosphoric acids, and half esters of phosphonic acids, an additional broad band appears at 1700–1630 cm^{-1}.[26] Phosphonic acids (with two OH groups) generally do not have this band.[26] This band has been assigned to the POH stretching vibration interacting with the overtone of the out-of-plane POH bending vibration which absorbs weakly near 900 cm^{-1} and is usually somewhat obscured by P—O bands.[34,35] A similar set of OH bands have been found for selenic and sulfinic acids.[34,35] Acid salts containing the POH group show broad bands at 2700–2560 cm^{-1} and 2500–1600 cm^{-1}.[36]

12.25 P—O—P

The P—O—P group gives rise to a strong band at 1000–870 cm^{-1} due to the asymmetric stretching of the P—O—P bonds.[29,33,37] A much weaker band near 700 cm^{-1} has also been associated with the P—O—P group.[29]

12.26 PO_2^-, POS^-, PO_3^{2-}, PO_4^{3-}

Salts of alkyl and aryl phosphinic $R_2PO_2^-$ and phosphonous acids $R(H)PO_2^-$ have two strong bands probably due to asymmetric and symmetric PO_2 stretching[33] at 1190–1100 and 1075–1000 cm^{-1}. When the substituent electro-negativity is higher the bands are higher. The salt $(RO)_2PO_2^-$ absorbs at 1282–1210 and ca. 1100 cm^{-1}.[38]

Inorganic salts containing the PO_2^- group absorb at 1300–1150 cm^{-1}.[29,36]

Salts containing the POS^- group absorb at 1163–1064 cm^{-1} (strong) and 652–560 cm^{-1} (medium strong).[33]

Salts containing the PO_3^{2-} group absorb at 1090–970 cm^{-1} (asymmetric stretch) and 990–920 cm^{-1} (symmetric stretch).[29,33]

The PO_4^{3-} ion absorbs strongly at 1100–1000 cm^{-1}.[29]

[34] S. Detoni and D. Hadži, *Spectrochim. Acta* **20**, 949 (1964).
[35] D. Hadži, *Pure Appl. Chem.* **11**, 435 (1965).
[36] J. V. Pustinger, Jr., W. T. Cave, and M. L. Neilsen, *Spectrochim. Acta* **15**, 909 (1959).
[37] R. A. McIvor, G. A. Grant, and C. E. Hubley, *Can. J. Chem.* **34**, 1611 (1956).
[38] J. R. Ferraro, *Pittsburg Conf. Appl. Spectrosc.*, 1962 Paper 162.

12.27 P—O—C (Aliphatic)

A very strong band[26,30,37] occurs at 1050–970 cm^{-1} in all compounds having the P—O—C (aliphatic) link. This band is probably due to an asymmetric P—O—C stretching vibration. In most ethoxy phosphorus compounds this band is a doublet. In methoxy and ethoxy phosphorus compounds a second strong band appears at 830–740 cm^{-1} probably due to symmetric P—O—C stretching. This band is usually absent in other alkoxy phosphorus compounds.

In the hydrogen stretching and deformation regions, methyl and ethyl phosphorus esters resemble carbonyl esters. A methoxy group on a phosphorus atom absorbs near 2960, 1460, and at 1190–1170 cm^{-1} (CH$_3$ rock), and an ethoxy group on a phosphorus atom absorbs near 2990 cm^{-1} (CH$_3$ asymmetric stretch), 1485 cm^{-1} (O—CH$_2$ deformation), 1450 cm^{-1} (CH$_3$ asymmetric deformation), 1395 cm^{-1} (O—CH$_2$ wag), 1375 cm^{-1} (CH$_3$ symmetric deformation), and 1165–1155 cm^{-1} (CH$_3$ rock).[37] The bands at 1190–1170 cm^{-1} for the methoxy group and 1165–1155 cm^{-1} for the ethoxy group on a phosphorus atom are probably the best distinguishing bands.[29] An isopropoxy group has the usual doublet near 1385 and 1370 cm^{-1} (symmetric CH$_3$ deformation), 1350 cm^{-1} (CH deformation), and a triplet of bands between 1200 and 1100 cm^{-1},[37] some of which involve the asymmetric motion of the branched carbon (asymmetric C—C—C, asymmetric C—C—O). A medium band at 900–870 cm^{-1} [32] is probably due to CH$_3$ rock.

12.28 P—O—C (Phenyl)

The P—O-phenyl linkage gives rise to two bands. Strong absorption[29] at 1260–1160 cm^{-1} is mainly due to the stretching of the C—O bond of the phenoxy group, and a strong absorption at 994–914 cm^{-1} in pentavalent phosphorus compounds and 875–855 cm^{-1} in trivalent phosphorus compounds[33,39] is mainly due to stretching of the P—O bond.

12.29 P—CH$_2$

In alkyl phosphines, the CH$_2$ deformation of the CH$_2$ next to the phosphorus absorbs at 1440–1405 cm^{-1}, where it can be seen in the presence of the remaining normal hydrocarbon groups. In a trialkyl phosphine, this is the only indication of the presence of phosphorus.

[39] A. C. Chapman and R. Harper, *Chem. Ind. (London)* p. 985 (1962).

12.30   P—CH₃

The symmetric deformation of the CH_3 group attached to a phosphorus gives rise to medium intensity absorption at 1310–1280 cm^{-1}.[29] The CH_3 rocking usually gives rise to a strong band near 940 cm^{-1} for $P(CH_3)_2$ compounds, at 920–860 cm^{-1} for P—CH_3 compounds, and 880–875 cm^{-1} for monomethyl tertiary phosphines.[40]

12.31   P—Phenyl

When a phenyl group is attached to a phosphorus, the bands due to the phenyl ring near 3050 cm^{-1} (weak), 1600 cm^{-1} (weak), 1500 cm^{-1} (weak), 1450–1425 cm^{-1} (strong), 750 cm^{-1} (strong), and 700 cm^{-1} (strong) all can be seen. Another strong band usually occurs at 1130–1090 cm^{-1} due probably to an aromatic vibration involving some P—C stretching. A medium, sharp phenyl band occurs at 1010–990 cm^{-1}.[29]

12.32   P=S

The P=S group appears[30,41] at 750–580 cm^{-1} but unfortunately, unlike the P=O vibration which always gives rise to a strong infrared band, the P=S band varies in intensity, making its identification difficult in the infrared. The P=S stretching vibration is expected to interact mechanically with the attached P—O and P—C stretching vibrations at 850–650 cm^{-1} to a greater extent than the P=O stretching vibration does. This probably is the reason why two regions 835–713 cm^{-1} (medium) and 675–568 cm^{-1} (variable) have been given as being characteristic for the P=S group.[33] The Raman spectrum is very valuable here, as its identification of the P=S vibration is more definite in some cases than the infrared. The band position varies when different groups are attached to the phosphorus, so it is necessary to be specific about atoms next to the phosphorus. The approximate positions of some examples are given in Table 12.5.[41]

[40] K. B. Mallion, F. G. Mann, B. P. Tong, and V. P. Wystrach, *J. Chem. Soc., London* p. 148 (1963).
[41] E. M. Popov, T. A. Mastryukova, N. P. Rodionova, and M. I. Kabachnik, *Zh. Obshch. Khim.* **29**, 1998 (1959).

TABLE 12.5

Approximate P=S Spectral Regions Identified by Both Infrared and Raman
Studies (in cm^{-1})

$Cl_3-P=S$	750	$(RO)_2$ R $\diagdown$ P=S	580
Cl_2 RO $\diagdown$ P=S	700	Cl_2 R_2N $\diagdown$ P=S	670
Cl $(RO)_2$ $\diagdown$ P=S	660	$(RO)_2$ RNH $\diagdown$ P=S	640
$(RO)_3-P=S$	610	$(RO)_2$ HS $\diagdown$ P=S	650
Cl_2 R $\diagdown$ P=S	665	$(RO)_2$ $R-S$ $\diagdown$ P=S	660
Cl $RO-$P=S R	620	$(RO)_2$ H $\diagdown$ P=S	630

The force constant of the P=S group is not affected as much as that of the P=O group by changes in the electronegativities of the substituents.[42] The frequency shifts are largely due to mechanical effects where the P=S stretching vibration (estimated to be at roughly 675 cm^{-1}) interacts with the in-phase PX_3 stretching vibration to some extent. Some examples of PX_3 stretching frequencies are listed in Table 12.6.[42]

If the attached PX_3 single bonds have frequencies below 675 cm^{-1} (P—Cl, P—S), out-of-phase interaction will raise the "P=S" frequency. If the PX_3 frequencies are higher than 675 cm^{-1} (P—OR, P—CH$_2$), in-phase interaction will lower the "P=S" frequency. The closer the PX_3 and P=S frequencies

[42] F. N. Hooge and P. J. Christen, *Rec. Trav. Chim. Pays-Bas* **77**, 911 (1958).

TABLE 12.6

FREQUENCIES FOR IN-PHASE X_3P STRETCH AND P=S STRETCH (IN cm^{-1})

Substituent X	X_3P sym. PX_3	X_3PO sym. PX_3	X_3PS sym. PX_3	X_3PS P=S
C_2H_5	669	692	686	535
OC_2H_5	731	734	785	610
SC_2H_5	—	558	526	685
Cl	510	486	435	745

are, the greater will be the magnitude of the shift.[42] The P=S frequency can be roughly calculated by adding or subtracting from the 675 cm^{-1} P=S frequency the following emperical corrections for each substituent: Cl, $+23$ cm^{-1}; S, $+10$ cm^{-1}; H, 0 cm^{-1}; OR, -23 cm^{-1}; and CH$_2$, -50 cm^{-1}. Sometimes the "P=S" absorption in R$_3$P=S and (RO)$_3$P=S compounds is weak in the infrared. In these cases the "P=S" vibration is actually the in-phase XPS stretch. When there is at least one Cl or S substituent, the "P=S" absorption is generally medium or strong in the infrared. In these cases the "P=S" vibration is actually the out-of-phase XPS stretch.

12.33   P—SH

The SH stretching vibration in dialkyl dithiophosphoric acids [(RO)$_2$PSSH] absorbs broadly in the liquid state at 2480–2440 cm^{-1}. A band at 865–835 cm^{-1} is probably due to SH bending. Dithiophosphinic acids (R$_2$PSSH) absorb broadly in the condensed state at 2420–2300 cm^{-1} due to SH $\cdot\cdot$ S bonding. In dilute solution the free SH in the above dithiophosphoric and phosphinic acids absorbs sharply[43,44] at 2590–2550 cm^{-1}.

12.34   P—N

Most compounds having a P—N bond have absorption[32,45,46] at 1110–930 cm^{-1}, which probably involves stretching of the P—N bond. Compounds

[43] A. Memefee, D. Alford, and C. Scott, *J. Chem. Phys.* **25**, 370 (1956).
[44] G. Allen and R. O. Colclough, *J. Chem. Soc., London* p. 3912 (1957).
[45] B. Holmstedt and L. Larson, *Acta Chem. Scand.* **5**, 1179 (1951).
[46] R. B. Harvey and J. E. Mayhood, *Can. J. Chem.* **33**, 1552 (1955).

with the P—NH$_2$ group absorb here, in addition to the NH$_2$ bands at 3330–3100 cm^{-1} (stretch), 1600–1535 cm^{-1} (deformation), and 840–660 cm^{-1} (wag).[36] The P—NH group absorbs at 3200–2900 cm^{-1}.[36]

Compounds involving the P—N—C (aliphatic) group have bands[32,45,46] at 1110–930 cm^{-1} (asymmetric P—N—C stretch) and[29,32,45,46] 770–680 cm^{-1} (symmetric P—N—C stretch). In addition to these bands, absorption occurs for P—N(CH$_3$)$_2$ at 1316–1270 cm^{-1}, near 1190 cm^{-1}, and near 1064 cm^{-1};[32,46] for P—N(C$_2$H$_5$)$_2$ near 1210 cm^{-1} and near 1175 cm^{-1};[32] and for P—N(i-Pr)$_2$ near 1200, 1183, 1160, 1139 (shoulder), and 1129 cm^{-1}.[32]

The P—N-(phenyl) group absorbs near 1290 cm^{-1} (phenyl-N) and near 932 cm^{-1} (P—N).

The above-suggested assignments indicate the basic similarity of P—N frequencies to comparable P—O frequencies as might be expected. The intensities are somewhat weaker.

12.35   P=N

Cyclic P=N compounds, such as phosphonitrilic chlorides, phosphonitrilic esters, and salts of phosphonitrilic acids, absorb strongly[26,29,47,48] at 1320–1100 cm^{-1} due to the stretching of the P=N bond. Compounds of the type (RO)$_3$P=N—C$_6$H$_5$ and (RO)$_2$(R)P=N—C$_6$H$_5$ give rise to a strong band at 1385–1325 cm^{-1}.[49]

12.36   P—F, P—Cl, P—C

The P—F group gives rise to absorption at 835–720 cm^{-1} for phosphorfluoridate salts.[29] Organic phosphorus–fluorine compounds absorb at 890–805 cm^{-1} for pentavalent phosphorus.[26,27,33]

The P—Cl group gives rise to bands at 587–435 cm^{-1}.[29,33]

The P—C linkage gives rise to bands at 770–650 cm^{-1}, but these are weak and not of too much value.[29]

[47] L. W. Daasch, J. Amer. Chem. Soc. **76**, 3403 (1954).

[48] D. E. C. Corbridge and E. J. Lowe, J. Chem. Soc., London p. 4555 (1954).

[49] M. I. Kabachnik, V. A. Gilyarov, and E. N. Tsvetkov, Izv. Akad. Nauk. SSSR p. 2135 (1959).

12.37 S—H

The SH stretch in mercaptans and thiophenols absorbs at 2590–2540 cm^{-1}.[50] The band is weak in the infrared but strong in the Raman spectra. Thiol acids C(=O)SH absorb in the same region and show only small hydrogen bonding shifts. The PSSH group absorbs broadly at 2480–2300 cm^{-1} in the bonded condensed state and sharply at 2590–2550 cm^{-1} in dilute solution.[43,44]

12.38 Sulfides and Disulfides

The C—S stretch (705–570 cm^{-1}, medium weak)[50,51] does not give rise to strong bands in the infrared nor does the S—S stretch (500 cm^{-1}) in disulfides, which makes these linkages difficult to detect in some cases. They are both better bands in the Raman effect. The most characteristic bands are those of the CH_2 or CH_3 groups attached to the sulfur.

12.39 CH_2—S

The CH_2 next to an unoxidized sulfur atom gives rise to the bands in Table 12.7.[51,52]

TABLE 12.7

CH_2—S SPECTRAL REGIONS (IN cm^{-1})

Asymmetric stretch	2948–2922
Symmetric stretch	2878–2846
Deformation	1435–1410
Wag	1270–1220 (strong)

[50] N. Sheppard, *Trans. Faraday Soc.* **46**, 429 (1950).
[51] I. F. Trotter and H. W. Thompson, *J. Chem. Soc., London* p. 481 (1946).
[52] S. E. Wiberley, S. C. Bunce, and W. H. Bauer, *Anal. Chem.* **32**, 217 (1960).

12.40 CH_3-S

The CH_3 next to an unoxidized sulfur atom gives rise to the bands in Table 12.8.[50-52]

TABLE 12.8

CH_3—S SPECTRAL REGIONS (IN cm^{-1})

Asymmetric stretch	2992–2955
Symmetric stretch	2897–2867
Asymmetric deformation	1440–1415
Symmetric deformation	1330–1290
Rock	1030– 960

The group SO_2CH_3 absorbs at 1325–1310 cm^{-1} (symmetric deformation) and at 976–964 cm^{-1} [53] (probably CH_3 rock).

12.41 $S-CH=CH_2$

The $R-S-CH=CH_2$ group gives rise to vinyl bands near 1590, 965, and 860 cm^{-1}.

12.42 $S-Aryl$

A band near 1090 cm^{-1} is usually characteristic for the aryl-S linkage. It is thought to be an aromatic ring vibration having some $C-S$ stretching character.[54]

12.43 $S-F$

The $S-F$ stretching frequency appears as a strong infrared band at 815–755 cm^{-1}.[54]

[53] E. Merian, *Helv. Chim. Acta* **43**, 1122 (1960).
[54] N. S. Ham, A. N. Hambly, and R. H. Laby, *Aust. J. Chem.* **13**, 443 (1960).

12.44 SO

The S=O stretching vibration in alkyl and aryl sulfoxides gives rise to strong absorption at 1065–1030 cm^{-1}.[55] The S=O stretching vibration in dialkyl sulfites gives rise to strong absorption at 1220–1195 cm^{-1}.[56,57] Alkyl chlorosulfites absorb at 1216–1210 cm^{-1}.[57] The S=O frequency has been correlated with the sum of the electronegativities of the attached groups.[58] Electronegative substituents reduce the resonance contribution from the polar form

$$\overset{+}{S}-\overset{-}{O}$$

resulting in a stiffer S=O bond and a higher frequency.

Dialkyl sulfites have bands due to the S—O—CH$_2$ group at 1050–850 cm^{-1} and two bands between 750 and 690 cm^{-1}.[56]

12.45 SO$_2$

The SO$_2$ group results in two strong bands due to symmetric (1200–1100 cm^{-1}) and asymmetric (1400–1300 cm^{-1}) stretch of the SO$_2$ group (see Table 12.9).[53–56,59–63]

The frequencies of both bands have been correlated with the sum of the electronegativities of the substituents.[58] Substituents with higher electronegativities reduce resonance contributions from polar forms which result in stiffer SO bonds and higher frequencies. The SO$_2$ bonds are sensitive to inductive effects but not to mesomeric effects.[64]

[55] D. Barnard, J. M. Fabian, and H. P. Koch, *J. Chem. Soc., London* p. 2442 (1949).

[56] S. Detoni and D. Hadži, *Spectrochim. Acta* **11**, 601 (1957).

[57] H. H. Szmant and W. Emerson, *J. Amer. Chem. Soc.* **78**, 454 (1956).

[58] L. W. Daasch, *Spectrochim. Acta* **13**, 257 (1958).

[59] P. M. G. Bavin, G. W. Gray, and A. Stephenson, *Spectrochim. Acta* **16**, 1312 (1960).

[60] E. A. Robinson, *Can. J. Chem.* **39**, 247 (1961).

[61] J. N. Baxter, J. Cymerman-Craig, and J. B. Willis, *J. Chem. Soc., London* p. 669 (1955).

[62] R. D. Guthrie and H. Spedding, *J. Chem. Soc., London* p. 953 (1960).

[63] G. Geiseler and K. O. Bindernagel, *Z. Elektrochem.* **64**, 421 (1960).

[64] L. J. Bellamy and R. L. Williams, *J. Chem. Soc., London* p. 863 (1957).

TABLE 12.9

SO$_2$ Spectral Regions

Compound	Functional group	Symmetric stretch (cm^{-1})	Asymmetric stretch (cm^{-1})	References
Sulfones	C—SO$_2$—C	1165–1120	1340–1290	55,59,60
Sulfonamides	C—SO$_2$—N	1180–1140	1380–1310	53,60,61
Anhydrous sulfonic acids	C—SO$_2$—OH	1165–1150	1352–1342	56,60
Sulfonates	C—SO$_2$—O—C	1195–1165	1375–1335	60,62
Sulfonyl chlorides	C—SO$_2$Cl	1185–1168	1390–1361	54,60,63
Dialkyl sulfates	C—O—SO$_2$—O—C	1200–1187	1415–1390	56,60
Sulfonyl fluorides	C—SO$_2$F	1213–1203	1412–1398	54,60

The approximate ratio of the wavenumbers of the two bands[64] is asymmetric stretch ÷ symmetric stretch = 1.16.

12.46   Sulfones

Sulfones absorb[55,59,60] strongly at 1340–1290 and 1165–1120 cm^{-1} due to the stretching of the SO$_2$ group. The higher frequency band as measured in CCl$_4$ solution very frequently consists of a triplet with the regions 1333–1318, 1316–1306, and 1302–1287 cm^{-1}, with the highest frequency band being most intense in solution.[59] These bands are shifted down 3–20 cm^{-1}, in nujol mulls.

Most sulfones have the lower frequency band in the range 1165–1120 cm^{-1}. Some sulfones give an additional band at 1130–1100 cm^{-1}, and a few have a band at 1185–1175 cm^{-1}.[59]

All sulfones have a medium-strong band at 610–499 cm^{-1} assigned as SO$_2$ scissors deformation, and all saturated sulfones have a medium-strong band at 525–463 cm^{-1} assigned as SO$_2$ wag.[65]

12.47   Sulfonamides

Sulfonamides absorb strongly at 1180–1140 and 1380–1310 cm^{-1} due to the stretching of the SO$_2$ group.[53,60,61]

[65] W. R. Feairheller, Jr. and J. E. Katon, *Spectrochim. Acta* **20**, 1099 (1964).

Primary sulfonamides have bands near 3330 and 3250 cm^{-1} due to asymmetric and symmetric stretching of the NH_2 group. The deformation absorbs near 1570 cm^{-1}. A band is found at 910–900 cm^{-1} which may be due to the $S—N(H_2)$ stretching vibration.

Secondary sulfonamides absorb at 3300–3270 cm^{-1} (NH stretch).[53]

12.48 Sulfonic Acids

Anhydrous sulfonic acids absorb at 1352–1342 cm^{-1} (asymmetric SO_2 stretch), 1165–1150 cm^{-1} (symmetric SO_2 stretch), and 910–895 cm^{-1} (S—O stretch). The OH gives rise to a strong broad band near 2900 cm^{-1} with a smaller band near 2400 cm^{-1}.[56]

Sulfonic acids hydrate very easily. Hydrated sulfonic acids are thought to exist as hydronium sulfonates $(R—SO_3^-\ H_3O^+)$ and therefore resemble sulfonate salts with strong absorption in the 1230–1120 cm^{-1} region. A very broad band at 2800–1650 cm^{-1} with diffuse minima near 2600, 2250, and 1680 cm^{-1} is thought to be due to the hydronium ion.[56]

12.49 Sulfonic Acid Salts

Sulfonic acid salts absorb strongly at 1230–1120 and weakly at 1080–1025 cm^{-1} due to the asymmetric and symmetric stretching of the SO_3 group. In the aromatic sulfonic acid salts there are frequently four bands here near 1230, 1190, 1130, and 1040 cm^{-1} (three SO and one S-phenyl vibrations interacting).

12.50 Sulfinic Acids

Sulfinic acids in the solid state (R—SOOH) have bands at 2790–2340 cm^{-1}, (OH stretch), 1090–990 cm^{-1} (S=O stretch), and 870–810 cm^{-1} (S—O stretch).[66]

[66] S. Detoni and D Hadži, J. Chem. Soc., London p. 3163 (1955).

12.51   HSO₄⁻

The $HO-SO_3^-$ ion resembles the sulfonic acid salts in that it has bands at 1190–1160 cm^{-1} (asymmetric SO_3^- stretch) and 1080–1015 cm^{-1} (symmetric SO_3^- stretch). It has a band the sulfonates do not have[67] at 870–850 cm^{-1} which might be loosely described as involving $S-O(H)$ stretch.

12.52   SO₄²⁻

The inorganic sulfate ion absorbs strongly at 1125–1080 cm^{-1} due to asymmetric SO_4 stretching. The symmetric SO_4 stretch is normally forbidden by symmetry but may occasionally be seen as a very weak sharp band near 1000 cm^{-1}.[67] An SO_4 bending band is seen near 620 cm^{-1}.

12.53   SO₃²⁻

The sulfite ion absorbs strongly at 980–920 cm^{-1}.[67]

12.54   S—O—CH₂

The asymmetric stretching of the $S-O-CH_2$ bonds gives rise to strong absorption at 1020–850 cm^{-1}. Absorption also occurs near 830–690 cm^{-1} (symmetric $S-O-C$ stretch).

12.55   Thionylamine

The thionylamine group ($-N=S=O$) absorbs at 1300–1230 cm^{-1} (asymmetric NSO stretch) and 1180–1110 cm^{-1} (symmetric NSO stretch).[68,69]

[67] F. A. Miller and C. H. Wilkins, *Anal. Chem.* **24**, 1253 (1952).
[68] W. K. Glass and A. D. E. Pullin, *Trans. Faraday Soc.* **57**, 546 (1961).
[69] G. Kresze and A. Maschke, *Chem. Ber.* **94**, 450 (1961).

12.56 C=S

The C=S group does not give rise to as characteristic a band as the C=O group does due to the complication of interaction.[70-84] Due to the greater mass of sulfur the C=S vibration is expected to occur at considerably lower frequencies than the C=O vibration, in the C—O or C—N region. This means that the C=S and the attached single bonds can interact with the result that more than one band involves C=S stretching. In thioamides the NH_2 rocking vibration can also interact.[72,84] Vibrational analysis reveals that thioformamide has a band at 843 cm^{-1} which is almost pure C=S stretching, but that in other thioamides much mixing occurs.[84] Bands have been assigned as involving C=S stretching at 1420–700 cm^{-1}.[70-84] Thioacetamide has bands at 718, 975, and 1306 cm^{-1} involving C=S stretching.[77,84] Monosubstituted thioamides have bands near 1000 cm^{-1} which involve C=S.[84,85] The cyclic thioamides and ureas as well as so-called mercapto-N-heteroaromatic compounds usually have bands[78] probably involving C=S at 1210–1045 cm^{-1} not present in the oxygen analogs.

In noncyclic thioamides and ureas the NH frequencies are similar to those in amides and ureas. In the solid state the NH_2 group absorbs near 3380, 3180, and 1630 cm^{-1}. In the solid state the NH stretch absorbs near 3170 cm^{-1} and the CNH group absorbs near 1530 and 1350 cm^{-1}.[85] The cyclic CS—NH—R group gives rise to absorption[79] near 1550 cm^{-1} due to the CNH group, similar to noncyclic cases and unlike cyclic lactams which have no band near 1550 cm^{-1}.

[70] E. Spinner, *J. Org. Chem.* **23**, 2037 (1958).

[71] R. Mecke, R. Mecke, Jr., and A. Luttringhaus, *Chem. Ber.* **90**, 975 (1957).

[72] A. Yamaguchi, R. B. Penland, S. Mizushima, T. J. Lane, C. Curran, and J. V. Quagliano, *J. Amer. Chem. Soc.* **80**, 527 (1958).

[73] M. Davies and W. J. Jones, *J. Chem. Soc., London* p. 955 (1958).

[74] J. I. Jones, W. Kynaston, and J. L. Hales, *J. Chem. Soc., London* p. 614 (1957).

[75] T. A. Scott and E. L. Wagner, *J. Chem. Phys.* **30**, 465 (1959).

[76] R. Mecke, R. Mecke, Jr., and A. Luttringhaus, *Z. Naturforsch. B* **10**, 367 (1955).

[77] L. J. Bellamy and P. E. Rogasch, *J. Chem. Soc., London* p. 2218 (1960).

[78] R. Mecke and R. Mecke, Jr., *Chem. Ber.* **89**, 343 (1956).

[79] H. M. Randall, R. G. Fowler, N. Fuson, and R. Dangl, "Infrared Determination of Organic Structures." Van Nostrand-Reinhold, Princeton, New Jersey, 1949.

[80] B. Bak, L. Hansen-Nygaard, and C. Pedersen, *Acta Chem. Scand.* **12**, 1451 (1958).

[81] C. S. Marvel, P. Radzitsky, and J. J. Brader, *J. Amer. Chem. Soc.* **77**, 5997 (1955).

[82] N. Lozac'h and G. Guillouzo, *Bull. Soc. Chim. Fr.* p. 1221 (1957).

[83] C. N. R. Rao and R. Venkataraghavan, *Spectrochim. Acta* **18**, 541 (1962).

[84] I. Suzuki, *Bull. Chem. Soc. Jap.* **35**, 1286, 1449, and 1456 (1962).

[85] D. Hadži, *J. Chem. Soc., London* p. 847 (1957).

The region 1250–1020 cm^{-1} has been assigned to vibrations involving C=S stretching in compounds where nitrogen is not attached to the C=S group.[83]

Dialkyl trithiocarbonates $(R-S)_2 C=S$ have a C=S band near 1070 cm^{-1}.

Noncyclic dithioesters have strong absorption[77,80,81] at 1225–1170 cm^{-1} which is thought to involve the stretching of the C=S bond.

Compounds with the structure $R-O-(C=S)-S-X$ such as O, S dialkyl xanthates, dixanthogens, and copper and zinc xanthates show bands involving the $R-O-C=S$ group at 1250–1200 cm^{-1}, strong; 1140–1110 cm^{-1}, medium; and 1070–1020 cm^{-1}, strong.[86] The sodium and potassium xanthates are characterized by a number of strong bands between 1180 and 1030 cm^{-1} involving C—O and CS bonds.[86] These probably differ from the heavy metal xanthates because the two CS bonds are more nearly equivalent, $R-O-CS_2^- Na^+$.

Derivatives of thiobenzophenone have a medium-weak band[82] at 1224–1207 cm^{-1} not present in the comparable benzophenones.

Some frequencies which are thought to involve the C=S vibration in different types of molecules are listed in Table 12.10.[71] The noncyclic compounds are dimethyl derivatives and the cyclic compounds are dimethylene derivatives.

TABLE 12.10

Some C=O and C=S Frequencies for Cyclic and Noncyclic "Compounds" (in cm^{-1})

Compound	C=O Five-membered ring	C=O Noncyclic	C=S Five-membered ring	C=S Noncyclic
O—(C=X)—O	1798	1742	—	1127
CH$_2$—(C=X)—O	1773	1734	—	—
CH$_2$—(C=X)—CH$_2$	1742	1706	—	—
O—(C=X)—NH	1724	1703	1171	—
CH$_2$—(C=X)—NH	1692	1658	1109	1097
NH—(C=X)—NH	1638	1626	1201	1183
S—(C=X)—S	1638	1655	1058	1076
S—(C=X)—NH	—	—	1047	—
CH$_2$—(C=X)—N—CH$_3$	—	—	1136	1122

[86] L. H. Little, G. W. Poling, and J. Leja, *Can. J. Chem.* 39, 745 (1961).

12.57 (FHF)⁻

The linear bifluoride ion $(FHF)^-$ has one of the strongest hydrogen bonds known. The asymmetric FHF stretching vibration absorbs broadly at 1700–1400 cm^{-1} and the FHF deformation absorbs at 1260–1200 cm^{-1}.[87,88]

12.58 FCH

A single hydrogen on a fluorine bearing carbon atom absorbs[89,90] near 3000 cm^{-1} with some indication of shifts with the number of fluorine atoms on the carbon. CF_3H absorbs at 3062 cm^{-1}, $-CF_2H$ at 3008 cm^{-1}, and $>CFH$ at 2990 cm^{-1}.[89]

12.59 FC=C

The C=C stretching frequency is unusually high when fluorine atoms are directly on the carbon atoms. The group $C=CF_2$ absorbs at 1755–1735 cm^{-1} and the group $-CF=CF_2$ absorbs[89,91–94] at 1800–1780 cm^{-1}.

12.60 FC=O

Acid fluoride groups attached to fluorinated alkane groups $(-CF_2-COF)$ absorb at 1900–1870 cm^{-1}.[89,95] In the gas phase, acetyl fluoride absorbs at 1869 cm^{-1} and benzoyl fluoride absorbs at 1820 cm^{-1}.[96]

[87] J. A. A. Ketelaar, C. Haas, and J. van der Elsken, *J. Chem. Phys.* **24**, 624 (1956).

[88] J. A. A. Ketelaar and W. Vedder, *J. Chem. Phys.* **19**, 654 (1951).

[89] D. G. Weiblen, *in* "Fluorine Chemistry" (J. H. Simons, ed.), Vol. 2, Chapter 7, Academic Press, New York, 1954.

[90] D. C. Smith, M. Alpert, R. A. Saunders, G. M. Brown, and N. B. Moran, *Nav. Res. Lab. Rep.* **3924** (1952).

[91] D. C. Smith, J. R. Neilsen, L. H. Berryman, H. H. Claassen, and R. L. Hudson, *Nav. Res. Lab. Rep.* **3567** (1949).

[92] P. Torkington and H. W. Thompson, *Trans. Faraday Soc.* **41**, 236 (1945).

[93] T. J. Brice, J. D. La Zerte, L. J. Hals, and W. H. Pearlson, *J. Amer. Chem. Soc.* **75**, 2698 (1953).

[94] R. N. Haszeldine, *J. Chem. Soc., London* p. 4423 (1952).

[95] R. N. Haszeldine, *Nature (London)* **168**, 1028 (1951).

[96] J. Overend and J. R. Scherer, *Spectrochim. Acta* **16**, 773 (1960).

12.61   F-Aryl

Fluorine atoms directly attached to an aromatic ring give rise to bands in the region 1270–1100 cm^{-1}. Many of these compounds including the simpler ones with one fluorine only on the ring absorb near 1230 cm^{-1}. Fluorine substitution does not abnormally affect the ring frequencies at 850–700 cm^{-1}.

12.62   CF Stretch

The groups CF$_3$ and CF$_2$ on the whole are difficult to differentiate in the infrared.[89,91,97] The CF$_3$ group absorbs strongly at 1350–1120 cm^{-1} [91] and the CF$_2$ at 1280–1120 cm^{-1}.[91] Four- or five-membered cyclic CF$_2$ compounds absorb at 1350–1140 cm^{-1}.[89] In the group CF$_3$(CF$_2$)$_n$X, a band is usually found at 1365–1325 cm^{-1} attributed to the CF$_3$ group.[98] When the CF$_3$ is absent in this type of compound, absorption here is usually not present. Another band involving the CF$_3$ group occurs at 780–680 cm^{-1}.[89] The narrower region of 745–730 cm^{-1} is characteristic for the group CF—CF$_3$.[99] The group CF=CF$_2$ absorbs strongly at 1340–1300 cm^{-1}.[89,91] The electronegative fluorinated alkane group causes some shifts in common functional groups when attached thereon. A summary of these positions is found in Table 12.11.[89]

TABLE 12.11

Common Functional Groups Attached to Fluorinated Alkane Groups (in cm^{-1})

—CF$_2$—C≡N	2280–2270
—CF$_2$—N=C=O	2300
—CF$_2$—CHO	1785–1755
—CF$_2$—CO—CH$_2$	1770
—CF$_2$—CO—CF$_2$	1800
—CF$_2$—COF	1900–1870
—CF$_2$—COCl	1820–1795
—CF$_2$—COOH (liquid)	1785–1750
—CF$_2$—CO—O—CO—CF$_2$—	1890 and 1820
—CF$_2$—COO—CH$_2$	1796–1784
—CH$_2$—COO—CH$_2$—CF$_2$—	1773–1761
—CF$_2$—COO—CH$_2$—CF$_2$—	1814–1802
—CF$_2$—CO$_2$$^-$ metal$^+$	1695–1615
—CF$_2$—CO—NH$_2$	1730–1700 and 1630–1610 (weaker)

[97] H. W. Thompson and R. B. Temple, *J. Chem. Soc., London* p. 1432 (1948).
[98] M. Hauptschein, C. S. Stokes, and E. A. Nodiff, *J. Amer. Chem. Soc.* **74**, 4005 (1952).
[99] L. J. Bellamy, "The Infrared Spectra of Complex Molecules." Wiley, New York, 1954.

12.63 C—Cl

The aliphatic C—Cl bond absorbs at 830–560 cm^{-1}.[100,101] The group C—CH$_2$—CH$_2$—Cl gives rise to two bands due to rotational isomers. The planar C—C—C—Cl *trans* zig-zag form absorbs near 726 cm^{-1}, and the *gauche* C—C—C—Cl form absorbs near 649 cm^{-1}.[101,102] Just as the *trans* form absorbs at higher wavenumbers than the *gauche*, the equatorial C—Cl absorbs higher than the axial C—Cl bond in cyclic rings. In steroids, the equatorial C—Cl absorbs at 755 cm^{-1} in the 2-position, at 782–750 cm^{-1} in the 3-position, and 749 cm^{-1} in the 7 position. Comparable axial C—Cl bands occur at 693, 730–617, and 588 cm^{-1}.[103] Monochloroalkanes, where the α-carbon atoms are branched with methyl groups, have C—Cl bands at the lower wavenumber end of the region.[101] The complexity of the C—Cl region is indicative of C—Cl interaction with C—C and other oscillators. In chloroalkanes where there is one chlorine per carbon of an aliphatic compound, close correlations have been made, specifying the rotational isomers by specifying the atom X "*trans*" to the chlorine on the β-carbon so that the atoms X—C—C—Cl zig-zag in one plane. In Table 12.12[104,105] P stands

TABLE 12.12

C—X Bands in Haloalkanes (in cm^{-1})

	Cl[104]	Br[105]	I[105]
P$_C$	730–723	649–635	602–584
P$_H$	657–648	565–557	513–500
P$'_H$	686–679	617–610	583–578
S$_{CC}$	~758	—	—
S$_{CH}$	674–655	617–608	581–575
S$'_{HH}$	637–627	588–578	550–521
S$_{HH}$	615–608	539–529	490–483
T$_{CHH}$	632–595	600–571	578–559
T$_{HHH}$	581–538	524–488	508–466

[100] J. K. Brown and N. Sheppard, *Trans. Faraday Soc.* **50**, 1164 (1954).

[101] N. Sheppard, *Trans. Faraday Soc.* **46**, 527 and 533 (1950).

[102] J. K. Brown and N. Sheppard, *Proc. Roy. Soc., Ser. A* **231**, 555 (1955).

[103] D. H. R. Barton, J. E. Page, and C. W. Shoppee, *J. Chem. Soc., London* p. 331 (1956).

[104] J. J. Shipman, V. L. Folt, and S. Krimm, *Spectrochim. Acta* **18**, 1603 (1962).

[105] F. F. Bently, L. D. Smithson, and A. L. Rozek, "Infrared Spectra and Characteristic Frequencies 730-300 cm^{-1}." Wiley (Interscience), New York, 1968.

for primary (CH_2Cl), S for secondary (CHCl), and T for tertiary. The subscript C or H is the "*trans*" atom X. In this nomenclature the planar skeleton *trans* form of *n*-propyl chloride would be noted as P_C and the *gauche* skeleton form as P_H. Isopropyl chloride would be S_{HH} (two "*trans*"-hydrogens). The symbol P'_H means that the β-carbon is branched and the symbol S'_{HH} means the carbon chain does not have the planar zig-zag configuration. Tertiary haloalkanes with two methyl groups α to the halogen are found in the upper half of the T_{CHH} and T_{HHH} regions.

When a carbon atom is *trans* to the chlorine, the "C—Cl stretching" vibration sharply bends the adjacent C—C—C bond angle. This forced interaction raises the frequency compared to the C—Cl stretching vibration when a hydrogen is *trans* to the chlorine, in which case the adjacent C—C—C bond angle is not sharply bent.[106]

Putting more than one chlorine on a carbon raises the CCl frequency. The CCl_3 group has an asymmetric stretching vibration which gives rise to a strong band at 830–700 cm^{-1}. The group CH_2—Cl has a strong CH_2 wag band at 1300–1240 cm^{-1}.[102]

12.64   C—Br

The aliphatic C—Br bond gives rise to bands in the region 680–515 cm^{-1}.[107] The group C—CH_2—CH_2—Br gives rise to two bonds near 644 cm^{-1} (*trans*) and 563 cm^{-1} (*gauche*)[100,102,107] in both infrared and Raman spectra due to two rotational isomers.[100,102] Branching of the α-carbon with methyl groups lowers the band to 560–515 cm^{-1}.[101] Putting two or three bromines on the same carbon atom results in bands at the high frequency end of the region. The CH_2—Br group has strong CH_2 wag bonds near 1230 cm^{-1}.[102]

12.65   C—I

The aliphatic C—I bond gives rise to bands about 610–485 cm^{-1}.[100–102] The group C—CH_2—CH_2—I gives rise to two bands near 594 and 503 cm^{-1}. The CH_2—I group has strong CH_2 wag bands near 1170 cm^{-1}.[102]

[106] N. B. Colthup, *Spectrochim. Acta* **20**, 1843 (1964).
[107] F. S. Mortimer, R. B. Blodgett, and F. Daniels, *J. Amer. Chem. Soc.* **69**, 822 (1947).

12.66   Aryl Halides

In aryl halides there are no bands obviously comparable to the aliphatic C—X stretching bands due to interaction with ring vibrations. One ring vibration which is X sensitive and involves some C—X stretching[108] is found approximately at 1270–1100 cm^{-1} for aryl-F compounds. The same band for aryl-Cl[108] is found at 1096–1089 cm^{-1} for *para*-, 1078–1074 cm^{-1} for *meta*-, and 1057–1034 cm^{-1} for *ortho*-substituted chlorobenzenes.[109] Bands are found at 1073–1068 cm^{-1} for *para*-, 1073–1065 cm^{-1} for *meta*-, and 1042–1028 cm^{-1} for *ortho*-substituted bromobenzenes. *Para*-substituted iodobenzenes absorb at 1061–1057 cm^{-1}.[109]

12.67   Organometallic Compounds

The hydrogen stretching vibration in Ge—H compounds absorbs at 2160–2010 cm^{-1}.[110] The Sn—H group absorbs at about 1900–1800 cm^{-1}.[111,112] Complexes containing AlH_3 and AlH_4 absorb at 1810–1674 cm^{-1}.[113]

The symmetric methyl deformation vibration absorbs strongly at 1240–1230 cm^{-1} in Ge—CH_3 compounds and 1200–1180 cm^{-1} in Sn—CH_3 compounds.[114] Methyl rocking vibrations absorb strongly at 900–700 cm^{-1}.

The CH_2-metal group absorbs at 1430–1415 cm^{-1} (CH_2 deformation) in compounds with such metals as mercury, zinc, cadmium, or tin.[115]

The CH_2=CH-(metal) group has bands at 1610–1565 cm^{-1} (C=C stretch), 1410–1390 cm^{-1} (CH_2 deformation), 1265–1245 cm^{-1} (CH rock), 1010–985 cm^{-1} (CH wag), and 960–940 cm^{-1} (CH_2 wag).[115,116]

In C_6H_5-(metal) compounds a band in the region 1120–1050 cm^{-1} is sensitive to a change in the metal. This is a benzene ring vibration interacting with C-(metal) stretching vibration.[108] Examples of this are tetraphenylsilane,

[108] A. Stojiljkovic and D. H. Whiffen, *Spectrochim. Acta* **12**, 47 (1958).

[109] A. R. Katritzky and J. M. Lagowski, *J. Chem. Soc., London* p. 2421 (1960).

[110] V. A. Ponomarenko, G. Y. Zueva, and N. S. Andreev, *Izv. Akad. Nauk SSSR* **10**, 1758 (1961),

[111] R. Mathis, M. Lesbre, and I. S. de Roch, *C.R. Acad. Sci.* **243**, 257 (1956).

[112] D. R. Lide, Jr., *J. Chem. Phys.* **19**, 1605 (1951).

[113] R. Dautel and W. Zeil, *Z. Elektrochem.* **64**, 1234 (1960).

[114] M. P. Brown, R. Okawara, and E. G. Rochow, *Spectrochim. Acta* **16**, 595 (1960).

[115] H. D. Kaesz and F. G. A. Stone, *Spectrochim. Acta* **15**, 360 (1959).

[116] M. C. Henry and J. G. Noltes, *J. Amer. Chem. Soc.* **82**, 555 (1960).

1100 cm^{-1}; tetraphenylgermane, 1080 cm^{-1}; tetraphenyltin, 1065 cm^{-1}; and tetraphenyllead, 1052 cm^{-1}.[116]

The Ge-O-Ge linkage has been tentatively assigned to the 900–700 cm^{-1} region where the Ge-CH$_3$ rocking vibrations also absorb.[114] The Sn-O-Sn linkage has been tentatively assigned to the 650–580 cm^{-1} region.[114] The Ti-O-Ti linkage usually absorbs[117,118] at 900–700 cm^{-1} and the Ti-O-Si linkage at about 950–900 cm^{-1}.[118]

Compounds with (metal)=O linkages where the oxygen is associated with only one metal generally absorb at 1100–900 cm^{-1} (example: VOCl$_3$, 1035 cm^{-1}).[117]

For a more detailed discussion of organometallic compounds, the books by Nakamoto[119] and Adams[120] should be consulted.

[117] C. G. Barraclough, J. Lewis, and R. S. Nyholm, *J. Chem. Soc., London* p. 3552 (1959).

[118] V. A. Zeitler and C. A. Brown, *J. Phys. Chem.* **61**, 1174 (1957).

[119] K. Nakamoto, "Infrared Spectra of Inorganic and Coordination Compounds." Wiley, New York, 1963.

[120] D. M. Adams, "Metal-Ligand and Related Vibrations." St. Martin's Press, New York, 1968.

CHAPTER 13

MAJOR SPECTRA–STRUCTURE
CORRELATIONS BY SPECTRAL REGIONS

13.1 Introduction

In previous chapters, the spectra of various chemical groups were discussed under functional group categories. In this chapter, the infrared spectrum will be considered in terms of spectral regions. This discussion will be briefer than that given for functional groups with only the major bands being mentioned.

13.2 3700–3100 cm^{-1} (OH, NH, and $\equiv$CH)

Bands in the region from 3700 to 3100 cm^{-1} are usually due to various OH and NH stretching vibrations. The NH$_2$ group gives rise to a doublet with approximately 70 cm^{-1} separation. The OH compounds include water, alcohols, and phenols but not phosphorus acids or carboxylic acid dimers which have bands at lower wavenumbers. The bonded OH group usually give rise to a broader band than NH groups. All these bands shift to higher wavenumbers and much become narrower and weaker in intensity when the hydrogen bond is broken by diluting the solute in a nonpolar solvent such as CCl$_4$. In concentrated solutions both bonded and unbonded forms may exist, giving rise to several bands.

The C$\equiv$C–H group has a CH stretching vibration which absorbs near 3300 cm^{-1}. This band is not nearly as broad as alcoholic bonded OH bands found in this region.

Overtones of lower bands occur here, notably overtones of carbonyls at twice the C=O frequency. These are weak bands in the 3500–3400 cm^{-1} region.

13.3   3100–3000 cm^{-1} (Aryl, Olefinic, and Three-Membered Ring CH)

The C—H stretching vibrations of olefins, aromatic rings, and three-membered rings absorb mainly in this region above 3000 cm^{-1}.

13.4   3000–2700 cm^{-1} (Aliphatic CH)

In this region below 3000 cm^{-1} are found aliphatic C—H stretching vibrations. Aliphatic CH$_2$ and CH$_3$ groups each give rise to a doublet approximately 80 cm^{-1} separation at slightly different frequencies (aliphatic CH$_3$ 2960 cm^{-1}, aliphatic CH$_2$, 2930 cm^{-1} for the higher band) and so can be differentiated. In tertiary amines, the CH$_2$ and CH$_3$ groups next to the nitrogen absorb near 2800 cm^{-1}. Many aldehydes absorb near 2730 cm^{-1}.

13.5   3100–2400 cm^{-1} (Acidic and Strongly Bonded Hydrogens)

In this region occur the broad bands of many acidic hydrogens. Carboxylic acid dimers absorb broadly near 3000 cm^{-1} with satellite bands near 2650 and 2550 cm^{-1}. Amine salts absorb in this region. Their bands have much fine structure. Phosphorus acids absorb broadly in this region. The strongly bonded hydrogen in the enol form of 1,3-diketones and related compounds absorbs broadly at 3100–2700 cm^{-1}.

13.6   2600–2100 cm^{-1} (SH, BH, PH, and SiH)

Mercaptans and thiophenols absorb at 2590–2540 cm^{-1} due to the SH stretching vibration. Compounds with a BH group absorb at 2630–2350 cm^{-1}. Compounds with a PH group absorb at 2440–2275 cm^{-1}. Compounds with an SiH group absorb at 2250–2100 cm^{-1}.

13.7   2300–1900 cm^{-1} (X≡Y and X=Y=Z)

In this region are found triple bonds (X≡Y) and cumulated double bonds (X=Y=Z). The bands of the latter tend to have considerably more intensity than the former. A combination band in primary amine salts occurs here. Metal carbonyl complexes (carbon monoxide complexes) absorb strongly in the lower part of the region.

13.8   2000–1700 cm^{-1} (Aryl and Olefinic Overtones)

Unusually intense overtones of aryl ring vibrations occur in this region. They form a pattern which can be used to differentiate substitution isomers. An unusually intense overtone of the CH_2 wag vibration in vinyl and vinylidine groups is an excellent check on the assignment of the fundamental near 900 cm^{-1}. These bands are somewhat weaker than fundamentals and are most clearly seen in spectra of moderately thick samples.

13.9   1900–1550 cm^{-1} (C=O)

Carbonyl compounds absorb strongly throughout this region. Those compounds absorbing above 1760 cm^{-1} are usually compounds with electronegative groups next to the carbonyl such as acid chlorides, compounds where the carbonyl is in a strained ring such as in cyclobutanone, or compounds with an anhydride group which gives rise to two strong bands. Very approximately, the carbonyl groups in saturated esters absorb near 1740 cm^{-1}, aldehydes near 1725 cm^{-1}, and ketones near 1715 cm^{-1}. Conjugation will lower these about 20 cm^{-1}. Carboxylic acid dimers absorb near 1700 cm^{-1} and amides near 1660 cm^{-1}. Increasing mesomerism increasingly lowers the carbonyl frequency. The extreme case is the carboxylic salt with two equivalent C=O bonds which absorbs near 1600 cm^{-1}. Also absorbing here is the enol form of 1,3-diketones where mesomerism is also extremely in evidence.

A convenient dividing line is 1600 cm^{-1}, where bands in the region 1800–1600 cm^{-1} are due to X=Y double bond stretching vibrations, and bands in the region 1600–1500 cm^{-1} are due to the asymmetric stretching of the X⁑Y⁑X "bond-and-a-half" bonds such as in carboxyl salts, nitro groups, and aromatic rings.

13.10   1700–1550 cm^{-1} (C=C and C=N)

Compounds containing the C=C group absorb in this region except for symmetric *trans*-disubstituted olefins where the band is forbidden by symmetry. Most olefins absorb at 1680–1600 cm^{-1}. Compounds containing the C=N group absorb ususally at 1690–1630 cm^{-1}, although some types such as guanidines absorb at lower wavenumbers.

13.11   1660–1450 cm^{-1} (N=O)

Organic nitrates absorb at 1660–1625 cm^{-1}. Organic nitrites have two bands due to rotational isomers at 1681–1648 and 1625–1605 cm^{-1}. Aliphatic nitro compounds absorb at 1590–1535 cm^{-1}, and aromatic nitro compounds absorb at 1530–1500 cm^{-1}. Monomeric C-nitroso monomers absorb at 1620–1488 cm^{-1}, and monomeric N-nitroso compounds absorb at 1490–1445 cm^{-1}.

13.12   1660–1500 cm^{-1} (NH$_2$, NH$_3{}^+$, and CNH)

The NH$_2$ group has its scissors deformation frequency at 1660–1590 cm^{-1}. The NH$_2{}^+$ group absorbs near 1600 cm^{-1} and the NH$_3{}^+$ group near 1600 and 1520 cm^{-1}. In a noncyclic monosubstituted amide or monosubstituted thioamide, the CHN group gives rise to a strong band near 1550 cm^{-1}. Liquid H$_2$O absorbs near 1640 cm^{-1}.

13.13   1620–1420 cm^{-1} (Aromatic and Heteroaromatic Rings)

Aromatic rings are characterized by sharp bands near 1600, 1580, 1500, and 1460 cm^{-1} which may vary in intensity with different substituents. Pyridines are closely related to the benzene compounds. Triazines have strong bands in the 1600–1500 cm^{-1} region. Heterocyclic compounds with two double bonds in a five-membered ring usually absorb at 1600–1530 and 1500–1430 cm^{-1}.

13.14   1500–1250 cm^{-1} (CH$_3$ and CH$_2$)

In the region 1500–1400 cm^{-1} are found the CH$_2$ scissors deformation and the CH$_3$ asymmetrical deformation which are near the same position when the substituents are the same. These bands are near 1460 cm^{-1} in hydrocarbons and near 1420 cm^{-1} when the group is on a sulfur, phosphorus or silicon or is next to a carbonyl or nitrile. The symmetric CH$_3$ umbrella deformation varies in frequency with the electronegativity of the substituent, being found near 1450 cm^{-1} for O—CH$_3$, 1375 cm^{-1} for C—CH$_3$, and 1265 cm^{-1} for Si—CH$_3$.

13.15 1470–1310 cm^{-1} (B—O, B—N, NO$_3^-$, CO$_3^{2-}$, and NH$_4^-$)

Compounds containing the B—O and B—N linkage usually absorb strongly in this region. The nitrate and carbonate ions absorb strongly and broadly near 1400 cm^{-1}, and the ammonium ion has a band near 1400 cm^{-1}.

13.16 1400–1000 cm^{-1} (SO$_2$, SO$_3^-$, SO, and SO$_4^{2-}$)

The SO$_2$ group gives rise to two bands separated by about 180 cm^{-1} in the regions 1400–1300 and 1200–1100 cm^{-1}, the exact position of the two bands varying with the electronegativity of the substituents. Sulfonic acid salts absorb broadly near 1200 cm^{-1}. Dialkyl sulfites absorb near 1200 cm^{-1}, and sulfoxides absorb near 1050 cm^{-1}. The SO$_4^{2-}$ group absorbs near 1120 cm^{-1}.

13.17 1300–1140 cm^{-1} (P=O)

Compounds containing the P=O group absorb in the region 1300–1140 cm^{-1}, the exact position depending on the sum of the electronegativities of the substituents. Trialkyl phosphates for example abosrb near 1290 cm^{-1}, and trialkyl phosphine oxides absorb near 1150 cm^{-1}. Cyclic P=N compounds also absorb in this region.

13.18 1350–1120 cm^{-1} (CF$_3$ and CF$_2$)

The fluorinated alkane groups CF$_3$ and CF$_2$ absorb strongly in the region 1350–1120 cm^{-1}.

13.19 1350–1150 cm^{-1} (CH$_2$ and CH Wag)

In the region 1340–1190 cm^{-1} occur the multiple weak bands due to CH$_2$ wagging vibrations in a normal hydrocarbon chain, the complexity of the band system increasing with chain length in the solid state spectra. The CH$_2$—Cl group absorbs strongly near 1275 cm^{-1}, CH$_2$—S near 1250 cm^{-1}, CH$_2$—Br near 1230 cm^{-1}, and CH$_2$—I near 1170 cm^{-1}. Comparable CH—Cl and CH—Br groups absorb near the CH$_2$ position but usually about 50 cm^{-1} lower.

13.20   1300–1000 cm^{-1} (C—O)

In this region occur strong bands involving the stretching of the C—O bonds. Bands in the region 1300–1150 cm^{-1} arise from those C—O bonds somewhat stiffened by resonance such as esters, phenols, phenyl ethers, and vinyl ethers. Saturated ethers absorb near 1125 cm^{-1}, and alcohols absorb in the region 1200–1000 cm^{-1}. Primary alcohols absorb at 1075–1000 cm^{-1}. Branching of the α-carbon tends to raise the frequency.

13.21   1100–830 cm^{-1} (Si—O and P—O)

The Si—O—alkyl group gives rise to a strong band at 1100–1000 cm^{-1} where the Si—O—Si group also absorbs. The Si—OH group absorbs at 915–830 cm^{-1}.

The P—O—alkyl group absorbs at 1050–970 cm^{-1}, and the P—O—P and P—OH groups absorb about 1000–900 cm^{-1}.

13.22   1000–600 cm^{-1} (Olefinic CH Wag)

Vinyl compounds absorb at 1000–940 and at 960–810 cm^{-1} (990 and 910 cm^{-1} for hydrocarbons). Vinylidene compounds absorb at 985–700 cm^{-1} (890 cm^{-1} for hydrocarbons). *trans*-Disubstituted ethylenes absorb at 980–890 cm^{-1} (965 cm^{-1} for hydrocarbons), and *cis* compounds absorb at 800–600 cm^{-1} (730–650 cm^{-1} for hydrocarbons). The position of the band within the regions can be predicted from the properties of the substituents.

13.23   900–700 cm^{-1} (Aromatic CH Wag)

Bands near 770–730 cm^{-1} are due to the 5 adjacent hydrogens of monosubstituted benzene. The 4 adjacent hydrogens of *ortho* disubstitution absorb near 750 cm^{-1}. The 3 adjacent hydrogens of *meta* and vicinal tri absorb near 780 cm^{-1}, the two adjacent hydrogens of *para* and unsymmetrical tri, near 820 cm^{-1}, and the isolated hydrogens of *meta*, unsymmetrical tri and symmetrical tri, near 870 cm^{-1}. Mono-, *meta*-, and symmetrical trisubstitution have a ring bending vibration which absorbs near 700 cm^{-1}. This classification by adjacent hydrogens can be extended to substituted napthalenes and pyridines.

13.24   830–500 cm^{-1} (CCl, CBr, and CI)

The CCl group gives rise to bands at 830–570 cm^{-1}, the CBr group to bands at 680–515 cm^{-1}, and the CI group to bands at 610–485 cm^{-1}.

13.25   Near Infrared Region Correlation Chart

For instrumental and theoretical reasons Kaye[1] has assigned the near infrared region from 0.7 to 3.5 μm (14,285–2860 cm^{-1}). Kaye has reviewed this region in two papers. The first paper[1] dealt with spectral identification and analytical applications and contained a correlation chart with many references. The second paper[2] reviewed instrumentation and technique. More recent correlation charts have been published by Goddu and Delker[3] and the Anderson Physical Laboratory, Champaign, Illinois. In 1960 a bibliography was published by Kaye.[4] A review by Whetsel was presented in 1968.[5]

Most of the absorption bands observed in this region are X—H stretching vibrations and overtone or combination bands of these vibrations. Since the C—H fundamental stretching vibrations as well as the O—H and N—H fundamental stretching vibrations are covered in more detail on subsequent charts (see the following sections), the main emphasis in the following discussion will concern the overtone bands and to a lesser degree the combination bands in the near infrared region.

A chart similar to that of Goddu and Delker is shown in Fig. 13.1. A few comments on this chart are in order. If one considers the paraffinic CH_3, CH_2, and aromatic C—H vibrations, Evans, Hibbard, and Powell[6] report that the most useful CH_3 band falls between 8375 and 8360 cm^{-1} while the CH_2 band falls between 8255 and 8220 cm^{-1} and the aromatic CH band between 8740 and 8670 cm^{-1}. The average value for the aromatic C—H was 8710 cm^{-1}, which is somewhat less than the 8755 cm^{-1} value previously reported by Hibbard and Cleaves.[7] The bands shown in the approximate range of 4075–4450 cm^{-1} are combination bands of saturated CH_2 and CH_3 groups.

Powers, Harper, and Tai[8] report that aromatic aldehydes are characterized by a band at 4525 cm^{-1}, which is possibly a combination of formyl C—H and of C=O, and two bands at 4445 and 8000 cm^{-1}.

Goddu[9] has reported that terminal methylene groups may be determined at 4740 or 6135 cm^{-1} with a sensitivity within 0.01 %. *Cis* double bonds may

[1] W. Kaye, *Spectrochim. Acta* **6**, 257 (1954).

[2] W. Kaye, *Spectrochim. Acta* **7**, 181 (1955).

[3] R. F. Goddu and D. A. Delker, *Anal. Chem.* **32**, 140 (1960).

[4] W. Kaye, *in* "The Encyclopedia of Spectroscopy" (G. L. Clark, ed.), p. 409. Van Nostrand-Reinhold, Princeton, New Jersey, 1960.

[5] K. B. Whetsel, *Appl. Spectrosc. Rev.* **2**, 1 (1968).

[6] A. Evans, R. R. Hibbard, and A. S. Powell, *Anal. Chem.* **23**, 1604 (1951).

[7] R. R. Hibbard and A. P. Cleaves, *Anal. Chem.* **21**, 486 (1949).

[8] R. M. Powers, J. L. Harper, and H. Tai, *Anal. Chem.* **32**, 1287 (1960); correction, **32**, 1598 (1960).

[9] R. F. Goddu, *Anal. Chem.* **29**, 1790 (1957).

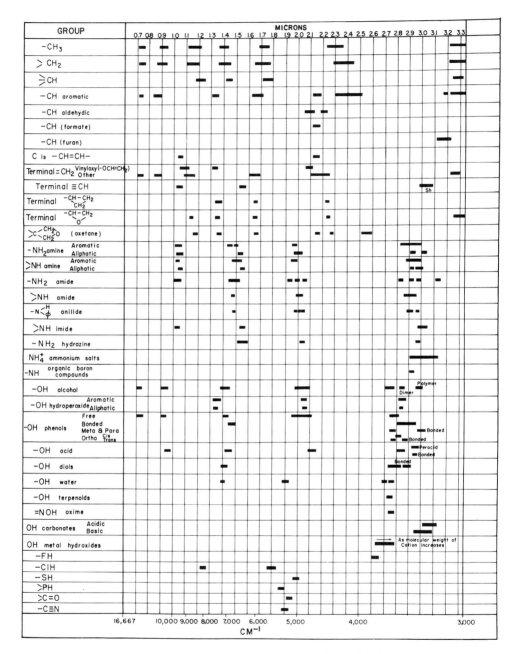

FIG. 13.1.   Correlation chart for the near infrared region.

be determined at 4675 cm^{-1} with a sensitivity of about 1%. The *trans* isomer has no unique band. Holman and Edmondson[10] have determined *cis* unsaturation in nonconjugated fatty acids at 4565, 4675, or 8475 cm^{-1}.

The chart published by Goddu and Delker[3] shows three bands for cyclopropyl compounds. Washburn and Mahoney[11] report for nine cyclopropyl compounds a band in the range 4505–4405 cm^{-1} and a second band in the range 6135–6060 cm^{-1}, undoubtedly the first overtone of the C—H fundamental at approximately 3080 cm^{-1}. The band between 6135 and 6060 cm^{-1} is about five times as intense as the first one mentioned. In a study of 16 simple cyclopropyl compounds, Gassman[12] found that the range was 6160–6060 cm^{-1} and the empirical absorptivity constant (extinction coefficient) is 0.30 for each unsubstituted cyclopropyl methylene group. For example, 11 compounds with two groups had absorptivity constants of 0.57–0.70 vs. the predicted 0.60. Seven nortricyclenes had peaks between 6010 and 6040 cm^{-1}, and although norbornenes (which lack a cyclopropyl group) absorb in the same region, they can be distinguished because the absorptivity constants are half those of the nortricyclenes.

Goddu and Delker[13] have determined terminal epoxides by measurement of the bands at 4545 and 6060 cm^{-1}. The combination band at 4545 cm^{-1} is more intense than the band at 6060 cm^{-1} which is the first overtone of the C—H stretching vibration of the terminal epoxide group. With regard to N—H bands, Whetsel, Roberson, and Krell,[14] have studied 40 primary aromatic amines and have reported intensities for a combination band near 5070 cm^{-1}, the first overtone asymmetric and symmetric stretching bands near 6915 and 6700 cm^{-1}, respectively, and the second overtone symmetric near 9810 cm^{-1}. These authors were able to correlate variations in band intensity and position with the electronic nature and position of the substituents on the ring. Solvent effects on the bands at 6700 and 5070 cm^{-1} of aniline, *m*-toluidine, *o*-chloroaniline, and *m*-chloroaniline have also been reported by the same authors.[15] Sauvageau and Sandorfy[16] comment that broad bands between 4600 and 4500 cm^{-1} serve to identify amine hydrohalides. They assign this band to a combination band of a

asymmetric bend with either a $\overset{+}{N}H_3$ $\overset{+}{N}H_3$

stretch or a CH_2 stretch or possibly both.

[10] R. T. Holman and P. R. Edmondson, *Anal. Chem.* **28**, 1533 (1956).
[11] W. H. Washburn and M. J. Mahoney, *J. Amer. Chem. Soc.* **80**, 504 (1958).
[12] P. G. Gassmann, *Chem. Eng. News* **40**, 49 (1962).
[13] R. F. Goddu and D. A. Delker, *Anal. Chem.* **30**, 2013 (1958).
[14] K. B. Whetsel, W. E. Roberson, and M. W. Krell, *Anal. Chem.* **30**, 1598 (1958).
[15] K. B. Whetsel, W. E. Roberson, and M. W. Krell, *Anal. Chem.* **32**, 1281 (1960).
[16] P. Sauvageau and C. Sandorfy, *Can. J. Chem.* **38**, 1901 (1960).

In connection with O—H absorptions, Kaye[4] has done some interesting work on cellulose and its derivatives. He has pointed out that the fundamental vibration at 3300 cm^{-1} is most advantageous for studying hydrogen bonded OH groups and the overtone region of the free OH band at approximately 7100 cm^{-1} is better for the study of unbonded compounds. The determination of hydroperoxides (O—O—H groups) at approximately 4810 and 6850 cm^{-1} has been discussed by Holman and Edmondson.[17,18] The first overtone of the OH stretching band of phenol has been assigned by Wulf, Jones, and Deming[19] to the band at 7050 cm^{-1}. In tri- and pentahalogenated phenols the overtone band is between 6890 and 6760 cm^{-1} and in *ortho*-halogenated phenols between 6910 and 6805 cm^{-1}. A combination of band medium intensity for mono-, tri-, and pentahalogenated phenols is located between 8250 and 8000 cm^{-1}.

13.26   Carbon—Hydrogen Stretching Region Correlation Chart

This region has been reviewed by Wiberley, Bunce, and Bauer[20] and the main correlations have been covered in previous chapters; so this discussion will be brief. As shown in Fig. 13.2 the CH$_3$ bands occur at 2960 and 2870 cm^{-1} and the CH$_2$ bands at 2930 and 2860 cm^{-1}. In many compounds only one band in the region 2870–2860 cm^{-1} can be resolved. In oxygenated and sulfur-containing compounds the methyl and methylene bands are approximately 7 cm^{-1} higher and the extinction coefficients are greater than for the corresponding hydrocarbons. In a given straight chain homologous series the asymmetric stretching CH$_3$ band at 2960 cm^{-1} is stronger than the corresponding CH$_2$ band when the ratio of methylene to methyl groups is 3 to 1 or less. With larger ratios the 2930 cm^{-1} band is more intense although exceptions may occur if the methyl group is adjacent to a carboxyl group.

Cyclopropyl CH$_2$ bands occur at approximately 3085 cm^{-1} for the unsymmetrical stretching frequency and at 3020 cm^{-1} for the symmetric stretch. Epoxides also have a distinct band above 2990 cm^{-1}. CH$_2$ groups in a four-membered ring have bands intermediate between the three- and five-membered ring systems. In five-membered rings the symmetric CH$_2$ band lies in the normal CH$_2$ range for straight chain hydrocarbons; however, the unsymmetric CH$_2$ band occurs at a higher value than the maximum CH$_2$

[17] R. T. Holman and P. R. Edmondson, *Anal. Chem.* **28**, 1533 (1956).

[18] R. T. Holman, C. Nickell, O. S. Privett, and P. R. Edmondson, *J. Amer. Oil Chem. Soc.* **35**, 422 (1958).

[19] O. R. Wulf, E. J. Jones, and L. S. Deming, *J. Chem. Phys.* **8**, 753 (1940).

[20] S. E. Wiberley, S. C. Bunce, and W. H. Bauer, *Anal. Chem.* **32**, 217 (1960).

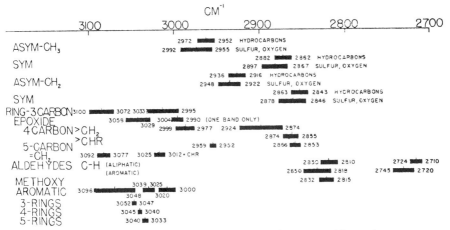

FIG. 13.2   Correlation chart for the carbon-hydrogen stretching region.

range shown for oxygenated and sulfur-containing compounds. The $=CH_2$ group in compounds containing $RHC=CH_2$ groups absorbs from 3192–3077 cm^{-1} while the $=CHR$ group absorbs in the 3025–3012 cm^{-1} range. The CH group in aldehydes gives rise to two bands at approximately 2820 and 2720 cm^{-1}. In aromatic aldehydes the bands occur at somewhat higher values.

Aromatic compounds contain one or more bands between 3100 and 3000 cm^{-1} and hence are readily distinguished from aliphatic compounds. As mentioned, the main exceptions are cyclopropyl derivatives, epoxides, alkenes, and compounds such as chloroform where the inductive effect of the chlorine atoms shifts the C—H stretching vibration to 3019 cm^{-1}. In general, for unsubstituted fused rings the aromatic C—H stretching vibration shifts to lower frequencies as the number of fused rings increases. For further details on the spectra of polynuclear aromatic compounds in the C—H stretching region, the paper by Wiberley and Gonzalez[21] should be consulted.

13.27   Sodium Chloride Region Correlation Charts

The correlation charts for this region are shown in Fig. 13.3(a)–(f). Since the previous chapters have discussed this region in great detail, additional comments on these correlations are not warranted.

[21] S. E. Wiberley and R. D. Gonzalez, *J. Appl. Spectrosc.* **15**, 174 (1961).

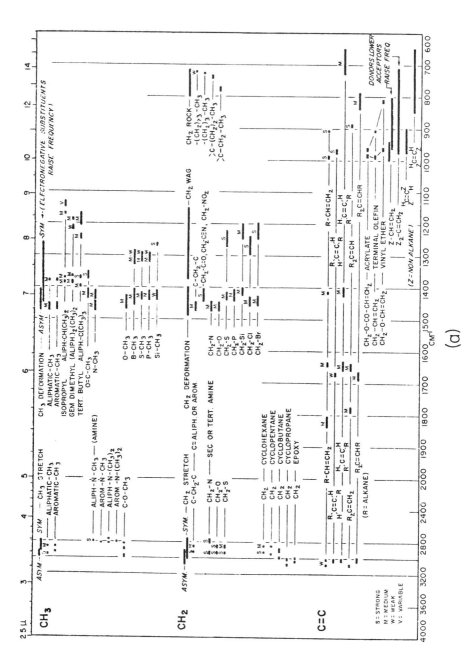

FIG. 13.3(a–f).   Correlation charts for the rock salt region.

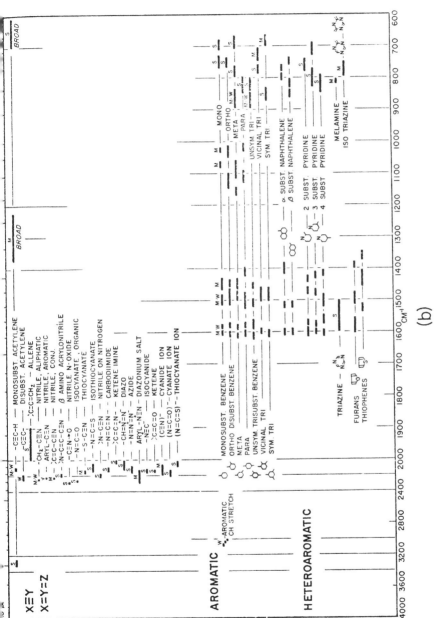

(b)

FIG. 13.3(b)

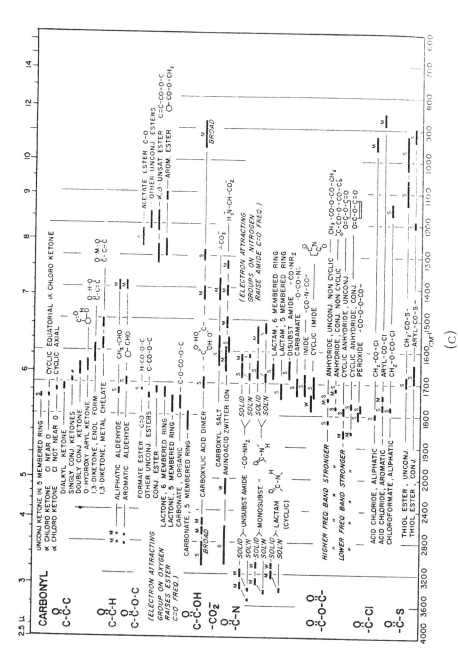

FIG. 13.3(c)

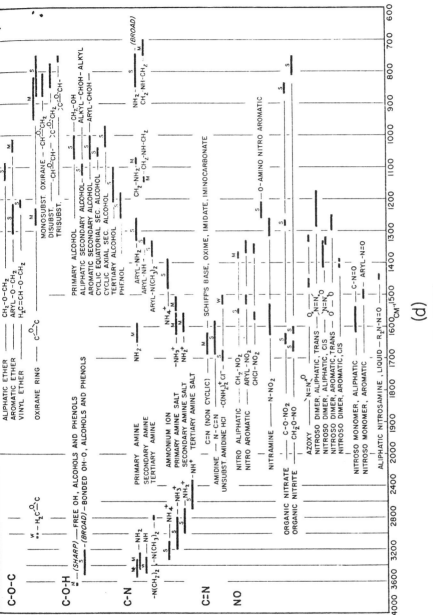

(d)

FIG. 13.3(d)

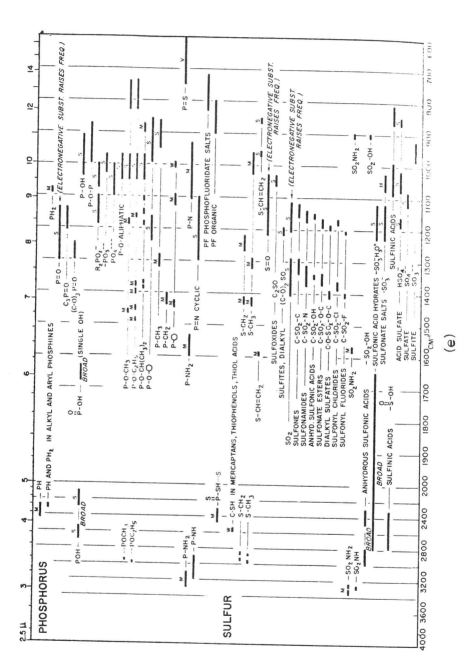

FIG. 13.3(e)

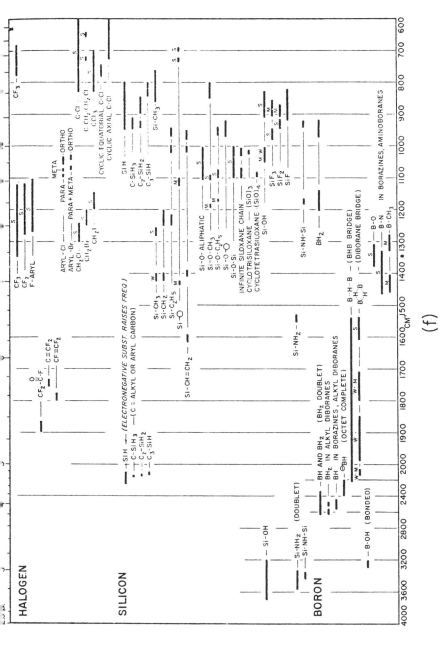

FIG. 13.3(f)

13.28   Low Frequency Infrared Region

Sodium chloride prism spectrometers have a lower limit of about 650 cm^{-1}, but grating instruments used with CsBr or CsI windows have extended this region to 250 or 200 cm^{-1}, respectively. The region beyond this, 200–10 cm^{-1}, is sometimes called the far infrared. The region beyond 10 cm^{-1} is studied by microwave techniques.

Group frequencies in the cesium bromide region have been extensively covered in the book by Bentley et al.[22] Group frequencies in this region include halogen stretching frequencies, aromatic ring bending frequencies, and various skeletal bending frequencies. Applications of low frequency infrared have been discussed in the book by Finch et al.[23] and in the review article by Brasch et al.[24] Some low frequency vibrations studied include the stretching of the hydrogen bond (O $\cdots$ H), methyl torsion vibrations, and crystal latice vibrations. Metal–ligand vibrations which occur in this region have also been covered in the book by Adams.[25]

13.29   Selected Infrared Spectra Illustrating Functional Group Frequencies

(a) PURPOSE OF THE SPECTRA

This collection of infrared spectra was run on an NaCl prism spectrometer. It does not pretend to be a complete library of spectra by any means. Its purpose is to provide one or two examples of most of the important functional group frequencies used in qualitative analysis. These are labeled directly on the spectra.

Since these spectra are not numerous enough to establish group frequencies by correlations in most cases, the order of the spectra has been arranged so the group frequency assignments given will seem reasonable by interspectral comparisons. The percent transmission coordinate has been condensed to

[22] F. F. Bentley, L. D. Smithson, and A. L. Rozek, "Infrared Spectra and Characteristic Frequencies ~ 700–300 cm^{-1}." Wiley (Interscience), New York, 1968.

[23] A. Finch, P. N. Gates, K. Radcliffe, F. N. Dickson, and E. F. Bentley, "Chemical Applications of Far Infrared Spectroscopy." Academic Press, New York, 1970.

[24] J. W. Brasch, Y. Mikawa, and R. J. Jakobsen, Appl. Spectrosc. Rev. 2, 187 (1968).

[25] D. M. Adams, " Metal-Ligand and Related Vibrations." St. Martin's Press, New York, 1968.

facilitate interspectral comparisons and to maximize the total number of spectra presented. The spectra are not intended to be used for quantitative analysis and are not suitable for this purpose.

In a few cases, enough spectra are presented of a group to illustrate a group frequency correlation. Examples of this would include groups such as methyl, tertiary butyl, phenyl, and a few others. Spectra 1–12 illustrate a spectra-structure correlation of a long chain aliphatic group.

(b) SPECTRAL PRESENTATION

The spectrum number is given in the upper left corner of each spectrum. In the lower left corner is given the sample thickness or preparation, explanations of which are given below.

0.015	liquid, 0.015 mm thick
0.17, 0.01	liquid, two thicknesses, 0.17 mm and 0.01 mm
cap	liquid between plates without spacers
smear	viscous liquid, between plates without spacers
10 cm, 25 mm,	gas run in 10 cm cell at 25 mm pressure
0.08 M, CCl_4-CS_2, 1 mm	solution, 0.08 M concentration in 1 mm cell run in CCl_4 from 4000–1300 cm^{-1} and in CS_2 from 1400–650 cm^{-1}, solvent bands compensated
N	solid run as Nujol mull, Nujol bands marked N
N, HC	solid run as a Nujol mull and a halocarbon mull, halocarbon used only, at 3100–2750 and 1490–1350 cm^{-1}, the rest of the spectrum being that of the Nujol mull
N, HCBD	same as above substituting hexachlorobutadiene for halocarbon (both are satisfactory)
N, melt	same as above substituting a melted and resolidified film for HC or HCBD
KBr	solid run in a KBr pressed pellet
molten	low melting solid run in the liquid molten state
melt	solid run as a melted and resolidified film

(c) BAND ASSIGNMENT

The assignments on the spectra are for the most part those discussed and referenced in the text. The description of the vibrations are necessarily brief. A description such as "C—O stretch" does not necessarily mean that the C—O bond stretches while the rest of the molecule stands still. It implies that the vibration giving rise to the band involves stretching of the C—O bond but may involve other bonds as well.

Certain vibrations have been drawn out with arrows representing in-plane motion and "plus" and "minus" signs indicating out-of-plane motion.

In some cases such as the phenyl group (ϕ), the whole group is indicated as being the origin of the band rather than always specifying the vibration within the group.

A dotted line in the description "OH def---C—O str" implies interaction so that at least to some extent both bands involve both OH deformation and C—O stretch.

In some cases where certain vibrations are inactive in the infrared, the Raman frequency of the vibration in question is indicated.

Significant infrared bands beyond the NaCl region such as C—Br and C—I bands are indicated on the spectra.

The abbreviations used include:

asym   asymmetric
sym   symmetric
str   stretch
def   deformation-scissors motion of an NH_2 or CH_2 group
wag   a bending motion out of the plane of the group
rk   rock—a bending motion in the plane of the group in the CH_2 case; in the CH_3 case, a rotation of the CH_3 group opposed by countermotion of the attached atom
adj   adjacent-applied to hydrogens on neighboring carbon atoms in an aromatic ring
ar   aromatic ring

(d) TABLE OF CONTENTS OF FUNCTIONAL GROUPS ILLUSTRATED IN THE SPECTRA

The number of the spectrum which is the primary example of a given functional group is given to the left of the group. Due to the fact that most molecules contain more than one functional group, secondary examples are given as cross references after the functional group.

1–39 Aliphatic Groups

17	methyl group on oxygen, cf. 44, 56, 150, 151
18	methyl group on nitrogen, cf. 57, 58
19	methyl group on iodine
—	methyl group on silicon, cf. 473, 474, 479
—	methyl group on phosphorus, cf. 495, 498
—	methyl group on sulfur, cf. 55, 61, 338, 518, 519, 536, 540, 556, 561
20–30	the group $Z-CH_2-CH_2-Z$ where Z represents different functional groups
31–33	cyclic CH_2 (6-, 5-, and 4-membered ring), cf. 20, 81, 84, 116–118, 303
34–39	cyclopropane ring

40–63 Triple Bonds

40	acetylene, monosubstituted, unconjugated, cf. 26
42, 43	acetylene, monosubstituted, conjugated with aryl group
44	acetylene, monosubstituted, conjugated with double bond
45	acetylene, monosubstituted, conjugated with carbonyl
41, 47	acetylene, symmetrically disubstituted
46	acetylene, unsymmetrically disubstituted
48–49	nitrile, unconjugated, cf. 27, 33, 34, 494–497, 543, 547, 620
48–49	$CH_2-C\equiv N$, cf. 27
50–52	nitrile with electronegative group in the α position
53–54	nitrile, conjugated, cf. 101–104, 343–346, 395, 396
55–57	nitrile, conjugated, with group in the β position, cf. 111, 112
58	nitrile on nitrogen cf. 343–346
59–60	isocyanide
61	thiocyanate, organic
62	cyanide, inorganic
63	diazonium group

64–72 Cumulated Double Bonds $X=Y=Z$

64	allene group
65	isocyanate, organic
66	isothiocyanate, organic
67	cyanate, inorganic
68	thiocyanate, inorganic
69	carbodiimide
70–71	azide
72	diazoacetate

73–84 Carbon–Carbon Double Bonds

73–74	vinyl group, cf. 25, 53, 79, 110, 198, 221, 252, 263, 264, 283, 311, 475, 541, 550, 572, 573
75	*trans*-disubstituted ethylene, cf. 57, 111, 200, 201, 378
76	*cis*-disubstituted ethylene, cf. 55, 56, 81, 112, 202
77	vinylidene group (1,1-disubstituted ethylene), cf. 54, 80, 113, 124, 199
78	trisubstituted double bond
79–80	double bond, aryl conjugated
81–84	cyclic double bond

85–104 Aromatic Rings

85–96	benzene ring, methyl substituted, *o*, *m*, *p*, etc.
85	monosubstituted benzene, cf. 32, 42, 43, 60, 63, 65, 66, 79, 80, 101, 114, 115, 135, 136, 148, 149, 160, 165, 167, 172, 203, 204, 222, 233, 305–309, 354, 465–472, 487–492, 621, 622
86	*o*-disubstituted benzene, cf. 102, 150, 159, 161, 286, 355, 358, 535, 574–578, 580, 581, 604, 605, 610
87	*m*-disubstituted benzene, cf. 103, 223, 356, 359, 582, 583, 606, 607, 611
88	*p*-disubstituted benzene, cf. 71, 104, 137, 151, 162, 275, 276, 357, 360, 366, 446, 448, 534, 558, 560, 564, 584, 585, 608, 609, 612
89–96	more highly substituted benzene ring, cf. 138
97–100	condensed rings, cf. 365
101–104	benzonitrile, methyl substituted

105–124 Ethers and Related C—O—C Groups

105–106	ether, aliphatic, unbranched and branched on α-carbon, cf. 3, 18, 21, 224, 225
107–109	orthoformate, orthoacetate, acetal
110–113	vinyl ether, cf. 44
114–115	aromatic ether, cf. 150, 151
116–118	cyclic ether, (6-, 5-, and 4-membered ring), cf. 21
119–124	epoxy ring

125–141 Alcohols, Phenols, and Related OH Compounds

125–126	primary alcohol, cf. 4, 229, 403, 586
127–128	secondary alcohol, cf. 226–228
129–130	tertiary alcohol, cf. 170
131–132	cyclic secondary alcohol

133–134	cyclic tertiary alcohol
135–138	phenol, cf. 161, 162
139–140	poly–ol, cf. 23
141	water

142–144 Peroxides

142	hydroperoxide
143	peroxide
144	perester

145–151 Aldehydes

145–146	aldehyde, aliphatic, cf. 5
147	aldehyde, electronegative substituent
148–151	aldehyde, conjugated
150	benzaldehyde, *ortho* substituted

152–177 Ketones

152–155	ketone, aliphatic, cf. 28
152–153	$CH_3-C=O$, cf. 155–157, 160–164, 171, 180–185
153–156	$CH_2-C=O$, cf. 145, 173–177
156	ketone, α-chloro
157	ketone, conjugated
158	ketone, doubly conjugated
159	ketone, doubly conjugated in strained ring
160	ketone, aromatic
161–169	ketone, hydrogen bonded and conjugated (hydroxy acetophenones, enols of 1,3-diketones, tropolone, etc.)
170	ketone-alcohol
171–172	1,2-diketone
173–177	ketone, cyclic (8–4-membered rings)

178–218 Esters

178–179	formate
178	$CH_3-O-C=O$, cf. 180, 186, 189, 190, 198–200, 205–207
179	$C_2H_5-O-C=O$, cf. 45, 181, 191, 194, 196, 197 201–203, 208
—	$n\text{-}C_3H_7-O-C=O$, cf. 187
180–181	acetate
182–183	acetate, branched substituent

184–185	acetate, election withdrawing substituent, cf. 52, 144
186–187	propionate
188	butyrate
189	isobutyrate
190–192	long chain ester, cf. 7
193	α-chloroester, election withdrawing substituent
194	oxalate
195	dithiooxalate
196	malonate
197	succinate
198	acrylate
199	methacrylate
200	crotonate
201	fumarate
202	maleate
203	benzoate
204	thiobenzoate
205	phthalate
206	isophthalate
207	terephthalate
208	carbonate (organic)
209–210	xanthate ester and salt
211	trithiocarbonate (organic)
212–215	cyclic lactones (6–4-membered rings)
216–217	cyclic carbonate
218	cyclic trithiocarbonate
—	propiolate, cf. 46
—	diazoacetate, cf. 72
—	peracetate, cf. 144
—	acetoacetate, cf. 169

219–241 Carboxylic Acids and Salts

219–223	carboxylic acid dimer, cf. 6, 46, 47
224–229	carboxylic acid with alternate hydrogen bonded forms, cf. 238, 239
230	carboxylic acid, electronegative substituent
231–234	carboxyl salt
235–237	amino acid zwitterion
238–239	amino acid hydrochloride
240	carbonate salt
241	bicarbonate salt

242–248 Anhydrides

242–243	anhydride, noncyclic, unconjugated
244–245	anhydride, noncyclic, conjugated
246	anhydride, cyclic, unconjugated
247–248	anhydride, cyclic, conjugated

249–265 Amides and Related Compounds

249–256	amide, unsubstituted, cf. 378, 587
254	thioamide, unsubstituted
257–259	amide, N-monosubstituted, noncyclic
260–261	amide, N-monosubstituted, cyclic lactam
262	amide, N-disubstituted
263–264	lactam, N-vinyl
265	hydroxamic acid

266–270 Carbamates

266–267	carbamate, N-unsubstituted
268	carbamate, N-monosubstituted
269	thiocarbamate, N-monosubstituted
270	carbamate, cyclic

271–281 Ureas

271	urea, monosubstituted
272–273	urea, N,N-disubstituted
274–277	urea, N,N'-disubstituted
276	urea, N,N'-disubstituted with electron withdrawing group
278–281	thiourea

282–291 Imides and Related Compounds

282–283	imide, noncyclic
284–286	imide, cyclic
287–288	hydantoin
289	N-acetyl benzamide
290	biuret
291	dithiobiuret

292–297 Chloro Carbonyl Compounds

292–293	acid chloride, unconjugated
294	acid chloride, aryl conjugated
295–296	chloroformate
297	chloroformamide

298–320 Amines and Amine Salts

298–301	primary amine, aliphatic, cf. 24
302–303	secondary amine, aliphatic, cf. 22
304	tertiary amine, aliphatic, cf. 17
305	aniline, *N*-unsubstituted, cf. 358–360, 535, 581, 583, 585
306–307	aniline, *N*-monosubstituted
308–309	aniline, *N*-disubstituted
310	ethylene imine
311	*N*-vinyl, aromatic amine
312	hydroxylamine
313	ammonium ion
314–315	primary amine salt, aliphatic, cf. 235–239
316	secondary amine salt, aliphatic
317	tertiary amine salt, aliphatic, cf. 234
318	quaternary amine salt, aliphatic
319–320	aniline salt, cf. 516, 517

321–353 C=N Compounds

321–323	C=N aliphatic substituents
324	C=N aromatic substituents
325–327	oxime
328–329	amide oxime
330–332	imidate
331	imidate salt
333	iminocarbonate
334–335	amidine salt
336	substituted formamidine
337	isourea salt
338	isothiourea salt
339–340	guanidines and salt
345	isourea, *N*-cyano
346	isothiourea, *N*-cyano
347–348	biguanide and salt
349–350	S₂C=N and salt
351–353	cyclic N=C—N

354–373 N=O Compounds

354–360	nitro, aromatic, cf. 276, 386, 387
361–362	nitro, aliphatic, cf. 367
363	nitrate, organic, cf. 8
364	nitramine
365–367	*C*-nitroso group

368	nitrite, organic, cf. 9
369–371	*N*-nitroso group
372	nitrate ion
373	nitrite ion

374–381 N=N Compounds

374–375	azo group
376	azoxy group
377–379	azocarbonyl compounds
380–381	N—N=N

382–423 Nitrogen Heteroaromatics

382–390	pyridine rings
383	pyridine hydrochloride
384–385	pyridine, *N*-oxide
391–394	quinolines and isoquinolines
395–396	pyrimidines
397	pyridazides
398	pyrazines
399–418	*s*-triazines, normal (3 double bonds in ring) and iso (one or more external double bonds), cf. 579
419–423	*s*-tetrazines, symmetrically and unsymmetrically substituted

424–441 Five-Membered Rings with Two Double Bonds

424–426	cyclopentadiene derivatives, cf. 167
427–430	furan and thiophene rings
431–441	rings containing 1, 2, 3, and 4-nitrogens

442–465 Boron Compounds

442–443	borane
442–443	B—H, cf. 453–455, 457, 461
444	boric acid
444–451	B—O, cf. 452, 454
445–446	boronic acid
447–448	boroxole
449–451	borate ester
452–455	coordinated boron and boron salt
456–461	borazine
456–464	B—N
459–460	B—Cl, cf. 464, 465
462–464	aminoborane
464–465	chloroborane

466–480 Silicon Compounds

466–468	SiH_3, SiH_2, and SiH
466–472	Si-phenyl
469	Si—OH
470	Si—NH_2
471	siloxane (mono)
472–473	Si—Cl
473–474	$Si(CH_3)_3$
474	Si—NH—Si
475	Si-vinyl
476	silicate ester
477	trisiloxane, cyclic
478	tetrasiloxane, cyclic
479	siloxane polymer
480	silica gel

481–531 Phosphorus Compounds

481–483	phosphine, aliphatic
481	PH_2, cf. 487, 494
482	PH, cf. 485–488, 500, 509, 516
481–486	CH_2—P, cf. 493–497, 507, 508, 510–515, 520, 521
484–485	phosphine oxide, aliphatic
484–485	P=O, cf. 490, 496, 498, 500–503, 505, 507–514, 526, 527
486	phosphine sulfide, aliphatic
486	P=S, cf. 491, 497, 506, 520, 522, 524, 525
487–489	phosphine, aromatic
487–492	P-phenyl, cf. 498, 509, 516, 526–528
490	phosphine oxide, aromatic
491	phosphine sulfide, aromatic
492	phosphonium salt
493	P-ethyl
495	P-methyl, cf. 498
494–497	cyanoethylphosphine, oxide, and sulfide
498	dimethylphenylphosphine oxide
499–500	CH_3—O—P, cf. 510, 519, 524, 525
501	C_2H_5—O—P, cf. 507, 616–619
502–503	n-C_4H_9—O—P, cf. 506
—	$(CH_3)_2CH$—O—P, cf. 522, 523
503	P—O—P, cf. 512
504–505	phenyl-O—P
500–503	C—O—P=O, cf. 505, 507
506	dithiophosphoric acid ester

507	phosphinate ester
508–512	phosphinic acid and other POOH acids
513–514	phosphonic acid
515–519	acid salts
518–519	CH_3-S-P
520–523	dithio acids and salts
524	thiophosphoryl chloride
525–528	$P-N$
529–531	inorganic salts

532–578 *Sulfur Compounds*

532	mercaptan
533–535	thiophenol
536–539	sulfides and disulfides, aliphatic
536	CH_3-S, cf. 540, 542, 556, 561
537–538	C_2H_5-S
540	sulfide, phenyl
541	sulfide, vinyl
542–544	sulfoxide
545–546	sulfite ester
545–546	alkyl-$O-S$, cf. 552, 555, 565
547–558	$-SO_2-$
547–548	sulfone
549–551	sulfonamide
550	vinyl on oxidized sulfur
552	sulfonate ester
553–554	sulfonyl chloride
555	sulfate ester
556–558	sulfonic acid, anhydrous
559–560	sulfonic acid, hydrate
561–564	sulfonic acid salt
565	alkyl sodium sulfate
566	sulfinic acid salt
567–570	inorganic sulfate, bisulfate, sulfite, and bisulfite
571–578	thiazole ring
571–572	$C=S$, cf. 209, 211, 218, 254, 269, 278–281, 291, 412, 415, 575

579–588 *Fluorine Compounds*

579–585	CF_3 group, cf. 230, 234, 256
586	CF_2 group, cf. 587
587	trifluorovinyl group
588	NF

—   F-aryl, cf. 610–612

589–614 Chlorine, Bromine, and Iodine Compounds

589–591	chloromethanes
592–594	*n*-alkyl halides, cf. 10–12, 19, 29, 30
595–597	isopropyl halides
598–600	benzyl halides
601–603	other C—Cl groups, cf. 36–39, 49, 50, 147, 156, 193, 255, 399–401
604–612	aryl halides, cf. 275, 375, 376, 580, 582, 584
613–614	chlorate and bromate ions

615–620 Metal Carbonyls

615	metal carbonyl
616–619	metal carbonyl–phosphorus complexes
620	bridged metal carbonyl

621–622 Arsenic Compounds

623–624 Tin Compounds

Inorganic Ions

62	$C{\equiv}N^-$
63	BF_4^-
67	NCO^-
68	NCS^-
240	CO_3^{2-}
241	HCO_3^-
313, 521	NH_4^+
372	NO_3^-
373	NO_2^-
455	BH_4^-
529	PO_4^{3-}
530	HPO_4^{2-}
531	$H_2PO_4^-$
567, 320, 348	SO_4^{2-}
568	HSO_4^-
569	SO_3^{2-}
570	HSO_3^-
613	ClO_3^-
614	BrO_3^-

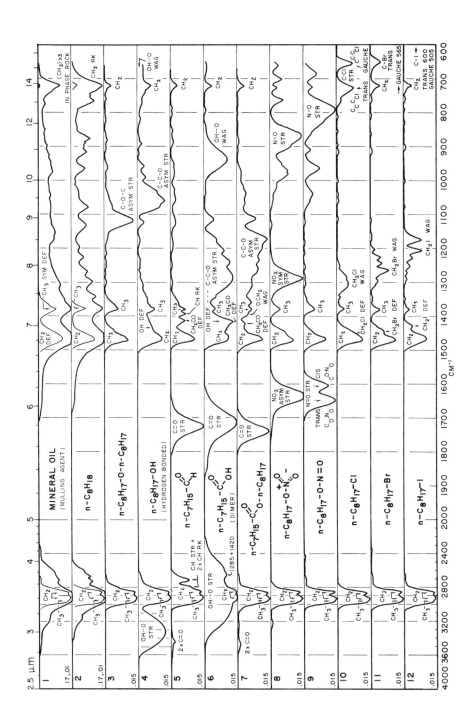

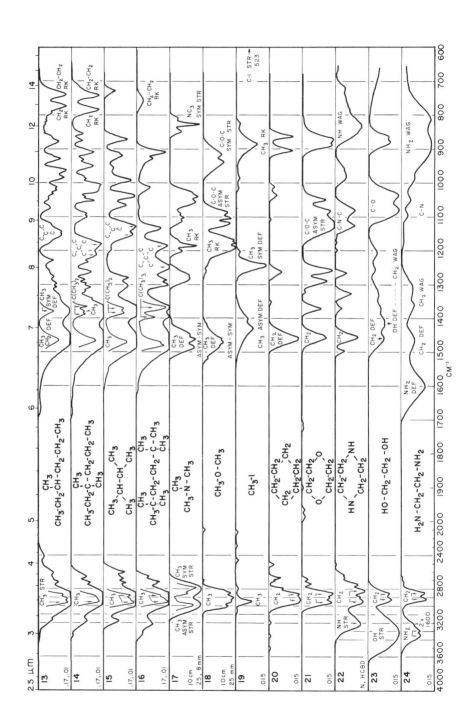

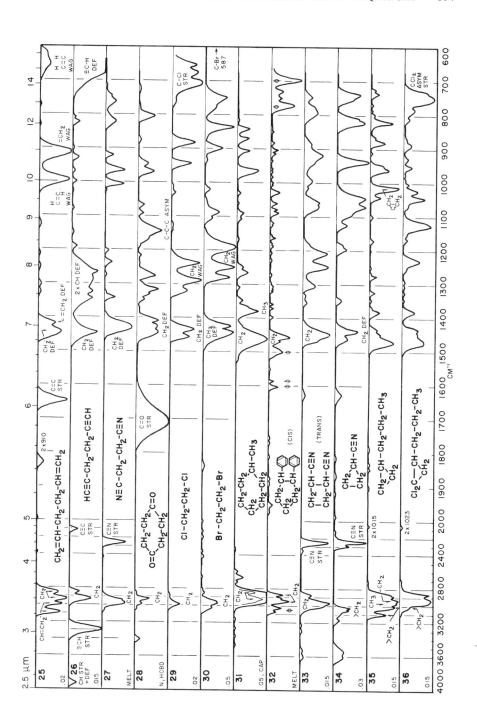

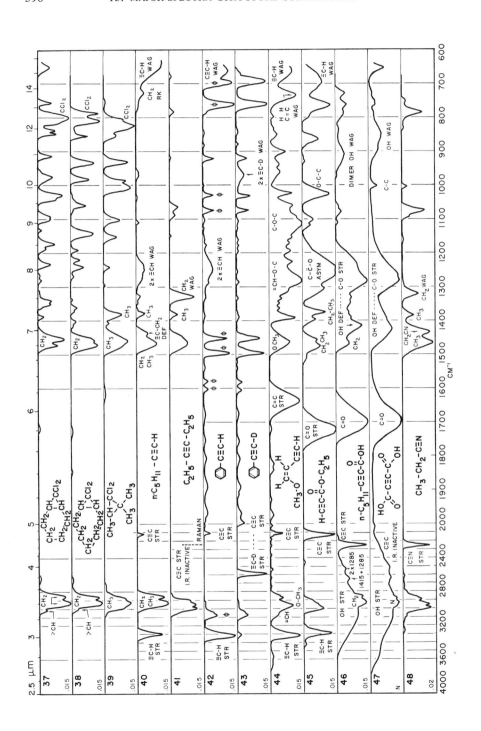

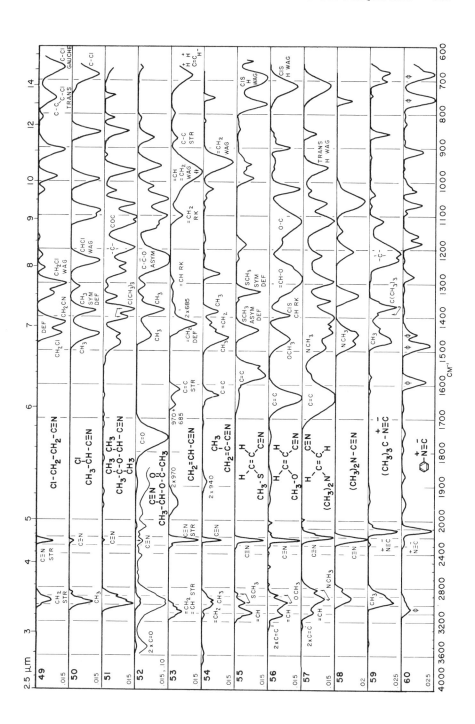

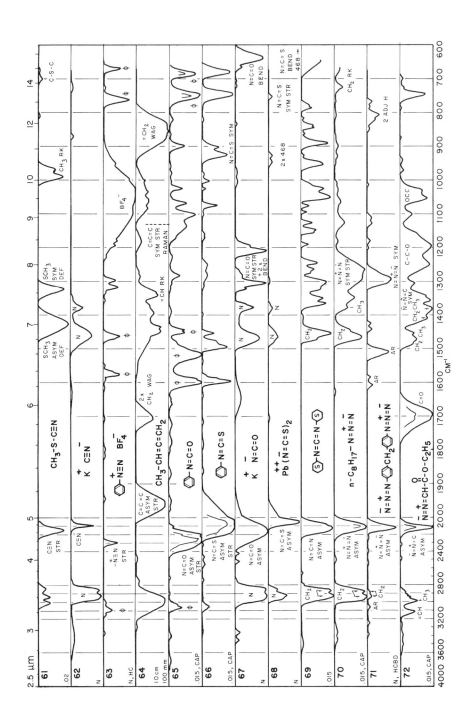

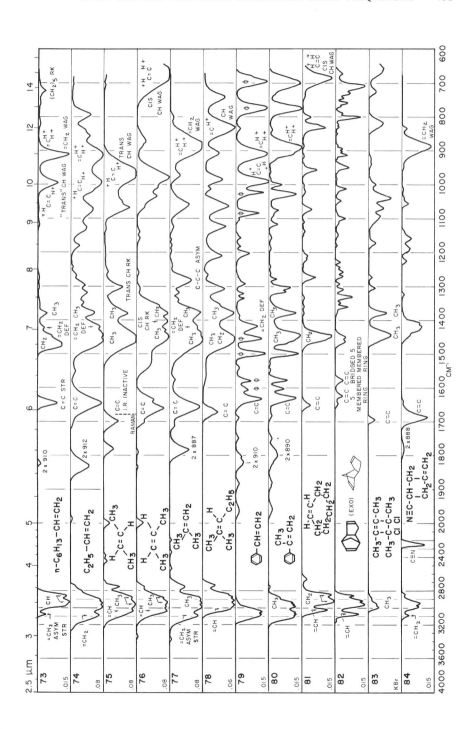

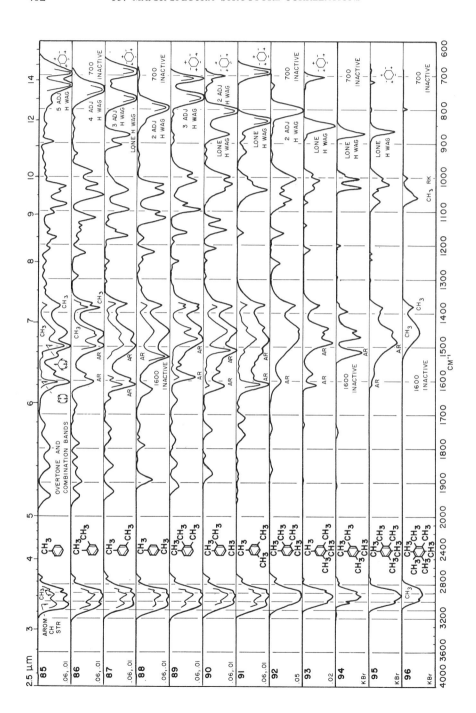

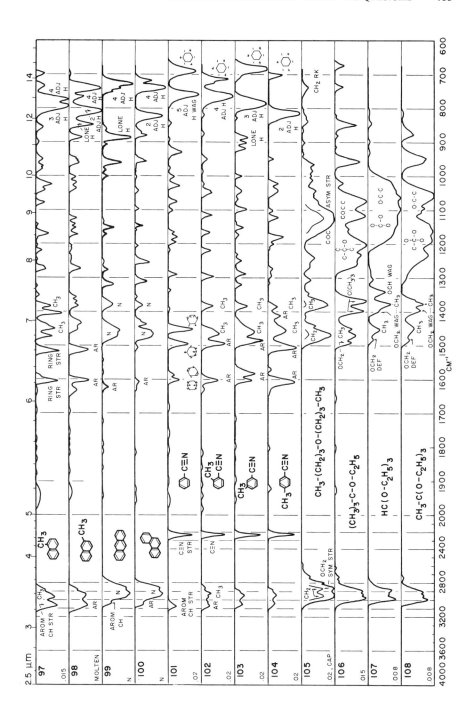

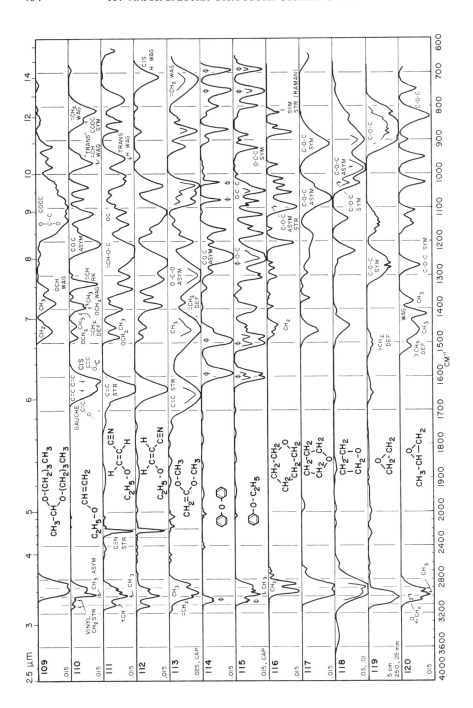

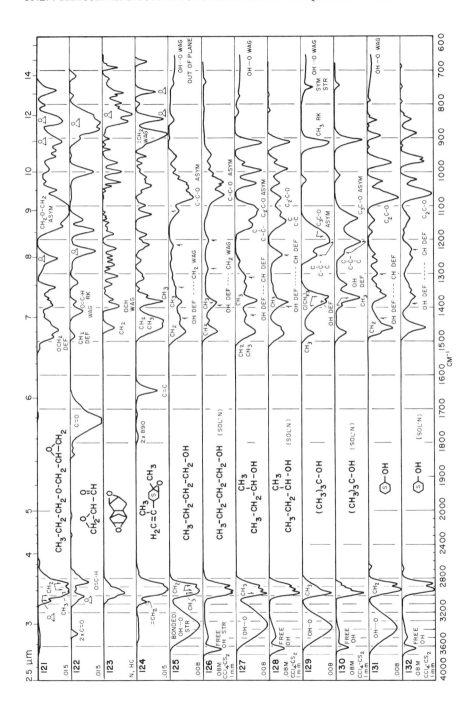

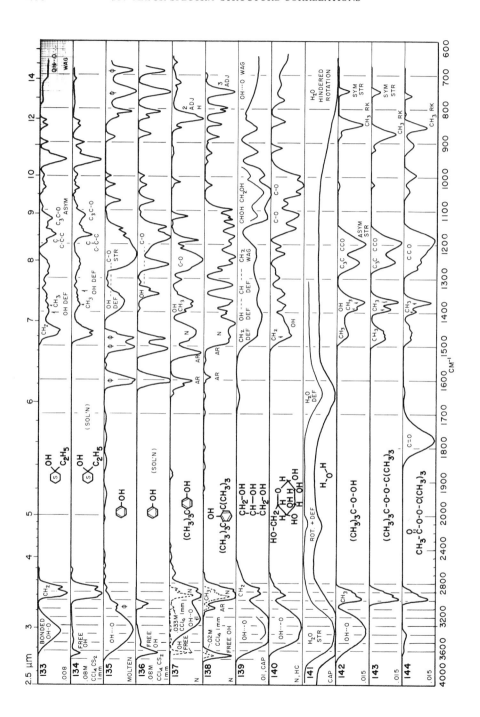

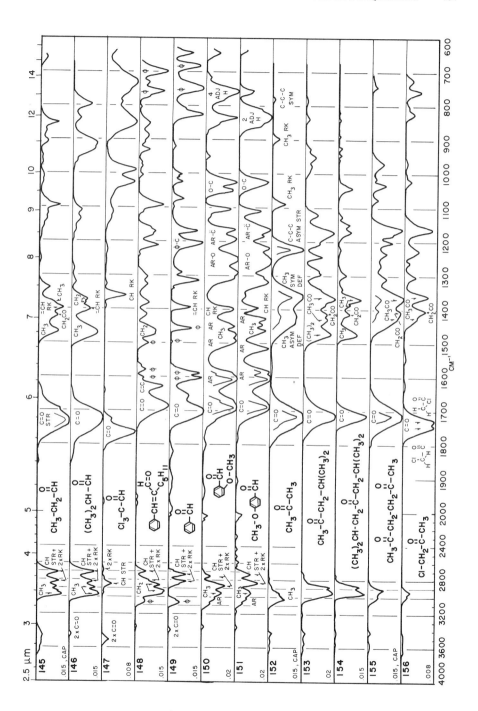

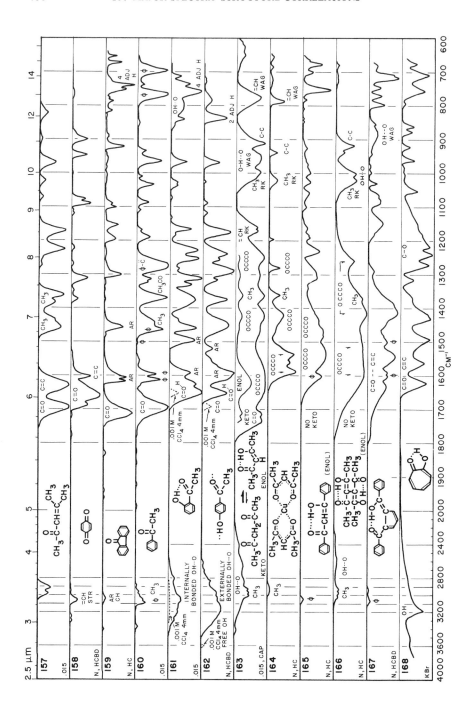

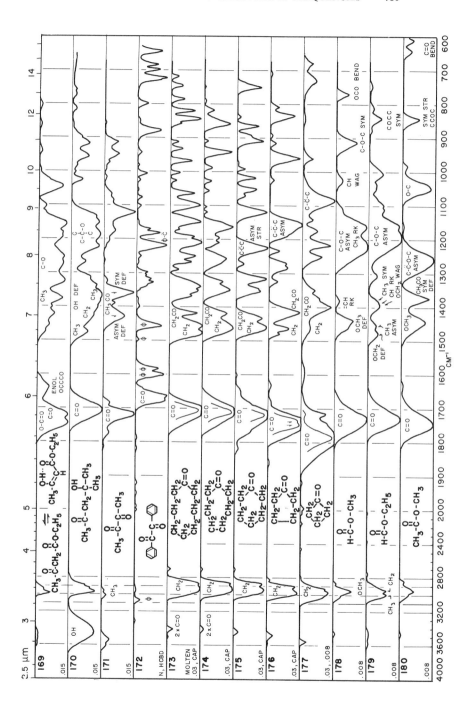

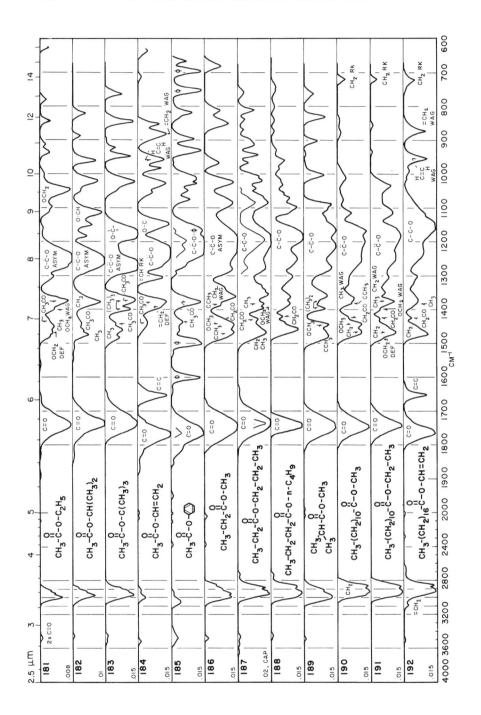

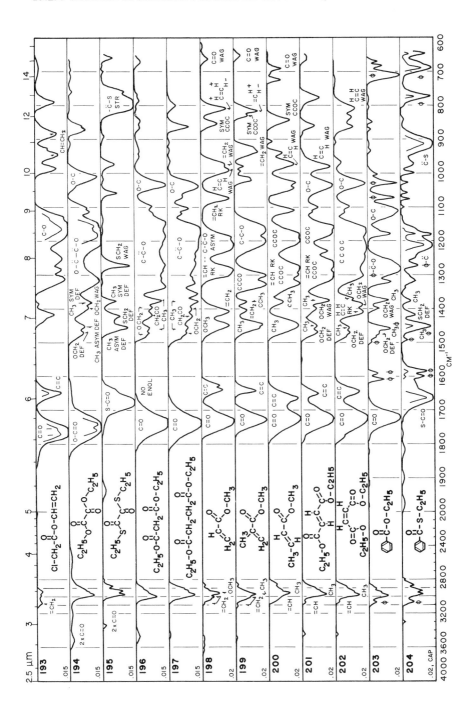

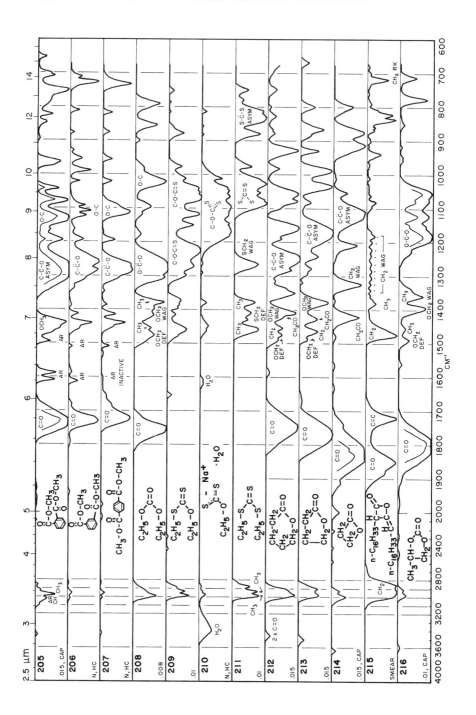

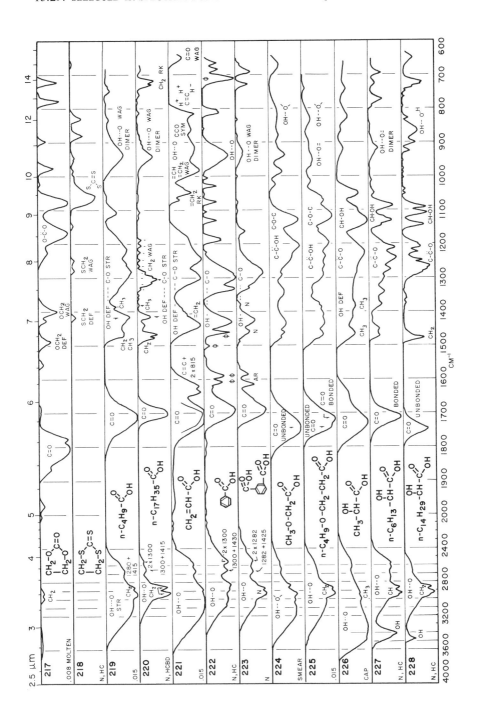

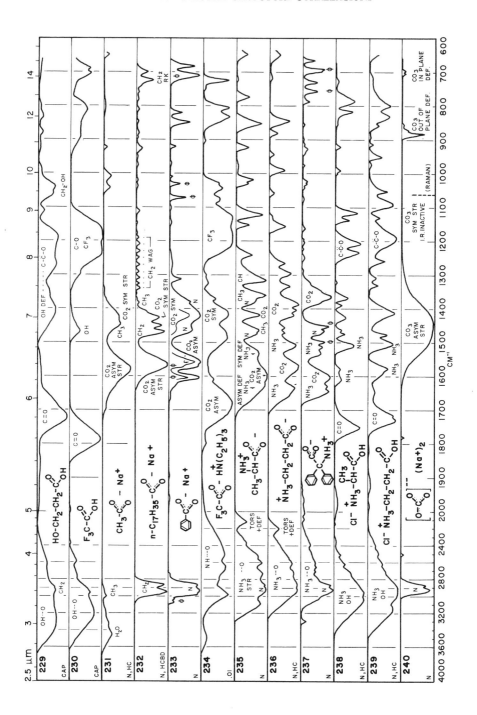

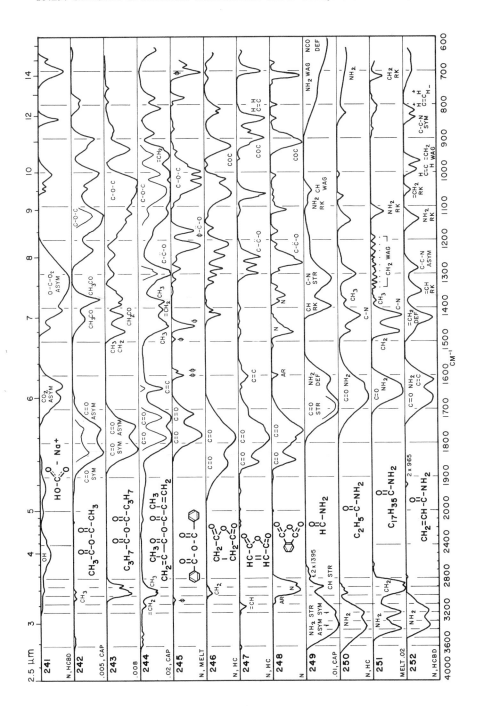

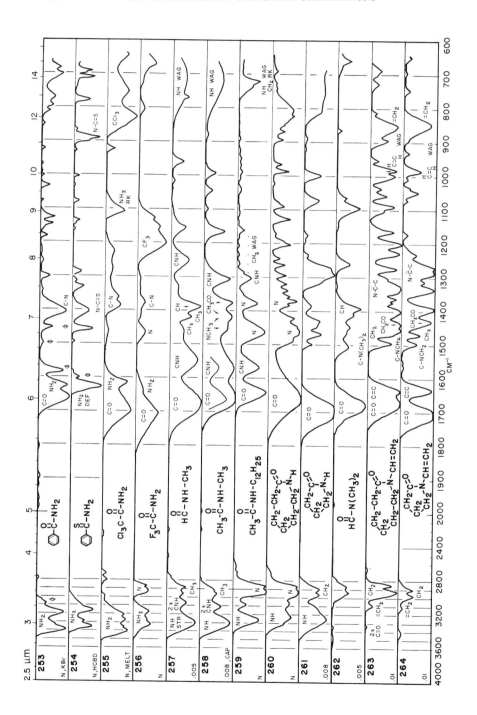

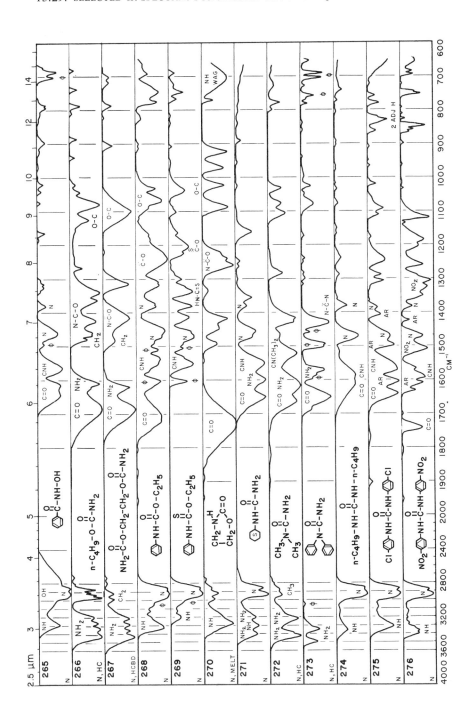

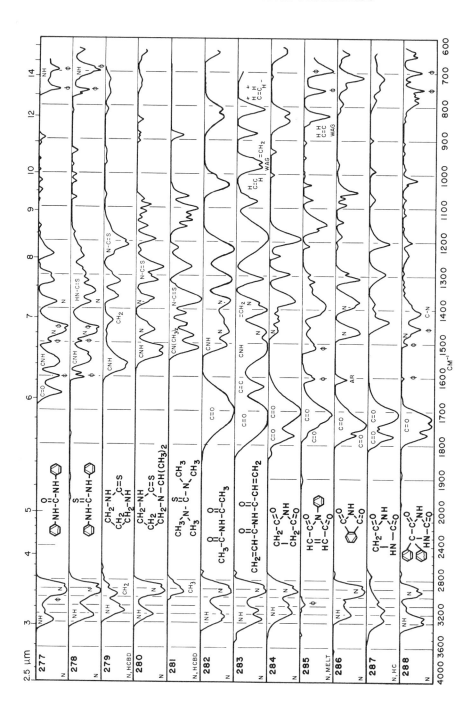

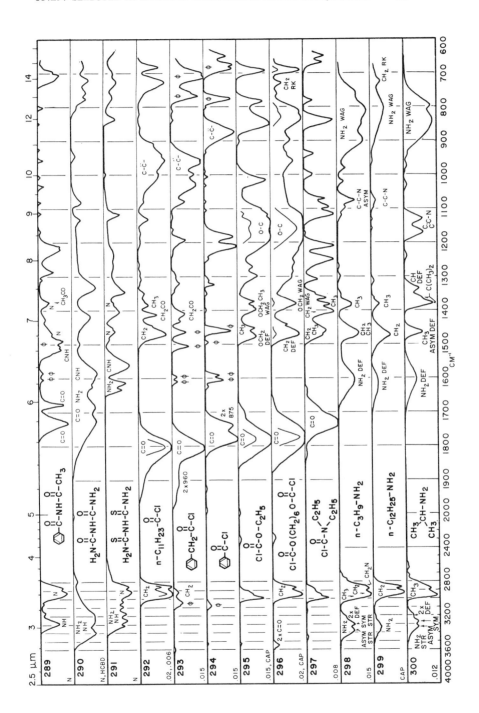

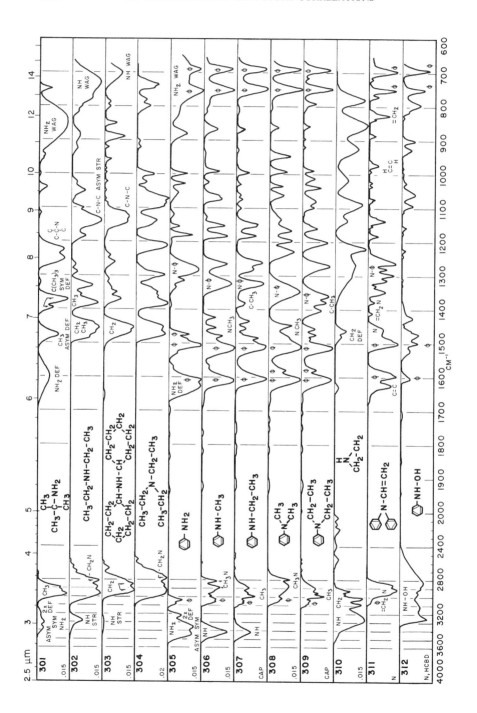

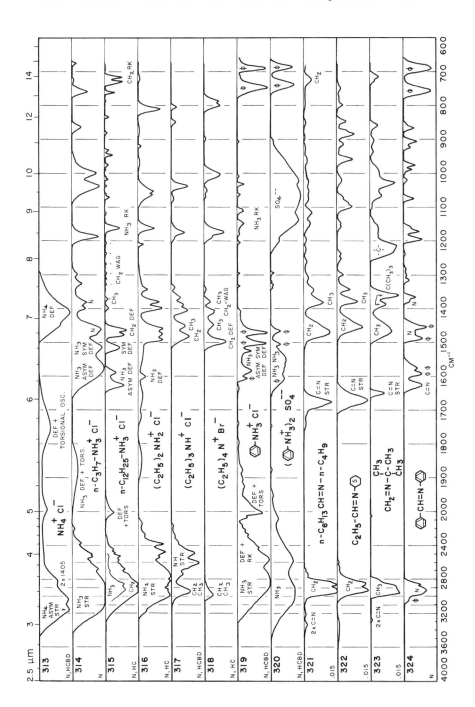

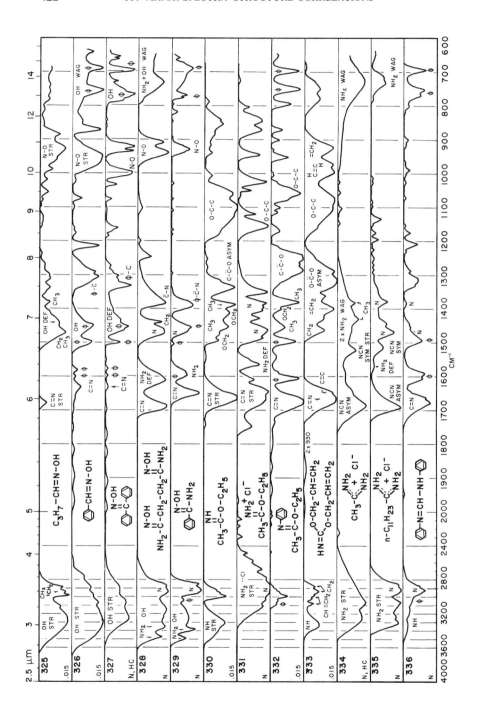

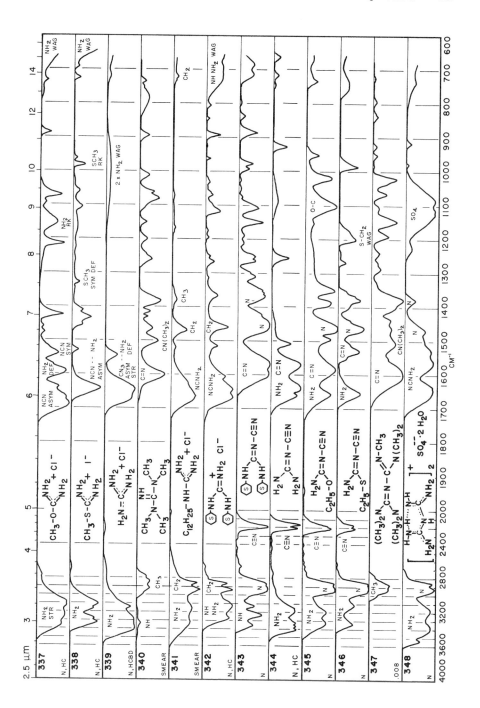

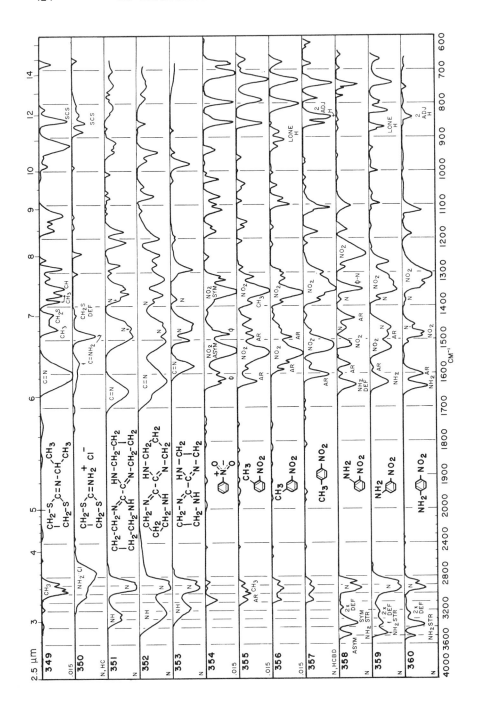

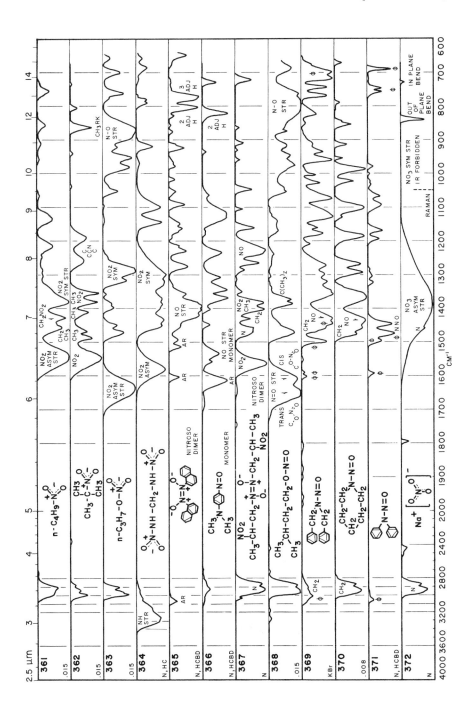

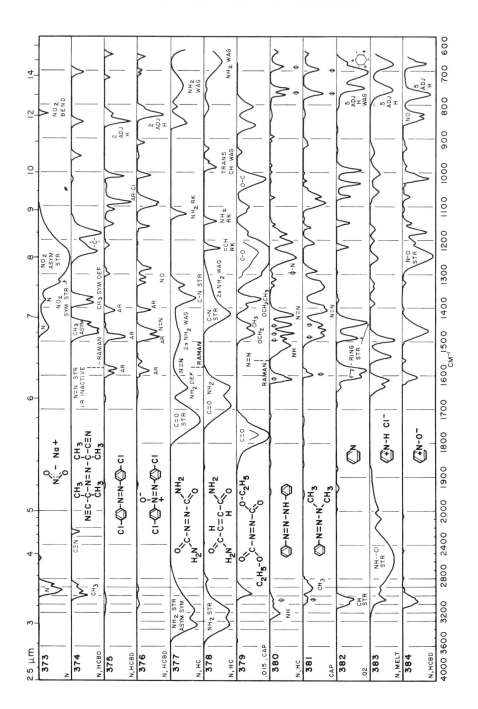

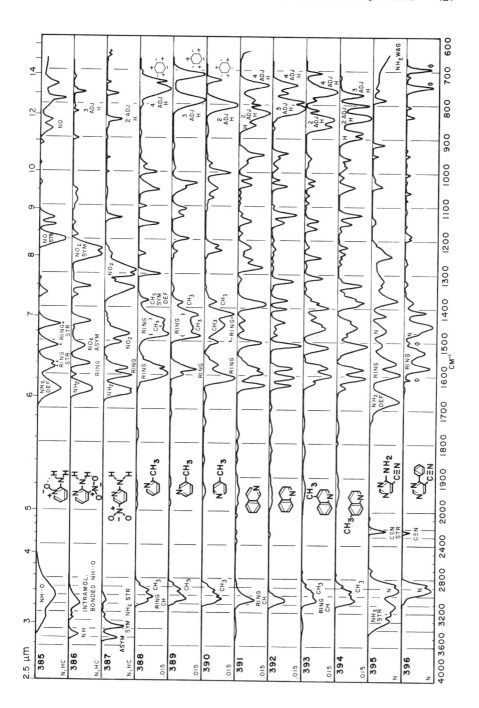

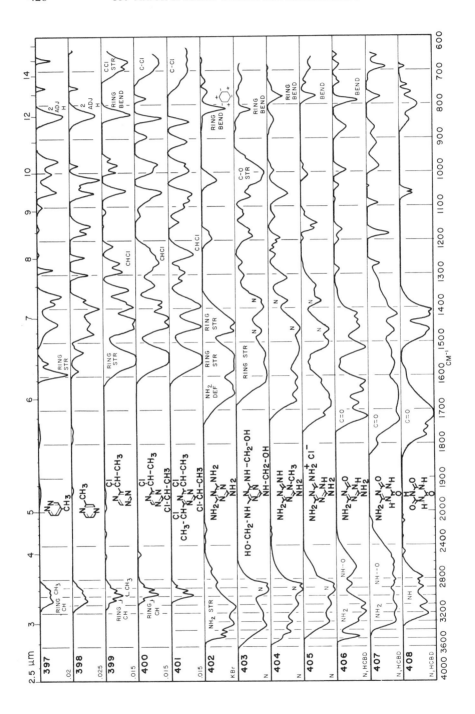

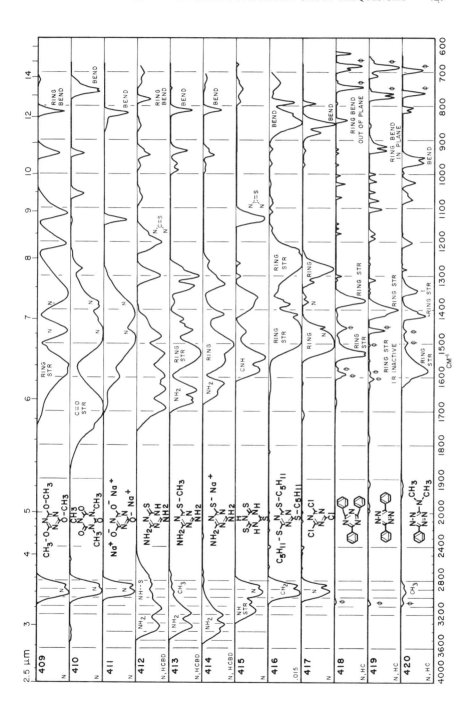

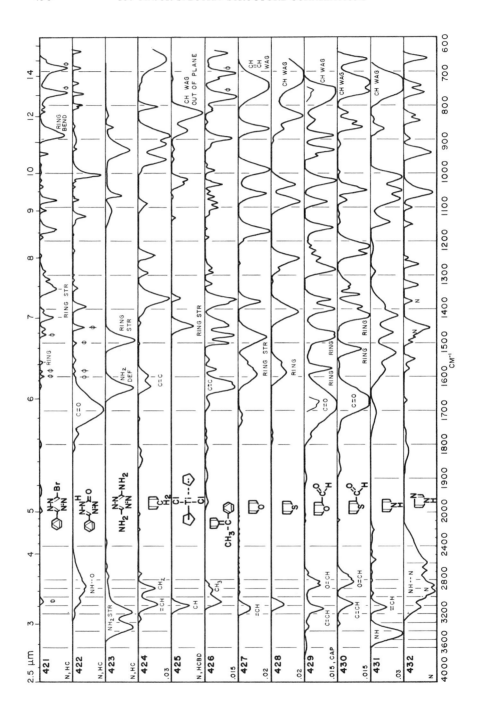

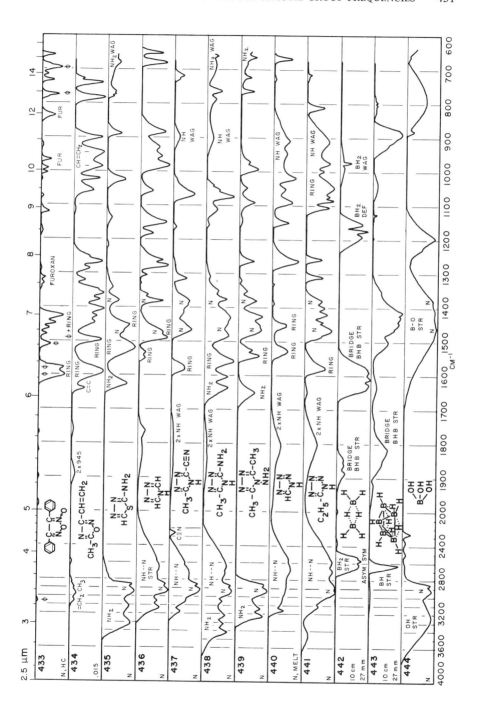

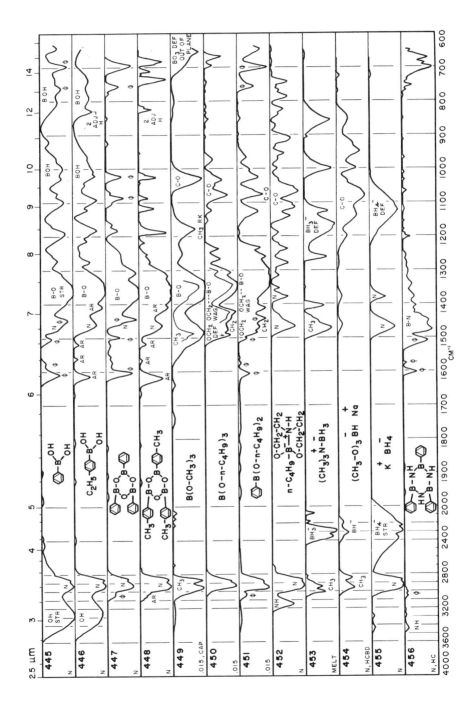

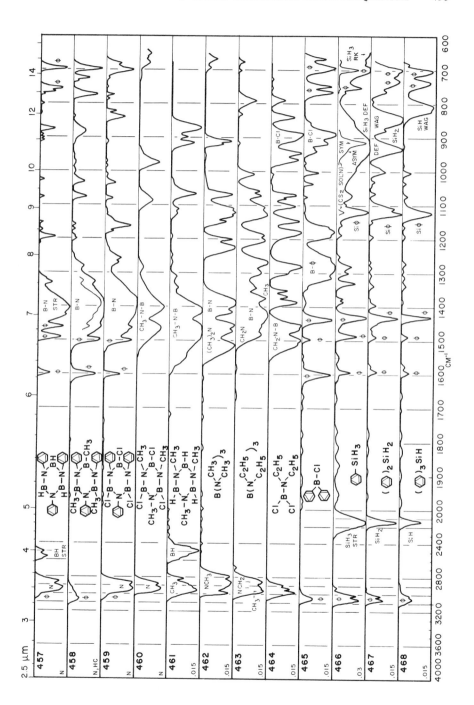

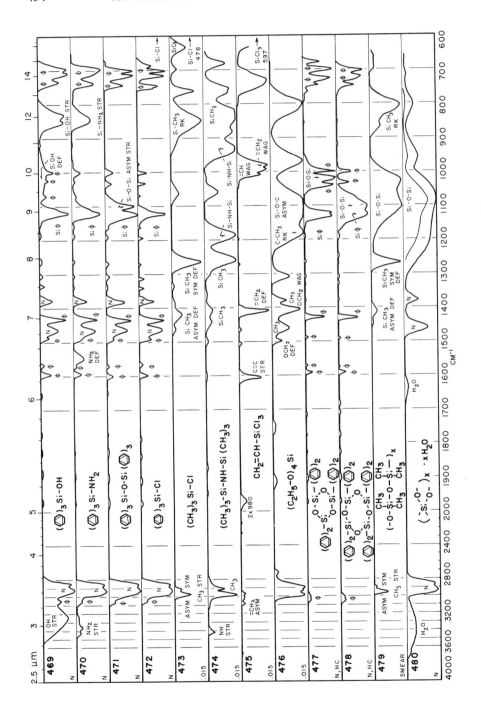

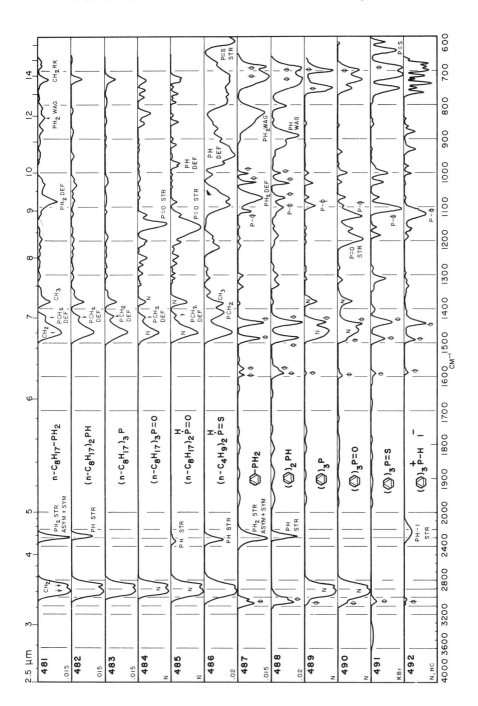

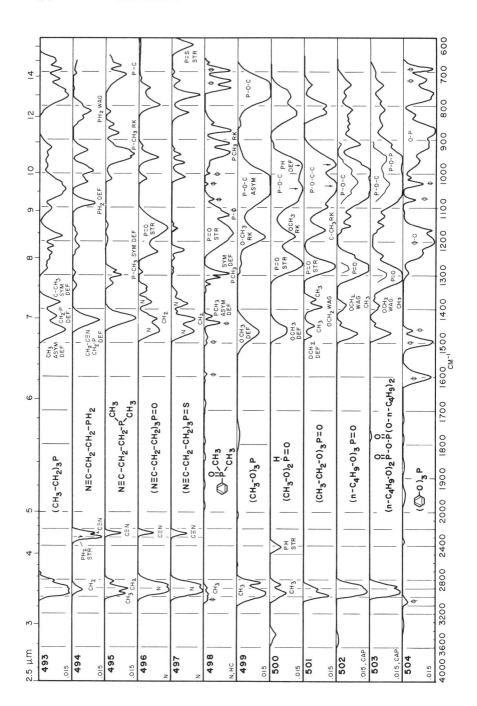

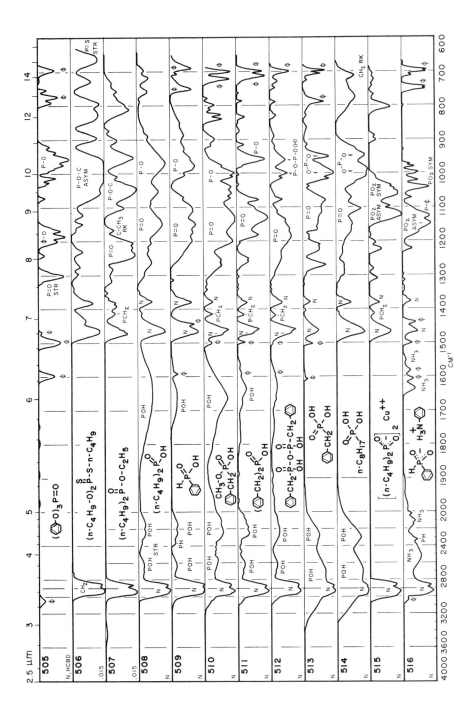

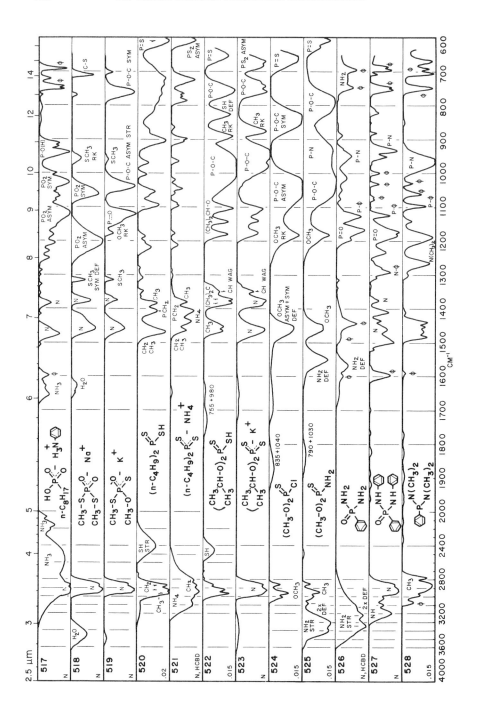

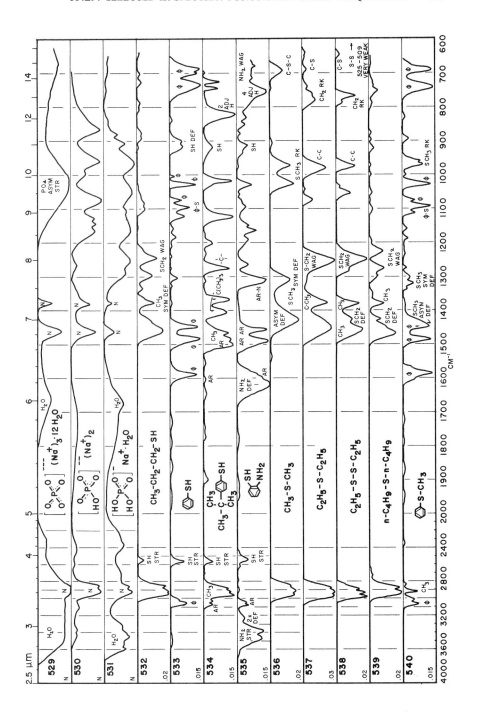

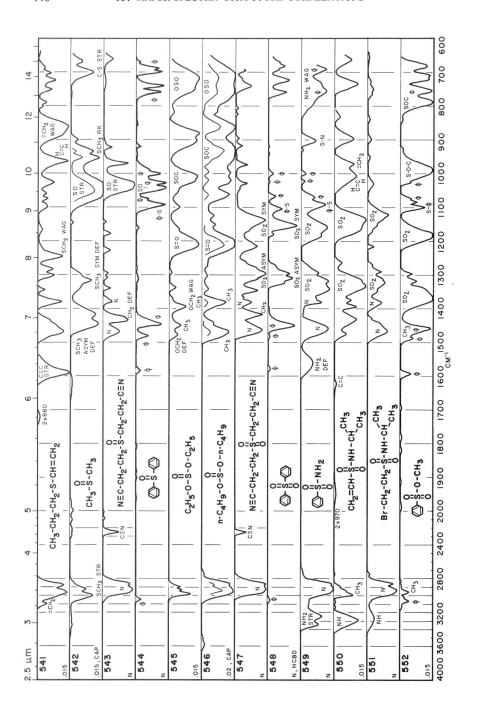

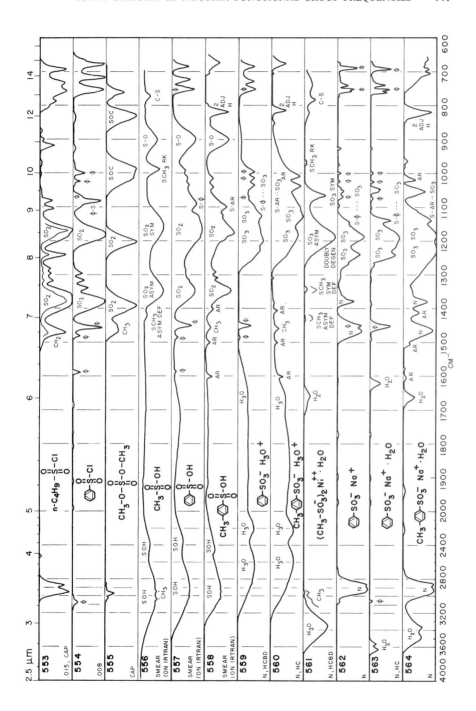

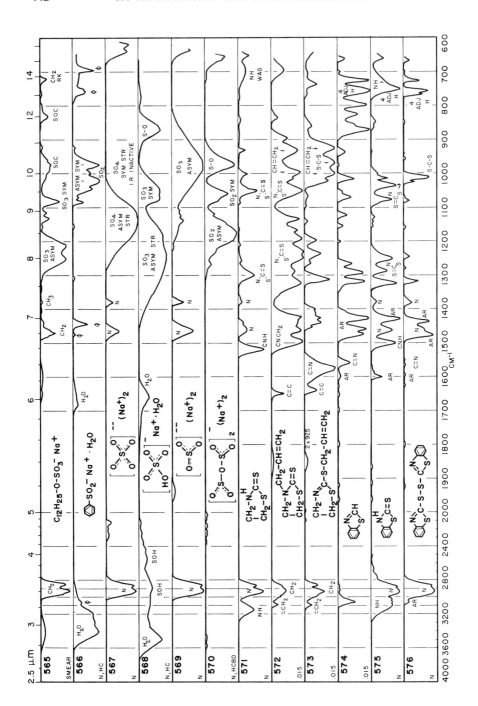

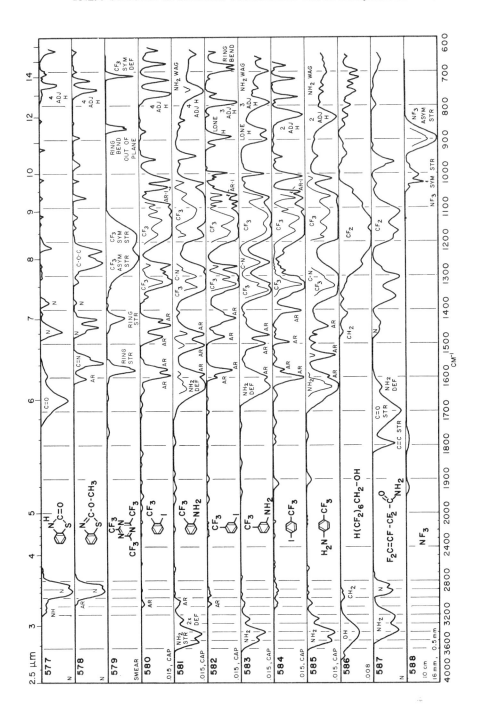

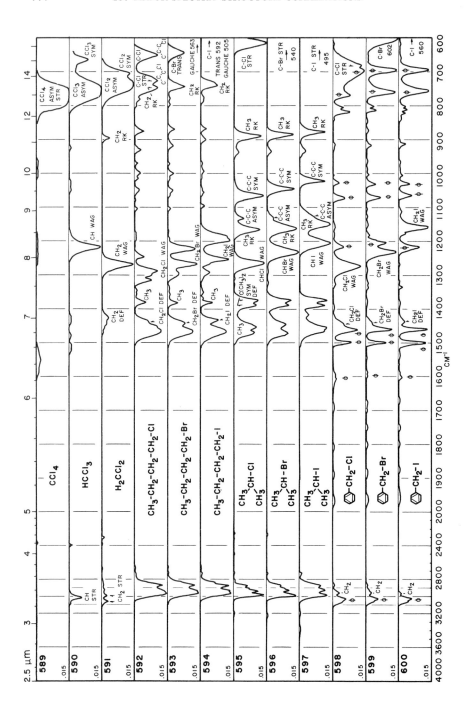

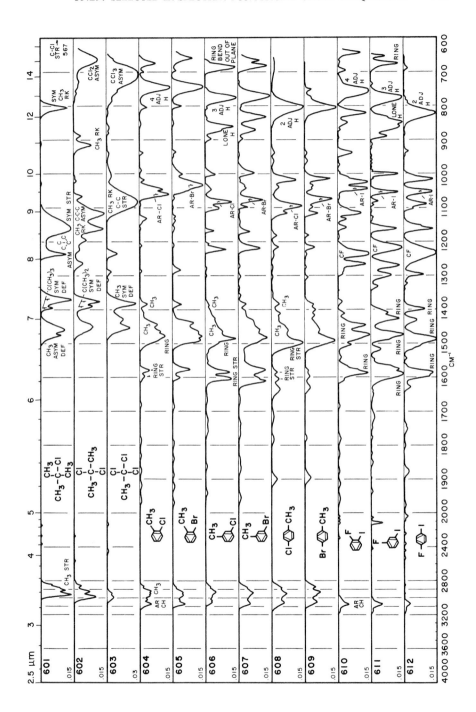

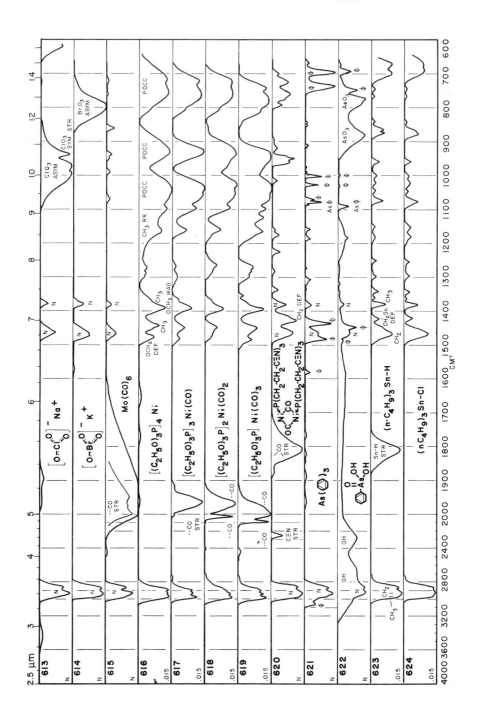

(e) MOLECULAR FORMULA INDEX OF INFRARED SPECTRA

Elements in each of the molecular formulas appear in the following sequence: B, Si, Bi, Sb, As, P, N, S, O, I, Br, Cl, F, C, H, then metals (in salts). Molecular formulas are indexed alphabetically.

$N_3C_8H_{11}$ 381
$N_3C_8H_{17}$ 70
$N_3C_{11}H_7$ 396
$N_3C_{12}H_{11}$ 380
$N_3C_{21}H_{15}$ 418
N_3ClCH_6 339
$N_3ClC_5H_6$ 399
$N_3ClC_{13}H_{26}$ 342
$N_3ClC_{13}H_{30}$ 341
$N_3Cl_2C_7H_9$ 400
$N_3Cl_3C_3$ 417
$N_3Cl_3C_9H_{12}$ 401
$N_3F_9C_6$ 579
$N_3OC_4H_7$ 345
$N_3O_2C_2H_5$ 290
$N_3O_2C_5H_5$ 386, 387
$N_3O_3C_3H_3$ 408
$N_3O_3C_6H_9$ 409, 410
$N_3O_3C_3Na_3$ 411
$N_3SC_2H_3$ 435
$N_3SC_4H_7$ 346
$N_3S_2C_2H_5$ 291
$N_3S_3C_3H_3$ 415
$N_3S_3C_{18}H_{33}$ 416
$N_4BrC_8H_5$ 421
N_4CH_2 440
$N_4C_2H_4$ 344
$N_4C_3H_6$ 438, 441
$N_4C_4H_4$ 437
$N_4C_4H_8$ 439
$N_4C_5H_4$ 395
$N_4C_6H_{10}$ 353
$N_4C_8H_{12}$ 374
$N_4C_8H_{14}$ 352
$N_4C_{10}H_{18}$ 351
$N_4C_{14}H_{10}$ 419
$N_4C_{14}H_{24}$ 343
$N_4OC_8H_6$ 422
$N_4O_2C_2H_4$ 377
$N_4O_2C_3H_4$ 407
$N_4O_2C_4H_{10}$ 328
$N_4O_5C_{13}H_{10}$ 276

$N_4O_4CH_4$ 364
$N_4O_6C_6H_{12}$ 367
$N_5C_9H_{21}$ 347
$N_5C_{10}H_{11}$ 420
$N_5OC_3H_5$ 406
$N_5SC_3H_4Na$ 414
$N_5SC_3H_5$ 412
$N_5SC_4H_7$ 413
$N_6C_2H_4$ 423
$N_6C_3H_6$ 402
$N_6C_4H_8$ 404
$N_6C_{13}H_{10}$ 71
$N_6ClC_3H_7$ 405
$N_6O_3C_6H_{12}$ 403
$N_{10}SO_4C_4H_{16}$ 348
OC_2H_4 119
OC_2H_6 18
OC_3H_6 118, 120, 145, 152
OC_4H_4 427
OC_4H_6 177
OC_4H_8 110, 117, 146
OC_4H_{10} 125–130
OC_5H_6 44
OC_5H_8 176
OC_5H_{10} 116
OC_6H_6 135, 136
OC_6H_{10} 157, 175
OC_6H_{12} 131, 132, 153
OC_6H_{14} 106
OC_7H_6 149
OC_7H_{12} 174
OC_8H_8 160
OC_8H_{10} 115
OC_8H_{14} 173
OC_8H_{16} 5, 133, 134
OC_8H_{18} 4, 105
OC_9H_{18} 154
$OC_{10}H_{14}$ 137
$OC_{10}H_{16}$ 124
$OC_{12}H_{10}$ 114
$OC_{13}H_8$ 159

$OC_{14}H_{18}$ 148
$OC_{14}H_{22}$ 138
$OC_{16}H_{34}$ 3
$OClC_3H_5$ 156
$OClC_7H_5$ 294
$OClC_8H_7$ 293
$OClC_{12}H_{23}$ 292
OCl_3C_2H 147
$OF_{12}C_7H_4$ 586
OH_2 141
$O_2C_2H_4$ 178
$O_2C_2H_6$ 23
$O_2C_3H_4$ 122, 214, 221
$O_2C_3H_6$ 179, 180
$O_2C_4H_6$ 171, 184, 198, 213
$O_2C_4H_8$ 21, 113, 181, 186
$O_2C_4H_{10}$ 142
$O_2C_5H_4$ 429
$O_2C_5H_6$ 45
$O_2C_5H_8$ 163, 199, 200, 212
$O_2C_5H_{10}$ 182, 189, 219
$O_2C_6H_4$ 158
$O_2C_6H_8$ 28
$O_2C_6H_{10}$ 155
$O_2C_6H_{12}$ 121, 170, 183, 187
$O_2C_7H_6$ 168, 222
$O_2C_8H_8$ 150, 151, 161, 162, 185
$O_2C_8H_{12}$ 46
$O_2C_8H_{16}$ 6, 188
$O_2C_8H_{18}$ 143
$O_2C_9H_{10}$ 203
$O_2C_{10}H_{12}$ 123
$O_2C_{10}H_{22}$ 109
$O_2C_{13}H_{26}$ 190
$O_2C_{14}H_{10}$ 172
$O_2C_{14}H_{28}$ 191

$SO_3C_7H_7Na$ 564	$S_2OC_5H_{10}$ 209	$(SiOC_2H_6)_x$ 479
$SO_3C_7H_8$ 552, 558, 560	$S_2O_2C_6H_{10}$ 195	$SiOC_{18}H_{16}$ 469
$SO_3C_8H_{18}$ 546	$S_3C_3H_4$ 218	$(SiO_2)_x$ 480
SO_3HNa 570	$S_3C_5H_{10}$ 211	$SiO_4C_8H_{20}$ 476
SO_3Na_2 569	SiC_6H_8 466	$Si_2NC_6H_{19}$ 474
$SO_4C_2H_6$ 555	$SiC_{12}H_{12}$ 467	$Si_2OC_{36}H_{30}$ 471
$SO_4C_{12}H_{25}Na$ 565	$SiC_{18}H_{16}$ 468	$Si_3O_3C_{36}H_{30}$ 477
SO_4HNa 568	$SiClC_3H_9$ 473	$Si_4O_4C_{48}H_{40}$ 478
SO_4Na_2 567	$SiClC_{18}H_{15}$ 472	$SnC_{12}H_{28}$ 623
$S_2C_4H_{10}$ 538	$SiCl_3C_2H_3$ 475	$SnClC_{12}H_{27}$ 624
$S_2OC_3H_5Na$ 210	$SiNC_{18}H_{17}$ 470	$TiCl_2C_{10}H_{20}$ 425

(f) UNKNOWNS FOR INTERPRETATION

Spectra A—X are presented without labels or interpretation as an exercise for the reader in functional group analysis. These are all pure liquids (no Nujol or solvent) 0.01–0.3 mm thick. For interpretations of these see Section 13.29 (g). The spectra are on pages 452 and 453.

(g) INTERPRETATION OF UNKNOWNS A—X

(A) Aromatic ring, 3100–3000 cm^{-1}, 1608 cm^{-1}, and 1500 cm^{-1}. Mono-substituted aromatic ring, 700, 767, 1032, 1085, and 2000–1700 cm^{-1} (overtones). Aliphatic group, 2960–2870 and 1470 cm^{-1}. Isopropyl group, 1388 and 1368 cm^{-1}. *Isopropylbenzene.*

(B) Carboxylic acid, 3100 cm^{-1} (broad) (OH), 1712 cm^{-1} (C=O), 1420, 1300–1200, and 930 cm^{-1} (broad). Aliphatic group, 2960–2860 and 1468 cm^{-1}. Methyl group on a carbon 1380 cm^{-1}. *Propionic acid.*

(C) Aromatic ring, 3100–3000, 1605, 1583, and 1490 cm^{-1}. *Ortho*-substituted aromatic ring, 740 cm^{-1}. Aliphatic group, 2960–2860, 1465, and 1380 cm^{-1} (CH$_3$). Conjugated ester, 1725 cm^{-1} (C=O). Aromatic ester, ca. 1280 and 1120 cm^{-1}. *Di (2-ethyl-hexyl)phthalate.*

(D) Monosubstituted aromatic ring, 3100–3000, 1592, 1505, 750, and 700 cm^{-1}. Aliphatic group, 2930–2880 cm^{-1}. Conjugated ester, 1727 cm^{-1} (C=O). Unsaturated ester, ca. 1190 cm^{-1}. Vinyl group, 1635, 1412, 985, and 968 cm^{-1}. Acrylate, 812 and 1620 cm^{-1} (overtone). *Benzyl acrylate.*

(E) Aromatic ring, 3100–3000, 1602, 1582, and 1520 cm^{-1}. Substituted napthalene, 810 cm^{-1} (3 adjacent hydrogens) and 780 cm^{-1} (4 adjacent hydrogens). Aldehyde, 2860 cm^{-1}, 2740 cm^{-1} (CH), and 1692 cm^{-1} (C=O). *I-Naphthaldehyde.*

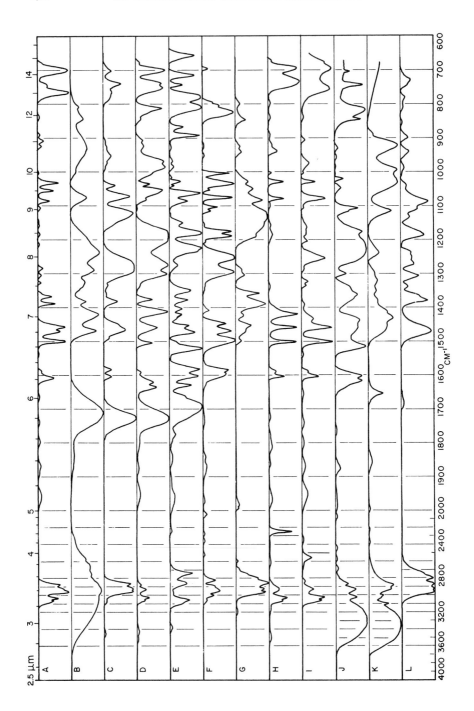

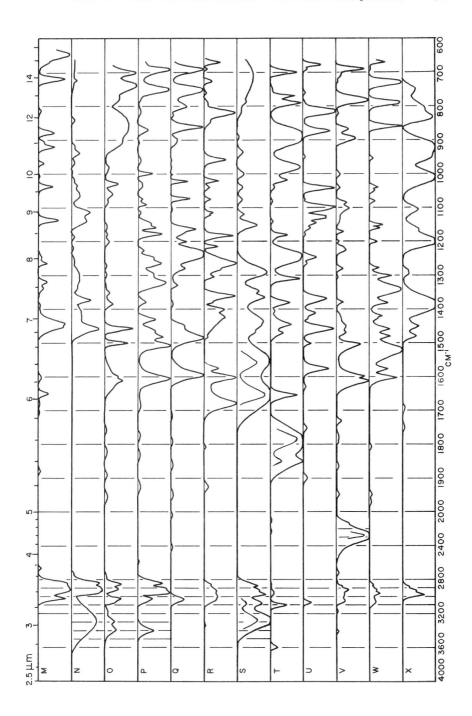

(F) Aromatic ring, 3100–3000, 1597, 1585, and 1497 cm^{-1}. *Para*-substituted aromatic ring, 825 cm^{-1}. Aliphatic, 2960–2830 cm^{-1}, methoxy, 2830 cm^{-1}. Aromatic-*o*-aliphatic ether, 1252 and 1035 cm^{-1}. *Para*-substituted bromo benzene, 1075 cm^{-1}. *p-Bromoanisole.*

(G) Aliphatic groups 2970–2850 cm^{-1}. Aliphatic ether, 1130 cm^{-1}. *Diethyl ether.*

(H) Monosubstituted aromatic ring, 3100–3000, 1605, 1590, 1502, 1458, 1069, 1029, 733, and 693 cm^{-1}. Aliphatic group, 2930 cm^{-1}. Unconjugated nitrile, 2250 cm^{-1}, $CH_2-C\equiv N$, 1420 cm^{-1}. *Phenyl acetonitrile.*

(I) Monosubstituted aromatic ring [see (H)]. Aliphatic group, 2930 cm^{-1}. SH, 2560 cm^{-1}, CH_2-S, 1433 and 1255 cm^{-1}. *Benzyl mercaptan.*

(J) Hydroxyl group, 3300 cm^{-1} (broad) (OH), aromatic OH, 1225 cm^{-1} *para*-substituted aromatic, 3030, 1617, 1604, 1515, and 817 cm^{-1} aliphatic group, 2930–2870 and 1460 cm^{-1}. *p-Cresol.*

(K) Hydroxyl group, 3350 cm^{-1}, primary alcohol, 1030 cm^{-1}, vinyl group, 3100–3000, 1855 (overtone), 1655 (C=C), 995, and 922 cm^{-1}. Aliphatic group, 2930 and 2880 cm^{-1}. *Allyl alcohol.*

(L) Aliphatic group, 2960–2860, 1470, and 1830 (CH_3) cm^{-1}. CH_2 next to tertiary amine nitrogen, 2800 cm^{-1}. *Tributylamine.*

(M) Olefinic CH, 3100–3000 cm^{-1}. Alkane CH_2, 3000–2800 and 1450–1440 cm^{-1}. C=C, 1650 cm^{-1}. *Cis* CH=CH, 640 cm^{-1}. (Note no 1375 or 2965 cm^{-1} $C-CH_3$.) *Cyclohexene.*

(N) OH, 3300 cm^{-1}. CH_2 3000–2800 and 1460 cm^{-1}. CH_3, 1370 cm^{-1}. Secondary alcohol $C-O$, 1110 cm^{-1}, COH bend, 1410 cm^{-1}, Long chain, $(CH_2)_x$, 724 cm^{-1}. *2-Octanol.*

(O) NH_2, 3400, 3300, 3200, 1615, and 860 cm^{-1}. Phenyl, 3100–3000, 1600, 1500, 740, and 700 cm^{-1}. CH_2, 3000–2800 cm^{-1}. *Benzylamine.*

(P) NH, 3400 cm^{-1}. CH_2, CH_3, 3000–2800 and 1375 cm^{-1}. Phenyl, 3030, 1610, 1505, 750, and 691 cm^{-1}. Aryl-N, 1220 cm^{-1}. *N-n-Butylaniline.*

(Q) Phenyl, 3030, 1595, 1490, 1075, 1025, 750, and 692 cm^{-1}. Aryl-O, 1240 cm^{-1}. (Note no OH, 3300, no CH_2 or CH_3, 2900 cm^{-1}). *Phenyl ether.*

(R) Aromatic, 3100–3000, 1605, 1575, and 1480 cm^{-1} (weak). *Para,* 816 cm^{-1}. Conjugated ketone, 1680 cm^{-1}, Aryl ketone, 1265 cm^{-1}. *p-Methyl acetophenone.*

(S) NH, 3300 cm^{-1}. Amide C=O, 1655 cm^{-1}. CNH, 1555, 3100 (overtone), 1290, and 715 cm^{-1}. $O=C-CH_3$, 1370 cm^{-1}. CH_2, CH_3, 3000–2800 and 1465–1430 cm^{-1}. *N-Ethyl acetamide.*

(T) Cyclic anhydride, 1850 (med.) and 1780 cm^{-1} (strong). C=C, 1650 cm^{-1}. Olefinic CH, 3100 cm^{-1}. C—CH$_3$, 2960, 1450, and 1375 cm^{-1}. Anhydride C—O, 1240 and 890 cm^{-1}. *Citraconic anhydride.*

(U) Aromatic, 3100–3000, 1580, 1500, and 1460 cm^{-1}. 1,2,4-Trisubstituted, 872 and 818 cm^{-1}. (Note no aliphatic, 2900 cm^{-1}.) *1,2,4-Trichlorobenzene.*

(V) N=C=O, 2285 cm^{-1}. Aromatic, 3100–3000, 1615, 1600, and 1505 cm^{-1}. *Meta*, 783 and 688 cm^{-1}. C—CH$_3$, 2925 and 1375 cm^{-1}. *m-Tolyl isocyanate.*

(W) Conjugated nitro, 1515 and 1345 cm^{-1}. Aromatic, 3100–3000, 1610, and 1580 cm^{-1}. CH$_3$, 3000–2800 and 1378 cm^{-1}. *o-Nitrotoluene.*

(X) CH$_2$, CH$_3$, 3000–2800, 1480, and 1450 cm^{-1}. —O—SO$_2$—O—, 1380 and 1200 cm^{-1}. C—O—S, 1010, 925, and 833 cm^{-1}. *Diethyl sulfate.*

13.30 Selected Raman Spectra

(a) ILLUSTRATING FUNCTIONAL GROUP FREQUENCIES

These spectra were plotted from runs on a Jarrell-Ash 25–300 Raman spectrophotometer with a 4880 Å argon ion laser. In some spectra the region from 4000 to 2000 cm^{-1} has been plotted so that the intensity is 0.5 times its true value compared to the rest of the spectrum. These are marked ×0.5. Like the infrared spectra, these Raman spectra illustrate a group frequencies which are labeled directly on the spectra. Groups illustrated include alkanes in spectra 1–6, cyclohexanes 7–8, aromatics 9–12, 15, 17, 18, 20, 21, 25, 32–34, double bonds 13, 14, 24, isocyanate 15, triple bond 16, nitrile 17, 18, carbonyls 19–26, alcohols 27–29, ether 30, amines 31, 32, nitro 33, C—Cl 34, C—Br 35, and mercaptan 36. A molecular formula index of the Raman spectra follows.

(b) MOLECULAR FORMULA INDEX OF RAMAN SPECTRA

BrC$_4$H$_9$ 35	Cl$_2$C$_2$H$_2$ 13	OC$_7$H$_6$ 20
C$_5$H$_{11}$ 1	NC$_6$H$_{15}$ 31	OC$_8$H$_{18}$ 30
C$_6$H$_{12}$ 7	NC$_7$H$_9$ 32	O$_2$C$_3$H$_6$ 22
C$_6$H$_{13}$ 2	NC$_8$H$_7$ 17, 18	O$_2$C$_6$H$_{12}$ 23
C$_7$H$_8$ 9	NOC$_3$H$_7$ 26	O$_2$C$_8$H$_8$ 21
C$_7$H$_{14}$ 8	NOC$_7$H$_5$ 15	O$_3$C$_6$H$_{10}$ 14
C$_7$H$_{15}$ 3	NO$_2$C$_6$H$_5$ 33	O$_4$C$_6$H$_6$ 16
C$_8$H$_{10}$ 10, 11, 12	OCH$_4$ 27	O$_4$C$_8$H$_{12}$ 25
C$_8$H$_{17}$ 4, 6	OC$_3$H$_8$ 28	O$_4$C$_{28}$H$_{46}$ 25
C$_9$H$_{19}$ 5	OC$_4$H$_8$ 19	SC$_{12}$H$_{26}$ 36
ClC$_7$H$_7$ 34	OC$_4$H$_{10}$ 29	

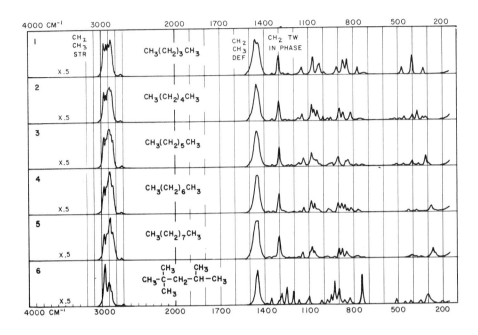

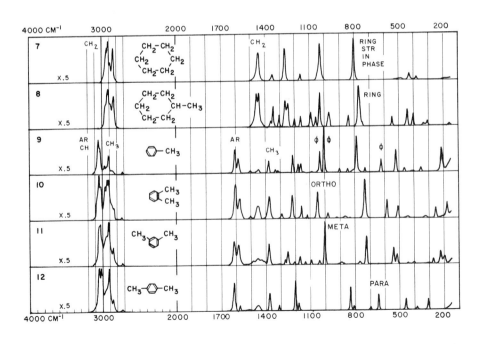

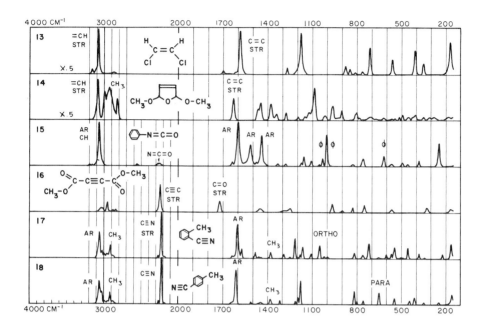

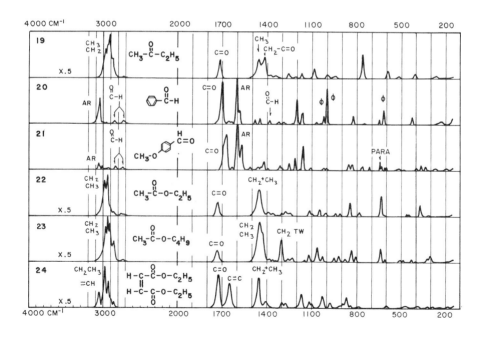

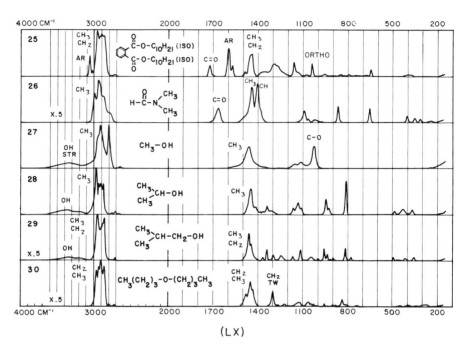

(LX)

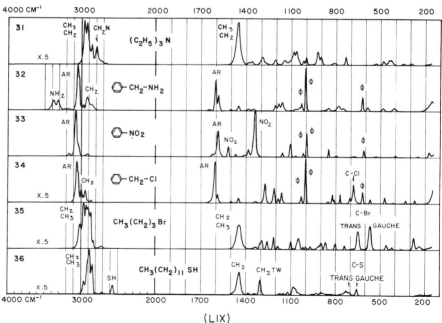

(LIX)

CHAPTER 14

THE THEORETICAL ANALYSIS OF MOLECULAR VIBRATIONS

As discussed in Chapter 1 a nonlinear molecule may have $3N - 6$ normal modes of vibration. The purpose of this chapter is to show how the frequencies and forms of these normal modes of vibration may be calculated mathematically. The calculations require a knowledge of interatomic forces (force constants) and the configuration (bond lengths and angles) of the molecule. The method is outlined and then discussed in detail using chloroform as an example. Some use is made of group theory (see Chapter 3) and matrix algebra (see Section 14.21) but little detailed previous knowledge of these fields is necessary. Other texts may be consulted for further discussion of vibrational theory and mathematical principles involved.[1-6]

14.1 Normal Modes of Vibration

Let the displacements from the equilibrium position of each of the N atoms be represented by a set of coordinates X_1, Y_1, Z_1, for atom 1, X_2, Y_2, Z_2 for atom 2, X_N, Y_N, Z_N for atom N, or more generally $q_1, q_2, q_3, q_4, \ldots, q_{3N}$.

[1] E. T. Whittaker, "Analytical Dynamics of Particles and Rigid Bodies," 3rd ed., Chapter 7. Cambridge Univ. Press, London and New York, 1927.

[2] S. Glasstone, "Theoretical Chemistry." Van Nostrand-Reinhold, Princeton, New Jersey, 1944.

[3] E. B. Wilson, Jr., J. C. Decius, and P. C. Cross, "Molecular Vibrations." McGraw-Hill, New York, 1955.

[4] G. Herzberg, "Infrared and Raman Spectra of Polyatomic Molecules." Van Nostrand-Reinhold, Princeton, New Jersey, 1945.

[5] H. Margenau and G. M. Murphy, "The Mathematics of Physics and Chemistry," 2nd ed. Van Nostrand-Reinhold, Princeton, New Jersey, 1956.

[6] D. Steele, "Theory of Vibrational Spectroscopy." Saunders, Philadelphia, Pennsylvania, 1971.

The $3N$ coordinates q_i where $i = 1, 2, 3, \ldots, 3N$ may be used to express the potential energy V and the kinetic energy T. The appropriate relations are

$$2V = \sum_{i=1}^{3N} \sum_{j=1}^{3N} f_{ij} q_i q_j \quad \text{and} \quad 2T = \sum_{i=1}^{3N} \sum_{j=1}^{2N} m_{ij} \dot{q}_i \dot{q}_j \quad (14.1)$$

where $\dot{q}$ is dq/dt, and q_i and q_j are the ith and jth coordinates and are not necessarily the same for the general case ($i = j$ or $i \neq j$). The f_{ij} values are force constants and the m_{ij} values are functions of the atomic masses. The kinetic energy expression is a generalization of $\frac{1}{2}mv^2$ for a single particle. If cartesian displacement coordinates are used, the m_{ij} values are simply atomic masses and cross terms where $\dot{q}_i \neq \dot{q}_j$ are zero. The potential energy expression is a generalization of $\frac{1}{2}fX^2$ for a single particle held in a harmonic oscillator type force field where the restoring force ($-fX$) is proportional to the displacement from the equilibrium position. If cartesian displacement coordinates are used, cross terms ($q_i \neq q_j$) are generally not zero.

Newton's equations of motion can be written in the Lagrange form. When T is only a function of $\dot{q}_i$ and V is only a function of q_i, then

$$\frac{d}{dt}\left(\frac{\partial T}{\partial \dot{q}_i}\right) + \frac{\partial V}{\partial q_i} = 0 \quad (14.2)$$

From Eq. (14.1) we obtain for each value of q_i,

$$\frac{\partial V}{\partial q_i} = \sum_{j=1}^{3N} f_{ij} q_j \qquad \frac{\partial T}{\partial \dot{q}_i} = \sum_{j=1}^{3N} m_{ij} \dot{q}_j \qquad \frac{d}{dt}\left(\frac{\partial T}{\partial \dot{q}_i}\right) = \sum_{j=1}^{3N} m_{ij} \ddot{q}_j \quad (14.3)$$

Substituting the values in Eq. (14.3) into Eq. (14.2) we get for each of the $3N$ values for i,

$$\sum_{j=1}^{3N} m_{ij} \ddot{q}_j + \sum_{j=1}^{3N} f_{ij} q_j = 0 \quad (14.4)$$

This is a generalization of the statement, mass times acceleration ($m\ddot{q}$) minus force ($-fq$) equals zero.

These $3N$ equations have the general solution

$$q_j = A_j \sin(\sqrt{\lambda} t + \alpha) \quad \text{where} \quad \sqrt{\lambda} = 2\pi v \quad (14.5)$$

which is an equation characteristic of simple harmonic motion with a frequency v equal to $\sqrt{\lambda}/(2\pi)$, a maximum amplitude A_j and a phase constant α. Differentiating Eq. (14.5) twice with respect to time we get

$$\dot{q}_j = \sqrt{\lambda} A_j \cos(\sqrt{\lambda} t + \alpha) \qquad \ddot{q}_j = -\lambda A_j \sin(\sqrt{\lambda} t + \alpha) \quad (14.6)$$

Substituting the values for q_j and $\ddot{q}_j$ in Eqs. (14.5) and (14.6) into Eq. (14.4) we get for each of the $3N$ values for i,

$$\sum_{j=1}^{3N} f_{ij} A_j \sin(\sqrt{\lambda} t + \alpha) - \sum_{j=1}^{3N} m_{ij} \lambda A_j \sin(\sqrt{\lambda} t + \alpha) = 0$$

$$(14.7)$$

$$\sum_{j=1}^{3N} (f_{ij} - m_{ij}\lambda)A_j = 0$$

When this equation is written out we get

$$(f_{11} - m_{11}\lambda)A_1 + (f_{12} - m_{12}\lambda)A_2 \cdots (f_{1,3N} - m_{1,3N}\lambda)A_{3N} = 0$$
$$(f_{21} - m_{21}\lambda)A_1 + (f_{22} - m_{22}\lambda)A_2 \cdots (f_{2,3N} - m_{2,3N}\lambda)A_{3N} = 0$$
$$\cdots \qquad \cdots \qquad \cdots \qquad \cdots \qquad (14.8)$$
$$(f_{3N,1} - m_{3N,1}\lambda)A_1$$
$$+ (f_{3N,2} - m_{3N,2}\lambda)A_2 \cdots (f_{3N,3N} - m_{3N,3N}\lambda)A_{3N} = 0$$

These are called the secular equations. This set of linear homogeneous equations has nontrivial (not zero) solutions for the A values only for certain specific values for λ. These can be found by making use of the fact that, in these types of equations, the determinant of the coefficients of the A values must equal zero.

$$\begin{vmatrix} f_{11} - m_{11}\lambda & f_{12} - m_{12}\lambda & \cdots & f_{1,3N} - m_{1,3N}\lambda \\ f_{21} - m_{21}\lambda & f_{22} - m_{22}\lambda & \cdots & f_{2,3N} - m_{2,3N}\lambda \\ \cdots & \cdots & \cdots & \cdots \\ f_{3N,1} - m_{3N,1}\lambda & f_{3N,2} - m_{3N,2}\lambda & \cdots & f_{3N,3N} - m_{3N,3N}\lambda \end{vmatrix} = 0 \quad (14.9)$$

Equation (14.9) is called the secular determinant. When it is expanded, a $3N$ order characteristic equation for λ will be obtained which can be solved for the $3N$ characteristic values for λ in terms of the f and m values. Each value of λ can be then put back into Eq. (14.8) to calculate the corresponding values for A_j. Actually only the ratios of the A_j values, that is, the ratio of the amplitudes for each λ, can be determined but this is sufficient to describe the vibration. The result indicates that each atom is oscillating about its equilibrium position with amplitude A_j, generally different for each coordinate, but with the *same* frequency $v = \lambda^{\frac{1}{2}}/(2\pi)$ and phase constant α which means all the atoms go through their equilibrium positions simultaneously. Such a motion is called a *normal mode of vibration*.

There are two difficulties with the use of cartesian displacement coordinates. The first is that the $3N$ solutions for λ include six nongenuine vibrations with zero frequency. These are the translations and rotations. This makes the secular determinant have six more rows and columns than necessary and increases the difficulty of the solution. The second difficulty is that the force constants (the f_{ij} values) have the most direct physical meaning in terms of internal coordinates (bond stretching and angle bending), and in terms of cartesian displacement coordinates these become complex and not easy to transfer to related molecules. For these reasons the vibrational problem is usually handled using internal coordinates. This excludes the translations and rotations and the secular determinant has $3N - 6$ rows and columns. However, before proceeding to the internal coordinate method it will be helpful to discuss a simple specific example to illustrate the general principles involved, using cartesian displacement coordinates.

14.2   The Linear Triatomic Model's Stretching Frequencies

Consider a model consisting of three masses, M_1, M_2, and M_3, held together by springlike bonds in a linear configuration, free to move only on the model axis, like beads on a wire for example [see Fig. 14.1(a)]. Let the displacement from the equilibrium position for each mass be measured in cartesian displacement coordinates X_1, X_2, and X_3. New coordinates S_1 and S_2 can be defined [see Fig. 14.1(b)] which characterize how much the bond lengths differ from their equilibrium lengths.

$$S_1 = X_2 - X_1 \qquad \text{and} \qquad S_2 = X_3 - X_2 \tag{14.10}$$

Referring to Eq. (14.1), the complete potential energy expression in terms of the two S internal coordinates is

$$2V = F_1 S_1^2 + F_2 S^2 + F_{12} S_1 S_2 + F_{21} S_2 S_1 \tag{14.11}$$

where $F_{12} = F_{21}$. The meaning of these terms may be clarified as follows. Since from classical physics the force is equal to the negative of the derivative of the potential energy with respect to the coordinate, we can write that the restoring force for the S_1 coordinate is

$$-\frac{\partial V}{\partial S_1} = -F_1 S_1 - F_{12} S_2 = \text{restoring force for } S_1 \tag{14.12}$$

This means that when the potential energy is written in the above form, then the force tending to restore S_1 to its equilibrium length depends on a force constant F_1 times S_1 (Hooke's law), but the S_1 restoring force also depends in the most general case on a second constant (F_{12}) times S_2. This second constant (F_{12}) is called an interaction force constant and is usually much smaller than the main force constant (F_1) and in one type of approximation is assumed to be zero. By substituting the values for S in Eq. (14.10) into

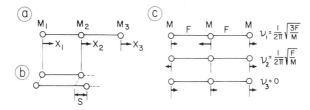

FIG. 14.1.   The axial motions of the $M_1 M_2 M_3$ linear triatomic molecule. In (a) the cartesian displacement coordinates are illustrated. In (b) a stretching internal coordinate is illustrated. In (c) are illustrated the forms of the axial motions of the special case where the masses and force constants are equal.

(14.11) and rearranging, the potential energy in terms of the cartesian coordinates X can be written in the format

$$
\begin{aligned}
2V = \quad & F_1 X_1^2 && + (F_{12} - F_1)X_1 X_2 && - F_{12} X_1 X_3 \\
& + (F_{12} - F_1)X_2 X_1 && + (F_1 - 2F_{12} + F_2)X_2^2 && + (F_{12} - F_2) X_2 X_3 \\
& - F_{12} X_3 X_1 && + (F_{12} - F_2)X_3 X_2 && + F_2 X_3^2
\end{aligned}
\tag{14.13}
$$

The kinetic energy is $\frac{1}{2}M_1 \dot{X}_1^2 + \frac{1}{2}M_2 \dot{X}_2^2 + \frac{1}{2}M_3 \dot{X}_3^2$ and is written in the same format as Eq. (14.13).

$$
\begin{aligned}
2T = \; & M_1 \dot{X}_1^2 && + (0)\dot{X}_1 \dot{X}_2 && + (0)\dot{X}_1 \dot{X}_3 \\
& + (0)\dot{X}_2 \dot{X}_1 && + M_2 \dot{X}_2^2 && + (0)\dot{X}_2 \dot{X}_3 \\
& + (0)\dot{X}_3 \dot{X}_1 && + (0)\dot{X}_3 \dot{X}_2 && + M_3 \dot{X}_3^2
\end{aligned}
\tag{14.14}
$$

Referring to Eq. (14.9), the secular determinant can be written directly from the energy expressions as

$$
\begin{vmatrix}
F_1 - M_1\lambda & F_{12} - F_1 & -F_{12} \\
F_{12} - F_1 & F_1 - 2F_{12} + F_2 - M_2\lambda & F_{12} - F_2 \\
-F_{12} & F_{12} - F_2 & F_2 - M_3\lambda
\end{vmatrix} = 0
\tag{14.15}
$$

When the determinant is expanded we get the characteristic equation, third order in this case,

$$\lambda^3 - \lambda^2\left[F_1\left(\frac{1}{M_1} + \frac{1}{M_2}\right) + F_2\left(\frac{1}{M_2} + \frac{1}{M_3}\right) - 2F_{12}\left(\frac{1}{M_2}\right)\right]$$

$$+ \lambda\left[(F_1F_2 - F_{12}^2)\left(\frac{M_1 + M_2 + M_3}{M_1M_2M_3}\right)\right] = 0 \qquad (14.16)$$

One characteristic value for λ is zero and, since $v = \lambda^{1/2}/(2\pi)$, this represents a nongenuine vibration with a zero frequency which turns out to be a translation of the model as a whole. This, plus the two vibrational frequencies obtained from the quadratic equation

$$\lambda^2 - \lambda\left[F_1\left(\frac{1}{M_1} + \frac{1}{M_2}\right) + F_2\left(\frac{1}{M_2} + \frac{1}{M_3}\right) - 2F_{12}\left(\frac{1}{M_2}\right)\right]$$

$$+ \left[(F_1F_2 - F_{12}^2)\left(\frac{M_1 + M_2 + M_3}{M_1M_2M_3}\right)\right] = 0 \qquad (14.17)$$

make up the three degrees of freedom in the X direction possessed by the three masses. If the masses are expressed in unified atomic mass units u and the force constants are expressed in millidynes per angstrom, then the frequency in cm^{-1} (v/c) is $\bar{v} = 1303 \lambda^{1/2}$.

The above treatment shows how to obtain the frequencies if the masses and force constants are given. If, on the other hand, the frequencies are known and the force constants are unknown (the usual situation), the procedure is more difficult. In a quadratic equation written in the same form as Eq. (14.17), the values in brackets are equal to the sum and product, respectively, of the specific roots. [Let $\lambda = a$ and $\lambda = b$, then $(\lambda - a)(\lambda - b) = 0$ and $\lambda^2 - \lambda(a + b) + ab = 0$.] Therefore we can write from Eq. (14.17)

$$\lambda_1 + \lambda_2 = F_1\left(\frac{1}{M_1} + \frac{1}{M_2}\right) + F_2\left(\frac{1}{M_2} + \frac{1}{M_3}\right) - 2F_{12}\left(\frac{1}{M_2}\right) \qquad (14.18)$$

$$\lambda_1\lambda_2 = (F_1F_2 - F_{12}^2)\left(\frac{M_1 + M_2 + M_3}{M_1M_2M_3}\right) \qquad (14.19)$$

where λ_1 and λ_2 are the specific roots and equal $(2\pi v_1)^2$ and $(2\pi v_2)^2$ where v_1 and v_2 are the vibrational frequencies. In this case using the *general valence force field* (retaining all interaction form constants) there are three force constants and these cannot be unambiguously evaluated from the two experimental frequencies. If all the interaction force constants are made zero (they are usually small) we have a *simple valence force field* which in our example has two force constants to determine from two frequencies. Unfortunately the problems are not over because when Eqs. (14.18) and (14.19) are solved for the F values when $F_{12} = 0$ we get

$$F_1^2 - F_1\left[\left(\frac{M_1 M_2}{M_1 + M_2}\right)(\lambda_1 + \lambda_2)\right]$$

$$+ \frac{(M_1^2 M_2)(M_2 + M_3)}{(M_1 + M_2 + M_3)(M_1 + M_2)}(\lambda_1 \lambda_2) = 0 \qquad (14.20)$$

$$F_2 = \frac{1}{F_1}\left(\frac{M_1 M_2 M_3}{M_1 + M_2 + M_3}\right)(\lambda_1 \lambda_2) \qquad (14.21)$$

Because these equations are nonlinear, we obtain two sets of the two force constants, *each set of which* will lead to the same two frequencies. For example, from the two stretching frequencies for FCN (2290 and 1077 cm^{-1}) we deduce two sets of force constants which can be roughly represented by F$-$C$\equiv$N and F$\equiv$C$-$N. The actual molecular force field is represented by one of these, but in this case the theory cannot say which and a choice must be made using other arguments. In this case the choice is clear but in more complex cases this is not so easy and the number of sets to choose from increases rapidly as the molecule becomes bigger. In practice it is usually easier to assume an approximate force field, calculate the frequencies, modify the force field, re-calculate the frequencies, continuing until the calculated frequencies match experimental frequencies as closely as possible.

If a set of force constants are assumed, then the characteristic equation is solved for the two characteristic values for λ. The form of the vibration characterized by λ_1 is obtained by putting the value for λ_1 into the secular equations which can be constructed from the secular determant, Eq. (14.15). By referring to Eq. (14.8) we can write

$$(F_1 - M_1\lambda)A_1 + (F_{12} - F_1)A_2 + (-F_{12})A_3 = 0$$
$$(F_{12} - F_1)A_1 + (F_1 - 2F_{12} + F_2 - M_2\lambda)A_2 + (F_{12} - F_2)A_3 = 0 \quad (14.22)$$
$$(-F_{12})A_1 + (F_{12} - F_2)A_2 + (F_2 - M_3\lambda)A_3 = 0$$

These can be solved for the ratios of the A values which are the amplitudes of the masses. For example, if $F_{12} = 0$ the first and third secular equation are used to get

$$\frac{A_1}{A_2} = \frac{F_1}{F_1 - M_1\lambda} \quad \text{and} \quad \frac{A_3}{A_2} = \frac{F_2}{F_2 - M_3\lambda} \qquad (14.23)$$

Both of these are solved using λ_1, and then both are solved using λ_2.

For the special case where $M_1 = M_2 = M_3 = M$ and $F_1 = F_2 = F$ and $F_{12} = 0$, the three characteristic values are $\lambda_1 = 3F/M$, $\lambda_2 = F/M$, and $\lambda_3 = 0$. The amplitude ratios for the λ_1 vibration are $A_1 : A_2 : A_3 = 1 : -2 : 1$ which is the out-of-phase stretch vibration [see Fig. 14.1(c)]. For the λ_2 vibration we get $A_1 : A_2 : A_3 = -1 : 0 : 1$ which is the in-phase stretch. For the λ_3 non-genuine vibration we get $A_1 : A_2 : A_3 = 1 : 1 : 1$ which is a translation of the model as a whole.

14.3   Internal Coordinates

Internal coordinates characterize the change in shape of a molecule from its equilibrium shape without regard for its position or orientation as a whole in space. The type most commonly used are changes in bond lengths and bond angles from their equilibrium values. This is appropriate because ideas about chemical forces indicate that such forces tend to resist bond length or bond angle distortions from their equilibrium configurations. A method for obtaining complete sets has been given by Decius[7] and is summarized below.

The four general types of internal coordinates and illustrated in Fig. 14.2.

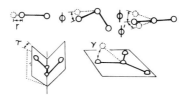

FIG. 14.2.   General types of internal coordinates.

A *bond stretching* coordinate, r, is a change from equilibrium in the interatomic bond distance. A *bond angle bending* coordinate, ϕ, is a change from equilibrium in an interbond angle where two bonds meet at an atom. If there are any linear subsections the molecule can be bent at each atom with 180° bond angles in two mutually perpendicular directions. The additional bond angle bending coordinate is designated as ϕ'. *Bond torsion* can occur when atoms 1, 2, 3, and 4 are connected in sequence nonlinearly. The bond torsion coordinate, τ, is a change from equilibrium in the dihedral angle between the two planes containing atoms 1, 2, and 3 and 2, 3, and 4. *Out-of-plane* bending can occur when three or more bonds meet at an atom and all the bonds are coplanar. The out-of-plane bending coordinate, γ, is a change from equilibrium in the angle between one of these bonds and the plane containing

[7] J. C. Decius, *J. Chem. Phys.* **17**, 1315 (1949).

any other two bonds. In the above discussion of bonds, the bond type such as single, double, or triple has no significance.

There are as many r coordinates as there are bonds. There are as many ϕ' coordinates as there are atoms with $180°$ bond angles in linear subsections. There are as many τ coordinates as there are nonterminal bonds, minus the number of ϕ' coordinates. The multiplicity, m, of an atom is equal to the number of bonds meeting at the atom, again ignoring bond type. At each nonterminal atom the number of independent ϕ coordinates is $2m - 3$ with the following exception. If all the bonds meeting at this atom are coplanar and m equals three or more, the number of γ coordinates will be $m - 2$ and the number of independent ϕ coordinates will be $m - 1$. If N is the number of atoms in the molecule the resultant set will consist of $3N - 6$ linearly independent internal coordinates for noncyclic molecules. If the molecule has cyclic rings then the resultant set will consist of $3N - 6$ independent internal coordinates plus 6μ cyclic redundant coordinates where μ is the number of bonds which must be deleted to make the molecule acyclic (benzene, $\mu = 1$; naphthalene, $\mu = 2$). The maximum number of ϕ coordinates at each nonterminal atom is $m(m - 1)/2$ but this number may include some redundant coordinates (so that the total is more than $3N - 6$). Sometimes it is desirable to retain a redundant coordinate of this "local" type in order to take advantage of the symmetry of the molecule.

14.4   Vibrational Solution Using Internal Coordinates

The general potential energy function of the molecule expressed in internal coordinates S is written

$$2V = \sum_{ij} F_{ij} S_i S_j \quad \text{or} \quad 2V = \mathbf{S}'\mathbf{F}\mathbf{S} \tag{14.24}$$

On the right the convenient matrix notation is used (see Section 14.21) where the bold face symbols stand for matrices written out as follows:

$$2V = \|S_1\ S_2\ \cdots\| \left\|\begin{matrix} F_{11} & F_{12} & \cdots \\ F_{21} & F_{22} & \cdots \\ \cdots & \cdots & \cdots \end{matrix}\right\| \left\|\begin{matrix} S_1 \\ S_2 \\ \cdots \end{matrix}\right\|$$

where $\mathbf{F}$ is

	S_1	S_2	$\cdots$
S_1	F_{11}	F_{12}	$\cdots$
S_2	F_{21}	F_{22}	$\cdots$
$\cdots$	$\cdots$	$\cdots$	$\cdots$

The symbol S' is the transpose of the S matrix which is a column matrix of internal coordinates. The force constants F_{ij} characterize the restoring forces for distorted internal coordinates S (bond lengths and bond angles). The kinetic energy expression which is easily expressed using cartesian displacement coordinates is more complicated when internal coordinate are used. Wilson [8,9] has shown that the kinetic energy for a nontranslating and nonrotating molecule is given by

$$2T = \sum_{ij} (G^{-1})_{ij} \dot{S}_i \dot{S}_j \quad \text{or} \quad 2T = \dot{S}'G^{-1}\dot{S}$$

or

$$2T = \|\dot{S}_1 \quad \dot{S}_2 \quad \cdots\| \left\|\begin{matrix} (G^{-1})_{11} & (G^{-1})_{12} & \cdots \\ (G^{-1})_{21} & (G^{-1})_{22} & \cdots \\ \cdots & \cdots & \cdots \end{matrix}\right\| \left\|\begin{matrix} \dot{S}_1 \\ \dot{S}_2 \\ \cdots \end{matrix}\right\| \quad (14.25)$$

Where the symbol G^{-1} stands for the inverse of the G matrix. (See Section 14.22 for the G matrix origin.) The G matrix elements will be discussed later in Section 14.5.

When these energy expressions are treated in as Section 14.1 the solutions are

$$S_j = L_j \sin(\sqrt{\lambda} t + \alpha) \quad (14.26)$$

where $\lambda = (2\pi\nu)^2$ and L_j is the maximum amplitude of the internal coordinate distortion. The secular determinant is

$$\begin{vmatrix} F_{11} - (G^{-1})_{11}\lambda & F_{12} - (G^{-1})_{12}\lambda & \cdots \\ F_{21} - (G^{-1})_{21}\lambda & F_{22} - (G^{-1})_{22}\lambda & \cdots \\ \cdots & \cdots & \cdots \end{vmatrix} = 0 \quad (14.27)$$

In matrix form this is written

$$|\mathbf{F} - \mathbf{G}^{-1}\lambda| = 0 \quad (14.28)$$

It is usually more convenient to multiply this by $|\mathbf{G}|$ written as

$$|\mathbf{G}| \ |\mathbf{F} - \mathbf{G}^{-1}\lambda| = 0 \quad \text{to get} \quad |\mathbf{GF} - \mathbf{GG}^{-1}\lambda| = 0.$$

Since any matrix times its inverse equals the identity matrix $\mathbf{I}$ (ones on the diagonal, zero elsewhere) this is equivalent to

$$|\mathbf{GF} - \mathbf{I}\lambda| = 0 \quad (14.29)$$

[8] E. B. Wilson, Jr., *J. Chem. Phys.* **7**, 1047 (1939).
[9] E. B. Wilson, Jr., *J. Chem. Phys.* **9**, 76 (1941).

Written out this becomes

$$\left\| \begin{array}{ccc} G_{11} & G_{12} & \cdots \\ G_{21} & G_{22} & \cdots \\ \cdots & \cdots & \cdots \end{array} \right\| \left\| \begin{array}{ccc} F_{11} & F_{12} & \cdots \\ F_{21} & F_{22} & \cdots \\ \cdots & \cdots & \cdots \end{array} \right\| - \left\| \begin{array}{ccc} \lambda & 0 & \cdots \\ 0 & \lambda & \cdots \\ \cdots & \cdots & \cdots \end{array} \right\| = 0$$

(14.30)

$$\left| \begin{array}{ccc} \sum_i G_{1i} F_{i1} - \lambda & \sum_j G_{1j} F_{j2} & \cdots \\ \sum_i G_{2i} F_{i2} & \sum_j G_{2j} F_{j2} - \lambda & \cdots \\ \cdots & \cdots & \cdots \end{array} \right| = 0$$

If the **F** and **G** matrices are given, the λ values, and the v values which equal $\lambda^{1/2}/(2\pi)$, can be calculated from the secular determinant.

14.5 The G Matrix

The relationship between cartesian displacement coordinates X and internal coordinates S is

$$S_j = \sum_{i=1}^{3N} B_{ji} X_j \qquad \text{or} \qquad \mathbf{S} = \mathbf{BX}$$

(14.31)

$$\left\| \begin{array}{c} S_a \\ S_b \\ \cdots \end{array} \right\| = \left\| \begin{array}{ccc} B_{aX1} & B_{aY1} & \cdots \\ B_{bX1} & B_{bY1} & \cdots \\ \cdots & \cdots & \cdots \end{array} \right\| \left\| \begin{array}{c} X_1 \\ Y_1 \\ \cdots \end{array} \right\|$$

where **B** is

	X_1	Y_1	Z_1	X_2
S_a	B_{aX1}	B_{aY1}	$\cdots$	$\cdots$
S_b	B_{bX1}	B_{bY1}	$\cdots$	$\cdots$
$\cdots$	$\cdots$	$\cdots$	$\cdots$	$\cdots$

A unit change in the cartesian coordinate X_k causes a B_{jk} change in the internal coordinate S_j. Wilson[8,9] has shown that the **G** matrix elements are given by

$$G_{jk} = \sum_{i=1}^{3N} \frac{1}{M_i} B_{ji} B_{ki} \qquad \text{or} \qquad \mathbf{G} = \mathbf{BM}^{-1}\mathbf{B}'$$

or

$$\left\| \begin{matrix} G_{11} & G_{12} & \cdots \\ G_{21} & G_{22} & \cdots \\ \cdots & \cdots & \cdots \end{matrix} \right\| = \left\| \begin{matrix} B_{aX1} & B_{aY1} & \cdots \\ B_{bX1} & B_{bY1} & \cdots \\ \cdots & \cdots & \cdots \end{matrix} \right\|$$

$$\times \left\| \begin{matrix} \dfrac{1}{M_1} & 0 & 0 & 0 & \cdots \\[2mm] 0 & \dfrac{1}{M_1} & 0 & 0 & \cdots \\[2mm] 0 & 0 & \dfrac{1}{M_1} & 0 & \cdots \\[2mm] 0 & 0 & 0 & \dfrac{1}{M_2} & \cdots \\[2mm] \cdots & \cdots & \cdots & \cdots & \cdots \end{matrix} \right\| \left\| \begin{matrix} B_{aX1} & B_{bX1} & \cdots \\ B_{aY1} & B_{bY1} & \cdots \\ \cdots & \cdots & \cdots \end{matrix} \right\| \qquad (14.32)$$

As an example consider a diatomic molecule $M_1 - M_2$ (see Fig. 14.3) where the bond makes angle α with the XY plane in a cartesian coordinate system. Furthermore, let the *projection* of the bond on the XY plane make an angle β with the X axis. If the masses are given small displacements in the X, Y, or Z

FIG. 14.3.   A diatomic molecule in a cartesian coordinate system.

directions then the bond stretching internal coordinate S will equal the sum of the projections of these displacements onto the bond with appropriate signs to indicate whether the bond is shortened or lengthened by the displacements. If $c = \cos$ and $s = \sin$ then

$$S = - X_1 c\beta c\alpha - Y_1 s\beta c\alpha - Z_1 s\alpha + X_2 c\beta c\alpha + Y_2 s\beta c\alpha + Z_2 s\alpha \qquad (14.33)$$

This equation in matrix form $\mathbf{S} = \mathbf{BX}$ is written out as

$$\|\mathbf{S}\| = \| -c\beta c\alpha \quad -s\beta c\alpha \quad -s\alpha \quad c\beta c\alpha \quad s\beta c\alpha \quad s\alpha \| \left\| \begin{matrix} X_1 \\ Y_1 \\ Z_1 \\ X_2 \\ Y_2 \\ Z_2 \end{matrix} \right\| \qquad (14.34)$$

The equation $\mathbf{G} = \mathbf{BM}^{-1}\mathbf{B}'$ written out is

$$\mathbf{G} = \|-c\beta c\alpha \quad -s\beta c\alpha \quad -s\alpha \quad c\beta c\alpha \quad s\beta c\alpha \quad s\alpha\|$$

$$\times \begin{Vmatrix} \dfrac{1}{M_1} & 0 & 0 & 0 & 0 & 0 \\ 0 & \dfrac{1}{M_1} & 0 & 0 & 0 & 0 \\ 0 & 0 & \dfrac{1}{M_1} & 0 & 0 & 0 \\ 0 & 0 & 0 & \dfrac{1}{M_2} & 0 & 0 \\ 0 & 0 & 0 & 0 & \dfrac{1}{M_2} & 0 \\ 0 & 0 & 0 & 0 & 0 & \dfrac{1}{M_2} \end{Vmatrix} \begin{Vmatrix} -c\beta c\alpha \\ -s\beta c\alpha \\ -s\alpha \\ c\beta c\alpha \\ s\beta c\alpha \\ s\alpha \end{Vmatrix} \qquad (14.35)$$

$$\mathbf{G} = \frac{c^2\alpha(c^2\beta + s^2\beta) + s^2\alpha}{M_1} + \frac{c^2\alpha(c^2\beta + s^2\beta) + s^2\alpha}{M_2} = \frac{1}{M_1} + \frac{1}{M_2} \qquad (14.36)$$

Notice that the **G** matrix element is independent of the bond's orientation in the cartesian coordinate system, and this is generally true for other **G** matrix elements. Since the potential energy is $\frac{1}{2}FS^2$, **F** is simply F, and the one by one secular determinant $|\mathbf{GF} - \mathbf{I}\lambda| = 0$ becomes

$$\left(\frac{1}{M_1} + \frac{1}{M_2}\right) F - \lambda = 0 \qquad v = \frac{1}{2\pi}\sqrt{\lambda} = \frac{1}{2\pi}\sqrt{F\left(\frac{1}{M_1} + \frac{1}{M_2}\right)} \qquad (14.37)$$

Wilson[8,9] introduced a convenient vectorial method for obtaining matrix elements which does not use cartesian coordinates. For a complete treatment of this type, see Wilson, Decius, and Cross,[3] Chapter 4, and also Meister and Cleveland.[10] This method will not be discussed here since general formulas have been tabulated for all the common **G** matrix elements.[3,11,12]

[10] A. G. Meister and F. F. Cleveland, *Amer. J. Phys.* **14**, 13 (1946).
[11] J. C. Decius, *J. Chem. Phys.* **16** (1948).
[12] S. M. Ferigle and A. G. Meister, *J. Chem. Phys.* **19**, 982 (1951).

Wilson, Decius, and Cross[3] have suggested a notation for **G** matrix elements which are given in terms of internal coordinates, such as bond stretching, r, and angle deformation, ϕ. The element $G_{r\phi}$, for example, involves a stretching and bending internal coordinate, but can be one of three common types depending on the structure involved. If both atoms involved in the stretching coordinate are also involved in the bending coordinate, the designation is $G_{r\phi}^2$. The superscript two indicates two common atoms. If, however, only one of the atoms involved in the stretching is also involved in the bending coordinate, there are two possible situations. The common atom may be either an end atom or a central atom in the bending coordinate. The two elements are called $G_{r\phi}^1\binom{1}{2}$ and $G_{r\phi}^1\binom{1}{1}$, respectively. Representations of the commonly used kinetic energy elements, similar to those in Wilson, Decius, and Cross,[3] are shown in Fig. 14.4. The subscripts r or ϕ give the internal

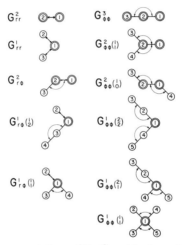

FIG. 14.4.   Schematic representations of the **G** matrix elements. (Redrawn from " Molecular Vibrations" by Wilson, Decius, and Cross. Copyright 1955, McGraw-Hill Book Company, Inc. Used by permission.)

coordinates involved and the superscript gives the number of atoms common to both coordinates. (Here they are restricted to stretching and bending coordinates.) The representations are schematic with atoms common to both coordinates shown as double circles on a horizontal line. Noncommon atoms for the first coordinate (first subscript) are on 45° lines above the horizontal and noncommon atoms of the second coordinate are on 45° lines below. A pair of numbers in parentheses, for example, $G_{r\phi}^1\binom{1}{2}$, gives the number of atoms on the top left line (top number) and on the bottom left line (bottom number). Each atom is identified by a number. By means of this scheme each type of **G** matrix element can be uniquely characterized and the general formulas have been tabulated in terms of these designations.[3,11,12]

14.6 The L Matrix

Once the frequencies have geen determined the next step is to define the form of the normal mode of vibration in terms of the internal coordinates. In order to do this the L matrix should be evaluated. As an example of the procedure let us imagine we have multiplied 3×3 G and F matrices to give an H matrix ($GF = H$) from which we can construct a secular determinant by subtracting λ from each diagonal element.

$$\begin{vmatrix} H_{11} - \lambda & H_{12} & H_{13} \\ H_{21} & H_{22} - \lambda & H_{23} \\ H_{31} & H_{32} & H_{33} - \lambda \end{vmatrix} = 0 \qquad (14.38)$$

This can be solved for the three values for λ. The secular equations from which this secular determinant arose can be reconstructed from the determinant and are given by

$$\begin{aligned} (H_{11} - \lambda_k)L_{1k} + H_{12}L_{2k} + H_{13}L_{3k} &= 0 \\ H_{21}L_{1k} + (H_{22} - \lambda_k)L_{2k} + H_{23}L_{3k} &= 0 \qquad (14.39) \\ H_{31}L_{1k} + H_{32}L_{2k} + (H_{33} - \lambda_k)L_{3k} &= 0 \end{aligned}$$

When one of the three specific λ values is substituted into these equations they can be solved for the L values that go with that λ value. Only the relative values, or L ratios, can be determined however, but this is sufficient to characterize the form of the vibrations by specifying the relative distortion of each internal coordinate at any time during the vibration. For example, the Eqs. (14.39) can each be divided by L_{1k} and then any pair of these can be solved for L_{2k}/L_{1k} and L_{3k}/L_{1k}.

A systematic general procedure for finding the relative L values is as follows: Select any one row of the secular determinant numerically evaluated using a specific λ value. Find the cofactor for each element in the selected row. The cofactor of an element is the determinant left after deleting the row and column containing the element, multiplied by plus or minus one depending on whether the sum of the H_{ij} element subscripts is even or odd, respectively. The cofactors of the first, second, and third elements of any one row in the secular determinant evaluated for a specific λ value (λ_k) are, respectively, L_{1k}, L_{2k}, and L_{3k}. These are unnormalized L values. If a different row is selected, a different set of L values will be obtained but the values will be in the same proportion to each other as the first set of values. The L values for a specific λ value are usually tabulated in a column and the complete array makes up the (unnormalized) **L** matrix.

$$\mathbf{L} = \begin{array}{c|ccc} & \lambda_j & \lambda_k & \cdots \\ \hline S_1 & L_{1j} & L_{1k} & \cdots \\ S_2 & L_{2j} & L_{2k} & \cdots \\ \cdots & \cdots & \cdots & \cdots \end{array}$$

14.7   Normal Coordinates

New coordinates Q can be defined which are related to internal coordinates by

$$S_i = \sum L_{ij} Q_j \qquad \text{or} \qquad \mathbf{S} = \mathbf{LQ}$$

$$\left\| \begin{array}{c} S_1 \\ S_2 \\ \cdots \end{array} \right\| = \left\| \begin{array}{ccc} L_{ij} & L_{1k} & \cdots \\ L_{2j} & L_{2k} & \cdots \\ \cdots & \cdots & \cdots \end{array} \right\| \left\| \begin{array}{c} Q_j \\ Q_k \\ \cdots \end{array} \right\| \tag{14.40}$$

where **L** is

$$\begin{array}{c|ccc} & Q_j & Q_k & \cdots \\ \hline S_1 & L_{1j} & L_{1k} & \cdots \\ S_2 & L_{2j} & L_{2k} & \cdots \\ \cdots & \cdots & \cdots & \cdots \end{array}$$

so that a unit change in the coordinate Q_j causes an L_{ij} change in the internal coordinate S_i. The factor L_{ij} is chosen so that the kinetic and potential energy can be written in terms of the coordinates Q_k and the constants λ_k as

$$2T = \sum_k \dot{Q}_k^2 \qquad 2V = \sum_k \lambda_k Q_k^2 \tag{14.41}$$

The coordinates Q in which terms the energy expressions can be written like this without cross terms are called *normal coordinates*.

The Lagrange equation written in terms of normal coordinates is

$$\frac{d}{dt}\left(\frac{\partial T}{\partial \dot{Q}_k}\right) + \frac{\partial V}{\partial Q_k} = 0 \tag{14.42}$$

when this is used with Eq. (14.41) we get

$$\ddot{Q}_1 + \lambda_1 Q_1 = 0 \qquad \ddot{Q}_2 + \lambda_2 Q_2 = 0 \cdots \ddot{Q}_k + \lambda_k Q_k = 0 \qquad (14.43)$$

the solutions of which are, respectively,

$$Q_1 = Q_1^0 \sin(2\pi v_1 t + \alpha_1)$$
$$Q_2 = Q_2^0 \sin(2\pi v_2 t + \alpha_2) \qquad (14.44)$$
$$Q_k = Q_k^0 \sin(2\pi v_k t + \alpha_k)$$

where Q_k^0 and α_k are constants. When these solutions are substituted into Eq. (14.43) it can be seen that $\lambda_k = (2\pi v_k)^2$ and that λ_k in the energy expression Eq. (14.41) has the same meaning as the λ values evaluated in the internal coordinate treatment. This means that each normal coordinate is vibrating with its own independent frequency v_k since λ_1 is only associated with Q_1, etc. Each normal coordinate can be independently activated because there are no cross terms in the energy expressions.

If only one normal coordinate Q_1 is excited, then Eq. (14.40) will become

$$S_i = L_{i1} Q_1 \qquad \text{(where } L_{i2} Q_2 + L_{i3} Q_3 \cdots = 0) \qquad (14.45)$$

Combining this with Eq. (14.44) we get

$$S_i = L_{i1} Q_1^0 \sin(2\pi v_1 t + \alpha_1) \qquad (14.46)$$

This means that when one normal coordinate Q_1 is excited, every internal coordinate S_i (and therefore every atom) vibrates with the same frequency v_1 and with the same phase constant α_1, which means that all the atoms move in-phase in the sense that they all go through the equilibrium position simultaneously. From previous analysis we have seen that this type of motion is a normal mode of vibration. This means that each normal mode of vibration can be characterized by a normal coordinate. As seen from Eq. (14.46) if only one normal coordinate Q_1 is excited then every internal coordinate S_i will vibrate with a relative amplitude proportional to L_{i1}, Q_1^0 being a constant common factor. Therefore, the relative values of the amplitudes of the internal coordinates vibrating during a normal mode of vibration (evaluated in Section 14.6) are the same as the relative values of the factors L_{i1} which relate internal to normal coordinates, in Eq. (14.40). The only step left is normalization.

Substituting the value for S_j when only one normal coordinate Q_k is excited, $S_j = L_{jk} Q_k$, into the potential energy expression using internal coordinates,

$2V = \sum_{ij} F_{ij} S_i S_j$, we get the potential energy V_k when only one normal coordinate Q_k is excited is

$$2V_k = Q_k^2 \sum_{ij} F_{ij} L_{ik} L_{jk} \tag{14.47}$$

If only one normal coordinate is excited Eq. (14.41) gives $2V_k = \lambda_k Q_k^2$. By combining this with Eqs. (14.47) we obtain the normalization condition

$$\lambda_k = \sum_{ij} F_{ij} L_{ik} L_{jk} \tag{14.48}$$

If L'_{ik} is an unnormalized L value obtained from the internal coordinate treatment (Section 14.6), then

$$NL'_{ik} = L_{ik} \tag{14.49}$$

where L_{ik} is a normalized L value, by which is meant an L value which relates internal and normal coordinates, and where N is the normalization factor which can be evaluated from

$$\lambda_k = N^2 \sum_{ij} F_{ij} L'_{ik} L'_{jk} \tag{14.50}$$

For example in a 3×3 case (see Section 14.6) N is given by

$$N = \{\lambda_k / [F_{11}(L'_{1k})^2 + F_{22}(L'_{2k})^2 + F_{33}(L'_{3k})^2 \\ + 2F_{12}(L'_{1k} L'_{2k} + 2F_{13} L'_{1k} L'_{3k} + 2F_{23} L'_{2k} L'_{3k})]\}^{1/2}$$

As we have shown previously, the properties of the normal coordinates (L and λ) can be evaluated from the internal coordinate treatment.

14.8 Potential Energy Distribution

Because the dimensions of L are different for stretching and bending coordinates it is sometimes preferable to use the potential energy distribution to characterize the form of the normal coordinates. In Eq. (14.47) the potential energy when only one normal coordinate is activated is

$$2V_k = Q_k^2 \sum_{ij} F_{ij} L_{ik} L_{jk} \tag{14.47}$$

The potential energy distribution is the fractional part of the potential energy of a normal mode of vibration contributed by each force constant F_{ij}. If each $F_{ij}L_{ik}L_{jk}$ term is divided by the total sum of $F_{ij}L_{ik}L_{jk}$ terms for this vibration which from Eq. (14.48) equals λ_k, then the potential energy distribution (P.E.D.) can be expressed as

$$\text{P.E.D.} = \frac{F_{ij}L_{ik}L_{jk}}{\lambda_k} \qquad (14.51)$$

Because the terms in Eq. (14.47) where $i = j$ are large compared to terms where $i \neq j$, an approximate potential energy distribution for internal (or symmetry) coordinates in each normal coordinate Q_k can be defined in percent using only diagonal terms ($i = j$).

$$\text{P.E.D.}_{\text{diag}} = \left(\frac{100 F_{ii} L_{ik}^2}{\sum_i F_{ii} L_{ik}^2} \right) \% \qquad (14.52)$$

14.9   The Form of the Normal Coordinates

If we want to draw a diagram of the atomic displacements during the vibration we must evaluate the cartesian coordinate displacements per unitary change in the normal coordinates.[13] The transformation from cartesian displacement coordinates to normal coordinates is given by

$$X_j = \sum_k A_{jk} Q_k \qquad \text{or} \qquad \mathbf{X} = \mathbf{A}\mathbf{Q}$$

$$
\begin{Vmatrix} X_1 \\ Y_1 \\ Z_1 \\ X_2 \\ \cdots \end{Vmatrix}
=
\begin{Vmatrix} A_{X1a} & A_{X1b} & \cdots \\ A_{Y1a} & A_{Y1b} & \cdots \\ A_{Z1a} & A_{Z1b} & \cdots \\ A_{X2a} & A_{X2b} & \cdots \\ \cdots & \cdots & \cdots \end{Vmatrix}
\begin{Vmatrix} Q_a \\ Q_b \\ \cdots \\ \cdots \\ \cdots \end{Vmatrix}
\qquad (14.53)
$$

where **A** is

	Q_a	Q_b	$\cdots$
X_1	A_{X1a}	A_{X1b}	$\cdots$
Y_1	A_{Y1a}	A_{Y1b}	$\cdots$
Z_1	A_{Z1a}	A_{Z1b}	$\cdots$
X_2	A_{X2a}	A_{X2a}	$\cdots$
$\cdots$	$\cdots$	$\cdots$	$\cdots$

[13] B. L. Crawford, Jr. and W. H. Fletcher, *J. Chem. Phys.* **19**, 141 (1951).

A unit change in the coordinate Q_a causes an A_{X1a} change in the coordinate X_1. It will be shown in Section 14.22 that $\mathbf{A}$ is given by a matrix product

$$\mathbf{A} = \mathbf{M}^{-1}\mathbf{B}'\mathbf{F}\mathbf{L}\mathbf{\Lambda}^{-1} \tag{14.54}$$

All these matrices have been illustrated previously except $\mathbf{\Lambda}^{-1}$ which is

$$\begin{Vmatrix} 1/\lambda_1 & 0 & 0 & \cdots \\ 0 & 1/\lambda_2 & 0 & \cdots \\ 0 & 0 & 1/\lambda_3 & \cdots \\ \cdots & \cdots & \cdots & \cdots \end{Vmatrix}$$

Where λ_1 is associated with Q_1 and so forth.

14.10 Symmetry Coordinates

If the molecules has some symmetry the solution of the secular determinant can be simplified by the use of symmetry coordinates. Symmetry coordinates are linear combinations of the internal coordinates. After a little experience one can frequently select symmetry coordinates intuitively by taking combinations of symmetry related internal coordinates. The symmetry coordinates picked out are not necessarily unique but must have definite properties. They must be normal, orthogonal, and they must transform properly as will be described below. A symmetry coordinate has the form

$$\mathscr{S}_j = \sum_k U_{jk} S_k \qquad \text{or} \qquad \mathscr{S} = \mathbf{U}\mathbf{S}$$

$$\begin{Vmatrix} \mathscr{S}_1 \\ \mathscr{S}_2 \\ \cdots \end{Vmatrix} = \begin{Vmatrix} U_{11} & U_{12} & \cdots \\ U_{21} & U_{22} & \cdots \\ \cdots & \cdots & \cdots \end{Vmatrix} \begin{Vmatrix} S_1 \\ S_2 \\ \cdots \end{Vmatrix} \tag{14.55}$$

where $\mathbf{U}$ is

	S_1	S_2	$\cdots$
$\mathscr{S}_1$	U_{11}	U_{12}	$\cdots$
$\mathscr{S}_2$	U_{21}	U_{22}	$\cdots$
$\cdots$	$\cdots$	$\cdots$	$\cdots$

When $\mathscr{S}_j$ is the jth symmetry coordinate, S_k is the kth internal coordinate and U_{jk} is a suitable coefficient for S_k.

Since symmetry coordinates are not necessarily unique, a perfectly general method for obtaining them cannot be given. Several methods are available for obtaining trial coordinates. For example, a coordinate may be given by[14]

$$\mathscr{S}_j = \sum \chi_i(R)RD_k \qquad (14.56)$$

where $\chi_i(R)$ is the character for the vibration species i (A_1, etc.) and covering operation $R(C_3$, etc.) and is obtained from the group character table. RD_k is the coordinate which the generating coordinate, D_k, becomes when each operation, R, is performed and the equation is summed over all operations of the group. D_k is the "generating coordinate" and may be a single internal coordinate or some linear combination of internal coordinates.

The trial coordinates must be shown to have the necessary properties. The condition for normality is

$$\sum_k U_{jk}^2 = 1 \qquad (14.57)$$

Where U_{jk}^2 is the square of a coefficient of the jth symmetry coordinate and the summation is carried over all the k internal coordinates involved in that symmetry coordinate. The condition therefore requires the sum of the squares of the coefficients in each symmetry coordinate be one. The coordinates will be orthogonal if

$$\sum_k U_{jk} U_{lk} = 0 \qquad (14.58)$$

where l and j refer to different symmetry coordinates. That is, the coefficient of an internal coordinate in one symmetry coordinate is multiplied by the corresponding (same internal coordinate) coefficient in another symmetry coordinate and these products are summed over all internal coordinates. The summation must be zero and the same result must hold true for every symmetry coordinate applied to all the others for them to be mutually orthogonal.

The corrections of transformation can be shown by use of the point group character table and a table of transformations of the internal coordinates for each symmetry operation. The table of transformation lists what happens to each of the internal coordinates as a result of each of the covering operations. The coordinates may remain unchanged or become one of the other internal coordinates. Each of the covering operations is performed on the symmetry coordinate and the result is either no change (represented by a factor $+1$) or a new coordinate differing by a factor (-1 for nondegenerate vibrations). The factor (1 or -1) must be the same as the character of the covering operation

[14] J. R. Nielson and L. H. Berryman, *J. Chem. Phys.* **17**, 659 (1949).

for the vibration species to which the symmetry coordinate belongs. The case for degenerate species is somewhat more involved and will be discussed with the chloroform example.

Since the $\mathbf{U}$ matrix in Eq. (14.55) is orthogonal, the inverse $\mathbf{U}^{-1}$ will equal the transpose $\mathbf{U}'$. Therefore the identity matrix $\mathbf{I}$ which equals $\mathbf{U}^{-1}\mathbf{U}$ by definition will also equal $\mathbf{U}'\mathbf{U}$. We can insert this last expression for the identity matrix into the potential energy expression

$$2V = \mathbf{S}'\mathbf{F}\mathbf{S} = \mathbf{S}'\mathbf{U}'\mathbf{U}\mathbf{F}\mathbf{U}'\mathbf{U}\mathbf{S} \qquad (14.59)$$

The transpose of $\mathscr{S}$ in Eq. (14.55) equals the transpose of $(\mathbf{US})$ which equals $\mathbf{S}'\mathbf{U}'$ since in matrix algebra the transpose of a product equals the product of the transposes taken in reverse order. Therefore we can write

$$2V = \mathscr{S}'\mathscr{F}\mathscr{S} \qquad \text{where} \qquad \mathscr{F} = \mathbf{U}\mathbf{F}\mathbf{U}' \qquad (14.60)$$

In a similar manner the kinnetic energy expression is

$$2T = \dot{\mathbf{S}}'\mathbf{G}^{-1}\dot{\mathbf{S}} = \dot{\mathscr{S}}'\mathscr{G}^{-1}\dot{\mathscr{S}} \qquad \text{where} \qquad \mathscr{G}^{-1} = \mathbf{U}\mathbf{G}^{-1}\mathbf{U}' \qquad (14.61)$$

When both sides of this last expression are premultiplied by $\mathbf{U}\mathbf{G}\mathbf{U}'$ and postmultiplied by $\mathscr{G}$, one obtains after removing identity matrices such as $(\mathbf{U}'\mathbf{U})$ and $(\mathbf{G}\mathbf{G}^{-1})$

$$\mathscr{G} = \mathbf{U}\mathbf{G}\mathbf{U}' \qquad (14.62)$$

When internal coordinates are used, the $\mathbf{F}$ and $\mathbf{G}$ matrices and the secular determinant are square as in Fig. 14.5(a). When symmetry coordinates which transform as described above are used, the $\mathscr{F}$ matrix obtained from $\mathbf{U}\mathbf{F}\mathbf{U}'$ and the $\mathscr{G}$ matrix obtained from $\mathbf{U}\mathbf{G}\mathbf{U}'$ and the secular determinant $|\mathscr{F}\mathscr{G} - \mathbf{I}\lambda| = 0$ all consist of smaller square blocks, one for each symmetry species, distributed along the diagonal as in Fig. 14.5(b). Since all terms in the nonshaded part of Fig. 14.5(b) are zero the problem factors into the symmetry species.

$$|\mathscr{F}_a\mathscr{G}_a - \mathbf{I}\lambda| = 0 \qquad |\mathscr{F}_b\mathscr{G}_b - \mathbf{I}\lambda| = 0 \quad \cdots \qquad (14.63)$$

where $\mathscr{F}_a$ and $\mathscr{G}_a$ are matrices for the (a) symmetry species, and so forth. Since the determinants are smaller they can be solved considerably more easily for the λ values for each symmetry species.

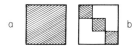

FIG. 14.5.   Factoring the secular determinant.

When symmetry coordinates are used an $\mathscr{L}$ matrix is derived which relates symmetry coordinates to normal coordinates

$$\mathscr{S} = \mathscr{L}\mathbf{Q} \qquad (14.64)$$

Since we know that $\mathscr{S} = \mathbf{US}$ and $\mathbf{S} = \mathbf{LQ}$ we can write $\mathscr{S} = \mathbf{ULQ}$ so $\mathbf{UL} = \mathscr{L}$ and

$$\mathbf{L} = \mathbf{U'}\mathscr{L} \qquad (14.65)$$

This relationship enables us to calculate the $\mathbf{L}$ matrix from the $\mathscr{L}$ matrix.

14.11 The CHCl$_3$ Molecule

The chloroform molecule will be used as an example for vibrational analysis. The CHCl$_3$ molecule has one threefold axis of symmetry and three σ_v planes of symmetry. Chloroform thus belongs to the C_{3v} point group as discussed in Chapter 3. The symmetry elements, symmetry operations, selection rules, and vibration types have been discussed in detail in Section 3.6 where it was shown that chloroform has nine fundamental vibrations, three A_1 vibrations, and three double degenerate type E vibrations. The results of the calculation to follow are presented in Fig. 14.6 where the forms of the vibrations are illustrated.

The observed infrared spectra of CHCl$_3$ and CDCl$_3$ are illustrated in Fig. 14.7 and in Tables 14.1 and 14.2. The observed Raman spectra are given in the same tables, and in Figs. 1.31 and 1.35. All wavenumbers are for the liquid state and are best estimates taken from the data of Plyler,[15] Cleveland,[16,17] and unpublished data.[18]

There are six strong Raman lines and these are taken as the six fundamentals. This procedure is by no means always correct but is supported by the rest of the data. The 3018, 668, and 367 cm^{-1} lines are polarized in the Raman spectrum and therefore are assigned to the completely symmetric A_1 type species and are listed as ν_1, ν_2, and ν_3 in order of decreasing wavenumber. The 1216, 758, are 261 cm^{-1} lines in the Raman spectrum are com-

[15] E. K. Plyler and W. S. Benedict, *J. Res. Nat. Bur. Stand.* **47**, 202 (1951).

[16] J. R. Madigan and F. F. Cleveland, *J. Chem. Phys.* **19**, 119 (1951).

[17] J. R. Madigan and F. F. Cleveland, *J. Chem. Phys.* **18**, 1081 (1950).

[18] Data from the Laboratories at Rensselaer Polytechnic Institute, Troy, New York (1963).

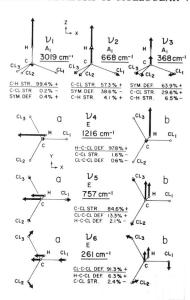

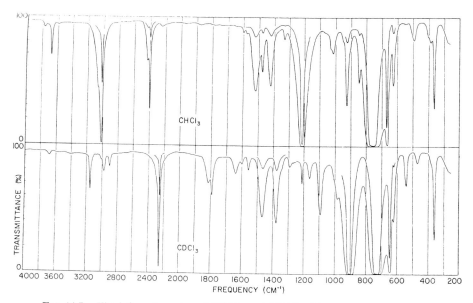

FIG. 14.6.   The cartesian coordinate displacements per unitary change in the normal coordinates for $CHCl_3$. The potential energy distribution for each mode is listed in table form. The plus and minus sign come from the $\mathscr{L}$ matrix showing the relative phases of the symmetry coordinates involved in the normal coordinates.

FIG. 14.7.   The infrared spectra of $CHCl_3$ and $CDCl_3$. These spectra were run on a PE 521 grating spectrometer in a 0.180 mm cell with some regions run in a 0.023 mm cell.

pletely depolarized and are the three type E vibrations listed as ν_4, ν_5, and ν_6. Essentially all the observed wavenumbers in Table 14.1 may be accounted for as resulting from allowed combinations or overtones of the selected six fundamentals, which gives additional support for their choice. The corresponding assignment is given for CDCl$_3$ in Table 14.2. It may be noted that ν_1 and ν_4 which chiefly involve CH stretching and CH bending are shifted to considerably lower wavenumbers in the deuterated chloroform spectrum, as expected, where as the other fundamentals are much less affected by deuterium substitution. See Section 3.16 for the product rule calculation for CHCl$_3$ and CDCl$_3$.

TABLE 14.1

INFRARED AND RAMAN SPECTRA OF CHCl$_3$ (LIQUID)[a]

ν	Infrared intensity	Type	Assignment	$\Delta\nu$	Raman intensity polarization
3683	w	A_1	$\nu_1 + \nu_2$		
3019	s	A_1	ν_1	3018	s (polarized)
2400	m	$A_1 + E$	$2\nu_4$		
1521	m	$A_1 + E$	$2\nu_5$	1518	w
1475	w	$A_1 + A_2 + E$	$\nu_4 + \nu_6$	1501	vw
1423	m	E	$\nu_2 + \nu_5$	1420	vw
1334	w	A_1	$2\nu_2$		
1216	vs	E	ν_4	1216	m (depolarized)
1032	w	A_1	$\nu_2 + \nu_3$	1024	vw
1019	w	$A_1 + A_2 + E$	$\nu_5 + \nu_6$		
929	m	E	$\nu_2 + \nu_6$		
849	w	E	$\nu_4 - \nu_3$		
757	vs	E	ν_5	758	s (depolarized)
668	s	A_1	ν_2	668	s (polarized)
627	m	E	$\nu_3 + \nu_6$	622	vw
497	w	$A_1 + A_2 + E$	$\nu_5 - \nu_6$	497	vw
407	w	E	$\nu_2 - \nu_6$		
392	w	$A_1 + E$	$\nu_5 - \nu_3$		
368	m	A_1	ν_3	367	s (polarized)
303	vw	A_1	$\nu_2 - \nu_3$		
261	w	E	ν_6	261	vs (depolarized)
230	w	$A_1 + A_2 - 2E$	$\nu_5 - 2\nu_6$		

[a] The frequencies given are in cm^{-1} and the band intensities are abbreviated as follows: vs—very strong; s—strong; m—moderate; w—weak; vw—very weak.

TABLE 14.2

INFRARED AND RAMAN SPECTRA OF CDCl$_3$ (LIQUID)[a]

ν	Infrared intensity	Type	Assignment	Raman frequencies $(\Delta\nu)$
3155	w	E	$\nu_1 + \nu_4$	
2983	w	E	$\nu_1 + \nu_5$	
2902	w	A_1	$\nu_1 + \nu_2$	
2255	s	A_1	ν_1	2254
1817	w	$A_1 + E$	$\nu_3 + 2\nu_5$	
1795	w	$A_1 + E$	$2\nu_4$	1796
1643	vw	$A_1 + A_2 + E$	$\nu_4 + \nu_5$	1642
1562	vw	E	$\nu_2 + \nu_4$	
1468	m	$A_1 + E$	$2\nu_5$	1463
1382	m	E	$\nu_2 + \nu_5$	
1295	w	A_1	$2\nu_2$	
1168	w	$A_1 + A_2 + E$	$\nu_4 + \nu_6$	
1097	m	E	$\nu_3 + \nu_5$	
989	w	$A_1 + A_2 + E$	$\nu_5 + \nu_6$	
905	vs	E	ν_4	907
740	vs	E	ν_5	736
650	s	A_1	ν_2	650
627	m	$A_1 + A_2 + E$	$\nu_3 + \nu_6$	
545	w	E	$\nu_4 - \nu_3$	
475	w	$A_1 + A_2 + E$	$\nu_5 - \nu_6$	365
366	m	A_1	ν_3	261

[a] The frequencies given are in cm^{-1} and the band intensities are abbreviated as in Table 14.5.

14.12   The Internal Coordinates for CHCl$_3$

Referring to Section 14.3 it can be seen that chloroform has only two types of internal coordinates, bond stretching, and angle bending. These are illustrated in Fig. 14.8. The four bond stretching internal coordinates are r, t_1', t_2,

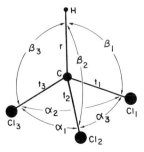

FIG. 14.8.   Internal coordinates for chloroform.

and t_3. Again referring to Section 14.3, chloroform has five independent bending coordinates but has a maximum number of six bending coordinates which are designated as α_1, α_2, α_3, β_1, β_2, and β_3. In order to take advantage of the symmetry of the molecule, all six will be used even though they are not all independent. Therefore, the ten selected internal coordinates will include one redundant coordinate since there are only nine fundamentals.

14.13 The Symmetry Coordinates for CHCl$_3$

Symmetry coordinates for chloroform may be obtained from Eq. (14.56). The characters [$\chi_i(R)$ values] are obtained from the C_{3v} character table, shown in Table 14.3, and the generating coordinates, D_k, from the internal coordinates.

TABLE 14.3

C_{3v} CHARACTER TABLE

C_{3v}	I	C_3	σ_v
A_1	1	1	1
A_2	1	1	-1
E	2	-1	0

For example, using r as D_k and the $\chi_i(R)$ values for the A_1 class, Eq. (14.56) is

$$\mathscr{S}_j^i = \sum_R \chi_i(R) R D_k$$
$$\mathscr{S}_1^{A_1} = (1)r + (1)r + (1)r + (1)r + (1)r \tag{14.66}$$

since none of the symmetry operations, R, effect r, and all the characters for A_1 vibrations are one. (See Table 14.4 to conveniently obtain the RD_k values). The correct coordinate, however, would be

$$S_1 = r \tag{14.67}$$

since Eq. (14.66) is obviously not normal $\left(\sum_k U_{jk}^2 \neq 1\right)$ and the A_1 superscript has been omitted. Using t_1 as D_k gives

$$\mathscr{S}_2 = t_1 + t_2 + t_3 + t_1 + t_2 + t_3 = 2(t_1 + t_2 + t_3) \tag{14.68}$$

which would be

$$\mathscr{S}_2 = 1/\sqrt{3}(t_1 + t_2 + t_3) \tag{14.69}$$

since $1/(3^{1/2}) + 1/(3^{1/2}) + 1/(3^{1/2})$ is one and the coordinate would then be normal. It can be seen that the symmetry coordinates from Eq. (14.56) have the correct ratio for the internal coordinates but the actual coefficient must be picked to give a normal coordinate. Two more A_1 symmetry coordinates are given by using $\alpha_1 - \beta_1$ and $\alpha_1 + \beta_1$ as D_k values to give

$$\begin{aligned}\mathscr{S}_3 = (\alpha_1 - \beta_1) + (\alpha_3 - \beta_3) + (\alpha_2 - \beta_2) + (\alpha_2 - \beta_2) \\ + (\alpha_3 - \beta_3) + (\alpha_1 - \beta_1)\end{aligned} \qquad (14.70)$$

and

$$\begin{aligned}\mathscr{S}_4 = (\alpha_1 + \beta_1) + (\alpha_3 + \beta_3) + (\alpha_2 + \beta_2) + (\alpha_2 + \beta_2) \\ + (\alpha_3 + \beta_3) + (\alpha_1 + \beta_1)\end{aligned} \qquad (14.71)$$

respectively. The two normalized coordinates would then be

$$\mathscr{S}_3 = 1/\sqrt{6}(\alpha_1 + \alpha_2 + \alpha_3 - \beta_1 - \beta_2 - \beta_3) \qquad (14.72)$$

and

$$\mathscr{S}_4 = 1/\sqrt{6}(\alpha_1 + \alpha_2 + \alpha_3 + \beta_1 + \beta_2 + \beta_3) \qquad (14.73)$$

There are only three A_1 vibrations for $CHCl_3$ and thus only three symmetry coordinates are necessary, indicating that one of the four is redundant and is either equal to zero or not independent of the others. This point will be discussed later in this section. The coordinates must first be shown to be orthogonal and transform properly. $\mathscr{S}_1$ is orthogonal with $\mathscr{S}_2$ since

$$\sum_k U_{jk} U_{lk} = (1)(0) + (0)(1/3) + (0)(1/3) + (0)(1/3) = 0 \qquad (14.74)$$

and $\mathscr{S}_1$ would also be orthogonal with $\mathscr{S}_3$ and $\mathscr{S}_4$ because no other internal coordinate appears in $\mathscr{S}_1$ and r is not present in any other symmetry coordinate. The same is also true of $\mathscr{S}_2$ since t_1, t_2, and t_3 are not present in any other symmetry coordinate and either U_{jk} or U_{lk} would always be zero. $\mathscr{S}_3$ and $\mathscr{S}_4$ are orthogonal because

$$\begin{aligned}\sum_k U_{jk} U_{lk} = (1/\sqrt{6})(1/\sqrt{6}) + (1/\sqrt{6})(1/\sqrt{6}) + (1/\sqrt{6})(1/\sqrt{6}) \\ + (-1/\sqrt{6})(1/\sqrt{6}) + (-1/\sqrt{6})(1/\sqrt{6}) + (-1/\sqrt{6})(1/\sqrt{6}) = 0\end{aligned} \qquad (14.75)$$

and thus all four coordinates are mutually orthogonal.

Correctness of transformation can be shown by use of Table 14.4 which is the table of transformations for the internal coordinates of chloroform. For example taking $\mathscr{S}_2$ and performing the indicated symmetry operations

$$(I)\mathscr{S}_2 = 1/\sqrt{3}(t_1 + t_2 + t_3) = (1)\mathscr{S}_2$$
$$(C_3^+)\mathscr{S}_2 = 1/\sqrt{3}(t_3 + t_1 + t_2) = (1)\mathscr{S}_2 \qquad (14.76)$$
$$(\sigma_{v_1})\mathscr{S}_2 = 1/\sqrt{3}(t_1 + t_3 + t_2) = (1)\mathscr{S}_2$$

which means the transformed coordinate is the same as the original or multiplication of the original coordinate by one gives the transformed coordinate. The characters for all covering operations for A_1 vibrations are one and thus $\mathscr{S}_2$ transforms as an A_1 vibration. Referring to Table 14.4 it can easily be shown that all four coordinates are unchanged by all transformations and thus all have the plus one character necessary for A_1 vibrations.

TABLE 14.4

TRANSFORMATION OF INTERNAL COORDINATES

	I	C_3^+	C_3^-	σ_1	σ_2	σ_3
r	r	r	r	r	r	r
t_1	t_1	t_3	t_2	t_1	t_3	t_2
t_2	t_2	t_1	t_3	t_3	t_2	t_1
t_3	t_3	t_2	t_1	t_2	t_1	t_3
α_1	α_1	α_3	α_2	α_1	α_3	α_2
α_2	α_2	α_1	α_3	α_3	α_2	α_1
α_3	α_3	α_2	α_1	α_2	α_1	α_3
β_1	β_1	β_3	β_2	β_1	β_3	β_2
β_2	β_2	β_1	β_3	β_3	β_2	β_1
β_3	β_3	β_2	β_1	β_2	β_1	β_3

The four coordinates which have been shown to have all the necessary characteristics for symmetry coordinates are

$$\mathscr{S}_1 = r$$
$$\mathscr{S}_2 = 1/\sqrt{3}(t_1 + t_2 + t_3)$$
$$\mathscr{S}_3 = 1/\sqrt{6}(\alpha_1 + \alpha_2 + \alpha_3 - \beta_1 - \beta_2 - \beta_3) \qquad (14.77)$$
$$\mathscr{S}_4 = 1/\sqrt{6}(\alpha_1 + \alpha_2 + \alpha_3 + \beta_1 + \beta_2 + \beta_3) = 0$$

Since only three coordinates are necessary for the A_1 vibrations, $\mathscr{S}_4$ may be picked as the redundant coordinate. This is permissible because the sum of all the changes around a point (the carbon atom) would be zero and thus not affect the kinetic or potential energies.

The E type vibrations introduces a somewhat different approach to the symmetry coordinates. There are also three E vibrations for chloroform but, since they are doubly degenerate, six symmetry coordinates are needed. The six will be three pairs, usually designated a and b. Generally each pair will involve the same type internal coordinate.

Taking t_1 as a generating coordinate and again using Eq. (14.56) of Section 14.10 with the characters for E vibrations gives

$$\mathscr{S}_{1a} = (2)t_1 + (-1)t_2 + (-1)t_3 + (0)t_1 + (0)t_3(0)t_2 = 2t_1 - t_2 - t_3 \quad (14.78)$$

which must be multiplied by $1/\sqrt{6}$ to give the normal coordinate

$$\mathscr{S}_{1a} = 1/\sqrt{6}(2t_1 - t_2 - t_3) \quad (14.79)$$

Similar coordinates may be obtained using α_1 and β_1 as generating coordinates to give

$$\mathscr{S}_{2a} = 1/\sqrt{6}(2\alpha_1 - \alpha_2 - \alpha_3) \quad (14.80)$$

and

$$\mathscr{S}_{3a} = 1/\sqrt{6}(2\beta_1 - \beta_2 - \beta_3) \quad (14.81)$$

respectively.

The second coordinate of the first pair may be obtained using $t_2 - t_3$ as D_k to give

$$\begin{aligned}
\mathscr{S}_{1b} &= 2(t_2 - t_3) + (-1)(t_1 - t_2) + (-1)(t_3 - t_1) + (0)(t_3 - t_2) \\
&\quad + (0)(t_2 - t_1) + (0)(t_1 - t_3) \\
&= 2t_2 - 2t_3 - t_1 + t_2 - t_3 + t_1 \\
&= 3(t_2 - t_3).
\end{aligned} \quad (14.82)$$

which multiplied by $1/2(2^{1/2})$ gives the normal coordinate

$$\mathscr{S}_{1b} = 1/\sqrt{2}(t_2 - t_3). \quad (14.83)$$

Following the same procedure with $\alpha_2 - \alpha_3$ and $\beta_2 - \beta_3$ as generating coordinates gives

$$\mathscr{S}_{2b} = 1/\sqrt{2}(\alpha_2 -)\alpha_3 \quad (14.84)$$

and

$$\mathscr{S}_{3b} = 1/\sqrt{2}(\beta_2 - \beta_3) \tag{14.85}$$

respectively.

The six coordinates must be proved orthogonal. $\mathscr{S}_{1a}$ and $\mathscr{S}_{1b}$ are orthogonal since

$$\sum_k U_{jk} U_{lk} = (2/\sqrt{6})(0) + (-1/\sqrt{6})(1/\sqrt{2}) + (1/\sqrt{6})(-1/\sqrt{2}) = 0 \tag{14.86}$$

$\mathscr{S}_{1a}$ and $\mathscr{S}_{1b}$ are obviously orthogonal with the other four coordinates because no t appears in $\mathscr{S}_{2a}$, $\mathscr{S}_{2b}$, $\mathscr{S}_{3a}$, or $\mathscr{S}_{3b}$. In the same way $\mathscr{S}_{2a}$ and $\mathscr{S}_{2b}$ and $\mathscr{S}_{3a}$ and $\mathscr{S}_{3b}$ can be shown to be orthogonal with each other.

The correctness of transformation for the coordinates of degenerate vibrations differs from the A_1 coordinates because a covering operation applied to one of a pair of E coordinates does not give the same coordinate or its negative. Instead a linear combination of the two coordinates forming the pair is produced. For example, when the identity operation, I, is applied to $\mathscr{S}_{1a}$

$$(I)\mathscr{S}_{1a} = 1/\sqrt{6}(2t_1 - t_2 - t_3) = A\mathscr{S}_{1a} + B\mathscr{S}_{1b} \tag{14.87}$$

or

$$1/\sqrt{6}(2t_1 - t_2 - t_3) = A(2t_1 - t_2 - t_3)1/\sqrt{6} + B(t_2 - t_3)1/\sqrt{2} \tag{14.88}$$

where A and B are constants. Doing the same for $\mathscr{S}_{1b}$ gives

$$(I)\mathscr{S}_{1b} = 1/\sqrt{2}(t_3 - t_3) = A'\mathscr{S}_{1a} + B'\mathscr{S}_{1b} \tag{14.89}$$

or

$$1/\sqrt{2}(t_2 - t_3) = A'(2t_1 - t_2 - t_3)1/\sqrt{6} + B'(t_2 - t_3)1/\sqrt{2} \tag{14.90}$$

where A' and B' are constants. By equating coefficients the constants may be found. For example, using t_1 in Eq. (14.88) gives

$$2A/\sqrt{6} = 2/\sqrt{6} \qquad A = 1 \tag{14.91}$$

or using t_2 in the same equations

$$-A/\sqrt{6} + B/\sqrt{2} = -1/\sqrt{6} \qquad (14.92)$$

and since A is one

$$B/\sqrt{2} = -1/\sqrt{6} + 1/\sqrt{6} = 0 \qquad B = 0 \qquad (14.93)$$

In the same way A' can be shown to be zero and B' to be one. Then, forming the matrix

$$\left\| \begin{matrix} A & B \\ A' & B' \end{matrix} \right\| = \left\| \begin{matrix} 1 & 0 \\ 0 & 1 \end{matrix} \right\| \qquad (14.94)$$

in which the sum of the elements along the principal diagonal is two, it can be seen that this is the character for E vibrations under the identity operation and thus $\mathscr{S}_{1a}$ and $\mathscr{S}_{1b}$ transform properly.

$$(C_3^+)\mathscr{S}_{1a} = (2t_3 - t_1 - t_2)1/\sqrt{6}$$
$$= A(2t_1 - t_2 - t_3)1/\sqrt{6} + B(t_2 - t_3)1/\sqrt{2} \qquad (14.95)$$

and

$$(C_3^+)\mathscr{S}_{1b} = (t_1 - t_2)1/\sqrt{2}$$
$$= A'(2t_1 - t_2 - t_3)1/\sqrt{6} + B'(t_2 - t_3)1/\sqrt{2} \qquad (14.96)$$

Equating coefficients for t_1 in Eq. (14.95)

$$2A/\sqrt{6} = -1/\sqrt{6} \qquad A = -1/2 \qquad (14.97)$$

and for t_2 in the same equation

$$-A/\sqrt{6} + B/\sqrt{2} = -1/\sqrt{6} \qquad (14.98)$$

and since A is $-1/2$

$$\tfrac{1}{2}/\sqrt{6} + B/\sqrt{2} = -1/\sqrt{6} \qquad B = -\sqrt{3}/2 \qquad (14.99)$$

In the same way, using Eq. (14.53), A' is $\sqrt{3}/2$ and B' is $-1/2$ to give the matrix

$$\begin{Vmatrix} A & B \\ A' & B' \end{Vmatrix} = \begin{Vmatrix} -1/2 & -\sqrt{3}/2 \\ \sqrt{3}/2 & -1/2 \end{Vmatrix} \qquad (14.100)$$

in which the principal diagonal is -1 in agreement with the character for E vibrations under C_3^+.

The matrix for the C_3^- operation is

$$\begin{Vmatrix} A & B \\ A' & B' \end{Vmatrix} = \begin{Vmatrix} -1/2 & \sqrt{3}/2 \\ -\sqrt{3}/2 & -1/2 \end{Vmatrix} \qquad (14.101)$$

again giving the correct -1 character. The matrices for the three σ_v reflections are

$$\begin{Vmatrix} A & B \\ A' & B' \end{Vmatrix} = \begin{Vmatrix} 1 & 0 \\ 0 & -1 \end{Vmatrix}$$

$$\begin{Vmatrix} A & B \\ A' & B' \end{Vmatrix} = \begin{Vmatrix} -1/2 & -\sqrt{3}/2 \\ -\sqrt{3}/2 & 1/2 \end{Vmatrix} \qquad (14.102)$$

$$\begin{Vmatrix} A & B \\ A' & B' \end{Vmatrix} = \begin{Vmatrix} -1/2 & \sqrt{3}/2 \\ \sqrt{3}/2 & 1/2 \end{Vmatrix}$$

in which the sum of the elements along the principal diagonal is zero for all three, which is the character for E vibrations under σ_v.

All members of a given set must transform in an identical way. This means that the transformation matrices for $\mathcal{S}_{2a}$ and $\mathcal{S}_{2b}$ and those for $\mathcal{S}_{3a}$ and $\mathcal{S}_{3b}$ must be identical to those for $\mathcal{S}_{1a}$ and $\mathcal{S}_{1b}$ [Eqs. (14.94), (14.100), (14.101), and (14.106)]. This can be verified by the reader.

Before leaving the discussion of symmetry coordinates it should be mentioned that the choice of generating coordinates has been somewhat arbitrary and others might have been tried first. For example, if $t_1 - t_2$ were used instead of $t_2 - t_3$, the $\mathcal{S}_{1b}$ coordinate would be

$$\mathcal{S}_{1b}^* = 1/\sqrt{2}(t_1 - t_2) \qquad (14.103)$$

instead of Eq. (14.40). However, this coordinate would not be orthogonal with $\mathscr{S}_{1a}$ since

$$\sum_k U_{jk} U_{1k} = (2/\sqrt{6})(1/\sqrt{2}) + (-1/\sqrt{6})(-1/\sqrt{2}) + (-1/\sqrt{6})(0) \neq 0$$
(14.104)

and the pair $\mathscr{S}_{1b}^*$ and $\mathscr{S}_{1a}$ would not transform properly since

$$\left\| \begin{array}{cc} A & B \\ A' & B' \end{array} \right\| = \left\| \begin{array}{cc} -1/2 & 0 \\ -3 & 1 \end{array} \right\|$$
(14.105)

is the matrix for the C_3^- operation and the sum of the elements along the principal diagonal is not -1 as it should be for an E vibration.

On the other hand, it is possible that several generating coordinates may give identical symmetry coordinates. The problem then is to try various generating coordinates until a sufficient number of proper symmetry coordinates is obtained. The symmetry coordinates are represented in Fig. 14.9.

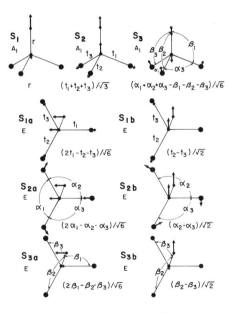

FIG. 14.9. Symmetry coordinates for chloroform. These symmetry coordinate drawings indicate schematically the major motions resulting from specified combinations of internal coordinates.

The complete **U** matrix is

$$\mathbf{U} = $$

		r	t_1	t_2	t_3	α_1	α_2	α_3	β_1	β_2	β_3
A_1	$\mathscr{S}_1$	1	0	0	0	0	0	0	0	0	0
	$\mathscr{S}_2$	0	$\frac{1}{\sqrt{3}}$	$\frac{1}{\sqrt{3}}$	$\frac{1}{\sqrt{3}}$	0	0	0	0	0	0
	$\mathscr{S}_3$	0	0	0	0	$\frac{1}{\sqrt{6}}$	$\frac{1}{\sqrt{6}}$	$\frac{1}{\sqrt{6}}$	$\frac{-1}{\sqrt{6}}$	$\frac{-1}{\sqrt{6}}$	$\frac{-1}{\sqrt{6}}$
	$\mathscr{S}_4$	0	0	0	0	$\frac{1}{\sqrt{6}}$	$\frac{1}{\sqrt{6}}$	$\frac{1}{\sqrt{6}}$	$\frac{1}{\sqrt{6}}$	$\frac{1}{\sqrt{6}}$	$\frac{1}{\sqrt{6}}$
$E_{(a)}$	$\mathscr{S}_{1a}$	0	$\frac{2}{\sqrt{6}}$	$\frac{-1}{\sqrt{6}}$	$\frac{-1}{\sqrt{6}}$	0	0	0	0	0	0
	$\mathscr{S}_{2a}$	0	0	0	0	$\frac{2}{\sqrt{6}}$	$\frac{-1}{\sqrt{6}}$	$\frac{-1}{\sqrt{6}}$	0	0	0
	$\mathscr{S}_{3a}$	0	0	0	0	0	0	0	$\frac{2}{\sqrt{6}}$	$\frac{-1}{\sqrt{6}}$	$\frac{-1}{\sqrt{6}}$
$E_{(b)}$	$\mathscr{S}_{1b}$	0	0	$\frac{1}{\sqrt{2}}$	$\frac{-1}{\sqrt{2}}$	0	0	0	0	0	0
	$\mathscr{S}_{2b}$	0	0	0	0	0	$\frac{1}{\sqrt{2}}$	$\frac{-1}{\sqrt{2}}$	0	0	0
	$\mathscr{S}_{3b}$	0	0	0	0	0	0	0	0	$\frac{1}{\sqrt{2}}$	$\frac{-1}{\sqrt{2}}$

$$(14.106)$$

This **U** matrix can be used to convert the **F** or **G** matrices based on internal coordinates to $\mathscr{F}$ and $\mathscr{G}$ matrices based on symmetry coordinates. When this is done both $\mathscr{F}$ and $\mathscr{G}$ matrices will break up as in Fig. 14.4 into blocks on the diagonal, a 4×4 for the A_1 species and a 6×6 block for the E species. Furthermore, the 6×6 block for the E species breaks up into two identical 3×3 blocks for the a and b pairs. All elements outside the blocks are zero.

The $\mathscr{F}_{A_1}$ matrix block for the A_1 species can be obtained separately by using only that part of the **U** matrix which contains the A_1 symmetry coordinates (the first four rows), so that $\mathscr{F}_{A_1} = \mathbf{U}_{A_1} \mathbf{F} \mathbf{U}'_{A_1}$. The same holds for the $\mathscr{G}$ matrix calculation. However, because of the redundant coordinate $\mathscr{S}_4$, one of the roots of the 4×4 A_1 secular determinant will be zero. Because $\mathscr{S}_4 = 0$ the fourth row and column of the $\mathscr{F}_{A_1}$ and $\mathscr{G}_{A_1}$ matrices may simply be deleted. The same effect is achieved by deleting the $\mathscr{S}_4$ row from the U_{A_1} matrix. The resulting 3×3 A_1 secular determinant will have the same roots as the nonzero roots of the 4×4 A_1 secular determinant retaining the redundant coordinate. Because the two 3×3 a and b blocks of the $\mathscr{F}_{E(a)}$ and $\mathscr{F}_{E(b)}$ matrices are identical, only one of the two E sets in the U matrix need be used (for example, the last three rows only). The same holds true for generating the $\mathscr{G}_{E(a)}$ and $\mathscr{G}_{E(b)}$ matrices.

14.14   The G Matrix for CHCl$_3$

The G matrix for chloroform is writtten

	r	t_1	t_2	t_3	α_1	α_2	α_3	β_1	β_2	β_3
r	G_{rr}	G_{rt_1}	G_{rt_2}	G_{rt_3}	$G_{r\alpha_1}$	$G_{r\alpha_2}$	$G_{r\alpha_3}$	$G_{r\beta_1}$	$G_{r\beta_2}$	$G_{r\beta_3}$
t_1		$G_{t_1t_1}$	$G_{t_1t_2}$	$G_{t_1t_3}$	$G_{t_1\alpha_1}$	$G_{t_1\alpha_2}$	$G_{t_1\alpha_3}$	$G_{t_1\beta_1}$	$G_{t_1\beta_2}$	$G_{t_1\beta_3}$
t_2			$G_{t_2t_2}$	$G_{t_2t_3}$	$G_{t_2\alpha_1}$	$G_{t_2\alpha_2}$	$G_{t_2\alpha_3}$	$G_{t_2\beta_1}$	$G_{t_2\beta_2}$	$G_{t_2\beta_3}$
t_3				$G_{t_3t_3}$	$G_{t_3\alpha_1}$	$G_{t_2\alpha_2}$	$G_{t_3\alpha_3}$	$G_{t_3\beta_1}$	$G_{t_3\beta_2}$	$G_{t_3\beta_3}$
α_1	Symmetric				$G_{\alpha_1\alpha_1}$	$G_{\alpha_1\alpha_2}$	$G_{\alpha_1\alpha_3}$	$G_{\alpha_1\beta_1}$	$G_{\alpha_1\beta_2}$	$G_{\alpha_1\beta_3}$
α_2		about				$G_{\alpha_2\alpha_2}$	$G_{\alpha_2\alpha_3}$	$G_{\alpha_2\beta_1}$	$G_{\alpha_2\beta_2}$	$G_{\alpha_2\beta_3}$
α_3			diagonal				$G_{\alpha_3\alpha_3}$	$G_{\alpha_3\beta_1}$	$G_{\alpha_3\beta_2}$	$G_{\alpha_3\beta_3}$
β_1								$G_{\beta_1\beta_1}$	$G_{\beta_1\beta_2}$	$G_{\beta_1\beta_3}$
β_2									$G_{\beta_2\beta_2}$	$G_{\beta_2\beta_3}$
β_3										$G_{\beta_3\beta_3}$

$$(14.107)$$

Each pair of internal coordinates has a G matrix element associated with it. However, because of the symmetry of the molecule many of the G matrix elements are identical. The G matrix, simplified by the use of a single symbol for equivalent elements, is written

$$G = \begin{Vmatrix} G_r & G_{rt} & G_{rt} & G_{rt} & G_{r\alpha} & G_{r\alpha} & G_{r\alpha} & G_{r\beta} & G_{r\beta} & G_{r\beta} \\ & G_t & G_{tt} & G_{tt} & G'_{t\alpha} & G_{t\alpha} & G_{t\alpha} & G_{t\beta} & G'_{t\beta} & G'_{t\beta} \\ & & G_t & G_{tt} & G_{t\alpha} & G'_{t\alpha} & G_{t\alpha} & G'_{t\beta} & G_{t\beta} & G'_{t\beta} \\ & & & G_t & G_{t\alpha} & G_{t\alpha} & G'_{t\alpha} & G'_{t\beta} & G'_{t\beta} & G_{t\beta} \\ & \text{Symmetric} & & & G_\alpha & G_{\alpha\alpha} & G_{\alpha\alpha} & G'_{\alpha\beta} & G_{\alpha\beta} & G_{\alpha\beta} \\ & \text{about} & & & & G_\alpha & G_{\alpha\alpha} & G_{\alpha\beta} & G'_{\alpha\beta} & G_{\alpha\beta} \\ & \text{diagonal} & & & & & G_\alpha & G_{\alpha\beta} & G_{\alpha\beta} & G'_{\alpha\beta} \\ & & & & & & & G_\beta & G_{\beta\beta} & G_{\beta\beta} \\ & & & & & & & & G_\beta & G_{\beta\beta} \\ & & & & & & & & & G_\beta \end{Vmatrix} \quad (14.108)$$

See Fig. 14.8 showing the internal coordinates for justifications of the equivalencies.

The $\mathscr{G}$ matrices for the A_1 and E vibrations may now be determined from $\mathscr{G} = UGU'$. The U matrix for the A_1 vibrations with the redundant coordinate deleted is

	r	t_1	t_2	t_3	α_1	α_2	α_3	β_1	β_2	β_3
$\mathscr{S}_1$	1	0	0	0	0	0	0	0	0	0
$U_{A_1} = \mathscr{S}_2$	0	$\dfrac{1}{\sqrt{3}}$	$\dfrac{1}{\sqrt{3}}$	$\dfrac{1}{\sqrt{3}}$	0	0	0	0	0	0
$\mathscr{S}_2$	0	0	0	0	$\dfrac{1}{\sqrt{6}}$	$\dfrac{1}{\sqrt{6}}$	$\dfrac{1}{\sqrt{6}}$	$\dfrac{-1}{\sqrt{6}}$	$\dfrac{-1}{\sqrt{6}}$	$\dfrac{-1}{\sqrt{6}}$,

$$(14.109)$$

and exchanging rows and columns gives the transpose, U':

$$U_{A_1} = \begin{Vmatrix} 1 & 0 & 0 \\ 0 & \dfrac{1}{\sqrt{3}} & 0 \\ 0 & \dfrac{1}{\sqrt{3}} & 0 \\ 0 & \dfrac{1}{\sqrt{3}} & 0 \\ 0 & 0 & \dfrac{1}{\sqrt{6}} \\ 0 & 0 & \dfrac{1}{\sqrt{6}} \\ 0 & 0 & \dfrac{1}{\sqrt{6}} \\ 0 & 0 & \dfrac{-1}{\sqrt{6}} \\ 0 & 0 & \dfrac{-1}{\sqrt{6}} \\ 0 & 0 & \dfrac{-1}{\sqrt{6}} \end{Vmatrix} \qquad (14.110)$$

The product $\mathbf{GU'}$ is first obtained

$$
\mathbf{GU'} = \left\|
\begin{array}{lll}
G_r & \dfrac{3}{\sqrt{3}}G_{rt} & \dfrac{3}{\sqrt{6}}G_{r\alpha} - \dfrac{3}{\sqrt{6}}G_{r\beta} \\[2ex]
G_{rt} & \dfrac{1}{\sqrt{3}}G_t + \dfrac{2}{\sqrt{3}}G_{tt} & \dfrac{1}{\sqrt{6}}G'_{t\alpha} + \dfrac{2}{\sqrt{6}}G_{t\alpha} - \dfrac{1}{\sqrt{6}}G_{t\beta} - \dfrac{2}{\sqrt{6}}G'_{t\beta} \\[2ex]
G_{rt} & \dfrac{1}{\sqrt{3}}G_t + \dfrac{2}{\sqrt{3}}G_{tt} & \dfrac{1}{\sqrt{6}}G'_{t\alpha} + \dfrac{2}{\sqrt{6}}G_{t\alpha} - \dfrac{1}{\sqrt{6}}G_{t\beta} - \dfrac{2}{\sqrt{6}}G'_{t\beta} \\[2ex]
G_{rt} & \dfrac{1}{\sqrt{3}}G_t + \dfrac{2}{\sqrt{3}}G_{tt} & \dfrac{1}{\sqrt{6}}G'_{t\alpha} + \dfrac{2}{\sqrt{6}}G_{t\alpha} - \dfrac{1}{\sqrt{6}}G_{t\beta} - \dfrac{2}{\sqrt{6}}G'_{t\beta} \\[2ex]
G_{r\alpha} & \dfrac{1}{\sqrt{3}}G'_{t\alpha} + \dfrac{2}{\sqrt{3}}G_{t\alpha} & \dfrac{1}{\sqrt{6}}G_\alpha + \dfrac{2}{\sqrt{6}}G_{\alpha\alpha} - \dfrac{1}{\sqrt{6}}G'_{\alpha\beta} - \dfrac{2}{\sqrt{6}}G_{\alpha\beta} \\[2ex]
G_{r\alpha} & \dfrac{1}{\sqrt{3}}G'_{t\alpha} + \dfrac{2}{\sqrt{3}}G_{t\alpha} & \dfrac{1}{\sqrt{6}}G_\alpha + \dfrac{2}{\sqrt{6}}G_{\alpha\alpha} - \dfrac{1}{\sqrt{6}}G'_{\alpha\beta} - \dfrac{2}{\sqrt{6}}G_{\alpha\beta} \\[2ex]
G_{r\alpha} & \dfrac{1}{\sqrt{3}}G'_{t\alpha} + \dfrac{2}{\sqrt{3}}G_{t\alpha} & \dfrac{1}{\sqrt{6}}G_\alpha + \dfrac{2}{\sqrt{6}}G_{\alpha\alpha} - \dfrac{1}{\sqrt{6}}G'_{\alpha\beta} - \dfrac{2}{\sqrt{6}}G_{\alpha\beta} \\[2ex]
G_{r\beta} & \dfrac{1}{\sqrt{3}}G_{t\beta} + \dfrac{2}{\sqrt{3}}G'_{t\beta} & \dfrac{1}{\sqrt{6}}G'_{\alpha\beta} + \dfrac{2}{\sqrt{6}}G_{\alpha\beta} - \dfrac{1}{\sqrt{6}}G_\beta - \dfrac{2}{\sqrt{6}}G_{\beta\beta} \\[2ex]
G_{r\beta} & \dfrac{1}{\sqrt{3}}G_{t\beta} + \dfrac{2}{\sqrt{3}}G'_{t\beta} & \dfrac{1}{\sqrt{6}}G'_{\alpha\beta} + \dfrac{2}{\sqrt{6}}G_{\alpha\beta} - \dfrac{1}{\sqrt{6}}G_\beta - \dfrac{2}{\sqrt{6}}G_{\beta\beta} \\[2ex]
G_{r\beta} & \dfrac{1}{\sqrt{3}}G_{t\beta} + \dfrac{2}{\sqrt{3}}G'_{t\beta} & \dfrac{1}{\sqrt{6}}G'_{\alpha\beta} + \dfrac{2}{\sqrt{6}}G_{\alpha\beta} - \dfrac{1}{\sqrt{6}}G_\beta - \dfrac{2}{\sqrt{6}}G_{\beta\beta}
\end{array}
\right\|
$$

$$(14.111)$$

and multiplication of $\mathbf{U}$ and $\mathbf{GU'}$ gives $\mathbf{UGU'}$ which is equivalent to the $\mathscr{G}$ matrix for the A_1 vibrations.

$$\mathbf{UGU'} = \begin{Vmatrix} G_r & \dfrac{3}{\sqrt{3}} G_{rt} & \dfrac{3}{\sqrt{6}} G_{r\alpha} - \dfrac{3}{\sqrt{6}} G_{r\beta} \\[3ex] \dfrac{3}{\sqrt{3}} G_{rt} & G_t + 2G_{tt} & \dfrac{1}{\sqrt{2}} G'_{t\alpha} + \dfrac{2}{\sqrt{2}} G_{t\alpha} \\[2ex] & & -\dfrac{1}{\sqrt{2}} G_{t\beta} - \dfrac{2}{\sqrt{2}} G'_{t\beta} \\[3ex] \dfrac{3}{\sqrt{6}} G_{r\alpha} - \dfrac{3}{\sqrt{6}} G_{r\beta} & \dfrac{1}{\sqrt{2}} G'_{t\alpha} + \dfrac{2}{\sqrt{2}} G_{t\alpha} & \dfrac{1}{2} G_\alpha + G_{\alpha\alpha} - G'_{\alpha\beta} \\[2ex] & -\dfrac{1}{\sqrt{2}} G_{t\beta} - \dfrac{2}{\sqrt{2}} G'_{t\beta} & -2G_{\alpha\beta} + \dfrac{1}{2} G_\beta + G_{\beta\beta} \end{Vmatrix}$$

$$(14.112)$$

The $\mathscr{G}$ matrix for the E vibrations is obtained in the same way except that the **U** and **U'** of the E symmetry coordinates are used. Either the a or b sets may be used since both give the same result for the $\mathscr{G}$ matrix. The **U** matrix from the b set of E coordinates is

	r	t_1	t_2	t_3	α_1	α_2	α_3	β_1	β_2	β_3
$\mathscr{S}_{1b}$	0	0	$1/\sqrt{2}$	$-1/\sqrt{2}$	0	0	0	0	0	0
$\mathbf{U}_{E(b)} = \mathscr{S}_{2b}$	0	0	0	0	0	$1/\sqrt{2}$	$-1/\sqrt{2}$	0	0	0
$\mathscr{S}_{3b}$	0	0	0	0	0	0	0	0	$1/\sqrt{2}$	$-1/\sqrt{2}$

$$(14.113)$$

and the **U'** matrix is

$$\mathbf{U}_{E(b)} = \begin{Vmatrix} 0 & 0 & 0 \\ 0 & 0 & 0 \\ 1/\sqrt{2} & 0 & 0 \\ -1/\sqrt{2} & 0 & 0 \\ 0 & 0 & 0 \\ 0 & 1/\sqrt{2} & 0 \\ 0 & -1/\sqrt{2} & 0 \\ 0 & 0 & 0 \\ 0 & 0 & 1/\sqrt{2} \\ 0 & 0 & -1/\sqrt{2} \end{Vmatrix} \qquad (14.114)$$

The matrix multiplication of $\mathbf{GU'}$ gives

$$\mathbf{GU'} = \begin{Vmatrix} 0 & 0 & 0 \\[4pt] 0 & 0 & 0 \\[4pt] \dfrac{1}{\sqrt{2}}(G_t - G_{tt}) & \dfrac{1}{\sqrt{2}}(G'_{t\alpha} - G_{t\alpha}) & \dfrac{1}{\sqrt{2}}(G_{t\beta} - G'_{t\beta}) \\[10pt] \dfrac{1}{\sqrt{2}}(G_{tt} - G_t) & \dfrac{1}{\sqrt{2}}(G_{t\alpha} - G'_{t\alpha}) & \dfrac{1}{\sqrt{2}}(G'_{t\beta} - G_{t\beta}) \\[10pt] 0 & 0 & 0 \\[4pt] \dfrac{1}{\sqrt{2}}(G'_{t\alpha} - G_{t\alpha}) & \dfrac{1}{\sqrt{2}}(G_\alpha - G_{\alpha\alpha}) & \dfrac{1}{\sqrt{2}}(G'_{\alpha\beta} - G_{\alpha\beta}) \\[10pt] \dfrac{1}{\sqrt{2}}(G_{t\alpha} - G'_{t\alpha}) & \dfrac{1}{\sqrt{2}}(G_{\alpha\alpha} - G_\alpha) & \dfrac{1}{\sqrt{2}}(G_{\alpha\beta} - G'_{\alpha\beta}) \\[10pt] 0 & 0 & 0 \\[4pt] \dfrac{1}{\sqrt{2}}(G_{t\beta} - G'_{t\beta}) & \dfrac{1}{\sqrt{2}}(G'_{\alpha\beta} - G_{\alpha\beta}) & \dfrac{1}{\sqrt{2}}(G_\beta - G_{\beta\beta}) \\[10pt] \dfrac{1}{\sqrt{2}}(G'_{t\beta} - G_{t\beta}) & \dfrac{1}{\sqrt{2}}(G_{\alpha\beta} - G'_{\alpha\beta}) & \dfrac{1}{\sqrt{2}}(G_{\beta\beta} - G_\beta) \end{Vmatrix} \tag{14.115}$$

and multiplication of $\mathbf{U}$ with $\mathbf{GU'}$ gives

$$\mathbf{UGU'} = \mathscr{G} \text{ for the } E \text{ vibrations}$$

$$\mathbf{UGU'} = \begin{Vmatrix} G_t - G_{tt} & G'_{t\alpha} - G_{t\alpha} & G_{t\beta} - G'_{t\beta} \\[6pt] G'_{t\alpha} - G_{t\alpha} & G_\alpha - G_{\alpha\alpha} & G'_{\alpha\beta} - G_{\alpha\beta} \\[6pt] G_{t\beta} - G'_{t\beta} & G'_{\alpha\beta} - G_{\alpha\beta} & G_\beta - G_{\beta\beta} \end{Vmatrix} \tag{14.116}$$

The bond angles for $CHCl_3$ are not exactly tetrahedral but will be assumed tetrahedral for convenience in calculation. The values listed by Wilson, Decius, and Cross[3] for tetrahedral bond angles are shown in Table 14.10.

TABLE 14.10^a

G ELEMENTS FOR $\phi = 109°28'$

G_{rr}^2	$\mu_1 + \mu_2$
G_{rr}^1	$-\dfrac{1}{3}\mu_1$
$G_{r\phi}^2$	$-\dfrac{2\sqrt{2}}{3}\rho_{23}\mu_2$
$G_{r\phi}^1\begin{pmatrix}1\\2\end{pmatrix}$	$\dfrac{2\sqrt{2}}{3}\rho_{13}\mu_1 \cos\tau$
$G_{r\phi}^1\begin{pmatrix}1\\1\end{pmatrix}$	$\dfrac{\sqrt{2}}{3}(\rho_{13} + \rho_{14})\mu_1$
$G_{\phi\phi}^3$	$\rho_{12}^2\mu_1 + \rho_{23}^2\mu_3 + \dfrac{1}{3}(3\rho_{12}^2 + 3\rho_{23}^2 + 2\rho_{12}\rho_{23})\mu_2$
$G_{\phi\phi}^2\begin{pmatrix}1\\1\end{pmatrix}$	$-\dfrac{1}{6}\{3\rho_{21}^2\mu_1 + [3\rho_{21}^2 + (\rho_{23} + \rho_{24})\rho_{21} - 5\rho_{23}\rho_{24}]\mu_2\}$
$G_{\phi\phi}^2\begin{pmatrix}1\\0\end{pmatrix}$	$-\dfrac{1}{3}\rho_{12}\cos\tau[(3\rho_{12} + \rho_{14})\mu_1 + (3\rho_{12} + \rho_{23})\mu_2]$
$G_{\phi\phi}^1\begin{pmatrix}2\\2\end{pmatrix}$	$-\dfrac{1}{3}(3\sin\tau_{25}\sin\tau_{34} - \cos\tau_{25}\cos\tau_{34})\rho_{12}\rho_{14}\mu_1$
$G_{\phi\phi}^1\begin{pmatrix}2\\1\end{pmatrix}$	$-\dfrac{1}{3}[(3\cos\tau_{35} + \cos\tau_{34})\rho_{14} + (3\cos\tau_{34} + \cos\tau_{35})\rho_{15}]\rho_{12}\mu_1$
$G_{\phi\phi}^1\begin{pmatrix}1\\1\end{pmatrix}$	$-\dfrac{2}{3}(\rho_{12} + \rho_{14})(\rho_{13} + \rho_{15})\mu_1$

^{a} From "Molecular Vibrations," by Wilson, Decius, and Cross. Copyright (1955). McGraw-Hill Book Company, Inc. used by permission.

These values can be used to evaluate the G elements as follows

$$G_r = [G_{rr}^2] = \mu_C + \mu_H \tag{14.117}$$

$$G_{rt} = [G_{rr}^1] = -\frac{1}{3}\mu_C \tag{14.118}$$

$$G_{r\alpha} = [G_{r\phi}^1(\begin{smallmatrix}1\\1\end{smallmatrix})] = \frac{2\sqrt{2}}{3}\rho_{CCl}\mu_C \tag{14.119}$$

$$G_{r\beta} = [G_{r\phi}^2] = -\frac{2\sqrt{2}}{3}\rho_{CCl}\,\mu_C \tag{14.120}$$

$$G_t = [G_{rr}^2] = \mu_C + \mu_{Cl} \tag{14.121}$$

$$G_{tt} = [G_{rr}^1] = -\frac{1}{3}\mu_C \tag{14.122}$$

$$G_{t\alpha}' = [G_{r\phi}^1(\begin{smallmatrix}1\\1\end{smallmatrix})] = \frac{2\sqrt{2}}{3}\rho_{CCl}\,\mu_C \tag{14.123}$$

$$G_{t\alpha} = [G_{r\phi}^2] = -\frac{2\sqrt{2}}{3}\rho_{CCl}\,\mu_C \tag{14.124}$$

$$G_{t\beta} = [G_{r\phi}^2] = -\frac{2\sqrt{2}}{3}\rho_{CH}\,\mu_C \tag{14.125}$$

$$G_{t\beta}' = [G_{r\phi}^1(\begin{smallmatrix}1\\1\end{smallmatrix})] = \frac{\sqrt{2}}{3}(\rho_{CH} + \rho_{CCl})\mu_C \tag{14.126}$$

$$G_\alpha = [G_{\phi\phi}^3] = \frac{8}{3}\rho_{CCl}^2\,\mu_C + 2\rho_{CCl}^2\,\mu_{Cl} \tag{14.127}$$

$$G_{\alpha\alpha} = [G_{\phi\phi}^2(\begin{smallmatrix}1\\1\end{smallmatrix})] = -\frac{1}{2}\rho_{CCl}^2\,\mu_{Cl} \tag{14.128}$$

$$G_{\alpha\beta}' = [G_{\phi\phi}^1(\begin{smallmatrix}1\\1\end{smallmatrix})] = -\frac{4}{3}(\rho_{CCl}^2 + \rho_{CH}\,\rho_{CCl})\mu_C \tag{14.129}$$

$$G_{\alpha\beta} = [G_{\phi\phi}^2(\begin{smallmatrix}1\\1\end{smallmatrix})] = -\left(\frac{2}{3}\rho_{CCl}^2 - \frac{2}{3}\rho_{CH}\,\rho_{CCl}\right)\mu_C - \frac{1}{2}\rho_{CCl}^2\,\mu_{Cl} \tag{14.130}$$

$$G_\beta = [G_{\phi\phi}^3] = \rho_{CH}^2\,\mu_H + \rho_{CCl}^2\,\mu_{Cl} + \left(\rho_{CH}^2 + \rho_{CCl}^2 + \frac{2}{3}\rho_{CH}\,\rho_{CCl}\right)\mu_C \tag{14.131}$$

and

$$G_{\beta\beta} = [G_{\phi\phi}^2(\begin{smallmatrix}1\\1\end{smallmatrix})] = -\frac{1}{2}\rho_{CH}^2\,\mu_H - \frac{1}{6}(3\rho_{CH}^2 + 2\rho_{CH}\,\rho_{CCl} - 5\rho_{CCl}^2)\mu_C \tag{14.132}$$

where μ_H, μ_C, and μ_{Cl} are the reciprocals of the atomic masses of hydrogen, carbon, and chlorine and ρ_{CH} and ρ_{CCl} are the reciprocals of the C—H and C—Cl bond distances. Making use of the above equations the $\mathscr{G}$ matrices for the A_1 and E species in Eqs. (14.112) and (14.116) become

$$\mathscr{G}_{A_1} = \begin{Vmatrix} \mu_H + \mu_C & -\dfrac{1}{\sqrt{3}}\mu_C & \dfrac{4}{\sqrt{3}}\rho_{CC}\mu_C \\[2ex] -\dfrac{1}{\sqrt{3}}\mu_C & \dfrac{1}{3}\mu_C + \mu_{Cl} & -\dfrac{4}{3}\rho_{CCl}\mu_C \\[2ex] \dfrac{4}{\sqrt{3}}\rho_{CCl}\mu_C & -\dfrac{4}{3}\rho_{CCl}\mu_C & \dfrac{16}{3}\rho^2_{CCl}\mu_C + 2\rho^2_{CCl}\mu_C \end{Vmatrix} \quad (14.133)$$

$$\mathscr{G}_E = \begin{Vmatrix} \dfrac{4}{3}\mu_C + \mu_{Cl} & \dfrac{4\sqrt{2}}{3}\rho_{CCl}\mu_C & -\left[\sqrt{2}\rho_{CH} + \dfrac{\sqrt{2}}{3}\rho_{CCl}\right]\mu_C \\[2ex] \dfrac{4\sqrt{2}}{3}\rho_{CCl}\mu_C & \dfrac{8}{3}\rho^2_{CCl}\mu_C + \dfrac{5}{2}\rho^2_{CCl}\mu_{Cl} & -\dfrac{2}{3}[\rho^2_{CCl} + 3\rho_{CH}\rho_{CCl}]\mu_C \\ & & +\dfrac{1}{2}\rho^2_{CCl}\mu_{Cl} \\[2ex] -\left[\sqrt{2}\rho_{CH}\right. & -\dfrac{2}{3}[\rho^2_{CCl} + \rho_{CH}\rho_{CCl}]\mu_C & \dfrac{3}{2}\rho^2_{CH}\mu_H + \rho^2_{CCl}\mu_{Cl} \\ \left.+\dfrac{\sqrt{2}}{3}\rho_{CCl}\right]\mu_C & +\dfrac{1}{2}\rho^{2'}_{CCl}\mu_{Cl} & +\left[\dfrac{3}{2}\rho^2_{CH} + \dfrac{1}{6}\rho^2_{CCl}\right. \\ & & \left.+\rho_{CH}\rho_{CCl}\right]\mu_C \end{Vmatrix}$$

$$(14.134)$$

The $\mathscr{G}$ matrix elements can be numerically evaluated using the following values for the mass reciprocals μ and bond length reciprocals ρ where convenient units are unified atomic mass units (u) and angstroms (Å).

$$\mu_C = 1/12.0115\ u = 0.083254\ u^{-1} \quad (14.135)$$

$$\mu_H = 1/1.00797\ u = 0.99209\ u^{-1} \quad (14.136)$$

$$\mu_{Cl} = 1/35.453\ u = 0.028206\ u^{-1} \quad (14.137)$$

$$\rho_{CH} = 1/1.093\ \text{Å} = 0.9149\ \text{Å}^{-1} \quad (14.138)$$

$$\rho_{CCl} = 1/1.77\ \text{Å} = 0.5650\ \text{Å}^{-1} \quad (14.139)$$

Using these values the $\mathscr{G}$ matrices are

$$\mathscr{G}_{A_1} = \begin{Vmatrix} 1.07534 & -0.048067 & 0.108631 \\ -0.048067 & 0.055957 & -0.062718 \\ 0.108631 & -0.062718 & 0.159751 \end{Vmatrix} \qquad (14.140)$$

$$\mathscr{G}_{E} = \begin{Vmatrix} 0.139211 & 0.088697 & -0.125893 \\ 0.088697 & 0.093381 & -0.099282 \\ -0.129893 & -0.099282 & 1.406629 \end{Vmatrix} \qquad (14.141)$$

14.15   The F Matrix for CHCl$_3$

The general **F** matrix for chloroform is

	r	t_1	t_2	t_3	α_1	α_2	α_3	β_1	β_2	β_3
r	F_r	F_{rt}	F_{rt}	F_{rt}	$F_{r\alpha}$	$F_{r\alpha}$	$F_{r\alpha}$	$F_{r\beta}$	$F_{r\beta}$	$F_{r\beta}$
t_1		F_t	F_{tt}	F_{tt}	$F'_{t\alpha}$	$F_{t\alpha}$	$F_{t\alpha}$	$F_{t\beta}$	$F'_{t\beta}$	$F'_{t\beta}$
t_2			F_t	F_{tt}	$F_{t\alpha}$	$F'_{t\alpha}$	$F_{t\alpha}$	$F'_{t\beta}$	$F_{t\beta}$	$F'_{t\beta}$
t_3				F_t	$F_{t\alpha}$	$F_{t\alpha}$	$F'_{t\alpha}$	$F'_{t\beta}$	$F'_{t\beta}$	$F_{t\beta}$
$\mathbf{F} = \alpha_1$	Symmetric				F_α	$F_{\alpha\alpha}$	$F_{\alpha\alpha}$	$F'_{\alpha\beta}$	$F_{\alpha\beta}$	$F_{\alpha\beta}$
α_2		about				F_α	$F_{\alpha\alpha}$	$F_{\alpha\beta}$	$F'_{\alpha\beta}$	$F_{\alpha\beta}$
α_3			diagonal				F_α	$F_{\alpha\beta}$	$F_{\alpha\beta}$	$F'_{\alpha\beta}$
β_1								F_β	$F_{\beta\beta}$	$F_{\beta\beta}$
β_2									F_β	$F_{\beta\beta}$
β_3										F_β

$$(14.142)$$

where each pair of coordinates has an F element associated with it. Since this is in the same form as the **G** matrix, Eq. (14.108), the $\mathscr{F}_{A_1}$ and $\mathscr{F}_E$ matrices can be written immediately by comparison with Eqs. (14.113) and (14.117).

$$\mathscr{F}_{A_1} = \begin{Vmatrix} F_r & \sqrt{3}F_{rt} & 3/\sqrt{6}(F_{r\alpha} - F_{r\beta}) \\ \sqrt{3}F_{rt} & F_t + 2F_{tt} & \begin{aligned} 1/\sqrt{2}(2F_{t\alpha} - F_{t\beta} \\ + F'_{t\alpha} - 2F'_{t\beta}) \end{aligned} \\ 3/\sqrt{6}(F_{r\alpha} - F_{r\beta}) & \begin{aligned} 1/\sqrt{2}(2F_{t\alpha} - F_{t\beta} \\ + F'_{t\alpha} - 2F'_{t\beta}) \end{aligned} & \begin{aligned} 1/2(F_\alpha + F_\beta + 2F_{\alpha\alpha} \\ + 2F_{\beta\beta} - 2F'_{\alpha\beta} - 4F_{\alpha\beta}) \end{aligned} \end{Vmatrix}$$

$$(14.143)$$

$$\mathscr{F}_E = \begin{Vmatrix} F_t - F_{tt} & F'_{t\alpha} - F_{t\alpha} & F_{t\beta} - F'_{t\beta} \\ F'_{t\alpha} - F_{t\alpha} & F_\alpha - F_{\alpha\alpha} & F'_{\alpha\beta} - F_{\alpha\beta} \\ F_{t\beta} - F'_{t\beta} & F'_{\alpha\beta} - F_{\alpha\beta} & F_\beta - F_{\beta\beta} \end{Vmatrix} \qquad (14.144)$$

The potential energy of a molecule has the form

$$2V = \sum F_{jk} S_j S_k \qquad (14.145)$$

In the cgs system a bond stretching coordinate, r, is expressed in cm and the associated F value in the term $F_r r^2$ is in dynes/cm. A convenient unit for this force constant in the molecular problem is mdynes/Å. For a bond angle bending coordinate, α, the angle is measured in radians. In the energy expression the force constant F_α in a term such as $F_\alpha \alpha^2$ must be expressed in mdynes Å/rad^2, if α is in radians, in order to have the same energy units as $F_r r^2$ type terms. For terms such as $F_{r\alpha} r\alpha$ where r and α are stretching and bending coordinates the force constant is in mdynes. Sometimes bending force constants are expressed in the same units as stretching force constants, mdynes/Å. These are obtained by dividing bending—stretching force constants such as $F_{r\alpha}$ by a scaling factor in Å and by dividing pure bending force constants such as F_α by the square of a scaling factor in Å. The scaling factor is arbitrary but becomes part of the characterization of bending force constants if they are expressed in mdynes/Å. It can be a bondlength included in the angle. The scaling factor (d) sometimes is included in the definition of a bending coordinate so that instead of α, $d\alpha$ is used. In this case the bending force constant unit is the same as the stretching force constant unit and the elements within the **F** and **G** and **L** matrices and the secular determinant are all in same units. However, the Decius **G** matrix elements (Table 14.10) must be modified since they are for unscaled internal coordinates. In the CHCl$_3$ problem we will not use scaling factors.

In order to evaluate the **F** matrix the force constants of the molecule must be known and this is usually a difficult requirement. In this section, however, we are primarily concerned with calculating frequencies from a given set of force constants. The force constants as determined by Zeitlow[19] will be used for the chloroform molecule and are given in Table 14.11. Using these values for the elements of the **F** matrices the numerical values for the $\mathscr{F}$ matrices are

$$\mathscr{F}_{A_1} = \begin{Vmatrix} 4.8540 & 0.1524 & 0.6765 \\ 0.1524 & 4.1834 & 0.4488 \\ 0.6765 & 0.4458 & 1.2774 \end{Vmatrix} \qquad (14.146)$$

$$\mathscr{F}_E = \begin{Vmatrix} 3.1256 & -0.6106 & 0.4884 \\ -0.6106 & 1.1588 & -0.1543 \\ 0.4884 & -0.1543 & 0.6437 \end{Vmatrix} \qquad (14.147)$$

[19] J. P. Zeitlow, M. S. Thesis, Illinois Institute of Technology, Chicago (1949).

TABLE 14.11

FORCE CONSTANTS ASSUMED FOR THE CHLOROFORM MOLECULE

	mdyne/Å		mdyne		mdyne Å
F_r	4.8540	$F_{r\alpha} - F_{r\beta}$	0.5524	F_α	1.6064
F_t	3.4782	$F_{t\alpha}$	0.4138	F_β	0.6527
F_{rt}	0.0880	$F_{t\beta}$	0.3243	$F_{\alpha\alpha}$	0.4476
F_{tt}	0.3526	$F'_{t\alpha}$	−0.1968	$F_{\beta\beta}$	0.00899
		$F'_{t\beta}$	−0.1641	$F_{\alpha\beta}$	0.1543
				$F'_{\alpha\beta}$	0

14.16   The Secular Determinants for CHCl$_3$

The secular determinants are obtained by getting the $\mathscr{G}\mathscr{F}$ matrix product, subtracting λ from the diagonal elements, and setting the determinant of this matrix equal to zero. The A_1 secular determinant is obtained using Eqs. (14.140) and (14.146).

$$\begin{vmatrix} 5.285864 - \lambda & 0.011552 & 0.844660 \\ -0.267218 & 0.198617 - \lambda & -0.087520 \\ 0.625808 & -0.174123 & 0.249407 - \lambda \end{vmatrix} = 0 \quad (14.148)$$

It can be determined by direct expansion that a determinant of the form

$$\begin{vmatrix} A - \lambda & B & C \\ D & E - \lambda & F \\ G & H & I - \lambda \end{vmatrix} = 0 \quad (14.149)$$

has a characteristic equation given by

$$\lambda^3 - \lambda^2(A + E + I) + \lambda\left[\begin{vmatrix} AB \\ DE \end{vmatrix} + \begin{vmatrix} AC \\ GI \end{vmatrix} + \begin{vmatrix} EF \\ HI \end{vmatrix}\right] - \begin{vmatrix} ABC \\ DEF \\ GHI \end{vmatrix} = 0 \quad (14.150)$$

The characteristic equation for the A_1 secular determinant is

$$\lambda^3 - 0.5.73389\lambda^2 + 1.87698\lambda - 0.115741 = 0 \quad (14.151)$$

The three roots are found to be

$$\lambda_1 = 5.389627 \qquad \lambda_2 = 0.262450 \qquad \lambda_3 = 0.0818236 \text{ mdyne Å}^{-1} u^{-1}$$

$$(14.152)$$

The units of λ are force constant units (mdyne/Å) divided by mass units (u). Since the wavenumber $\bar{\nu}$ is $\bar{\nu} = \sqrt{\lambda}/(2\pi c)$, in the present system of units

$$\bar{\nu} = 1303\sqrt{\lambda} \qquad (14.153)$$

$$1303 = \sqrt{N_A \times 10^5}/(2\pi c)$$

where N_A is Avogadro's number 6.02252×10^{23}. From this the A_1 wavenumbers can be calculated as

$$\bar{\nu}_1 = 3025 \text{ cm}^{-1} \qquad \bar{\nu}_2 = 668 \text{ cm}^{-1} \qquad \bar{\nu}_3 = 373 \text{ cm}^{-1} \qquad (14.154)$$

A similar procedure is used for the E species using Eqs. (14.143) and (14.149).

$$\begin{vmatrix} 0.317520 - \lambda & 0.037822 & -0.029307 \\ 0.171724 & 0.069371 - \lambda & -0.034997 \\ 0.341627 & -0.252782 & 0.857327 - \lambda \end{vmatrix} = 0 \quad (14.155)$$

$$\lambda^3 - 1.24422\lambda^2 + 0.348389\lambda - 0.012021 = 0 \qquad (14.156)$$

$$\lambda_4 = 0.851772 \qquad \lambda_5 = 0.352400 \qquad \lambda_6 = 0.0400481 \quad (14.157)$$

$$\bar{\nu}_4 = 1203 \text{ cm}^{-1} \qquad \bar{\nu}_5 = 774 \text{ cm}^{-1} \qquad \bar{\nu}_6 = 261 \text{ cm}^{-1} \quad (14.158)$$

The solution of a larger order secular determinant is much more difficult but the availability of computers has largely solved this problem. See Wilson, Decius, and Cross[3] and Steele[6] for a comprehensive discussion.

14.17 The Frequencies of the $CDCl_3$ Molecules

Within the limits of the Born–Oppenheimer approximation, the force field for a molecule is identical to that of another molecule where certain atoms have been replaced with different isotopes of that same atom. Therefore for the deuterated chloroform molecule $CDCl_3$ the $\mathscr{F}$ matrix will be identical to that for $CHCl_3$. In the $\mathscr{G}$ matrices for $CHCl_3$ [Eqs. (14.133) and (14.134)] μ_H is replaced by μ_D which has the value $0.496342u^{-1}$. In the A_1 species this only modifies $\mathscr{G}_{11}$ and the first row of the secular determinant. In the E species this only modifies $\mathscr{G}_{33}$ and the last row of the secular determinant.

The A_1 calculated wavenumbers are 2259, 641, and 369 cm^{-1} and the E calculated wavenumbers are 900, 746, and 259 cm^{-1}.

14.18   Comparison of Experimental and Calculated Wavenumbers

The calculated and observed wavenumbers for $CHCl_3$ and $CDCl_3$ are tabulated in Table 14.12.

TABLE 14.12

WAVENUMBERS FOR $CHCl_3$ AND $CDCl_3$ (IN cm^{-1})

Assignment	CHCl₃		CDCl₃	
	Observed	Calculated	Observed	Calculated
ν_1	3019	3025	2255	2259
ν_2	668	668	650	641
ν_3	368	373	366	369
ν_4	1216	1203	905	900
ν_5	757	774	740	746
ν_6	261	261	262	269

The assumed force field has led to calculated wavenumbers in reasonable agreement with the observed wavenumbers assumed to be fundamentals for both molecules. Since the force field was a reasonable one, this tends to verify the assignments of the fundamentals. However, since many force constants were utilized, it is possible that an equally good (or better) fit could be obtained using some other reasonable set of values for the force constants. See Shimanouchi,[20] Nakagawa,[21] and Lecuyer et al.[22] for other vibrational analyses for chloroform.

14.19   The Form of the Chloroform Normal Coordinates

The normalized $\mathscr{L}$ matrices for the A_1 and E species are obtained by the procedure listed in Section 14.7. These are combined into a complete $\mathscr{L}$ matrix (see Table 14.13).

[20] T. Shimanouchi, *J. Chem. Phys.* **17**, 245 (1949).

[21] I. Nakagawa, *J. Chem. Soc. Jap., Pure Chem. Sect.* **75**, 535 (1954).

[22] A. Lecuyer, M. M. Denairiez, and B. Barchewitz, *J. Chim. Phys. Physicochim. Biol.* **30**, 412 (1963).

TABLE 14.13

THE $\mathscr{L}$ MATRIX

	Q_1	Q_2	Q_3	Q_{4a}	Q_{5a}	Q_{6a}	Q_{4b}	Q_{5b}	Q_{6b}
$\mathscr{S}_1$	1.03518	-0.05382	0.03223	0	0	0	0	0	0
$\mathscr{S}_2$	-0.05544	-0.21763	-0.07430	0	0	0	0	0	0
$\mathscr{S}_3$	0.12791	0.32305	-0.19756	0	0	0	0	0	0
$\mathscr{S}_4$	0	0	0	0	0	0	0	0	0
$\mathscr{S}_{1a}$	0	0	0	-0.06938	-0.36617	0.01981	0	0	0
$\mathscr{S}_{2a}$	0	0	0	-0.06789	-0.23806	-0.19959	0	0	0
$\mathscr{S}_{3a}$	0	0	0	1.17735	0.12856	-0.07002	0	0	0
$\mathscr{S}_{1b}$	0	0	0	0	0	0	-0.06938	-0.36617	0.01981
$\mathscr{S}_{2b}$	0	0	0	0	0	0	-0.06789	-0.23806	-0.19959
$\mathscr{S}_{3b}$	0	0	0	0	0	0	1.17735	0.12856	-0.07002

A useful check is[6]

$$\mathscr{L}\mathscr{L}' = \mathscr{G} \qquad (14.159)$$

During the excitation of a normal coordinate the potential energy increase is distributed among the symmetry coordinates in which terms the normal coordinates are described. As was described in Section 14.8 the potential energy distribution can be defined as the ratios of the squares of the elements of the $\mathscr{L}$ matrix, each multiplied by the corresponding diagonal element of the $\mathscr{F}$ matrix. The potential energy distribution can be reported in percent if each $\mathscr{F}_{ii}\mathscr{L}^2_{ik}$ value is divided by the sum of the three such values in each specie and multiplied by 100. See Table 14.14 and Fig. 14.6.

TABLE 14.14

POTENTIAL ENERGY DISTRIBUTION FOR CHLOROFORM

A_1	Q_1	Q_2	Q_3
$\mathscr{S}_1$	99.4%	4.1%	6.5%
$\mathscr{S}_2$	0.2%	57.3%	29.6%
$\mathscr{S}_3$	0.4%	38.6%	63.9%
E	Q_4	Q_5	Q_6
$\mathscr{S}_{1a}$	1.6%	84.6%	2.4%
$\mathscr{S}_{2a}$	0.6%	13.3%	91.3%
$\mathscr{S}_{3a}$	97.8%	2.1%	6.3%

In order to evaluate the cartesian displacements per unitary change in the normal coordinates, the **B** matrix must be evaluated [see Eq. (14.13)]. This is given in Table 14.15. For the bond stretching coordinates r and t each entry is the projection of a unit displacement in the $+X$, Y, or Z direction onto the bond with the sign indicating whether the bond is lengthened ($+$) or shortened ($-$). For the angle bending coordinates α and β each entry for an end atom of the bond angle is the projection of a unit displacement in the $+X$, Y, or Z direction onto a line going through the end atom in a direction perpendicular to the bond and in the plane defined by the two bonds in the angle. This displacement projection is given a positive sign if it tends to increase the angle and a negative sign if it tends to decrease the angle. If the middle atom of the angle is displaced, the effect is the same as if instead, the two end atoms had been displaced equally, but in the opposite direction. Therefore each entry in the **B** matrix for a middle atom for a bending coordinate is the sum of comparable end atom entries already worked out but with a reversed sign.

TABLE 14.15

THE $\mathbf{B}$ MATRIX FOR $CHCl_3$

	X_H	Y_H	Z_H	X_C	Y_C	Z_C	X_{Cl_1}	Y_{Cl_1}	Z_{Cl_1}	X_{Cl_2}	Y_{Cl_2}	Z_{Cl_2}	X_{Cl_3}	Y_{Cl_3}	Z_{Cl_3}
r	0	0	1	0	0	-1	0	0	0	0	0	0	0	0	0
t_1	0	0	0	$-A$	0	$-B$	A	0	B	0	0	0	0	0	0
t_2	0	0	0	C	D	$-B$	0	0	0	$-C$	$-D$	B	0	0	0
t_3	0	0	0	C	$-D$	$-B$	0	0	0	0	0	0	$-C$	D	B
α_1	0	0	0	$-(N+N)$	0	$-(L+L)$	0	0	0	N	$-R$	L	N	R	L
α_2	0	0	0	$-(H-P)$	$-(S-K)$	$-(L+L)$	H	$-K$	L	0	0	0	$-P$	S	L
α_3	0	0	0	$-(H-P)$	$-(K-S)$	$-(L+L)$	H	K	L	$-P$	$-S$	L	0	0	0
β_1	$-E$	0	0	$-(J-E)$	0	M	J	0	$-M$	0	0	0	0	0	0
β_2	G	F	0	$-(G-Q)$	$-(F-S)$	M	0	0	0	$-Q$	$-S$	$-M$	0	0	0
β_3	G	$-F$	0	$-(G-Q)$	$-(S-F)$	M	0	0	0	0	0	0	$-Q$	S	$-M$

$A = \sin \phi$
$B = \cos \phi$
$C = \sin \phi \sin 30°$
$D = \sin \phi \cos 30°$
$E = 1/d_{CH}$
$F = \cos 30°/d_{CH}$
$G = \sin 30°/d_{CH}$
$H = (\sin 30° + \cos \phi)/d_{CCl}$

$J = \cos \phi/d_{CCl}$
$K = \cos 30°/d_{CCl}$
$L = (\cos^2 \phi - \cos \phi)/(d_{CCl} \sin \phi)$
$M = \sin \phi/d_{CCl}$
$N = (1 - \cos \phi) \sin 30°/d_{CCl}$
$P = (\cos \phi \sin 30° + 1)/d_{CCl}$
$Q = \cos \phi \sin 30°/d_{CCl}$
$R = (1 + \cos \phi) \cos 30°/d_{CCl}$

$S = \cos \phi \cos 30°/d_{CCl}$
$\sin \phi = 0.94281$
$\cos \phi = -0.33333$
$\sin 30° = 0.50000$
$\cos 30° = 0.86603$
$1/d_{CH} = 0.9149 \text{ Å}^{-1}$
$1/d_{CCl} = 0.5650 \text{ Å}^{-1}$

TABLE 14.16

THE A MATRIX

	Q_1	Q_2	Q_3	Q_{4a}	Q_{5a}	Q_{6a}	Q_{4b}	Q_{5b}	Q_{6b}
X_H	0	0	0	-0.958	0.187	0.128	0	0	0
Y_H	0	0	0	0	0	0	0.958	-0.187	-0.128
Z_H	0.939	-0.287	0.139	0	0	0	0	0	0
X_C	0	0	0	0.074	0.254	0.130	0	0	0
Y_C	0	0	0	0	0	0	-0.074	-0.254	-0.130
Z_C	-0.096	-0.233	0.107	0.006	-0.057	0	0	0	0
X_{Cl_1}	0	-0.045	-0.075	0	0	0.081	0	0	0
Y_{Cl_1}	0	0	0	0	0	0	0.005	0.004	0.113
Z_{Cl_1}	0.002	0.029	-0.014	-0.013	0.023	0.009	0	0	0
X_{Cl_2}	0	0.022	0.037	-0.002	-0.017	-0.064	-0.005	0.023	-0.084
Y_{Cl_2}	0	0.039	0.065	0.005	-0.023	0.084	-0.004	0.044	-0.033
Z_{Cl_2}	0.002	0.029	-0.014	0.006	-0.012	-0.004	-0.011	0.020	0.007
X_{Cl_3}	0	0.022	0.037	-0.002	-0.017	-0.064	0.005	-0.023	0.084
Y_{Cl_3}	0	-0.039	-0.065	-0.005	0.023	-0.084	-0.004	0.044	-0.033
Z_{Cl_3}	0.002	0.029	-0.014	0.006	-0.012	-0.004	0.011	-0.020	-0.007

The **B** matrix elements for the bending and stretching coordinates are related.[3] Once the **B** matrix elements for the stretching coordinates have been determined, the **B** matrix element $B_{\beta X}$ for the X (or Y or Z) coordinate for an end atom of a triatomic bond angle bending coordinate β can be determined from the equation $B_{\beta X} = (B_{SX} \cos \phi - B_{S'X'})/(d_S \sin \phi)$. Here B_{SX} is the element for the X coordinate of the end atom and its bond stretching coordinate S included in the triatomic unit, $B_{S'X'}$ is the corresponding element for the other end atom, d_S is the equilibrium bond length of the bond with the coordinate S, and ϕ is the equilibrium bond angle of the triatomic unit.[3]

The cartesian displacement per unitary change in the normal coordinates can be evaluated by the matrix multiplication

$$\mathbf{A} = \mathbf{M}^{-1}\mathbf{B'FU'}\mathscr{L}\mathbf{\Lambda}^{-1} \qquad (14.160)$$

as described in Eq. (14.54) where $\mathbf{U'}\mathscr{L}$ has been substituted for **L** [Eq. (14.65)]. The product $\mathscr{L}\mathbf{\Lambda}^{-1}$ is obtained by dividing each $\mathscr{L}$ matrix element by the appropriate calculated λ value. The $\mathbf{M}^{-1}\mathbf{B'}$ product is obtained by dividing each element in the transposed **B** matrix by the appropriate atomic mass. The $\mathbf{U'}$ matrix is obtained from the transpose of Eq. (14.106). The **F** matrix is obtained from Eq. (14.142) with values from Table 14.11. The complete **A** matrix is given in Table 14.16 from which a cartesian displacement per normal coordinate diagram can be drawn (see Fig. 14.6).

14.20   The Potential Function Problem

In previous sections we have traced the procedure used in determining the frequencies and forms of the molecular vibrations for a given molecular configuration when the force field is known. Usually, however, the frequencies are known and the force field is not known. A commonly used procedure is to assume a force field using force constants from closely related molecules. These are used to calculate rough frequencies and then the assumed force constants are systematically refined to make the frequency fit as close as possible. If isotopic analogs are available these will provide additional frequencies which must be properly calculated using the same force field which puts further restraints on the force field. However, frequently even with isotopic analogs the data are still insufficient to uniquely define all the interaction force constants and many may have to be left out. Crawford and Brinkley[23] have suggested methods for choosing the potential function, for example,

[23] B. E. Crawford and S. R. Brinkley, *J. Chem. Phys.* **9**, 69 (1941).

interaction force constants for two widely separated frequencies can usually be ignored and set equal to zero. Some models based on chemical bond theory can be helpful. For example, in $O=C=O$ one can write a resonance form $^-O-C\equiv O^+$ which will be increasingly favored if the left-hand bond is lengthened. This will generate forces tending to shorten the right-hand bond which suggests a positive sign for the stretching interaction force constant. In H_2O an increase in the bond angle (104°27′ at equilibrium) takes the H_2O molecule further from the pure p bond model with a 90° angle and closer to a pure sp^3 hybrid model with a tetrahedral bond angle. Since increasing s character tends to shorten bondlengths, this suggests a positive sign for the stretching–bending interaction force constant.[24]

In addition to simple, modified, or general valence force fields with no, some, or all interaction force constants retained, a force field referred to as the Urey–Bradley[25] force field is sometimes used[20] to reduce the number of force constants needed. It uses simple valence type force constants but also includes some force constants between nonbonded atoms representing van der Waals forces. For further discussion, see Herzberg,[4] Chapter 2; Wilson, Decius, and Cross,[3] Chapter 8; and Steele,[6] Chapter 5.

14.21   A Review of Matrix Notation

Matrices[5] are an array of numbers in table form with a certain number of rows and columns and may be represented by a bold force symbol.

$$\mathbf{F} = \begin{Vmatrix} F_{11} & F_{12} & \cdots \\ F_{21} & F_{21} & \cdots \\ \cdots & \cdots & \cdots \end{Vmatrix} \tag{14.161}$$

A square matrix has an equal number of rows and columns. A column matrix has any number of rows but only one column. A diagonal matrix is a square matrix with values on the diagonal (upper-left to lower-right) and zeros elsewhere.

In matrix multiplication the order of the matrices is important as it is not necessarily true that $\mathbf{AB} = \mathbf{BA}$. Let the matrices to be multiplied be

$$\mathbf{A} = \begin{Vmatrix} a & b & c \\ d & e & f \\ g & n & i \end{Vmatrix} \qquad \mathbf{B} = \begin{Vmatrix} j & k \\ l & m \\ n & o \end{Vmatrix} \tag{14.162}$$

[24] D. I. Heath and J. W. Linnett, *Trans. Faraday Soc.* **44**, 556 (1948).
[25] H. C. Urey and C. A. Bradley, *Phys. Rev.* **38**, 1969 (1931).

The product $\mathbf{AB} = \mathbf{C}$ is

$$
\begin{Vmatrix} a & b & c \\ d & e & f \\ g & h & i \end{Vmatrix} \begin{Vmatrix} j & k \\ l & m \\ n & o \end{Vmatrix} = \begin{Vmatrix} aj + bl + cn & ak + bm + co \\ dj + el + fn & dk + em + fo \\ gj + hl + in & gk + hm + io \end{Vmatrix} = \mathbf{C}
$$

(14.163)

Each element in a given row and column of the $\mathbf{C}$ matrix (the $\mathbf{AB}$ product) is obtained as follows. Select the corresponding row of the left-hand matrix $\mathbf{A}$ and the corresponding column of the right-hand matrix $\mathbf{B}$. Multiply the first element of the row and the first element of the column together, then multiply the second element of the row and the second element of the column together, and so forth. The sum of all the products for this row and column selected will be the element of the $\mathbf{AB}$ product matrix $\mathbf{C}$. The number of elements in the rows of the left-hand matrix must be the same as the number of elements in the column of the right-hand matrix in order for matrix multiplication to be possible.

The transpose of a matrix $\mathbf{A}$ is represented by $\mathbf{A}'$ and is obtained by exchanging rows and columns.

$$
\mathbf{A} = \begin{Vmatrix} a & b & c \\ d & e & f \\ g & h & i \end{Vmatrix} \qquad \mathbf{A}' = \begin{Vmatrix} a & d & g \\ b & e & h \\ c & f & i \end{Vmatrix}
$$

(14.164)

A useful relationship is that the transpose of a matrix product $(\mathbf{AB})'$ is equal to the product of the two matrices each transposed and taken in reverse order.

$$
\mathbf{C} = \mathbf{AB} \qquad \mathbf{C}' = (\mathbf{AB})' \qquad \mathbf{C}' = \mathbf{B}'\mathbf{A}'
$$

(14.165)

In matrix algebra when two matrices are equal such as $\mathbf{A} = \mathbf{B}$, it means that every element of the $\mathbf{A}$ matrix is equal to the corresponding element in the $\mathbf{B}$ matrix. When matrices are added or subtracted such as $\mathbf{A} - \mathbf{B}$ it means that every element of the $\mathbf{B}$ matrix is added or subtracted from the corresponding element of the $\mathbf{A}$ matrix. If a matrix is multiplied by a constant then each element of the matrix is multiplied by that constant.

Determinants are similar to square matrices in appearance

$$
|A| = \begin{vmatrix} a & b & c \\ d & e & f \\ g & h & i \end{vmatrix}
$$

(14.166)

but are defined in such a way that they have unique numerical values. One way to evaluate a determinant is to multiply each element in any one selected row by its cofactor and then taking the sum of these products for this one row. The cofactor of an element is the determinant left after deleting the row and column containing that element, multiplied by plus or minus one depending on whether the sum of the number of the row (first, second, third) and the number of the column containing the element is even or odd, respectively. For example for the determinant given above

$$|A| = a\begin{vmatrix} e & f \\ h & i \end{vmatrix} - b\begin{vmatrix} d & f \\ g & i \end{vmatrix} + c\begin{vmatrix} d & e \\ g & h \end{vmatrix}$$ (14.167)

$$\begin{vmatrix} e & f \\ h & i \end{vmatrix} = ei - fh \qquad \begin{vmatrix} d & f \\ g & i \end{vmatrix} = di - gf \qquad \begin{vmatrix} d & e \\ g & h \end{vmatrix} = dh - ge$$

The rules for multiplying two determinants together are the same as those for matrices.

The inverse of a matrix $\mathbf{A}$ is represented by $\mathbf{A}^{-1}$ and is defined by

$$\mathbf{A}^{-1}\mathbf{A} = \mathbf{A}\mathbf{A}^{-1} = \mathbf{I}$$ (14.168)

where $\mathbf{I}$ is the identity matrix which is a diagonal matrix with ones on the diagonal (zeros elsewhere). The identity matrix in matrix algebra corresponds to the digit one in scalar algebra. The identity matrix can be inserted or removed anywhere in a matrix equation

$$\mathbf{IA} = \mathbf{AI} = \mathbf{A}$$ (14.169)

Not all matrices possess inverses. A matrix whose determinant is zero is said to be singular. Only nonsingular (necessarily square) matrices possess inverses. The inverse of a matrix is found by replacing each element by its cofactor, then the resulting matrix is transposed and divided by the determinant of the original matrix. It can be seen from the definition that the inverse of a diagonal matrix is obtained simply by replacing each diagonal element by its reciprocal. If a matrix is orthogonal its inverse is equal to its transpose.

14.22   The Normal Coordinate Problem in Matrix Form

In order to properly introduce the $\mathbf{G}$ matrix (see Section 14.5 and 14.6) and the $\mathbf{A}$ matrix (see Section 14.9) we will review the normal coordinate problem in matrix form. Use will be made of matrix algebra but all the matrices used will have been introduced in detail in earlier sections.

The kinetic energy T of a molecule can be expressed in cartesian displacement coordinates where $\dot{\mathbf{X}}$ is a column matrix of the time derivatives thereof and $\mathbf{M}$ is a diagonal matrix with atomic masses as diagonal elements (zeros elsewhere).

$$2T = \dot{\mathbf{X}}'\mathbf{M}\dot{\mathbf{X}} \tag{14.170}$$

The potential energy V can be expressed in internal coordinates where $\mathbf{S}$ is a column matrix thereof and $\mathbf{F}$ is a square matrix of force constants.

$$2V = \mathbf{S}'\mathbf{FS} \tag{14.171}$$

Let the internal coordinates be transformed to cartesian displacement coordinates by

$$\mathbf{S} = \mathbf{BX} \tag{14.172}$$

We can write that $\mathbf{S}' = \mathbf{X}'\mathbf{B}'$ since in matrix algebra the transpose of a product is equal to the product of the transposes taken in reverse order. Using these values for $\mathbf{S}$ and $\mathbf{S}'$ we get from Eq. (14.171)

$$2V = \mathbf{X}'\mathbf{B}'\mathbf{FBX} \tag{14.173}$$

Let the cartesian displacement coordinates be transformed to normal coordinates by

$$\mathbf{X} = \mathbf{AQ} \tag{14.174}$$

where $\mathbf{Q}$ is a column matrix of normal coordinates in terms of which the kinetic and potential energies can be expressed as

$$2T = \dot{\mathbf{Q}}'\mathbf{I}\dot{\mathbf{Q}} \quad \text{and} \quad 2V = \mathbf{Q}'\mathbf{\Lambda}\mathbf{Q} \tag{14.175}$$

Here $\mathbf{I}$ is the identity matrix and $\mathbf{\Lambda}$ is a diagonal matrix where the diagonal elements are $\lambda_k = 4\pi^2 v_k^2$ (zero elsewhere). In this diagonal form there are no cross terms in the energy expressions and each normal coordinate vibrates with a frequency v_k and is independent of other normal coordinates. Since we do not know the forms ($\mathbf{A}$) or frequencies (from $\mathbf{\Lambda}$) of the normal coordinates, the problem is to express these as functions of terms we do know which are $\mathbf{F}$, $\mathbf{M}$, and $\mathbf{B}$.

Since from Eq. (14.174) we can write $\dot{\mathbf{X}} = \mathbf{A}\dot{\mathbf{Q}}$ and $\dot{\mathbf{X}}' = \dot{\mathbf{Q}}'\mathbf{A}'$, Eqs. (14.170) and (14.173) become

$$2T = \dot{\mathbf{Q}}'\mathbf{A}'\mathbf{M}\mathbf{A}\dot{\mathbf{Q}} \qquad \text{and} \qquad 2V = \mathbf{Q}'\mathbf{A}'\mathbf{B}'\mathbf{F}\mathbf{B}\mathbf{A}\mathbf{Q} \qquad (14.176)$$

By combining these with Eq. (14.175) we get

$$\mathbf{I} = \mathbf{A}'\mathbf{M}\mathbf{A} \qquad \text{and} \qquad \mathbf{A}'\mathbf{B}'\mathbf{F}\mathbf{B}\mathbf{A} = \boldsymbol{\Lambda} \qquad (14.177)$$

Since $\mathbf{A}'\mathbf{M}\mathbf{A}$ is the same as the identity matrix, it can be inserted into right-hand Eq. (14.177) to give

$$\mathbf{A}'\mathbf{B}'\mathbf{F}\mathbf{B}\mathbf{A} = \mathbf{A}'\mathbf{M}\mathbf{A}\boldsymbol{\Lambda} \qquad (14.178)$$

After cancelling $\mathbf{A}'$ and premultiplying by $\mathbf{M}^{-1}$ we get

$$\mathbf{M}^{-1}\mathbf{B}'\mathbf{F}\mathbf{B}\mathbf{A} = \mathbf{A}\boldsymbol{\Lambda} \qquad (14.179)$$

This is the secular equation in matrix form using cartesian displacement coordinates. It is a series of linear homogeneous simultaneous equations [similar to Eq. (14.8)] from which $\boldsymbol{\Lambda}$ and $\mathbf{A}$ can be determined.

At this point we remind ourselves there are $3N$ values for Q_k and λ_k and six of these are translations and rotations with zero λ values. We can designate that the column matrix $\mathbf{Q}$ has a submatrix $\mathbf{Q}_v$ within it consisting of $3N - 6$ normal coordinates associated only with vibration and that the diagonal matrix $\boldsymbol{\Lambda}$ has a submatrix $\boldsymbol{\Lambda}_v$ within it consisting of $3N - 6$ nonzero values for λ_k on the diagonal. From Eqs. (14.172) and (14.174) we get that the transformation from internal coordinates to normal coordinates is

$$\mathbf{S} = \mathbf{B}\mathbf{A}\mathbf{Q} \qquad (14.180)$$

The transformations from internal coordinates to *vibrational* normal coordinates is given by

$$\mathbf{S} = \mathbf{L}\mathbf{Q}_v \qquad (14.181)$$

where $\mathbf{L}$, with $3N - 6$ columns, is a submatrix of the $\mathbf{BA}$ matrix product with $3N$ columns (see Fig. 14.10). The equation holds because the six remaining columns in the $\mathbf{BA}$ matrix product must be zero since a change in rotational and translational normal coordinates Q_{RT} do not cause a change in internal coordinates S.

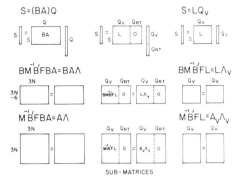

FIG. 14.10. Matrix equations in text in block form. The matrices on the left involve all the normal coordinates, including rotation and translation. The matrices on the right only involve vibrational normal coordinates and are submatrices of those on the left.

Let us premultiply both sides of Eq. (14.179) by **B** to get

$$\mathbf{BM^{-1}B'FBA = BA\Lambda} \qquad (14.182)$$

Then, if we substitute **L** for **BA** and $\mathbf{\Lambda}_v$ for **Λ** into Eq. (14.182) we will get an equation which *only* involves the vibrational normal coordinates.

$$\mathbf{BM^{-1}B'FL = L\Lambda}_v \qquad (14.183)$$

The equality holds because each side of this equation is a comparable submatrix of each side of Eq. (14.182) (see Fig. 14.10).
The matrix product $\mathbf{BM^{-1}B'}$ is called the **G** matrix so Eq. (14.183) becomes

$$\mathbf{GFL = L\Lambda}_v \quad \text{where} \quad \mathbf{G = BM^{-1}B'} \qquad (14.184)$$

This is a series of simultaneous equations (the secular equations) which determine **L** (except for normalization) when **Λ** is known. In order to define **Λ** we substitute for it **I**λ to get

$$(\mathbf{GF - I}\lambda)\mathbf{L} = 0 \qquad (14.185)$$

Setting the determinant of the coefficient of **L** equal to zero, we get for the secular determinant.

$$|\mathbf{GF - I}\lambda| = 0 \qquad (14.186)$$

from which the $3N - 6$ values for λ can be determined if the **G** and **F** values are known. Notice that this same secular determinant resulted when the kinetic energy was expressed in Section 14.5 as

$$2V = \mathbf{\dot{S}'G^{-1}\dot{S}} \qquad (14.187)$$

Having determined the λ values, the **L** can be evaluated from Eq. (14.185). If we want to determine **A** we substitute **L** for **BA**, $\mathbf{A}_v$ for $\mathbf{A}$, and $\mathbf{A}_v$ for **A** into Eq. (14.179).

$$\mathbf{M}^{-1}\mathbf{B}'\mathbf{FL} = \mathbf{A}_v\,\mathbf{\Lambda}_v \qquad (14.188)$$

The equality holds because each side of this equation is a comparable submatrix of each side of Eq. (14.179) (see Fig. 14.10). Postmultiplying both sides of Eq. (14.188) by $\mathbf{\Lambda}_v^{-1}$ we get

$$\mathbf{M}^{-1}\mathbf{B}'\mathbf{FL}\mathbf{\Lambda}_v^{-1} = \mathbf{A}_v \qquad (14.189)$$

This equation defines the form of the vibrational normal coordinates in terms of cartesian displacement coordinates

$$\mathbf{X} = \mathbf{A}_v\,\mathbf{Q}_v \qquad (14.190)$$

where $\mathbf{A}_v$ is a submatrix of **A** in Eq. (14.173). Each $\mathbf{A}_v$ element defines a cartesian displacement coordinate change per unitary change in a vibrational normal coordinate Q_v. This enables us to draw a picture of the atomic displacements in each vibration.

INDEX*

A

Absorbance, 97
Absorptivity, 97
Acetals, 319
Acetates, 290, 304
Acetonitrile, 54
Acetylene(s), 54, 64, 207, 236, 237
Acid chloride, 298
Acid hydrazide, 309
Acrylamide, 109
Acrylates, 291, 292
Alcohols, 316, 317
Aldehydes, 29, 287
Allenes, 182, 237
Amides, 211, 305
Amidines, 326
Amido acids, 305
Amines, 321, 322
Amine salts, 324
Amino acids, 304
Ammeline, 274
Ammonium ion, 324
Anharmonicity, 13
Anharmonic oscillator, 25
Anhydrides, 215, 297
Anilines, 321, 323
Anthracenes, 271
Anti-Stokes lines, 57
Aromatic CH wag, 209, 262
Aromatic summation bands, 266
Asymmetrical tops, 40, 41, 55, 56
Attenuated total reflectance, 92
Avogadro's number, 180
Azides, 242
Azo groups, 331
Azoxy groups, 333

B

Base line absorbance, 100, 101
Beer's law, 97
Benzene, 54
Benzene rings, 257
Benzoates, 292
Benzophenones, 283
B–H, 337
BHB bridge, 217, 337
Biflouride ion, 359
Bisulfates, 356
Blackbody radiation, 70
B–N, 336
B–O, 335
Bolometers, 81, 82
Boltzmann constant, 23
Boltzmann distribution function, 23, 25, 39
Born–Oppenheimer approximation, 14, 505
Boron groups, 335
Butatriene, 214, 245
Butyrates, 290

C

Calibration, 84
Carbamates, 308
Carbodiimides, 240
Carbohydrates, 319
Carbonates, 295
    trithio, 358
Carbon suboxide, 245
Carbon subsulfide, 245
Carbonyl group, 278
Carboxylic acids, 299
Carboxylic acid salts, 303

* See indexes for spectra on pp. 384, 447, and 455.

519